Principles of
Industrial Metalworking Processes

Principles of Industrial Metalworking Processes

by GEOFFREY W. ROWE

M.A., Ph.D., D.Sc., C.Eng., F.I.Mech.E., F.Inst.P., F.R.S.A.
Professor of Mechanical Engineering,
University of Birmingham

CBS Publishers & Distributors Pvt. Ltd.

New Delhi • Bengaluru • Chennai • Kochi • Kolkata • Mumbai
Hyderabad • Nagpur • Patna • Pune • Vijayawada

CBS Pubs ISBN: 81-239-0428-2
Ed. Arnold ISBN: 0-7131-3381-3

First Indian Reprint: 1996
Reprint: 2000, 2004, 2005

Copyright © 1977, G.W. Rowe

This edition has been reprinted in India by arrangement with Edward Arnold (Publishers) Limited, UK

All rights reserved. No part of this book may be reproduced or transmitted in any form or by any means, electronic or mechanical, including photocopying, recording, or any information storage and retrieval system without permission, in writing, from the publisher.

Published by **Satish Kumar Jain** and produced by **Varun Jain** for
CBS Publishers & Distributors Pvt. Ltd.,
4819/XI Prahlad Street, 24 Ansari Road, Daryaganj, New Delhi - 110002
delhi@cbspd.com, cbspubs@airtelmail.in • www.cbspd.com
Ph.: 23289259, 23266861, 23266867 • Fax: 011-23243014

Corporate Office: 204 FIE, Industrial Area, Patparganj, Delhi - 110 092
Ph: 49344934 • Fax: 011-49344935
E-mail: publishing@cbspd.com • publicity@cbspd.com

Branches:

- ***Bengaluru:*** 2975, 17th Cross, K.R. Road, Bansankari 2nd Stage, Bengaluru - 70 • Ph: +91-80-26771678/79 • Fax: +91-80-26771680 E-mail: cbsbng@gmail.com, bangalore@cbspd.com
- ***Chennai:*** No. 7, Subbaraya Street, Shenoy Nagar, Chennai - 600030 Ph: +91-44-26681266, 26680620 • Fax: +91-44-42032115 E-mail: chennai@cbspd.com
- ***Kochi:*** Ashana House, 39/1904, A.M. Thomas Road, Valanjambalam, Ernakulum, Kochi • Ph: +91-484-4059061-65 Fax: +91-484-4059065 • E-mail: cochin@cbspd.com
- ***Kolkata:*** 6-B, Ground Floor, Rameshwar Shaw Road, Kolkata - 700014 Ph: +91-33-22891126/7/8 • E-mail: kolkata@cbspd.com
- ***Mumbai:*** 83-C, Dr. E. Moses Road, Worli, Mumbai - 400018 Ph: +91-9833017933, 022-24902340/41 • E-mail: mumbai@cbspd.com

Representatives:

- Hyderabad: 0-9885175004
- Patna: 0-9334159340
- Vijayawada: 0-9000660880
- Nagpur: 0-9021734563
- Pune: 0-9623451994

Printed at:
J.S. Offset Printers, Delhi (India)

Preface

In the ten years since this book was first published, significant changes have been seen in the general world situation, as well as in metalworking. Efficiency of production remains, as always, a dominant industrial theme, but it is increasingly being interpreted in terms of social and environmental factors in addition to its strictly economic sense. This can have considerable repercussions on metalworking practice. For example, the noise produced by a drop hammer may now be unacceptable. It can, to some extent, be reduced at source and its transmission can be minimised, but thought is also being given to replacement of drop forging by inherently quieter processes where possible. Major changes in lubrication and cooling systems may be needed in some other processes, to avoid potential dangers to health by contact or ingestion, and to reduce disposal problems.

Superimposed upon these considerations is the need to conserve energy and material resources. More efficient utilisation of direct and indirect energy supplies has become essential, and the temporary and permanent shortages of certain raw materials demand greater flexibility in plant operation, as well as calling for substitute materials.

An attempt has been made to widen the background of this book, in the hope that some contribution may be made to the solution of these problems. A new chapter on materials has been added, dealing mainly with metals but including a brief introduction to polymers. The chapter on lubricants has been expanded to include more specific information, and brief descriptions of new process developments have been added to each chapter.

The main theoretical development in this decade has been in the widespread use of digital computers, and in the recognition that material properties should be included in analytical models. A new chapter outlining the numerical methods for solution of metalworking problems has been written, with critical guidance on their relative merits and applicability.

I am indebted to many colleagues in metallurgy and engineering for helpful comments and discussions. Among these it is a pleasure to record particularly Professor W. Johnson, Dr. J. A. Newnham, Professor P. L. B. Oxley, Professor A. A. Hendrickson, Professor M. C. Shaw, Mr. T. F. Li, Dr. A. N. Bramley, Professor K. Lange and Professor T. Wanheim. Finally I wish to thank my wife and family for their forbearance during the preparation of the manuscript.

G. W. R.

Preface to original edition

All metal objects, except castings, have at some time in their manufacture been subjected to at least one metalworking operation. Several different operations may often be necessary. Thus the steel used for tubes to make a simple office chair is forged, hot rolled several times, cold rolled into strip, slit, cold formed into tubular shape, welded, machined along the weld, and sometimes cold drawn as well, apart from all the subsidiary treatments. The theory of metalworking can assist in determining the most efficient utilisation of the machines, and so in improving productivity.

The main purpose of this book is to present an account of the basic principles of metalworking theory, which can be understood and applied by engineers and metalurgists directly concerned with improving production in the metalworking industry. It should also be of use to students in Universities and Colleges of Technology. It covers all the theory of plasticity necessary for this purpose, but it is not a mathematical treatise on general plastic deformation, and it contains no mathematics beyond VIth form level. On the other hand it is not a technological handbook containing descriptive details of all the complexities of modern processes.

Plasticity theory has not yet advanced sufficiently to provide accurate answers to all problems of metal deformation, but full treatments are available from simplified versions of practical processes, using idealised working materials. These can give valuable insight into the effects of various parameters such as tool-profile, speed, amount of deformation, and lubrication, on the energy consumed in the operation, the temperatures generated, the flow of the metal and its final properties.

The numerical examples are provided as an integral part of the purpose of the book, to encourage the reader to apply the methods personally, and to foster the attitude that the techniques are of practical as well as academic interest. The final chapter shows how the theory can be applied very conveniently with the aid of a small and relatively inexpensive analogue computer, so that optimum schedules can be assessed rapidly.

Grateful acknowledgement is made to many former and present colleagues in industry, colleges of technology and universities, particularly to Mr. A. P. Green, Dr. P. R. Lancaster, D. A. T. Male, Mr. D. R. Milner, Dr. J. A. Rogers and Professor E. C. Rollason for their comments leading to improvements in the manuscript. Mr. A. J. Collins and Messrs. Electronic Associates Ltd. have been most helpful in the production of Chapter XII. Various organisations including Messrs. W. H. A. Robertson Ltd., Messrs. Tube Investments Ltd., Messrs. Henry Wiggin Ltd., Messrs. Acheson Industries Ltd., the Ministry of Aviation and the Department of Scientific and Industrial Research have been helpful. I am also indebted to the individuals and organisations quoted for permission to reproduce illustrations, and to my wife for her cooperation, including typing and correcting the manuscript.

G. W. R.

Contents

Part II. EXAMINATION OF PROCESSES

LIST OF SYMBOLS AND ABBREVIATIONS

(Symbols used for more than one purpose are marked *)

Dimensions

h height or thickness
l length
L length
A area
*V volume
b platen or tool breadth
w strip width
D diameter
R radius
r radius
h_0 initial height in compression
D_b, h_b entry diameter or thickness in rolling or drawing
D_a, h_a exit diameter or thickness in rolling or drawing $(h_b > h_a)$
l_0 initial length
l_1 drawn length
A_0 initial area
A_1 final area

Tensile properties of materials

d_u diameter at M.T.S.
L_p limit of proportionality
L_e elastic limit
E Young's modulus (typically 200 kN/mm² (30×10^6 lb/in²) for mild steel, 120 kN/mm² (18×10^6 lb/in²) for copper
G Rigidity modulus, or Shear modulus
ν Poisson's ratio
m strain-hardening exponent
n strain-rate exponent
*s nominal stress
ϵ_u strain at M.T.S.
e linear strain
e_t tensile strain (positive)
e_c compressive strain (negative)
el% percentage elongation of gauge length
U.T.S. or M.T.S. ultimate or maximum tensile stress
Y.S. yield-point stress, equal to Y'
R.A.% percentage reduction in area to fracture

Metalworking properties of materials

σ true stress (Note that $\sigma \neq s$)
*$\bar{\sigma}$ mean stress
*$\bar{\sigma}$ generalised stress
ϵ natural or logarithmic strain $\left(\int \frac{dl}{l}\right)$. $\epsilon_t = \ln(1 + e_t)$

$$\epsilon_c = \ln\left(\frac{1}{1 - e_c}\right)$$

$*\bar{\epsilon}$ generalised strain
$*\bar{\epsilon}$ mean strain
$\dot{\epsilon}$ strain rate
γ shear strain ($\gamma/2$ is commonly used)
$\dot{\gamma}$ shear strain rate
$*B$ constant in yield stress equation
k yield stress in pure shear $k = \frac{S}{2} = \frac{Y}{\sqrt{3}}$
Y yield stress in uniaxial tension
S yield stress in plane strain. $S = 1{\cdot}55\,Y$
$*H$ Brinell Hardness Number (B.H.N.), expressed in kg/mm²
r fractional reduction of area in an operation
$R\%$ percentage reduction of area in an operation
$*c$ specific heat
ρ density

Metalworking analysis

log, ln logarithms to base 10 and base e respectively
$*P$ indentation or extrusion pressure kN/mm²
$*P$ working force or load, usually kN
$*F$ force, e.g. in drawing, usually kN
$\bar{Y}$ mean yield stress, assumed constant. $\bar{Y} = \frac{1}{\epsilon_2 - \epsilon_1} \int_{\epsilon_1}^{\epsilon_2} \sigma\, d\epsilon$
$\bar{S}$ mean yield stress in plane strain
$*t$ time
$*W$ work
$\dot{E}_b, \dot{E}_d, \dot{E}_f, \dot{E}_T$ boundary, deformation, frictional and total energy dissipation rates.
F_s shearing force
t_1 depth of cut (undeformed chip thickness)
t_2 chip thickness
T temperature
V_s shear plane velocity
$*\alpha$ cutting tool rake angle
Δb width of a zone boundary
χ roll load factor
$*\phi$ shear plane angle
σ_i flow stress in zone i

Stress analysis

σ_x, σ_y direct stresses on planes normal to X- and Y-axes respectively
τ_{xy} shear stress on a plane normal to the X-axis, in direction OY
$\sigma_\theta, \tau_\theta$ direct and shear stresses on a plane inclined at an angle θ to an axis of reference e.g. OY
$\sigma_1, \sigma_2, \sigma_3$ principal stresses; major, intermediate, and minor
θ inclination of an arbitrary plane to an axis of reference
inclination of the tangent to a slip line
$\theta*$ inclination of a principal plane to an axis of reference
$\theta**$ inclination of a plane of maximum shear to an axis of reference
$\epsilon_x, \epsilon_y, \epsilon_z$ and $\epsilon_1, \epsilon_2, \epsilon_3$ logarithmic strains
μ coefficient of friction

- $*2\alpha$ included angle of tool or die
- B a constant or variable equal to $\mu \cot \alpha$
- l, m, n direction cosines
- r_m maximum reduction of area

Slip-line field theory

- $*p$ hydrostatic pressure in a plastic metal
- k yield stress in pure shear
- α-line, β-line mutually perpendicular slip lines, defined by the algebraically greatest principal stress lying in the first quadrant
- $*\alpha, \beta$ distances along slip lines
- ϕ inclination of a slip line
- u, v velocities along α- and β-slip-lines respectively
- $*V$ velocity of an element of metal following a streamline
- $*s$ distance along a slip-line, or along a line in an upper-bound solution
- t_x drawing stress derived from slip-line field (t_x' includes friction)
- t_H drawing stress for work-hardening metal
- q die stress (q' includes friction)
- $f\left(\frac{c}{d}\right)$ and $*\phi\left(\frac{c}{d}\right)$ the redundant work factors of geometrical parameters
- $*\phi$ anticlockwise rotation of a slip line
- $*R$ extrusion ratio

Rolling theory

- $*R, R'$ undeformed and deformed roll radius
- L length of projected arc of contact, $L = \sqrt{R\Delta h}$
- Δh draft, or decrease in strip thickness
- $*c$ an elastic constant, in the Hitchcock equation
- $*s$ springback
- g roll gap
- $*H$ a function in Bland and Ford solution, $H = 2\sqrt{\frac{R'}{h_a}} \tan^{-1}\sqrt{\frac{R'}{h_a}} \cdot \alpha$
- w width
- P roll load (kN)
- T roll torque (Nm)
- t_b, t_a back and front tensions
- k_b, k_s deflection constants of the rolls, due to bending and to shear
- α general angular position, measured from the exit
- α_N the angular position of the neutral point
- α_b* angle of bite, equal to $\tan^{-1}\mu$
- p_r radial roll pressure
- $*p$ component of roll pressure normal to the strip
- p^+, p^- roll pressures on exit and entry sides of the neutral point, respectively
- λ lever arm
- λ' lever arm for deformed rolls
- Q a factor in Sims' equation

Friction theory

- A_x area of intimate asperity contact

A_0 apparent area of contact
$*\bar{s}$ mean shear strength of junctions, related to k
$\bar{p}$ mean yield stress of junctions, related to hardness, and so to k
$*F$ friction force
$*W$ load in a friction experiment
s_i shear strength at an interface

Metallurgy

A_1 lower critical point for hypoeutectoid steels
A_3 upper critical point
b Burger's vector
d grain diameter
D coefficient of diffusion
E activation energy
f_1, f_2 workability factors
F, G, H constants found from tensile tests (equation 12.7)
I second moment of area
k factor in Petch equation
L, M, N constants found from shear tests (equation 12.5)
R gas constant (8.315 joules/degree)
$*T$ temperature (K)
$*T$ relaxation time
T_m melting point (K)
W workability
α constituent phase (lower left on equilibrium diagram)
α-iron ferrite
γ-phase in steels, austenite
η coefficient of viscosity
$*\theta$ diffraction angle
$*\theta$ temperature
λ X-ray wavelength
σ_i glide resistance
σ_y yield stress (= Y)
σ'_A residual stress

Numerical methods

a, b height, width, in upper-bound solutions
a_i, b_i, c_i matrix coefficients
A_i weighting parameter
$[A]$ position coordinate matrix
$[B]$ displacement matrix
C_{ijkl} constant in a constitutive equation
$[D]$ elastic constant matrix
$D[\ ]$ differential operator
$E(x)$ error function
f^* postulated function of discrete coordinates
F body force
F_i force associated with displacement u_i
$[K]; [K_C], [K_G]$ stiffness matrix; material and geometric
n_1, etc. constant matrix coefficients
$N(u)$ function of spatial derivatives of u
$R(u^*)$ residuals; $R(u^*) = N(u^*) - \partial u^*/\partial t$
s_{ij} initial stress on finite element (contrast t_{ij})
S surface area

t_{ij} stress increment in deformed system
T traction (tangential surface force)
T_i traction associated with displacement u_i
$*u, v$ displacement components (finite elements)
$*u, v$ velocity components (visioplasticity, upper bounds and slip-line fields)
u^* trial solution for discrete displacements
u_i discrete displacement function
$*V$ volume
$*V; V_0, V_s$ velocity; initial and final
W_H, W_F, W_R, W_T homogeneous, frictional, redundant and total work
$*W_I, W_S$ internal and surface work
$*W_T, W_F, *W_I$ work due to surface traction, body forces and inhomogeneous deformation
W rate of performing work
α_{ij} weighting function
$*\beta$ fraction of platen contact length
$*\gamma$ angle in a slip-line field
δ displacement
δ_{ij} Kronecker delta
$\delta(\Delta u_i)$ hypothetical change in Δu_i
ΔE_{ij} incremental strain tensor component
$\Delta\epsilon_{ij}, \Delta\eta_{ij}$ linear and non-linear strain increments
Δf a force integral;

$$\Delta f_j = \int_V \frac{\Delta F_i}{V_0} \alpha_{ij} \, dV$$

$\Delta F_i, \Delta T_i$ force and traction increments associated with a hypothetical increase in Δu_i
Δt force integral;

$$\Delta t_j = \int_S \frac{\Delta T_i}{S_0} \alpha_{ij} dS$$

Δu_i increment of displacement (deformation)
e_{ii}, e_{ij} direct and shear strain tensor-components
$*\phi$ stress function
$*\Phi$ integral expression (equation 13.119)
$*\Phi$ flow function
χ stress function
$d\lambda$ constant in Levy-Mises equation
λ angle in a slip-line field
ψ scalar flow function
σ_m spherical or mean stress
σ' deviatoric stress
$*\sigma_{ii}, \sigma_{ij}$ direct and shear stress tensor-components
$*\sigma_{ij}$ stress after a deformation increment (contrast s_{ij})

USEFUL APPROXIMATE CONVERSIONS FROM f.p.s. TO S.I. UNITS

(It is seldom useful in metalworking theory to attempt to obtain accuracy better than 1%.)

Length

1 in. = 25·4 mm ≃ 25 mm; 10 mm ≃ 0·4 in.
$\frac{1}{8}$ in. ≃ 3·2 mm
$\frac{1}{16}$ in. ≃ 1·6 mm
10^{-3} in. (= 1 thou.) ≃ 0·025 mm; 10 μm ≃ 0·4 thou.
1 ft = 305 mm ≃ 0·3 m

Area

1 sq. in. = 645 mm^2

Mass

1 lb = 0·454 kg; 1 kg ≃ 2·2 lb
1 ton (= 1·016 tonnes) ≃ 1 tonne = 1,000 kg

Force

1 lbf ≃ 4·45 N
1 tonf (= 9,964 N) ≃ 10,000 N

Stress or Pressure

1,000 lb/in.^{2} (i.e. 1,000 lbf/in.^{2}) = 6·89 N/mm^2 ≃ 7 N/mm^2
1 ton/in.^{2} (i.e. 1 tonf/in.^{2}) = 15·44 × 10^6 N/m^2 ≃ 15·4 N/mm^2

Speed

100 ft/min = 0·51 m/s ≃ 0·5 m/s
10 rev/min = 1·05 rad/s ≃ 1 rad/s

Young's Modulus (for rough estimates)

Steels	$E \simeq 30 \times 10^6$ lbf/in.^{2};	$E \simeq 200$ kN/mm^2
Copper	$E \simeq 18 \times 10^6$ lbf/in.^{2};	$E \simeq 100$ kN/mm^2
Aluminium	$E \simeq 12 \times 10^6$ lbf/in.^{2};	$E \simeq 70$ kN/mm^2

Part 1: Basic Methods

1

The Nature and Purpose of Metalworking Theory

1.1 Introduction

It would be largely unnecessary to work metals at all if they could be cast, with good mechanical properties, in the wide variety of shapes and sizes required by modern industry. This is, however, far from being technically possible, apart from questions of economic production, though a trend in this direction can be seen in recent continuous casting processes for bar and tube[1.1, 1.2]. A serious difficulty is that most cast ingots require considerable working to refine the crystal structure and to homogenise the metal or alloy. Such working, which often amounts to 80% reduction in cross-sectional area and sometimes much more, greatly improves the mechanical properties, especially ductility and impact strength. It is usual to carry out this deformation at a high temperature, which gives two main advantages; the metal is softer, requiring less force and consequently lighter equipment, and the recrystallisation required for crystal grain refinement proceeds rapidly. The tensile strength and hardness of the homogenised metal can subsequently be increased by work-hardening in a cold-working operation performed at a temperature below the recrystallisation range, usually at room temperature.

Both hot- and cold-working processes are of great industrial importance today. In order to make the best economic use of metalworking plant it is necessary, of course, to know the capacity of the plant in terms of maximum safe loads for all the different operations it may be called upon to perform. As will be explained in Chapter 9, the maximum load for a given mill rolling 0·25 m (10 in.) wide strip, for example, may be appreciably less than that for the same mill rolling 1·25 m (50 in.) wide strip. A drawbench may be limited by maximum pull for one operation, and by stresses causing die fracture for another. In hot-working, the rate of mechanical straining plays an important part, while in cold-working the load is more dependent on the rate of work-hardening of the material. It is the function of metalworking theory to provide a background from which reasonable judgements may be made in the selection of equipment and the planning of individual passes for maximum productivity, or desirable product properties, on existing plant. Knowledge of the stresses and deformation may assist in diagnosing causes of fracture and suggest ways of overcoming the limitations. Theory can also contribute substantially to the design of equipment, though it must be admitted that many highly successful designs are far too complex for any theoretical treatment. In wire-drawing, for example, the drawing force or die load can easily be increased by 30% by incorrect choice of die angle. The precise influence of lubricant is also by no means obvious. For a very heavy reduction in area, the value of the coefficient of friction may make practically no difference to the maximum possible pass, but in a light sizing pass the friction may contribute half the drawing stress, and greatly influence the homogeneity of the final product. It is not even always desirable to reduce the coefficient of friction as far as possible. In cold strip-rolling, the maximum pass is often determined by the roll bite. If the lubrication is too good, the rolls will slip and the strip will not feed through.

Metalworking theory allows these factors to be evaluated. It has not yet advanced far enough to give accurate answers to all metalworking problems; or even to a majority, in precise quantitative terms. Nevertheless, the full treatment which is available for simplified versions of practical processes, and idealised forms of work material, gives an insight which can be of considerable value in predicting the influence of practical variables on working load and metal flow, and in choosing appropriate conditions to produce a desired result. The answers are unlikely to be final, but will almost certainly reduce the amount of empirical trial necessary before a process is established.

On the other hand, original publications in metalworking theory tend, necessarily, to be somewhat specialised and difficult reading for those engaged in practical metalworking. The results are consequently not as widely applied as would be useful to industry. It is the purpose of this book to present a simplified and condensed version of the important methods and some of the relevant conclusions, starting with elementary concepts and using no advanced mathematics. It is hoped that the reader will be able to apply some of the results directly, and will be stimulated to use the more advanced treatises which are available[1.3–1.6], and some of the original papers.

1.2 Yielding, and the simplification of stress combinations

Most metalworking theory is concerned finally with prediction of the stresses acting during metal deformation, and consequently of the forces which must be applied. The working load determines the power requirement, and the size of the equipment necessary to perform a particular operation. Measurement of forces also gives a good indication of the efficiency of operation of the plant under particular conditions. Until recently, it was the exception rather than the rule to see load meters on metalworking plant, but many manufacturers now consider load meters of some sort as an integral part of modern equipment. Even in olden days, there was usually at least an ammeter to measure armature current in the main motor.

In an approach to theoretical determination of working loads, two factors are of primary importance. The first is that when the stress applied to a metal is large enough, the metal will *yield*, that is, begin to deform plastically, so that it retains a permanent deformation after the load has been removed. This is considered a criterion of failure in engineering structures, all of which are designed so that the maximum loads which are likely to be imposed upon them will not raise the stress to this danger level. In metalworking a quite different view is taken. Any elastic deformation, which disappears as soon as the load is removed, is of no use to the metalworker, who wishes to produce a large permanent change in cross-section. Metalworking theory consequently begins exactly where structural theory ends, with the stress level which is just sufficient to cause plastic deformation, and frequently ignores the small amount of elastic recovery which occurs when the load is released.

Secondly, in a simple example, such as deformation by a tension applied to a wire, yielding is caused by a direct stress acting in only one direction, and can therefore be predicted easily. In most practical operations, however, there are at least two stresses acting on any element of metal in the deformation zone. Thus, in wire-drawing (Chapter 6, Figure 6.2) there is a pull on the wire, producing a tensile stress and a compressive stress at the die face, together with a surface shear due to the frictional drag on the die. It is not then possible to state, immediately, at what load plastic flow within the die region will occur. Yield depends on the combination of all the stresses acting, and can be predicted only on the basis of some relationship between them. This is the function of a *yield criterion*. Various types of yield criterion, or elastic failure limit in structural analysis, have been suggested. The best of these, proposed by von Mises in 1913, states that plastic flow commences when the shear strain-energy reaches a particular value. The magnitude of this energy can be expressed in various ways, but the yield criterion is most simply used in a form containing only the *principal stresses.* These important stresses are defined as the stresses acting on certain *principal planes* in the workpiece, which are the planes which are so oriented to the applied forces that no shear stress acts upon them. A secondary characteristic is that two of the three principal stresses are the greatest and least stresses acting on the workpiece. They always act in mutually perpendicular directions. The use of principal stresses greatly simplifies metalworking theory. General analysis of the combined influence of stress requires nine variables: three direct stresses acting normal to three mutually-perpendicular planes at the point considered, and six shear stresses acting parallel to those planes. The equations are so cumbersome to handle that it is common practice to use the shorthand notation of tensor analysis for them. This usually appears at the start of a textbook on plasticity, and not infrequently deters the less specialist reader from pursuing the subject further. However, the shear stresses always occur in equal pairs, because any shear process causes complementary shear in the opposite sense, to maintain rotational equilibrium. Thus only six of the nine stresses are independently variable. Moreover, if the three planes at the point are chosen so that they are principal planes, the three pairs of shear stresses are separately equal to zero, from the definition of principal planes, leaving only the three direct stresses. This simplification is very useful. Provided that the analysis is expressed in terms of principal stresses, it is necessary to consider only three stresses, all direct stresses. It is, indeed, often possible, because of symmetry or certain other conditions, to consider only two, the greatest and least direct stresses. Most metalworking theory, as distinct from general elasticity and plasticity, can be expressed in terms of principal stresses without requiring any tensor analysis. This non-mathematical approach will be followed throughout this book.

Chapter 2 describes the stress–strain curve or yield-stress curve which is the basis of all metalworking calculations, together with different means of determining the curve for any metal. This will probably be familiar to many readers, and may be omitted by them, but it is again important to emphasise the difference between metalworking and structural analysis. Whereas the stress–strain curve for many engineers ceases to be of interest beyond the elastic limit, except in so far as ultimate tensile strength and overall reduction of area give an indication of properties, the metalworker can usually ignore the elastic region. He is interested in plastic strains well beyond the 'ultimate' tensile elongation, and these are much better expressed in terms of *logarithmic strain* $\epsilon = \int dl/l$ discussed in Chapter 2, § 2.2, rather than the usual strain $e = (l - l_0)/l_0$ employed in a tensile test. One advantage can be seen in an elementary way by imagining a wire stretched to double its length. The linear strain is just $e = (2l_0 - l_0)/l_0 = 1$. The corresponding compressive operation is forging a cylinder, but an equal strain of unity implies reducing the height to h' given by $e = (h_0 - h')/h_0 = 1$; $h' = 0$. Clearly this is a much more difficult operation than stretching the wire, and a more realistic assessment is given by the logarithmic strain which equates doubling the length to halving the height (see Chapter 2, Table 1). Other advantages are discussed in Chapter 2. Logarithmic strain is used throughout this book.

1.3 Mohr's circles and the yield criterion

Because of the important simplifications resulting from the use of principal stresses, and the extent to which they are therefore applied, it is necessary to have a clear understanding of them, and of their relationship to the other stresses acting at a point. This is developed in Chapter 3, from elementary considerations of equilibrium by resolution of forces. It is then shown that these relationships can be quickly derived, and easily remembered, by use of a diagram suggested by Mohr in 1914. This takes the form of a circle plotted on axes of direct stress and shear stress (Chapter 3, Figure 3.3), showing the values of these stresses on any plane of known orientation. Once the principal stresses are known, this diagram may be drawn immediately, and from it the stresses on a plane inclined at any angle to the principal planes can be found directly. This construction is used in later development of the theory.

Having seen how the stresses on one plane can be related to those on other planes, the next stage is to discuss the particular combinations of stresses which will cause the metal to flow plastically.

Two yield criteria are mentioned in Chapter 3. The first, proposed by Tresca in 1865, has the advantages of simplicity and easy physical interpretation. This criterion suggests that yielding occurs when the maximum shear stress reaches a critical value. Unfortunately, this is not quite as accurate as the later criterion, due to von Mises (1913) and Hencky (1924), which suggests that shear-strain energy is the important factor, rather than shear stress.

In all metalworking theory, it is assumed that yielding of an element occurs sharply when the appropriate stress combination is reached. This is not strictly true for a real specimen, as may be shown by measuring stress–strain curves (Chapter 2, Figure 2.1). There is a good linearity over the elastic range, but this is followed not by a sharp kink but by a bend of fairly small radius, leading into a smooth curve, of much lower slope, in the fully plastic zone. The knee of the curve occupies a small but finite range of stress and of strain, so that the yield stress cannot be determined with precision, but the error can be neglected for practical purposes at least for operations involving severe deformation. This small error in practical determination should not be confused with the larger steady increase in yield stress with deformation due to strain-hardening with progressive cold-working. This can be allowed for in calculations by considering the stresses on a small element of the work-piece, and using the yield stress appropriate to metal which has undergone the strain of that particular element. A much easier method, which gives reasonable accuracy for the majority of practical operations, is to assume the yield stress constant at some mean value for the particular working pass. It is important to recognise that, however these small inaccuracies may be treated, the important stress combination in metalworking is that which will just cause plastic flow at the particular strain reached. This is still the 'elastic limit' for metal which has been appropriately strained, whatever the severity of the deformation, even approaching 100% reduction of area, though the value may then of course be very much greater than the elastic limit found in the common tensile test on annealed metal. A tensile test on the deformed metal will, however, show this high value of yield stress. (When reductions of area are very large, it is preferable to use area ratio, rather than reduction of area. For example extrusion, and some forms of rolling, may give changes in cross-sectional area of say 20:1 in one pass. This corresponds to 95% reduction of area, but 98% reduction of area gives little impression of the greatly increased difficulty in producing 40:1 area change.)

1.4 Simple estimation of working load from yield stress

In the most simple type of operation, in which the stress is the same throughout the deforming body, the force required to produce flow, which is always the load in which we are interested, is the product of the yield stress and the appropriate area. In tensile elongation of a wire, this area is the cross-sectional area of the wire; in compression of a cylinder, it is the contact area of the cylinder and a platen. This approach can be elaborated to some extent. Thus, to a first approximation, cold rolling can be regarded as simple compression and the roll load assessed from the mean yield stress of the metal between the rolls, and the projected area of contact (Chapter 4, equations 4.2, 4.3).

In many operations, however, the applied force is not related in this simple way to the cross-section. For example, in wire-drawing the drawing force is applied to the drawn wire, as in stretching, but the stress is that necessary to cause yielding in the die zone, not yielding of the drawn wire. An estimate of the working load may be obtained by considering the work done in plastic deformation of unit volume element of the metal. As will be shown in equation 4.6 (Chapter 4) this is equal to the area of the stress-strain curve appropriate to the strain undergone by that element. In the absence of friction and any other energy consuming process, the total of this work must be supplied by the external force. In this way a 'work-formula' may be derived giving the load for a particular process. In wire-drawing from cross-sectional area A_0 to A_1, for example, the drawing force F is related to the mean yield stress $\overline{Y}$ by the *work-formula* of equation 4.9:

$$F = A_1 \overline{Y} \ln \frac{A_0}{A_1}$$

Although this type of solution allows for different strain at different parts of the deformation zone, it assumes *homogeneous deformation*, in which elements which were originally rectangular in section remain geometrically similar, and do not distort. The principal stresses are then uniform across the section. Because friction is also neglected, this type of solution always gives a lower limit for the working load.

1.5 Stress-evaluation[1.3] to allow for friction

Allowance can be made for friction by considering the local equilibrium of forces acting on small elements of the deforming metal. This is shown in detail in Chapter 4. Taking wire-drawing as an example again, the forces acting on a cross-sectional lamina, at some diameter D in the die, are the drawing force, the radial (or near-radial) die force, and the frictional drag at the die surface (Chapter 6, Figure 6.2). By resolving these in an axial direction and equating their sum to zero, a differential equation can be obtained for the equilibrium drawing stress σ_x in wire (Chapter 6, equation 6.19), thus:

$$D\, d\sigma_x + 2[\sigma_x + p(1 + \mu \cot \alpha)]\, dD = 0$$

This can be integrated directly, for constant coefficient of friction μ and die angle 2α, because σ_x is related to the die pressure p and the simple tensile yield stress Y, by the yield criterion. The resulting equation (6.22) gives the drawing stress for a diameter reduction from D_b to D_a:

$$\frac{\sigma_{xa}}{Y} = \frac{1+B}{B}\left[1 - \left(\frac{D_a}{D_b}\right)^{2B}\right]$$

The parameter B is an abbreviation for the product $\mu \cot \alpha$.

The same approach can be followed, as will be shown in the appropriate chapters of Part II, for other processes: rolling, forging, and extrusion. It is reasonably straightforward to apply, though the calculations may sometimes become lengthy, particularly when curved tools are used, as in rolling and occasionally in drawing. However, this type of load determination often gives results in good agreement with practical values for these two processes.

It must, nevertheless, be applied with caution, because under many circumstances the load is seriously underestimated for drawing and forging, and is even more in error for piercing and extrusion. The error arises when the external constraint causes appreciable internal distortion of the workpiece beyond that strictly necessary for the shape change. To take the example of wire-drawing again, an element of the wire near the surface initially moves towards the die in a direction parallel to the axis, but on entering the die it is compelled to move with an inward, radial, velocity component (Figure 4.4). This is accomplished by shearing within the wire. As it passes through the exit from the die, the element is sheared back again, to proceed once more in an axial direction. The steeper the die angle, the more this shear exceeds the minimum shear required for elongation of the wire. Both these shear processes consume energy, though they do not contribute directly to the external form of the product. The corresponding amount of additional work is known as the *redundant work.* This is greater when the die has a large included angle than when it is a long die of small angle. On the other hand, long dies give more frictional drag, and since both redundant work and frictional work are superimposed on the work of homogeneous deformation, there is found to be an optimum die-angle for wire-drawing, giving the least drawing force for a particular reduction of area.

Extrusion dies are nearly always very steep, and the redundant work in extrusion is much greater than in wire drawing, so that the stress-evaluation technique has little value for practical extrusion.

1.6 Slip-line field theory[1.4], allowing for redundant work

There is unfortunately no way of allowing for redundant work in all the applications covered by stress evaluation. For the special conditions of plane-strain deformation, however, there is a method known as *slip-line field analysis* which gives strikingly close agreement with experiment. Though mathematically rigorous, this depends on drawing an initial field based on experience. *Plane strain* implies that there is no strain in one direction ($\epsilon_z = 0$); all elongation or contraction occurs on planes (*XOY*) perpendicular to this direction and is the same on all such planes, and can therefore be represented on a two-dimensional diagram. Deformation in plane strain is pure shear strain, and it is assumed that this is produced by pure shear stress. A slip-line field is, in fact, a two-dimensional vector diagram which shows the directions of maximum shear stress, identified with the directions of slip, at any point. There are always two such directions since, as already mentioned, shear is always accompanied by complementary shear at 90°. Thus a concrete cube, for example, if compressed to failure, will shear along its diagonals, leaving two pyramids (see also Figure 5.1). Similarly, if a long metal bar of square cross-section is laterally compressed in one direction, it will start to shear along lines at 45° to the sides. The appropriate slip-line field is a corresponding diagonal cross. In most real metals, this pattern becomes obscured if deformation is continued, but it can be recognised in lightly-deformed specimens by surface markings, or by grain structure revealed by etching in appropriate reagents. It is quite clear in brittle alloys such as 10% aluminium bronze or Mg 4% Al.

The slip-line field is always a network of lines crossing each other at right angles. It is built up by a trial and error process based on previous experience, starting with a knowledge of the equilibrium conditions at boundaries, as shown in Chapter 5. It must satisfy continuity and velocity conditions within the deforming metal. Because the deformation is pure shear strain, at the instant of yielding, the shear stress along any slip-line, anywhere, has a value k equal to the yield stress measured in a pure-shear test. The stress system at any point in the zone of plastic deformation can be considered as a combination of this stress with a hydrostatic pressure $-p$ which does not influence the yield stress (Chapter 5, equation 5.2). In fact, the principal stresses σ_1 and σ_3 have the values $-(p + k)$ and $-(p - k)$, respectively, and act at 45° to the directions of maximum shear, which are given by the directions of the slip-lines. Therefore, if p can be determined, the magnitude and direction of both principal stresses can be found from the slip-line field and a knowledge of the yield stress. Consideration of equilibrium, and an application of the Mohr circle, show that the change in p between two points is directly related to the angle between the slip-line tangents at these points. In the simple compression example quoted above, the slip-lines were straight, and for this condition the hydrostatic pressure is everywhere constant. For most operations, the field is more complex and involves lines of constant or varying curvature. The procedure in evaluating the stress from a given slip-line field is to establish a starting point at a boundary where the magnitude of the hydrostatic pressure is easily found; for example from the condition that no net normal or tangential stress acts on a free surface. The slip-line rotation from such a point to a point on the tool face is then determined from the diagram, and this gives the value of p and hence of σ_1 and σ_3 at the tool face.

The stress values found in this way are sometimes more than twice as great as would be predicted with the assumption of homogeneous deformation (cf. equations 5.23 and 5.24), but slip-line field solutions have always been found to agree well with experiments designed to test them. Their disadvantage, apart from relative complexity in some examples, is that they are restricted to plane-strain deformation, which seldom occurs in metalworking processes. Rolling of wide, flat, strip (Chapter 9) approximates to plane-strain working, since the deformation zone is elastically constrained by the undeformed strip on entry and exit sides of the parallel rolls, and thus prevented from major spreading in the lateral direction. Lateral strain can be ignored if the strip width is more than about ten times its thickness. Otherwise, few practical processes even approximate to plane-strain conditions. Ironically, cold-rolling usually involves very little redundant work (the equivalent 'die-angle' is small for large-diameter rolls) and the stress-evaluation results agree well with practice. Nevertheless, because slip-line field theory is mathematically rigorous, and the experimental verification of its predictions is so good, even when the predicted redundant work factors are large, the somewhat artificial condition of plane-strain deformation has attracted considerable attention.

It has recently been shown[1.7] that the slip-line field theory can be used to predict the type of result to be expected with axially-symmetrical conditions which are far more common. Thus, the slip-line field solution for plane-strain extrusion pressure P can be expressed in the form:

$$\frac{P}{2k} = a + b \ln \frac{1}{1-r},$$

where a and b are constants, and r is the reduction in area. The same relationship holds remarkably well for extrusion of round rod, if the two constants are determined empirically;

$$\frac{P}{Y} = 0{\cdot}8 + 1{\cdot}5 \ln \frac{1}{1-r}$$

This approach considerably extends the potential applications of slip-line field theory.

1.7 Load-bounding technique

Johnson and his colleagues[1.5, 1.7] have extensively applied methods which are simpler than the slip-line field technique and can be applied to a range of metalworking problems to obtain estimates of working loads. Their procedure is to determine a lower limit, or bound, which is certainly too low a load to deform the metal, and also to determine an upper bound which is certainly too high. The real load must then be between these limits, and the skill in application of the method lies in choosing deformation patterns which make the difference as small as possible.

The *lower bound* is found from consideration of stress conditions, paying no attention to internal flow restrictions. The *upper bound* is found from consideration of strains only, ignoring stress conditions. In metalworking the upper bound is of great interest, since it ensures that a particular operation can actually be carried out without exceeding the load predicted.

In a sense, the slip-line field solutions are upper-bound solutions, since they are based on the flow or slip of metal within the body. For general upper-load bounds, however, it is not necessary to use the full elaboration of slip-line fields. An approximation may be made by dividing the plastic zone into straight-sided areas, ignoring the orthogonality of the network formed. It is then assumed that all elements of metal within one of these areas move together without shearing, and that shearing is localised to the boundaries between the areas. Frequently the slip-line field solution may be used as a guide to the type of division to choose (Chapter 5, Figures 5.13*a* and *b*). A corresponding velocity diagram is then drawn, to show the velocity at any point in the plastic zone (Figure 5.13*c*). From these two diagrams, the rate of doing work by shearing can be assessed for the postulated deformation pattern. The shear force on any shear line is the product of its length s and the shear yield stress k. The distance moved in unit time is the velocity u, so the work done in unit time is the product $(k.u.s)$. The configuration giving the least value of the sum of such products for all shear lines, that is, the one calling for the least rate of working, is the most probable solution. The lowest value chosen is then equated to an expression for the rate at which the external force does work. The necessary load is then found from this simple equation (e.g. equations 5.33 and 5.35). Where solutions have been compared with slip-line field results, the accuracy of the best upper-bound load has been found quite good. The technique has also recently been applied to axially-symmetrical conditions of extrusion.

1.8 Situation in 1965

The work formula gives the least possible load for any operation, by considering the work necessary for homogeneous deformation only. The additional work necessary to overcome friction can be allowed for by stress evaluation. For cold-rolling, and some drawing and forging the results are then found to agree well with practical measurements of load. Redundant work, arising from excess internal distortion, requires analysis of the internal flow pattern. This can be done with considerable precision by slip-line field theory, for the special conditions of plane-strain deformation. In some instances it is possible to apply the type of solution obtained to conditions of axial symmetry, but this leaves constants to be determined empirically. A simpler form of slip analysis has been exploited more recently, which permits estimation of an upper bound to the load by considering simplified, straight, lines of block shearing. This approach gives numerical results much more readily than rigorous slip-line field analysis, though the results are less accurate, and the method cannot be used to find general solutions. It does not give a valid representation of the real flow pattern, as slip-lines do. Upper bound technique has been applied to load determination in axially-symmetrical extrusion. Slip-line field analysis can take account of friction, by modifying the angles at which the slip lines meet the interference. Upper-bound theory currently deals only with full sticking friction (no relative movement) or perfect lubrication ($\mu = 0$). Until recently, slip-line field theory considered only metal of constant yield stress, but modifications to the Hencky equations (which relate the hydrostatic pressure to slip-line rotation) have recently been proposed[1.8], to allow for strain-hardening. The stress evaluations on the other hand, do allow in detail for strain hardening. Examples can be given where this is conveniently done with the aid of a small analogue computer.

An advantage of using the slip-line field approach, apart from the gross inaccuracies in load determination which may arise if redundant work is ignored, is that it provides some insight into the actual flow pattern and the distribution of strain and stress within the deforming metal. This can be extended to give a pattern of temperature distribution for example, but caution should be exercised in this when strain-hardening is appreciable, since relatively little has yet been done with slip-line fields modified to allow for strain-hardening.

All these solutions apply to relatively simple metalworking operations, but they often provide a valuable guide to the more complex conditions frequently found in industrial practice.

1.9 Developments since 1965

Metalworking theory has been greatly influenced by the introduction of digital computer techniques, which have extended the scope of slip-line field and upper-bound solutions as well as providing new fully-numerical procedures (see Chapter 13).

Slip-line fields[1.9] still give the most accurate and detailed information about local deformation, the extent and shape of dead zones, and changes in free-surface profile, though the rigorous theory is restrictive in its assumptions. It is now possible to draw slip-line fields by computer, ensuring conformity to boundary conditions and continuity. This enables progressive solutions to be formulated and the effects of changes in geometric and frictional conditions to be evaluated quickly. The general form of the initial slip-line field must however be postulated first. This can be done from prior experience but a field can conveniently be established by the use of *visioplasticity*[1.10]. A model, preferably of pre-strained aluminium, is prepared in two halves with carefully ground and polished plane faces, one of which is etched with a fine grid pattern. The two faces are held in close contact while the specimen is deformed by a few percent, and the grid is then examined to reveal the local shear-strain directions, from which an approximate slip-line field can be built up, and then adjusted slightly to conform to the theoretical constraints.

The strict mathematical slip-line field theory still requires the assumptions of plane strain, and homogeneous, isotropic, non-hardening material[1.11]. Nevertheless, for most primary metalforming, as distinct from sheet-metal working, the effect of inhomogeneity and anisotropy on major forces and flow patterns can be ignored, though they may greatly influence local cracking, and other defects. Acceptable approximate solutions can be obtained for many axi-symmetric problems[1.9], and there has been a continuing interest at some laboratories in the remaining serious limitation of non-hardening material. Progress has been made towards a method of accounting for strain hardening, especially in metal cutting, so slip-line field theory is approaching a state of general applicability to cold metal working. It should be realised that for heavily pre-strained metal, the assumption of a constant mean yield stress value gives good accuracy.

In hot working, the slip-line field approach is less reliable. A good general picture of the deformation can be obtained for simple shapes, but in detail this is greatly influenced by strain-rate and temperature distributions[1.12]. It seems likely that this is the main area in which the newer fully-numerical techniques will find most application.

The most detailed of these is *finite-element analysis.* This has been used extensively since about 1960 for solving complex problems of elastic stress and strain. The equations and computer programs look daunting, but in fact the equations are all simple algebraic ones involving no high-level mathematics. The elements considered are very small, to provide good accuracy, so large numbers of simultaneous force-displacement equations need to be solved. A digital computer is therefore essential, but the solutions then give complete distributions of stress, strain and displacement. As applied to plasticity, the method relates the stresses acting on the faces of an element to the incremental displacements produced, using the principle of virtual work. A restriction to calculations of both elastic and plastic deformation is that linear relationships need to be assumed, either as Hooke's Law or as the Levy-Mises equations. Anisotropic properties and non-linear strain-hardening can be incorporated, but only by some form of optimisation. For such problems, and for large strains in non-steady forging and other operations, very large computer storage is needed.

A matrix method can be used to reduce the demands on the computer. The equations are formulated with a set of point velocities and a set of multipliers representing the redundant-work factors in inhomogeneous working. The problem is then to find suitable sets that will give a stationary value, according to the variational principle, for the total rate of dissipation of homogeneous and inhomogeneous deformation energy, together with the external surface energy. An iterative procedure is needed, so there is still a need for extensive computer facilities, but the calculations become feasible for quite detailed solutions, including assessment of the dead-zone configurations.

Another method depending on the calculus of variations utilises *weighted residuals.* A function of the discrete variables is postulated, approximating to the continuous stress and flow functions. The method of least squares is then used to find the best approximation, by minimising the residual errors through a choice of appropriate weighting factors for each separate value.

All these numerical processes are very long, as is also the slip-line field procedure. For many purposes it suffices to obtain approximate solutions by *limit analysis,* and usually the upper bound is much more important than the lower. The very simple plane-strain upper bounds previously described have now been considerably extended. An important contribution is the introduction of the spherical velocity-field for application to problems of axially-symmetric workpieces[1.13]. Kinematically-admissible velocity fields are developed with suitable spherical surfaces of velocity discontinuity. These may be effectively dictated by the geometric boundaries, as in simple wire drawing, or found by minimising the total energy dissipation. A recent development of this approach allows it to be applied to regular polygonal sections, for example square bars.

Upper-bound solutions to all but the simplest problems are now normally found by computer. Quite complicated models, with flow in more than one major direction, can be handled. For this purpose the workpiece is first supposed to be divided into separate but coherent deformation zones, which are then subdivided in an optimal way. According to one approach, the zones are subdivided into rigid triangles, so that energy is dissipated along the discontinuities and at the boundaries only. The other method considers each zone to deform homogeneously, but, as in the earlier upper-bound theory, the best solution is found by minimising the overall rate of performing work. The deformation zones can have curved boundaries, and unsymmetrical shapes can be included.

An important feature of the newer methods is that they make it possible to discard most of the restrictive assumptions of rigorous plasticity theory, and to produce good approximations to the flow behaviour of real materials. Basic material properties can be incorporated in solutions relating to both stress and strain distributions and hence to homogeneity and fracture in the final product. To a lesser extent, allowance can be made for temperature and strain-rate variations for hot working. These features are much more valuable in practice than a precise knowledge of the working load.

REFERENCES

1.1 Mort, E. R., 'Production of sheet and bar from molten metal', *Metal Ind.* (*Lond.*), **54**, 41–46, 1939.

1.2 Fenton, G., Littlewood, G. and Zaeytydt, T. J., 'Continuous casting research at BISRA and CNRM', *Iron and Coal Trades Review*, **182**, 681–685, 1961.

1.3 Hoffman, O. and Sachs, G., *Introduction to the Theory of Plasticity for Engineers*, McGraw-Hill, 1953.

1.4 Hill, R., *The Mathematical Theory of Plasticity*, Oxford Univ. Press, 1956.

1.5 Johnson, W. and Mellor, P. B., *Plasticity for Mechanical Engineers*, Van Nostrand, 1972.

1.6 Alexander, J. M. and Brewer, R. C., *Manufacturing Properties of Materials*, Van Nostrand, 1963.

1.7 Johnson, W. and Kudo, H., *The Mechanics of Metal Extrusion*, Manchester Univ. Press, 1962.

1.8 Palmer, W. B. and Oxley, P. B., 'Mechanics of orthogonal machining', *Proc. Inst. Mech. Engrs*, **173**, 623–654, 1960.

1.9 Johnson, W. and Mellor, P. B., *Engineering Plasticity*, van Nostrand, Reinhold, London 1973

1.10 Thomson, E. G., Yang, C. T., and Kobayashi, S., *Mechanics of plastic deformation in metal processing*, McMillan, New York, 1965.

1.11 Johnson, W., Sowerby, R., and Haddow, J. B., *Plane-strain slip-line fields*, Arnold, London, 1970.

1.12 Lange, K. (Ed.), *Lehrbuch der Umformtechnik Band 2 – Massivumformung*, 1974.

1.13 Avitzur, B., *Metal forming: processes and analysis*, McGraw-Hill, New York, 1968.

1.14 Hoffmanner, A. L., (Ed.), *Metal forming: interrelation between theory and practice*, *Metall. Soc. A.I.M.E. Symp.*, Plenum Press, New York, 1971.

2

Stress–Strain Curves

In all metalworking operations, the workpiece is permanently deformed, sometimes very severely. A major object of metalworking theories is to permit prediction of the amount of deformation, and the forces required to produce this. First, it is necessary to know the deformation characteristics of the particular metal, which may be determined from a test with a simple, calculable, stress system. The theory is then used to apply this information to the more complex system actually operating.

The simplest form of stress application is produced by a single tensile force, and this is the basis of the most widely used strength test for common engineering materials. This simple tensile test, described in §2.1, has certain disadvantages in the context of metalworking, especially that it is limited to relatively small shape changes. Other tests, of a compressive or shear type, are consequently necessary to obtain deformation characteristics applicable to severe metalworking operations.

2.1 The tensile test

The specimens used for a tensile test are characteristically simple in form, with a central uniform region of circular or rectangular cross-section. Two small marks are usually made in this region, to indicate the gauge length used for measurement of extension under load. The ends of the specimens increase smoothly in cross-section to a larger size, convenient for gripping in jaws without causing significant deformation.

The jaws may be serrated or threaded, to hold the ends of the specimen securely, and they are often self-aligning to ensure axial loading. A load is applied steadily by a hydraulic ram or a motor-driven screw, and its magnitude is determined continuously by the flexure of a beam, or by the movement of a dead-weight necessary to restore balance. The extension of the gauge length is magnified by some optical, mechanical, or electrical system, and is also measured continuously. The results are presented in the form of a load-extension graph, obtained by manual plotting or, in more modern machines, automatically.

2.1.1 *A typical load–extension curve and stress–strain curve*

Figure 2.1*a* shows the general form of load–extension curve obtained for most metals and alloys, tested in the annealed condition, without prior working. For engineering design purposes, it is more convenient to generalise the curves by plotting them in terms of stress and strain (Figure 2.1*b*), defined in the following way:

$$\text{Nominal stress } s = \frac{\text{Applied load}}{\text{Original cross-sectional area}} = \frac{P}{A_0} \tag{2.1}$$

$$\text{Linear strain } e = \frac{\text{Extension of gauge length}}{\text{Original gauge length}} = \frac{l - l_0}{l_0} \tag{2.2}$$

The stress is usually expressed in tons per square inch, Newtons per square millimetre or pounds per square inch, in English-speaking countries, or in kilogrammes per square millimetre. The strain is, of course, dimensionless.

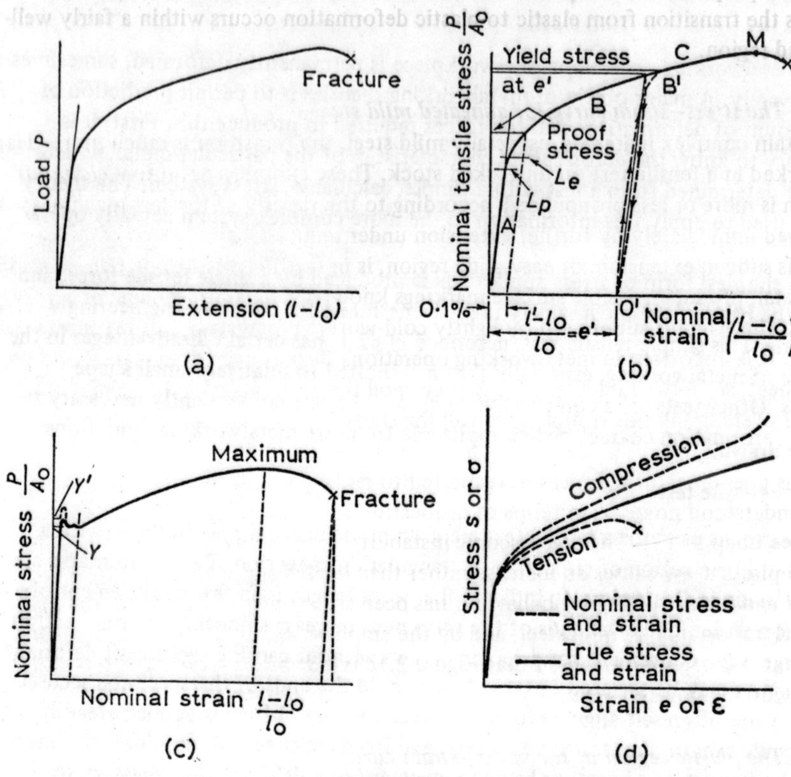

Figure 2.1

(*a*) A representative load–extension curve for a ductile material, such as alloy steel.

(*b*) The same data, replotted in terms of stress and strain.

(*c*) A stress–strain curve for mild steel, showing the yield point peculiarity.

(*d*) Stress–strain curves for idealised compression and tension.

When the strain is small, the stress is found to be directly proportional to the strain (Hooke's Law). The graph is correspondingly straight, up to a point L_p known as the *limit of proportionality*. Its slope is the *Young's modulus E*:

$$E = \frac{\text{Stress in uniaxial tension}}{\text{Corresponding elastic strain}} \tag{2.3}$$

This is a constant of the material, having the dimensions of stress. Typical approximate values, which are convenient to remember for rough calculations, are: 200 kN/mm² (30 x 10⁶ lb/in²) for mild steel; 120 kN/mm² (18 x 10⁶ lb/in²) for copper. The range of elastic strain is defined as the range within which the testpiece will return to its original dimensions when the load is removed. Beyond this *elastic limit* L_e, the metal will retain a permanent set. In other words it will deform plastically.

The actual positions ascribed to L_p and L_e will depend on the sensitivity of the measuring apparatus. With annealed metals, the true values may be very low; but for practical purposes the two points can be considered coincident, and with worked metals the transition from elastic to plastic deformation occurs within a fairly well-defined region.

2.1.2 *The stress–strain curve for annealed mild steel*

In certain complex materials, especially mild steel, the transition is much more clearly marked in a tensile test on unworked stock. These exhibit a definite *yield point* (which is more or less pronounced, according to the rigidity of the test machine), followed immediately by further extension under reduced load.

This sudden extension, or easy-glide region, is in fact localised to certain zones in the metal, producing visible surface markings known as 'stretcher-strains' or Lüders bands. The effect is suppressed by lightly cold-working mild steel, but reappears on ageing. It is important in metalworking operations such as deep-drawing of mild steel sheet, since the Lüders markings may spoil the surface appearance. The usual way of overcoming this trouble is to roll the sheet lightly, not more than 24 hours before drawing.

This phenomenon is, however, restricted to relatively few materials. It is quite well understood now[2.1], in terms of dislocation locking, but need not concern us yet (see Chapter 12). The macroscopic instability does not occur in compression, where plastic flow causes an increase rather than a decrease in cross-sectional area. Upper and lower yield-point behaviour has been so much accentuated by the commercial importance of mild steel, and by the emphasis on small strains in engineering, that it is often not realised that Figure 2.1*c* is exceptional, not typical, from a metalworking point of view.

2.1.3 *The plastic region of the stress–strain curve*

If the load in a tensile test is increased to a value at B' (Figure 2.1*b*), greater than the elastic limit L_e, and is then removed, the specimen will be plastically deformed. It will not, however, remain at a length l' corresponding to the point B', but will contract along the line $B'O'$ as the load is released. To a first approximation, this line is straight and parallel to the original elastic-deformation curve, with slope E. If the load is re-applied, the specimen will extend elastically, following $O'B'$ again until the point B' is reached, at which it will start to deform plastically again. In fact, $O'B'$ is slightly curved in opposite senses for loading and unloading and the point B' does not repeat precisely, so that the cycle of removal and reapplication of stress encloses a small loop and involves expenditure of unrecoverable work. This elastic hysteresis is, however, small enough to be ignored for metalworking purposes, though it is important in structures subjected to cyclic stressing and fatigue failure.

The curve $O'B'C$ can be considered as a new stress–strain curve for prestrained material. The transition from elastic to plastic deformation at B' is more clearly defined (except for mild steel, which now follows the general pattern, with no yield-point phenomenon) and occurs at a higher stress. Far from failing at the elastic limit L_e, the elastic strength of the material has been improved by stretching, a simple form of cold working. The nature of this *work-hardening* or *strain-hardening* process can be explained in terms of pile-up of dislocations in the metal[2.1], but in metal-working theory it is necessary to consider only the change in macroscopic properties.

In principle, the stress–strain curve could be built up over its whole range by straining the material a small known amount, and then subjecting it to a separate tensile test to determine the stress at which the transition from elastic to plastic stress occurs. This stress is known as the *yield stress*. For the purposes of metal-working, the stress–strain curve should be considered as a graph showing the value of the yield stress after any chosen degree of prior deformation. Before discussing this further, brief mention should be made of the technological features of the tensile stress–strain curve.

2.1.4 *Special characteristics and results of tensile testing*

The improvement in strength of metals due to work-hardening has long been known and utilised. It is accompanied by a decrease in ductility, and does not continue indefinitely in the tensile test. If the load is steadily increased beyond the point B', the load–extension curve will slowly rise, until at M it flattens out and then begins to fall. This fall coincides with the formation of a local neck in the testpiece. The load thus acts on a diminishing cross-sectional area, and this unstable situation terminates in fracture. This marks the final limit of use of the tensile specimen, or of any component subjected to tension. It is thus of importance in engineering; and in metal-working operations, such as drawing, which involve a major tensile stress. In the standard tensile test, the two parts of the fractured specimen are fitted accurately together again, and measurements are made of the smallest diameter and the final separation between the gauge marks. For structural design purposes, the results of the test are usually given in numerical form rather than as a complete graph, as follows:

Young's modulus

$$E = \frac{\text{Load/Original area}}{\text{(Length–Original length)/Original length}}$$

The measurement is made in the elastic region of the load–elongation curve. The value is little influenced by purity, grain size, strain rate, or temperature, and may usually be taken as a known constant of the material.

Yield point

$$Y' = \frac{\text{Load at elastic limit}}{\text{Original area}}$$

This value is clearly marked for annealed mild steel, to which the definition first applied, but for most annealed metals it is difficult to decide the exact position on the stress–strain curve. For design purposes it is therefore often preferable to specify the proof stress. This practice is becoming increasingly common.

Proof stress. The stress quoted is that required to produce a stated, small, permanent deformation, usually 0·1% strain. The strain measured during application of the load would, of course, exceed this value by an amount equal to the elastic recovery from this stress. Proof stress has no particular value in metalworking, where large strains are involved, except to specify, the properties of the final product.

Ultimate tensile stress

$$\text{U.T.S.} = \frac{\text{Maximum load}}{\text{Original area}}$$

The U.T.S. is widely quoted in engineering handbooks as a measure of strength, but it has no fundamental significance. It is not a real stress, since the actual area of the specimen will normally be very much smaller than its original size by the time the maximum load is reached, especially when annealed metals are tested. The same criticism can be applied to the deduced values of E and Y', but the strains there are too small (less than 0·1%) to alter the results. The cross-sectional area of annealed copper may, on the other hand, be reduced to half by the time the maximum load is reached.

The U.T.S. is also in no sense ultimate. It has been proposed that the term should be replaced by M.T.S.—maximum tensile stress, since it is evaluated from the maximum load. Even so, as will be seen in § 2.3, equation 2.13, the stress at M is determined by the rate of work-hardening rather than by a fracture criterion. The real stress at fracture, given by the quotient of load at fracture and area at fracture, will be much greater. Even this is governed by the hydrostatic tension induced by the neck formation, and the appearance of non-axial stresses. If the load is removed during the early stages of necking, and the specimen is remachined to a true cylindrical form, considerable further deformation and work-hardening is possible, even up to 80% total elongation.

In general, it is preferable to consider yield stress in design and in metalworking. The significance of U.T.S. in the past has been due largely to the reasonably constant ratio between ultimate tensile stress and yield stress, for most traditional materials. However, in several newer materials, including high-alloy steels, U.T.S. and Y.S. are much more nearly equal, and indiscriminate use of U.T.S. could be misleading.

Elongation

$$e = \frac{\text{Overall final length between gauge marks} - \text{Original gauge length}}{\text{Original gauge length}} \times 100\%$$

The elongation measured from the fractured parts has two components: a uniform extension up to M, which is proportional to gauge length; and a local extension in the neck, which depends on diameter but not on gauge length. Thus this quantity also has no fundamental significance, since it depends on specimen dimensions. For specified diameter and gauge length, it gives comparative results for application to known conditions, but it is not always a true guide to behaviour in service.

Reduction of area

$$\text{R.A.} = \left(\frac{\text{Original cross-sectional area} - \text{Final area}}{\text{Original area}}\right) \times 100\%$$

Because the final area, or fracture area, is largely determined by the necking parameters, the R.A. also has no fundamental significance. It is generally useful, in much the same way as the elongation, as an indication of ductility. A high reduction of area usually indicates a low rate of work hardening, and consequently delayed onset of necking.

2.2 True stress and natural strain

2.2.1 *True stress*

The nominal stress s defined for the tensile test in terms of original area is not really a stress, because the cross-sectional area A at the instant of load measurement is less than the original area A_0 used in the evaluation $s = P/A_0$. For the region of greatest interest to structural designers, where the strains are elastic, the difference between A and A_0 is negligibly small, but this is far from true when the maximum load is reached. Neither is the difference negligible for the great majority of metalworking processes. Whenever the strain exceeds a few per cent, it is preferable to consider the actual, or true stress operating at a given instant:

$$\text{True stress } \sigma = \frac{P}{A} \tag{2.4}$$

$$\text{cf. Nominal stress } s = \frac{P}{A_0} \tag{2.1}$$

It is not so easy to obtain values of σ from the test, since the force and the cross-sectional area must be measured simultaneously; but because of its more fundamental significance, σ is of importance for all metalworking calculations. The true stress, as here defined, is of course equally accurate in tension and compression, with appropriate regard to sign convention.

2.2.2 *Natural strain*

The linear strain e defined for the tensile test is of less basic significance than the incremental strain experienced by the test piece at a particular stage. This natural strain ϵ is defined in terms of the strain increment:

$$\text{Natural strain } \mathrm{d}\epsilon = \frac{\mathrm{d}l}{l}\,;\ \epsilon = \int_{l_0}^{l} \frac{\mathrm{d}l}{l} = \ln\left(\frac{l}{l_0}\right) \tag{2.5}$$

$$\text{cf. Linear strain } e = \frac{l - l_0}{l_0}\,;\ \mathrm{d}e = \frac{\mathrm{d}l}{l_0} \tag{2.2}$$

The natural strain ϵ obtained by integration is thus a logarithmic function, and is often referred to as the logarithmic strain.

It may be noted that logarithmic strains have the useful property of being additive in sequential processes, which is not true of linear strains. Thus,

$$\epsilon_{01} + \epsilon_{12} = \ln\left(\frac{l_1}{l_0}\right) + \ln\left(\frac{l_2}{l_1}\right) = \ln\left(\frac{l_2}{l_0}\right) = \epsilon_{02} \tag{2.6}$$

$$e_{01} + e_{12} = \frac{l_1 - l_0}{l_0} + \frac{l_2 - l_1}{l_1} \neq \frac{l_2 - l_0}{l_0}$$

2.2.3 *Relationships between nominal and natural strain*

Because linear strains are easily measured and recorded, but logarithmic strains are required for metalworking calculations, it is convenient to relate the two. In tension, according to the above definitions,

$$e_t = (l - l_0)/l_0 = l/l_0 - l$$

$$e_t = \ln (l/l_0) \quad = \ln (1 + e_t) \tag{2.7}$$

In compression, the corresponding relationships are expressed in terms of an initial height h_0:

$$e_c = \frac{h - h_0}{h_0} = \frac{h}{h_0} - 1$$

$$\epsilon_c = \int_{h_0}^{h} \frac{dh}{h} = \ln \frac{h}{h_0} = \ln (1 + e_c) \tag{2.8}$$

Because h is less than h_0, e_c is negative. This is the usual convention for elastic stress analysis. In metalworking, however, the predominant stresses are usually compressive, and it is common to reverse the convention so that compressive stress or strain becomes positive. Thus, with this metalworking convention:

$$e_c = \frac{h_0 - h}{h_0} = 1 - \frac{h}{h_0}$$

$$\epsilon_c = \int_{h}^{h_c} \frac{dh}{h} = \ln \frac{h_0}{h} = \ln \frac{1}{1 - e_c} \tag{2.9}$$

The strain in metalworking is frequently given as the fractional or percentage reduction in cross-sectional area:

$$r = \frac{A_0 - A}{A_0}\,;\, R = \frac{A_0 - A}{A_0} \times 100\% \tag{2.10}$$

Because the volume remains constant throughout a metalworking operation this can be related to the logarithmic strain.

$$Al = A_0 l_0.$$

$$r = 1 - \frac{A}{A_0},\ \frac{A}{A_0} = 1 - r.$$

$$\epsilon = \ln \frac{l}{l_0} = \ln \frac{A_0}{A} = \ln \frac{1}{1 - r} \tag{2.11}$$

The relationships between the different strain expressions can be seen in the following table.

The numerical value of the logarithmic strain is the same for doubling the length and halving the height. The linear strain, however, implies an equivalence between doubling the length, not a very difficult process, and compressing to negligible thickness; this is intuitively a much less realistic representation.

	e	ϵ	r
Extension to $l = 1{\cdot}1\, l_0$ (10% elongation)	+0·1	+0·095	0·1
Compression to $h = 0{\cdot}9\, h_0$ (10% compression)	−0·1	−0·104	−0·11
Extension to $2l_0$ (doubling length)	+1·0	+0·693	+0·5
Compression to $\frac{1}{2} l_0$ (halving height)	−0·5	−0·603	−1
Compression to zero	−1·0	−∞	

2.3 True stress–strain curves

2.3.1 *Tension*

If the load–extension curve (Figure 2.1*a*) is plotted in terms of true stress, it gives a real value of the yield stress after any chosen degree of prior straining. Because the cross-sectional area of the tensile specimen decreases continuously throughout the test, the true stress will always be greater than the nominal stress, as in Figure 2.1*d*. Conversely in a compressive test, the nominal stress would be the larger, because the area increases.

The true-stress curve exhibits no maximum, though the load itself reaches a maximum value and the nominal stress curve then begins to fall, as seen in §2.1.4. The reason for the maximum may, however, be found by considering the incremental conditions.

If the cross-sectional area of the specimen at any instant is A, and the load is then P, the true stress will be given by equation 2.4:

$$\sigma = P/A$$

The rate of increase of load with strain will thus be

$$\frac{dP}{d\epsilon} = \frac{d}{d\epsilon}(\sigma A) = A\frac{d\sigma}{d\epsilon} + \sigma\frac{dA}{d\epsilon} \tag{2.12}$$

If the volume $V = Al$ remains constant, which is very nearly true for all practical deformation processes,

$$\frac{dV}{d\epsilon} = \frac{d(Al)}{d\epsilon} = 0; \; l\frac{dA}{d\epsilon} + A\frac{dl}{d\epsilon} = 0$$

But, by definition,

$$d\epsilon = \frac{dl}{l} \tag{2.5}$$

so

$$\frac{dA}{d\epsilon} = -\frac{A}{l}\frac{dl}{d\epsilon} = -A$$

Equation 2.12 then becomes

$$\frac{dP}{d\epsilon} = A\frac{d\sigma}{d\epsilon} - \sigma A$$

This has a stationary value, which is a maximum, when

$$A\frac{d\sigma}{d\epsilon} = \sigma A, \text{ or } \frac{d\sigma}{d\epsilon} = \sigma \tag{2.13}$$

This shows that the load will reach a maximum value when the slope of the stress–strain curve, which is the *rate of work-hardening*, equals the magnitude of the applied stress. Physically this means that the increase in strength due to work-hardening just balances the decrease in area. Further strain causes instability, leading to formation of a neck, and finally to fracture. The work-hardening nevertheless continues; there is no change in the true stress–strain curve at this point.

2.3.2 *Compression*

The compressive stress–strain curve shows no necking, but even the true-stress curve, if plotted against nominal strain, shows a rapid increase at very high strain values, beyond about 80%. This does not, however, indicate a rapid increase in work-hardening, but arises from the nature of the nominal-strain scale. Thus 90% reduction in area corresponds to a ratio between original and final areas of 10:1, 95% to 20:1, and 98% to 50:1. As was seen in §2.2.3, the linear scale gives a poor indication of the increasing severity of very heavy passes. If the stress–strain curve is plotted in terms of logarithmic (natural) strain, the work-hardening is seen to continue steadily at a slightly decreasing rate.

In all metalworking it is desirable to refer to the true-stress/logarithmic-strain curves, and their use will be assumed throughout this book.

2.4 Simplified forms of stress–strain curve

It is sometimes convenient to express the stress–strain curve by an analytical expression, though the need for this has largely disappeared with the availability of computers which can handle the stress–strain curves, or corresponding numerical data, directly. Some simple expressions may still, however, be of use. They are represented graphically in Figure 2.2.

(*a*) It is found that the work-hardening of many metals approximates to a parabolic form. For annealed metals with a cubic lattice, the stress–strain curve can be represented with fair accuracy by a simple power law:

$$\sigma = B\epsilon^m, \tag{2.14}$$

where B and m are constants, the latter being known as the strain-hardening exponent.

(*b*) For cold-worked metals with an initial yield stress Y, the results are better represented by an equation of the form

$$\sigma = Y + B\epsilon^m \tag{2.15}$$

This ignores the elastic strain, which is very small (less than 0·1%), whereas plastic strains in metalworking are seldom less than 10%, and usually much greater. Neglect of elastic strain is sometimes considered equivalent to assuming an infinite Young's Modulus E, but this is a less tangible assumption.

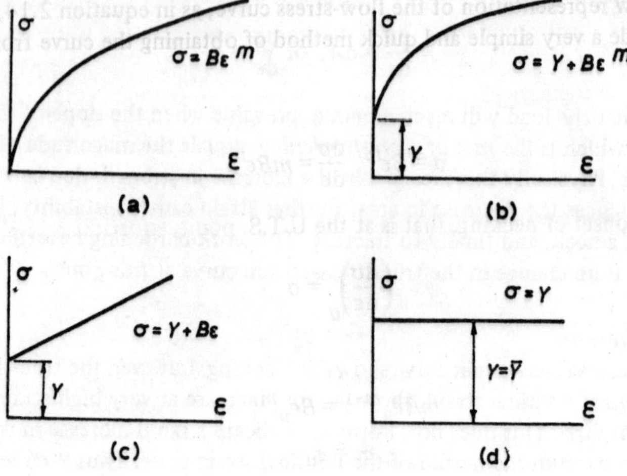

Figure 2.2. Approximate forms of stress–strain curve.

(*c*) A further simplification is often made by assuming that the strain-hardening rate is uniform

$$\sigma = Y + B\epsilon \tag{2.16}$$

This is a fair approximation for heavily-worked metals, but is no use for annealed or lightly-worked stock.

(*d*) The first approach to a problem is often to assume a material in which there is no work-hardening. Then the yield stress is constant, independent of strain. Elastic strain is again ignored, and the stress–strain curve is reduced to the horizontal straight line

$$\sigma = Y \tag{2.17}$$

This simple form is of basic importance in the majority of slip-line field theory (Chapter 5), though no real metal actually behaves in this way. However, in lead, for example, the processes of work-hardening and recrystallisation nearly balance at moderate strain rates, so that the yield stress is nearly independent of strain. This balance is critically dependent on temperature for pure lead, but tellurium-lead provides a fair model of a non-hardening material, and both have been extensively used in verifying and developing metalworking theory. Plasticine, though not a metal at all, approximates very well to the hypothetical material of constant-yield stress and negligible elastic-strain range[2.2]. Plasticine has been widely used for experiments and is particularly valuable also for studying flow patterns, since two different colours may be laminated or chequered in the billet. It is easy to impress fine grid patterns on it for more detailed flow examination.

In normal practical operations, the assumption of equation 2.17 is justifiable for very heavily worked metals, which are not appreciably hardened by further working.

2.4.1 *A simple method of determining a flow-stress curve*

The power-law representation of the flow-stress curve, as in equation 2.14, can be used to provide a very simple and quick method of obtaining the curve from a tensile test[2.8].

Thus if

$$\sigma = B\epsilon^m, \ \frac{d\sigma}{d\epsilon} = mB\epsilon^{m-1}$$

But at the onset of necking, that is at the U.T.S. point, equation 2.13 shows that

$$\left(\frac{d\sigma}{d\epsilon}\right)_u = \sigma$$

So

$$mB\epsilon_u{}^{m-1} = B\epsilon_u{}^m;$$

$$m = \epsilon_u$$

It is not even necessary to record the strain continuously, since the gauge length outside the neck will cease to deform once the necking commences. The value of ϵ_u is thus found directly from the diameters before the test, d_0, and outside the neck after the test, d_u

$$\epsilon_u = 2 \ln (d_0/d_u)$$

The constant B is found by measuring the maximum load L_u

$$\sigma_u = \frac{L_u}{\frac{\pi}{4} \cdot d_u{}^2}$$

The flow stress curve is represented by

$$\ln \sigma = \ln B + m \ln \epsilon$$

so a straight line of slope $m = \epsilon_u$ can be drawn through the point (σ_u, ϵ_u) on a $\ln \sigma$, $\ln \epsilon$ graph to determine the curve completely, up to the limiting strain at the onset of necking. Experimental comparisons with compression tests show that the power law can usually be extrapolated at least up to $\epsilon = 1$ (63% reduction) which is sufficient for wire drawing and most single-pass operations other than extrusion.

2.5 Selection of stress–strain curves for cold- and hot-working

For metalworking theory, the most important value obtained from the stress–strain curve is the yield stress. Ductility is important in tensile processes, such as drawing, and may determine the type of process to be used in some instances. For example, some rather brittle materials can be worked only by extrusion. Indeed, some materials are so brittle that even extrusion fails to prevent cracking unless a high hydrostatic pressure is superimposed.

There is a broad separation of metalworking into hot- and cold-working ranges[2.3]. The parameters governing the yield stress in these two types of process differ, and the tests must be made accordingly. The ranges are divided by the recrystallisation of the metal. As a rough guide, it is found that the recrystallisation temperature is approximately equal to half the melting point, expressed in degrees absolute. For the majority of metals, this is well above room temperature. Lead, however, recrystallises at room temperature, and super-pure aluminium and copper will recrystallise at only slightly elevated temperatures.

In the lower temperature range, including room temperature for most metals, the characteristic feature of the plastic portion of the stress–strain curve is strain-hardening. As the temperature is raised, recrystallisation, which causes softening, begins to compete with the strain-hardening. The increase in yield stress with strain is thus reduced, and the values become time-dependent. The strain rate in ordinary tensile testing is of the order 10^{-3} sec^{-1} (10^{-3} in./in./sec), and in this region the yield stress at room temperature shows practically no variation with strain rate. This is also true of most practical cold-working, where the rate may be say 1 sec^{-1} (1 in./in./sec). Some modern techniques, such as impact extrusion, or even explosive forming, may reach much higher rates (10^3 or 10^5 sec^{-1}) and the yield stress and work-hardening behaviour are then dependent on strain rate, but we shall not consider these specialised conditions here.

Above the recrystallisation temperature range, which must itself be defined with due attention to the times involved, strain-hardening becomes insignificant, but the yield stress is strongly dependent on time. If the times are sufficiently long, the specimen will continue to deform under a constant stress. This *creep process* is important in all high-temperature structural applications, but usually involves times measured in days, or hours at least, which are much longer than those encountered in metalworking. However, the value of the yield stress itself will depend on both the temperature and the strain rate in hot-working, so the test procedure must be made to match these parameters as closely as possible. This is often difficult because different parts of the workpiece may be subjected to different strain-rates, and even a single element of the workpiece may be strained at different rates as it passes through the process. The temperature will also vary, heat being lost by radiation and conduction to the tools, counterbalanced to some extent by the heat produced by deformation. Some reasonable mean values have therefore to be chosen.

Another factor which may be important in cold-working is the anisotropy of the workpiece, which may become quite pronounced after severe rolling or drawing for example. Test specimens taken across and along the rolling direction then give somewhat different stress–strain curves. This feature is of considerable importance in deep-drawing and spinning, but is very difficult to treat in a general way. Thus again a reasonable mean value is assumed for most purposes.

Tensile tests are usually unsuitable for hot-working, since the rate of work-hardening is so low that necking starts at a low stress (equation 2.13) and only small tensile strains are possible.

2.6 Compression tests for yield-stress determination

2.6.1 *Axially-symmetrical compression*

A serious limitation of the tensile test even for cold-working is that fracture occurs at a moderate strain, so that it is not possible to use this test for determination of yield stress after very heavy deformations. The fracture is most easily avoided by adopting some compressive form of test.

The simplest of these is axial compression of a cylinder between smooth platens. If the platens are well lubricated, this gives essentially the same yield stress as the tensile test, with small strains. As the strain is increased, however, the specimen spreads and the lubricant thins, so that the frictional component at the die face increases. This increases the load required to cause yielding, as shown in Chapter 10. It can be overcome to a large extent by removing the sample after small increments in strain and relubricating each time. To keep the friction between specimen and platens as low as possible, it is preferable to use cylinders of small diameter: height ratio. This test is somewhat slower than the tensile test, but with care can give useful results up to considerable strains.

A further complication in this test is that the cylinder tends to deform into a barrel shape. This interferes with the stress distribution, and may eventually lead to circumferential cracking due to secondary tensile stresses. To maintain uniform stress, as assumed, the barrelling may be machined away at intervals. By this means, Polakowski[(2.4)] was able to obtain an accurate stress–strain curve up to very heavy reductions, starting with a cylinder whose diameter was half its height. Unfortunately, the method is tedious, resembling the procedure for machining away the neck in a tensile test (§2.1.4).

A much simpler method combines measurement of the compressive load with a conventional ring test, described in §11.6[(2.9)]. The friction contribution, in the form of a friction factor $\tau/Y = m$, is deduced from the change in internal diameter of the ring, and this value is then inserted in an analytical equation that gives the average pressure p for forging in terms of the yield stress Y[(2.10)]. Unfortunately this equation is complicated, but it presents no problem if a computer is suitably programmed. A calibration chart showing p/Y has been devised.

2.6.2 *Plain-strain compression. (Ford test)*

The condition of plane-strain deformation is of great theoretical importance, as will be explained in Chapter 5. This condition implies that the strain in one dimension is zero; that is, that one dimension remains unchanged. This is very nearly true, for example, of most cold strip-rolling, where the change in thickness is entirely compensated by elongation in the rolling direction, with practically no lateral spread. Watts and Ford (1952) developed a compression test under plane-strain conditions[(2.5)]. A strip of thickness h is compressed between platens of breadth b, which overlap the strip as shown in Figure 2.3.

The width w of the strip should be at least 5 times and preferably 10 times the platen breadth, to ensure that lateral spread is negligible. The ratio h/b is chosen to lie between $\frac{1}{4}$ and $\frac{1}{2}$. If the thickness is too great, then the indentation pressure will exceed the simple yield stress, as explained in Chapter 5, §5.6. In the limit, when the thickness h is infinite, the process becomes one of indentation by a single punch, for which the indentation pressure is 2·57 times the yield stress. If, however, h does not exceed $\frac{1}{2}b$, the maximum error arising from this geometrical constraint will not be greater than about 2%. If h is very small, this error will be even less, but low h/b ratios imply a relatively large platen breadth b, which will increase the frictional contribution. Even when the dimensions are correctly chosen, it is necessary to ensure that the friction is as low as possible. For this purpose, the surfaces are well lubricated. Light rubbing with a soft pencil to form a graphite coating, together with a thin film of machine oil, is found to be best for copper, but a molybdenum-disulphide grease is most effective for steels. After each thickness reduction of a few per cent, the specimen is removed and relubricated.

This test has proved useful and reproducible, and can be recommended as a standard procedure. It should, however, be emphasised that the yield stress measured is S, the yield stress in plane strain; not Y, the yield stress in uniaxial tension, as measured in the tensile test. These are related, and it will be shown in Chapter 3, §3.6.4 that

$$S = \frac{2}{\sqrt{3}} Y = 1{\cdot}155\, Y \tag{3.27}$$

The disadvantage of the test is that it takes appreciably longer than a tensile test, but this is counterbalanced to some extent by machining time, since the specimens do not need to have a special shape. The time involved in the actual test depends on the frequency with which the specimens are relubricated, and this can be reduced

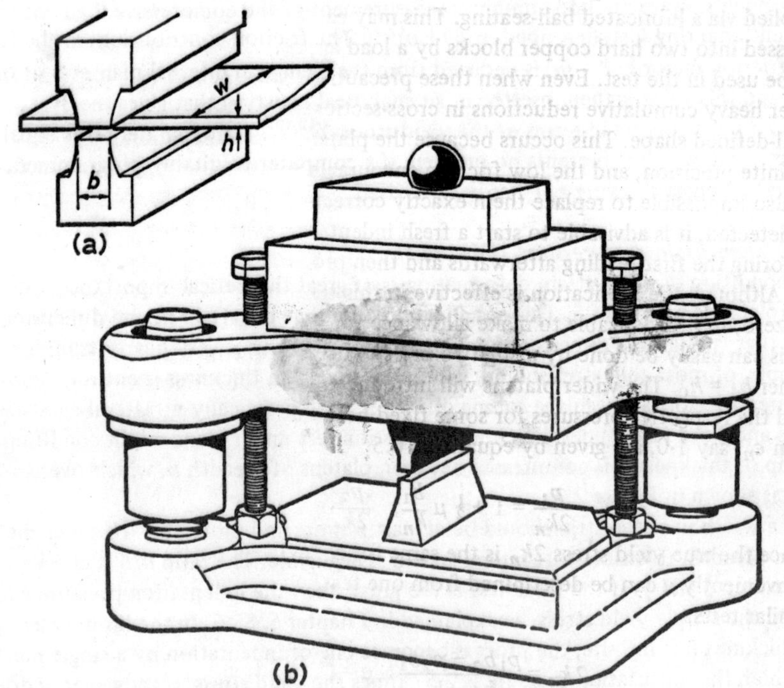

Figure 2.3. The plane-strain indentation test.

(*a*) Schematic arrangement showing dimensions.
(*b*) Outline drawing of an actual sub-press.

appreciably if a slightly lower accuracy is tolerable. Further time may often be saved in practice by making the first indentation large, so that the stress–strain curve starts effectively at say 20% reduction of area. Calculations based on annealed or lightly-worked metal are likely to be inaccurate, as we have seen, because the yield stress changes rapidly. It may be useful to tabulate the sequence of the test:

1. Choose a platen breadth b between 2 and 4 times the strip thickness h.
2. Cut the strip to a width w at least 5 times b.
3. Lubricate the strip.
4. Align the strip carefully so that it is square to the platens.
5. Estimate the load required to make say 2% or 5% reduction in h. Apply and measure the load accurately.
6. Remove the strip and measure the indentation accurately with a micrometer fitted with tips small enough to enter the narrow indentation.
7. Relubricate.
8. Replace the strip, with the dies in exactly the same position, and repeat the sequence 5–8.

The platens must be accurately ground square and parallel, and mounted rigidly in the sub-press. A special design is given by Watts and Ford, but a standard die-set provided with running ballrace sleeves performs excellently if the load is centrally applied via a lubricated ball-seating. This may easily be made with a 25 mm ball, pressed into two hard copper blocks by a load appreciably greater than the largest to be used in the test. Even when these precautions are taken, it will be found that after heavy cumulative reductions in cross-sectional area, the indentation loses its well-defined shape. This occurs because the platens cannot be ground parallel with infinite precision, and the low friction encourages any tendency to sideways slip. It is also impossible to replace them exactly correctly each time. As soon as this error is detected, it is advisable to start a fresh indentation with a single heavy reduction, ignoring the first reading afterwards and then proceeding as before.

Although the lubrication is effective in this test, the friction can never be reduced to zero, so it is desirable to make allowance for the finite coefficient of friction. This can easily be done by using two pairs of platens, one of width $b_1 = 2h_0$ and the other $b_2 = h_0$. The wider platens will introduce a larger contribution from friction and the respective pressures for some fixed height h_n corresponding to a large reduction ϵ_n, say 1·0, are given by equation 10.5:

$$\frac{p_1}{2k_n} = 1 + \tfrac{1}{2}\,\mu\,\frac{b_1}{h_n}; \quad \frac{p_2}{2k_n} = 1 + \tfrac{1}{2}\,\mu\,\frac{b_2}{h_n}$$

Since the true yield stress $2k_n$ is the same for both, μ can be eliminated, or more conveniently μ can be determined from one test and assumed to be the same in similar tests:

$$2k_n = \frac{p_1 b_2 - p_2 b_1}{b_2 - b_1}; \quad \mu = \frac{2h_n\,(p_1 - p_2)}{p_1 b_2 - p_2 b_1} \tag{2.18}$$

2.7 Torsion test

The plane-strain compression test is satisfactory for most cold-working applications, but for practical reasons it becomes unreliable for very heavy reductions. Cold-working is seldom carried beyond 80–90% reduction of area before re-annealing the workpiece. Because no limit is set by work-hardening and consequent loss of ductility, hot-working may be continued well beyond this. It is common to describe hot-working deformation in terms of area reduction ratios rather than percentage reductions (§2.3.2), and 20:1 or 40:1 reductions are quite common.

These very large strains are most conveniently simulated by a torsion test, in which the external dimensions of the testpiece remain essentially unchanged. Torsion testing is thus valuable for hot-working, and is widely used. The yield stress measured is the yield in pure shear, k, which, as shown in Chapter 3, §3.6.3, is related to the yield stress Y in uniaxial tension

$$k = Y/\sqrt{3} \tag{3.22}$$

The strain recorded is the shear strain. For structural calculations, shear strain is defined as the lateral displacement γ of an element, measured at unit distance from the base of the element. In the torsion tests, the shear strain is the circumferential rotation at unit distance from one end of the cylindrical specimen. Thus for a specimen of length l and diameter $2r$, twisted by an angle θ radians, the shear strain γ is

$$\gamma = r\theta/l$$

The number of revolutions possible before failure occurs is a measure of the ductility, or more usually, since this is primarily a high-temperature test, of hot-workability. It is often used at temperatures of 1,000–1,200°C, and for some metals, such as molybdenum, even higher. A detailed description of a torsion test is given by Hodierne[(2.6)].

2.8 Yield-stress determination at high strain rates

2.8.1 *Plane-strain compression*

The test described in §2.6.2 is slow, because the specimen must be incrementally deformed, and removed for measurement and relubrication after each compression. A less accurate but less tedious method involves simultaneous measurement of load and reduction in height, by suitable transducers. The force-displacement diagram can then be displayed directly on an XY plotter, in the form of a stress-strain curve if desired. This method is applicable at very high strain rates if an oscilloscope is used for presentation of the results. The load can conveniently be applied by a cam loading device, suitably profiled to produce constant strain rate as the specimen is reduced in thickness. The instrument used, which has to be robust in construction to avoid spurious deflections, is known as a *cam plastometer*[(2.11)].

This type of measurement is suitable for strains up to about 1·6 (80%) and strain rates between about 1 sec^{-1} and 1,000 sec^{-1}. The specimen can be heated but the temperatures become uncertain because of the variations due to cooling of the thin strip and to heating by the deformation itself.

2.8.2 *Twisted-bar test*

For hot working, the simple torsion test described in §2.7 is satisfactory provided that the strain rate is moderately low. At very high strain rates, the shearing approaches adiabatic conditions and a reliable knowledge of the stress/strain relationship becomes more important than the ductility measured in terms of revolutions before failure.

Simple drop-forging or high-speed press compression of cylindrical billets can be used, but the results are unreliable because of the high friction contribution and the unknown strain-rate and temperature distributions. Compression of a ring allows the friction factor to be eliminated but leaves the other deficiencies (see §2.6.1).

The most accurate method of testing at high strain-rates utilises a *Hopkinson bar*(2.12). The test specimen is a thin-walled cylinder, rigidly attached to two long cylindrical bars. One of these is initially loaded in torsion by a suitable hydraulically-rotated chuck, and held in a strained state by a notched high-tensile bolt. The other bar is clamped at its free end but is initially unstrained, as is the specimen. To perform the test, the bolt is fractured by over-tightening it, so that a high torque is suddenly applied to the specimen causing it to deform plastically at a very high rate, which may be 5,000 $\sec^{-1}$, depending on the stored energy of the bar. The torque transmitted through the specimen to the measuring bar is recorded electronically, and both torques are displayed on an oscilloscope. The stress, strain-rate and strain at any instant can be calculated, but the whole test is completed in about 1 ms, before the reflected stress wave can complicate the results.

The equipment and the testing are sophisticated, but the stress system is simple and the specimen can easily be heated uniformly. The results can be plotted directly as τ/γ curves without introducing a yield criterion or any other important assumption.

2.8.3 *Machining as a high-strain-rate property test*

In contrast to the test described above, stress/strain curves can be derived from a very simple cutting test on a lathe. Strain rates up to 10^5 $\sec^{-1}$ can be produced, but the interpretation of the results is semi-empirical(2.13).

The end of a tube is cut with a tool set squarely across the lathe so that the cutting is effectively in plane strain, or, as it is usually described, orthogonal cutting. It is assumed that the chip or swarf is formed continuously by simple shearing along a single plane inclined at an angle ϕ to the direction of cutting. Then if the chip thickness is t_2 and the depth of cut is t_1, the length L of the shear plane is given geometrically by

$$L = \frac{t_1}{\sin \phi} \quad \text{or} \quad L = \frac{t_2}{\cos(\phi - \alpha)} \tag{2.19}$$

if the rake angle of the tool is α, the angle of inclination of the tool face to the normal to the workpiece surface. Thus if t_1 is set by the lathe gears, and t_2 is measured, as an average vlaue, ϕ can be determined:

$$\tan \phi = \frac{t_1 \cos \alpha}{t_2 - t_1 \sin \alpha} \tag{2.20}$$

From consideration of the geometry, the shear strain γ can be found:

$$\gamma = \tfrac{1}{2} \frac{\cos \alpha}{\sin \phi \cos(\phi - \alpha)} \tag{2.21}$$

Similarly the stress acting along the shear plane is found from the resolved force component F_s and the area. If the width of the tube face is w,

$$\tau = \frac{F_s}{Lw} \tag{2.22}$$

These can thus be recorded and plotted for a given maximum strain rate $\dot{\gamma}$, which is assumed to follow the empirical equation

$$\dot{\gamma} = C\frac{V_s}{L} \tag{2.23}$$

where V_s is the resolved component of velocity along the shear plane. C is related to the spread of the shear zone about the assumed plane in a real material, and is taken to be a constant for a given metal.

This test can easily be undertaken with a wide range of materials at various cutting speeds.

2.9 **Hardness test**

Hardness testing does not directly provide a stress–strain curve, and it is mainly used for quality control. However, it is possible to obtain an approximate stress–strain curve, for rough calculation of cold-working loads, quickly and simply by hardness testing. The principle depends on the semi-empirical relationship[2.7] between the Brinell hardness number H and the yield stress Y

$$H = cY \tag{2.24}$$

If both H and Y are expressed in kg/mm² the constant c is roughly 3. Though the Brinell hardness number is in fact in kg/mm² units, Y is usually required in tons/in² or N/mm². This involves the conversion 1 ton/in² = 1·57 kg/mm², giving

$$Y \text{ tons/in}^2 = (\text{B.H.N.})/4{\cdot}7; \;\; Y \text{ N/mm}^2 = \text{B.H.N.}/0{\cdot}3 \tag{2.25}$$

A flat-sided off-cut of any size or shape can be compressed by known amounts, for example in a simple forging press with limit stops, without needing a knowledge of the deformation load. The yield stress after each deformation is then found by a hardness test. It is not reliable for small strains, because the hardness test itself involves some 8–10% strain[2.7], and the greater rate of working-hardening for this condition also impairs the accuracy, but the larger strains are usually of more interest. This method provides an easy assessment of yield stress, without using expensive load-measuring machines, which is often of sufficient accuracy for preliminary calculations.

EXAMPLES

2.1 The following data were obtained in a tensile test using a 15 mm diameter mild steel specimen with a 50 mm gauge length:

Load:	42·05	41·85	47·43	51·32	54·80	57·59	59·98	62·28	63·77 kN
Length:	51·18	51·59	52·37	53·16	53·92	54·71	55·50	56·29	57·05 mm
Load:	64·86	66·16	69·35	70·74	70·55	68·95	58·69	48·33 kN	
Length:	57·83	58·62	61·95	68·78	71·12	71·52	72·31	72·64 mm	

.Plot the stress/strain curve in terms of

(*a*) nominal stress and strain

(*b*) true stress and logarithmic strain

What would be the yield stress of this material after compression by 20% reduction of area?

Solution. (*a*) The nominal stress *s* is found by dividing the load *P* by the original cross-sectional area, $A_0 = 176{\cdot}7$ mm² for each value given. The nominal strain *e* is found from $e = (l - l_0)/l_0$, equation 2.2.

(*b*) The true stress σ is found by dividing each load value *P* by the actual cross-sectional area *A* at that stage of the test. Because the volume remains constant in plastic deformation (at $V_0 = 8836$ mm³), $A = A_0 l_0/l$, and $\sigma = Pl/8836$ N/mm², up to the maximum stress.

The logarithmic strain $\epsilon = \ln l_1/l_0$, equation 2.5.

At 20% reduction of area by compression $\epsilon = \ln \dfrac{1}{1 - e_c} = 0{\cdot}223$, and the appropriate yield stress is 0·492 kN/mm².

This assumes that the true stress is identical for yielding in tension and in compression.

2.2 The following data were obtained in plane-strain compression tests on (1) annealed 60/40 brass and (2) the same material after cold rolling to 41% reduction of area. The platens were 5·0 mm broad and were well lubricated. The initial dimensions of the test piece were chosen to be about 40 mm wide and 5 mm thick.

(1) Load:	0	5	50	70	80	89	100	120	149	175	199	kN
Thickness:	5·00	5·00	4·88	4·62	4·47	4·29	4·09	3·61	2·51	1·17	9·66	mm
Width:	39·4	39·4	39·4	39·4	39·4	39·4	39·6	39·6	40·1	41·4	42·2	mm

(2) Load:	0	75	120	130	135	140	145	150	170	kN
Thickness:	5·61	5·59	5·46	5·05	4·60	4·04	3·43	2·51	1·30	mm
Width:	36·3	36·3	36·3	36·5	36·5	36·6	36·6	37·1	38·1	mm

Plot the stress-strain curves of these specimens using true stress and (*a*) nominal strain, (*b*) logarithmic strain.

What is the yield stress of the material after 80% reduction of area?

How could the accuracy of the results be improved?

Solution. (1) The true stress is found by dividing the load *P* by the area of contact, which is the product of the strip width *w* and the platen breadth *b*. For example, at *P* = 149 kN, $\sigma = 149/(5{\cdot}0 \times 40{\cdot}1) = 0{\cdot}743$ kN/mm². It is quite common to assume no change in strip width, which would predict a stress $\sigma' = 0{\cdot}756$ kN/mm² in this instance. It is instructive to plot both in this example.

The nominal strain is $(h_0 - h)/h_0$; for example at $P = 149, h = 2{\cdot}51, e = 0{\cdot}498$. The true strain is $\epsilon = \ln h_0/h$, which at $P = 149$ is 0·689, (equation 2·8). At 80% reduction of area, $\epsilon_C = \ln 1/(1 - e_C) = \ln 5{\cdot}0 = 1{\cdot}61$. From the curve $S \approx 0{\cdot}86$ kN/mm² so $Y \approx 0{\cdot}75$ kN/mm².

The accuracy could be improved by taking more readings in the range 150 – 200 kN, which covers a large range of strains. It will be noticed that at the end of the test the *h*/*b* ratio has become small: at $P = 175, h/b = 0{\cdot}234$; at $P = 199, h/b = 0{\cdot}132$. The last point will therefore be high, because of friction on the platens. These should strictly be changed for narrower ones and the coefficient of friction μ evaluated, from equation 2.18. An allowance can however be made by assuming a reasonable value of μ. Thus, from equation 10.5

$$\bar{p} = 2k\left(1 + \tfrac{1}{2}\mu\frac{b}{h}\right)$$

For the last point of the curve, assuming $\mu = 0{\cdot}02$ or $0{\cdot}05$,

$$\frac{\bar{p}}{2k} = 1 + \tfrac{1}{2}\,.\,\mu\,.\,7{\cdot}6 = 1{\cdot}08 \text{ or } 1{\cdot}20$$

bringing the true stress to 0·87 or 0·79 instead of 0·94 kN/mm². Similar corrections can also be applied to the measurements at 175 and 149 kN.

(2) The true stress and true strain are evaluated as above, but the stress-strain curve should be plotted from an initial strain ϵ_{01} = ln 1(1 – 0·41) = 0·528. It will be found that this curve superimposes quite closely on that obtained from the data of part 1 of this example, particularly when allowance is made for friction in both, assuming μ = 0·02. With this correction the yield stress at 80% reduction is found to be $S \approx$ 0·80 kN/mm².

2.3 Plot the true stress/logarithmic strain curves for 0·15% C steel, for commercial purity aluminium, and for high-conductivity copper from the following data:

(a) Mild steel

Load:	0	10	20	41	60	81	100	120	141	150	159	170
Thickness:	5·16	5·16	5·16	5·10	5·05	4·90	4·52	3·99	3·18	2·74	2·29	1·80
Width:	37·3	37·3	37·3	37·3	37·3	37·5	37·5	37·8	38·1	38·1	38·3	38·6

(*b*) Aluminium

Load:	0	5	10	12	14	16	18	20	22	24	26	28
Thickness:	5·18	5·18	5·13	5·08	5·00	4·93	4·80	4·65	4·39	4·09	3·73	3·38
Width:	40·1	40·1	40·2	40·2	40·4	40·5	40·6	40·6	40·7	40·8	40·9	41·1

(*c*) Copper

Load:	0	20	30	40	50	60	65	70	75	80	85	90
Thickness:	5·03	4·80	4·62	4·47	4·19	3·84	3·63	3·40	3·20	3·02	2·84	2·59
Width:	38·6	38·6	38·6	38·8	38·8	38·9	39·1	39·4	39·4	39·6	39·8	40·1

The equipment used is the same as in Example 2.2.

Solution. The graphs are plotted in the same way as in the preceding example. They will be needed in later examples.

2.4 What are the 0·1% Proof Stress, the Percentage Elongation, the Reduction of Area, the Maximum Tensile Stress and the stress at fracture for the mild steel used in Example 2.1? The diameter at the fracture was 7·42 mm.
Why can Young's Modulus not be determined from the data? What proportion of the % elongation is due to necking? What would be the % elongation on a 20 mm gauge length?

Solution. From the graph plotted in Example 2.1, the proof stress at e = 0·001 is about 0·24 kN/mm², ignoring the Upper Yield behaviour. The final % elongation is 100 x (72·64 – 50)/50 = 45%. The final reduction of area, in the neck, is 1 – (7·42)²/(15·0)² ≈ 75%. The maximum tensile stress, which occurs at ϵ = 0·35, is 0·40 kN/mm². The stress at fracture is 48·33/0·785 x (7·42)² = 1·12 kN/mm².

To determine Young's Modulus, more accurate strain readings would be necessary in the elastic region (below 0·1% elongation). The specimen is assumed to extend uniformly until the maximum load is reached (at l = 68·78) and thereafter in the neck only. The proportion of the extension due to necking is thus (72·64 – 68·78)/(72·64 – 50) = 0·05. On a 100 mm gauge length the percentage elongation would be (37·56 + 3·86)/100 = 41·4%.

2.5 Plot the rate of strain hardening ($d\sigma/d\epsilon$) from Example 2.1 as a function of the true strain. What is this rate at a stress level equal to the M.T.S.?

Solution. The slope of the σ/ϵ curve at the M.T.S. point is roughly (0·568 – 0·551)/(0·352 – 0·319) = 0·52 kN/mm². According to equation 2.13 this should be equal to the true stress at that point, σ = 0·55 kN/mm², if the elongation has been strictly uniform. The values of $d\sigma/d\epsilon$ are less accurate than the values of σ.

2.6 What would be the total strain experienced by a block compressed under ideal conditions in five successive passes, each giving 20% reduction of area?

Solution. The separate passes are:

1 – 0·8; 0·8 – 0·64; 0·64 – 0·512; 0·512 – 0·410; 0·410 – 0·328

Thus $r_{05} = 0{\cdot}672$.

Considering logarithmic strain, each pass imparts

$$\epsilon_{01} = \ln \frac{1}{1 - e_c} = \ln \frac{1}{0{\cdot}8} = 0{\cdot}223$$

The total strain is $\epsilon_{05} = 5, \epsilon_{01} = 1{\cdot}116$;

$$1/(1 - e_c) = \exp(1{\cdot}116) = 3{\cdot}05; e_c = 0{\cdot}672$$

For complex schedules the logarithmic strain is more convenient, and it is useful to prepare a conversion graph relating ϵ_c and e_c as in Example 2.8.

2.7 What would be the yield stress of the block after five successive 20% passes, assuming the material to be the same as that tested in Example 2.2?

Solution. At $\epsilon = 1{\cdot}116, S = 0{\cdot}76$ kN/mm², so $Y = 0{\cdot}66$ kN/mm² after correction for friction in the test.

2.8 Plot a graph showing the relationship between ϵ_c and e_c, or r, up to 98% reduction of area.

Solution. This graph is useful for many metalworking calculations. It is plotted using equation 2.9 or 2.11. Some typical values are:

r	20%	30%	40%	50%	60%	63.2%	80%
ϵ	0·223	0·357	0·511	0·693	0·916	1·00	1·61

The most convenient form of this graph is in two parts, 0 – 65% and 65 – 100%. These can be plotted on A4 paper.

2.9 Approximate to the true stress/natural strain curve found in Example 2.2(*a*) by a linear hardening law, (*b*) by an exponential law. What errors are involved in those approximations for passes from 0 to 42% reduction of area, 0 to 18%, 18% to 42% and 42% – 64%?

Solution. (*a*) Using the graph corrected for spread, and assuming $\mu = 0{\cdot}05$, a suitable linear approximation at high reductions is from $S = 0{\cdot}80$ at $\epsilon = 2{\cdot}0$ to $S = 0{\cdot}74$ at $\epsilon = 0{\cdot}8$, which extrapolates to $S = 0{\cdot}7$ at $\epsilon = 0$. This involves no error for strains beyond $\epsilon = 0{\cdot}8$, 55% reduction of area. For light reductions, large errors would be introduced. Thus for the pass 0 – 42%, $\epsilon_{01} = 0{\cdot}545$, the value of $\overline{S}$ deduced would be 0·715 compared with the value derived from the area of the true curve, $\overline{S} = 0{\cdot}52$.

Straight lines can be used to approximate to selected portions of the curve. Thus from 0 – 18% a line from $S = 0{\cdot}22$ at $\epsilon = 0$ to $S = 0{\cdot}50$ at $\epsilon = 0{\cdot}198$ gives $\overline{S} = 0{\cdot}36$ compared with $\overline{S} = 0{\cdot}37$ from the areas.

The accuracy over any selected range depends only on drawing the appropriate line, and can usually be within about 2%, but a single line approximation to the whole curve is reliable only for very heavy reductions.

(*b*) To make the exponential approximation, log 10S is plotted against log 10ϵ. The best straight line over the larger reductions is found to be one from log $10S = 0{\cdot}9$ at log $10\epsilon = 1{\cdot}20$ to log $10S = 0{\cdot}72$ at log $10\epsilon = 0{\cdot}20$. Over the whole range a straight line can be drawn from (–0·6, 0·45) to (1·2, 0·93).

If this is replotted as S against ϵ, there is a satisfactory agreement up to $\epsilon = 1{\cdot}2$. From 0 – 18% r/a, the error is negligible. From 18% to 42% this exponential underestimates $\overline{S}$ by about 5%. From 42% to 64% there is a comparable underestimate.

The equation is sometimes useful in the form $S = 0{\cdot}753\ \epsilon^{0{\cdot}267}$.

2.10 The hardness of a cold-rolled strip is found to be 193 in a Brinell test. What is its approximate yield stress in N/mm²?

Solution. From equation 2.20, $Y \approx 193/0{\cdot}3 = 643$ N/mm².

REFERENCES

2.1 Smallman, R. E., *Modern Physical Metallurgy*, Butterworth, 1962.

2.2 Green, A. P., 'Use of Plasticine models to simulate plastic flow of metals', *Phil. Mag.*, 42, 365–373, 1951.

2.3 Rollason, E. C., *Metallurgy for Engineers*, 4th ed., Edward Arnold, 1973.

2.4 Polakowski, N. H., 'The compression test in relation to cold rolling', *J. Iron and Steel Inst.*, 163, 250–276, 1949.

2.5 Watts, A. B. and Ford H., 'An experimental investigation of the yielding of strip between smooth dies', *Proc. Inst. Mech. Engrs.*, B1, 448–453, 1952.

2.6 Hodierne, F., 'A torsion test for use in metalworking studies', *J. Inst. Metals*, 91(8), 267–273, 1963.

2.7 Tabor, D., *The Hardness of Metals*, Oxford Univ. Press, 1951.

2.8 Reihle, M., 'A simple method of obtaining the flow curves of steel at room tempeature', *Arch. für das Eisenhüttenwesen*, 32, 331–336, 1961.

2.9 de Pierre, V., Male, A. T., and Saul, G., U.S. Patent 3,693,419, 1971.

2.10 Bramley, A. N., and Abdul, N. A., 'Stress-strain curves from the ring test', *Proc. 15th Mach. Tool Des. Res. Conf.*, 431–436, 1974.

2.11 Adler, J. F., and Phillips, K. A., 'The effect of strain rate and temperature on the resistance of aluminium, copper and steel to compression', *J. Inst. Metals*, 83, 80–86, 1954.

2.12 Stevenson, M. G., 'Torsional Hopkinson-bar tests to measure stress-strain properties relevant to machining and high-speed forming', *Proc. III N. Amer. Metalwkg. Conf.*, Carnegie Press, Pittsburg, 291–304, 1975.

2.13 Stevenson, M. G., and Oxley, P. L. B., 'An experimental investigation of the influence of strain rate and temperature on the flow stress properties of a low carbon steel using a machining test', *Proc. Inst. Mech. Engrs.*, 185, 741–754, 1971.

3

Principal Stresses and Yielding

3.1 Introduction

The tensile test described in Chapter 2 imposes a very simple stress system. The force is applied in one direction only, and the specimen is free to contract in the other directions, so that, except at the ends, no constraint is applied. This free system requires the least possible expenditure of energy to obtain a particular change of cross-sectional area, as will be seen in Chapter 4. Although this appears to be desirable, pure tensile stretching is not used in metalworking, except for the rather trivial application of straightening a wire by pulling it until a small amount of plastic deformation has occurred. Stretch-forming of thin sheet by applying a radial tension over a former is used, but this involves a frictional constraint which increases the load, and it is subject to the limitation of all tensile processes, that fracture occurs before really large changes in cross-sectional area can be produced.

The nearest approach to the tensile test, in real metalworking, is wire drawing through a well-lubricated conical die of low angle. Much greater total reduction of cross-sectional area is then possible without fracture, because the compressive stress introduced by the die contributes to the deformation.

It is a simple matter to predict the load required to cause plastic flow in a wire stretched to a chosen elongation, once the stress–strain curve for that metal has been obtained in a tensile test. It is more difficult in general metal-working, because of the need to take account of the combined effects of at least two stresses, acting in different directions.

In principle, it would be possible to devise tests and to measure the stresses to produce yielding under all conceivable conditions of loading, finally tabulating the results. Fortunately this is not necessary, because all stress systems can be expressed in terms of nine direct-stress and shear-stress components, six of which must be known to specify the system completely. In fact, the stresses can be replaced by three direct stresses, acting in directions normal to three mutually perpendicular planes. No shear stresses act on these particular planes, which are known as principal planes. Then all considerations of plastic yielding can be expressed in terms of these three principal stresses alone. This concept greatly simplifies metalworking theory, and is of fundamental importance in practical operations.

3.2 Principal stresses in two dimensions

We shall consider first a system in which all the stresses act in directions parallel to one plane. This two-dimensional stress system can be represented by the stresses acting on a right angled prism, as in Figure 3.1.

Then σ_x is the direct stress acting on the plane $ABB'A'$, which is normal to the X-axis. This notation will be used throughout. To define the shear stresses, two

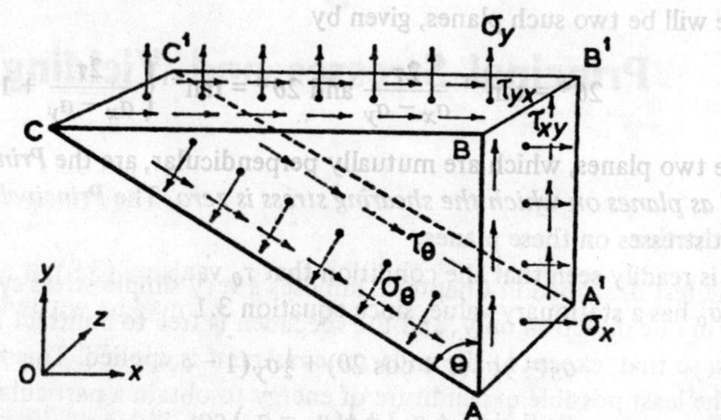

Figure 3.1 Stresses acting on an arbitrary plane in a two-dimensional stress system

suffices are required. The first defines the plane: τ_x acts on a plane normal to the X-axis. The second defines the direction. Thus τ_{xy} represents a shear stress acting on the plane normal to the X-axis, and in the direction of the Y-axis. (τ_{xz} would also act on this plane, but stresses in the z direction are zero in this example.) In fact τ_{xy} and τ_{yx} in Figure 3.1 are complementary shear stresses, which must be equal, to avoid rotation of the prism. The stresses acting on the plane $ACC'A'$, which is inclined at an arbitrary angle θ to the Y-axis, may be labelled σ_θ and τ_θ. The system must be in equilibrium under the action of the external forces, if it is not to change its velocity, so we can resolve and equate the forces (not the stresses) in any desired direction.

Resolving and equating forces in the direction perpendicular to $ACC'A'$:

$$\sigma_\theta AC.CC' = (\sigma_x AB.AA')\cos\theta + (\sigma_y BC.CC')\sin\theta + (\tau_{xy}AB.AA')\sin\theta + (\tau_{yx}BC.CC')\cos\theta.$$

This can be simplified, since $AA' = CC'$, $AB = AC\cos\theta$, and $BC = AC\sin\theta$.

$$\sigma_\theta AC = (\sigma_x AC\cos\theta)\cos\theta + (\sigma_y AC\sin\theta)\sin\theta + (\tau_{xy}AC\cos\theta)\sin\theta + (\tau_{yx}AC\sin\theta)\cos\theta$$

$$\sigma_\theta = \sigma_x\cos^2\theta + \sigma_y\sin^2\theta + 2\tau\sin\theta\cos\theta \tag{3.1}$$

Parallel to the plane $ACC'A'$

$$\tau_\theta AC.CC' = (\sigma_x AB.AA')\sin\theta - (\sigma_y BC.CC')\cos\theta - (\tau AB.AA')\cos\theta + (\tau BC.CC')\sin\theta$$

$$\tau_\theta = \sigma_x\sin\theta\cos\theta - \sigma_y\cos\theta\sin\theta - \tau(\cos^2\theta - \sin^2\theta).$$

$$= \tfrac{1}{2}(\sigma_x - \sigma_y)\sin 2\theta - \tau\cos 2\theta \tag{3.2}$$

Equation 3.2 shows that there will be no shear stress τ_θ on the plane AC when θ^* is chosen so that

$$\tan 2\theta^* = \frac{2\tau}{\sigma_x - \sigma_y} \tag{3.3}$$

There will be two such planes, given by

$$2\theta^* = \tan^{-1}\frac{2\tau}{\sigma_x - \sigma_y} \text{ and } 2\theta^* = \tan^{-1}\left\{\frac{2\tau}{\sigma_x - \sigma_y} + 180\right\}$$

These two planes, which are mutually perpendicular, are the *Principal Planes, defined as planes on which the shearing stress is zero.* The *Principal Stresses* are the direct stresses on these planes.

It is readily seen that the condition that τ_θ vanishes (3.3), is also the condition that σ_σ has a stationary value, since equation 3.1 may be written:

$$\sigma_\theta = \tfrac{1}{2}\sigma_x(1 + \cos 2\theta) + \tfrac{1}{2}\sigma_y(1 - \cos 2\theta) + \tau \sin 2\theta.$$

$$\sigma_\theta = \tfrac{1}{2}(\sigma_x + \sigma_y) + \tfrac{1}{2}(\sigma_x - \sigma_y)\cos 2\theta + \tau \sin 2\theta \qquad (3.4)$$

Differentiating to find the stationary value,

$$\frac{d\sigma_\theta}{d\theta} = -(\sigma_x - \sigma_y)\sin 2\theta + 2\tau\cos 2\theta$$

$$= 0, \text{ when } \tan 2\theta^* = \frac{2\tau}{\sigma_x - \sigma_y} \qquad (3.5)$$

Consequently the principal stresses are the maximum and minimum direct stresses in the material.

The magnitudes of the principal stresses may be found by substituting condition 3.3 in equation 3.4. (See example 3.1.) They are readily derived directly, and their significance is more clearly seen, by considering the equilibrium of a prism chosen so that the plane AC is a principal plane (Figure 3.2). For brevity let the length AA' equal unity. Note that $\tau_{\theta^*} = 0$, $\tau_{xy} = \tau_{yx} = \tau$. Resolving and equating the forces acting on the prism:

Horizontally: $\sigma_x \,.\, AB + \tau.BC = (\sigma.AC)\cos\theta^*.$

$$\sigma_x \cos\theta^* + \tau\sin\theta^* = \sigma\cos\theta^*.$$

$$(\sigma - \sigma_x)\cos\theta^* = \tau\sin\theta^*.$$

Vertically: $\sigma_y.BC + \tau.AB = (\sigma.AC)\sin\theta^*$

$$\sigma_y \sin\theta^* + \tau\cos\theta^* = \sigma\sin\theta^*$$

$$(\sigma - \sigma_y)\sin\theta^* = \tau\cos\theta^*$$

Multiplying these, $(\sigma - \sigma_x)(\sigma - \sigma_y) = \tau^2$

$$\sigma^2 - (\sigma_x + \sigma_y)\,\sigma + \sigma_x\sigma_y - \tau^2 = 0 \qquad (3.6)$$

The roots of this equation give the two principal stresses. By convention these are labelled σ_1 and σ_2, σ_1 being numerically the greater.

$$\sigma = \tfrac{1}{2}(\sigma_x + \sigma_y) \pm \tfrac{1}{2}\sqrt{\sigma_x{}^2 + 2\sigma_x\sigma_y + \sigma_y{}^2 - 4\sigma_x\sigma_y + 4\tau^2}$$

$$\sigma_1 = \tfrac{1}{2}(\sigma_x + \sigma_y) + \tfrac{1}{2}\sqrt{(\sigma_x - \sigma_y)^2 + 4\tau^2} \qquad (3.7)$$

$$\sigma_2 = \tfrac{1}{2}(\sigma_x + \sigma_y) - \tfrac{1}{2}\sqrt{(\sigma_x - \sigma_y)^2 + 4\tau^2}$$

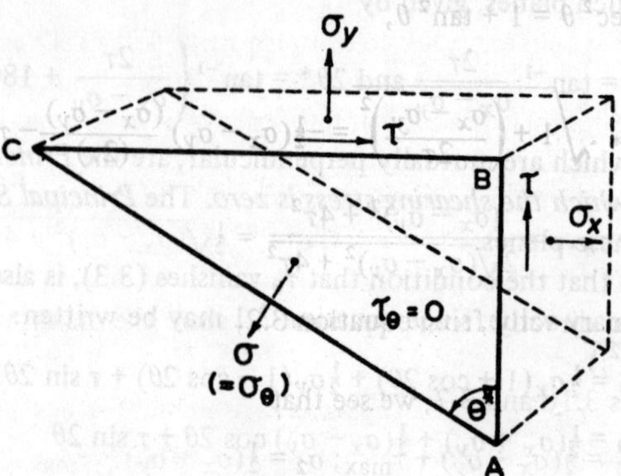

Figure 3.2 Stress system when θ^* is chosen so that AC is a principal plane.

We shall use the principal stresses extensively in the later chapters, because of the simplifications they offer.

3.3 Maximum shearing stresses

Before proceeding further, it is important to consider shearing stresses. As the maximum values of the direct stresses were found by differentiating equation 3.4, so the maximum shear-stresses may be found by differentiating equation 3.2 and equating to zero:

$$\tau_\theta = \tfrac{1}{2}(\sigma_x - \sigma_y)\sin 2\theta - \tau\cos 2\theta \tag{3.2}$$

$$\frac{d\tau_\theta}{d\theta} = (\sigma_x - \sigma_y)\cos 2\theta + 2\tau\sin 2\theta$$

$$= 0, \text{ when } \tan 2\theta^{**} = -\frac{\sigma_x - \sigma_y}{2\tau} \tag{3.8}$$

Again there will be two such planes which are mutually perpendicular.

Comparing the angle θ^* of the principal planes (equation 3.3) with this angle θ^{**} of the planes of maximum shear,

$$\tan 2\theta^* = \frac{2\tau}{\sigma_x - \sigma_y} = -\cot 2\theta^{**}$$

This condition is satisfied if

$$2\theta^{**} = 2\theta^* + \frac{\pi}{2};\ \theta^{**} = \theta^* + \frac{\pi}{4} \tag{3.9}$$

Thus the *maximum shear occurs on planes which are inclined at 45° to the principal planes*. The magnitude of the maximum shear-stress may be determined by substituting the condition 3·8 in equation 3.2:

$$\frac{\tau_{\theta^{**}}}{\cos 2\theta^{**}} = \tfrac{1}{2}(\sigma_x - \sigma_y)\tan 2\theta^{**} - \tau;$$

Remembering that $\sec^2 \theta = 1 + \tan^2 \theta$,

$$\pm \tau_{\theta**} \cdot \sqrt{1 + \left(\frac{\sigma_x - \sigma_y}{2\tau}\right)^2} = -\tfrac{1}{2}(\sigma_x - \sigma_y)\frac{(\sigma_x - \sigma_y)}{(2\tau)} - \tau.$$

$$\tau_{max} = \tau_{\theta**} = \frac{(\sigma_x - \sigma_y)^2 + 4\tau^2}{2\sqrt{(\sigma_x - \sigma_y)^2 + 4\tau^2}} = \tfrac{1}{2}\sqrt{(\sigma_x - \sigma_y)^2 + 4\tau^2} \quad (3.10)$$

(This can also be seen directly from equation 3.2, since the maximum value of $A \cos \theta + B \sin \theta$ is $\sqrt{A^2 + B^2}$.)

Comparing equations 3.10 and 3.7, we see that

$$\sigma_1 = \tfrac{1}{2}(\sigma_x + \sigma_y) + \tau_{max};\ \sigma_2 = \tfrac{1}{2}(\sigma_x + \sigma_y) - \tau_{max}$$

Thus the maximum shear-stress is equal to half the difference between the principal stresses.

$$\tau_{max} = \tfrac{1}{2}(\sigma_1 - \sigma_2) \quad (3.11)$$

It should be noted that though the shear stress is zero on the planes of maximum and minimum direct stress (the principal planes), the converse is not true. The normal stress on the planes of maximum shear is finite, and is easily shown to be equal to the mean of the principal stresses. (See Example 3.2.)

$$\sigma_{\theta**} = \tfrac{1}{2}(\sigma_x + \sigma_y) = \tfrac{1}{2}(\sigma_1 + \sigma_2) \quad (3.12)$$

3.4 Principal stresses in three dimensions

It can be shown by similar consideration of equilibrium in a three-dimensional stress system, that there are always three mutually perpendicular planes on which the shearing stress is zero. (See, for example, references 3.1 or 3.2.) There are thus three principal stresses $\sigma_1, \sigma_2, \sigma_3$, which are the direct stresses acting on these planes. By convention, $\sigma_1 > \sigma_2 > \sigma_3$.

The maximum shearing stress is found to be equal to half the difference between the greatest principal stress and the least.

$$\tau_{max} = \tfrac{1}{2}(\sigma_1 - \sigma_3) \quad (3.11a)$$

(in two dimensions the principal stresses are really σ_1, σ_2, 0. If, as usually occurs, σ_1 and σ_2 are of opposite sign, equation 3.11 is valid, but if they are both of the same sign, as might happen in a two-dimensional closed-die forging for example, the least principal stress is zero, and the maximum shearing stress would then be $\frac{1}{2}\sigma_1$.)

In three dimensions, as in two, the greatest shear stress occurs across a plane bisecting the angle between the planes on which the greatest and least principal stresses act.

3.5 Mohr's circle representation of stress states

The same equilibrium conditions at a point in a body subjected to two dimensional stress, can be considered somewhat differently. Mohr[3.3] (1914) suggested a graphical representation of the state of stress, which takes the form of a circle in a 'stress plane'. This approach is useful when rotation of the system of axes is required, and is of considerable practical value in simplifying the derivation of certain facts and equations required later. It is very helpful in the construction of slip-line fields (Chapter 5). The results obtained in the foregoing sections of this chapter can also be seen much more quickly and directly by reference to the appropriate circle. The construction of the circle is most easily followed for the simple example of two-dimensional stress already used. The results may then be derived from it immediately, simply by inspection.

3.5.1 *Mohr's circle for two-dimensional stress*

We consider again the stress system of Figure 3.1, reproduced here for convenience in Figure 3.3.

As before, all stresses act in directions parallel to the plane *ABC*. The plane *ACC'A'* is inclined at an arbitrary angle θ to the *Y*-axis. Resolution of forces perpendicular to the plane *AC* gives:

$$\sigma_\theta AC = \sigma_x AB \cos\theta + \sigma_y BC \sin\theta + \tau_{xy} AB \sin\theta + \tau_{yx} BC \cos\theta$$

$$\sigma_\theta = \sigma_x \cos^2\theta + \sigma_y \sin^2\theta + 2\tau_{xy} \sin\theta\cos\theta$$

Resolution parallel to *AC* gives:

$$\tau_\theta AC = \sigma_x AB \sin\theta - \sigma_y BC \cos\theta - \tau_{xy} AB \cos\theta + \tau_{yx} BC \sin\theta$$

$$\tau_\theta = (\sigma_x - \sigma_y)\sin\theta\cos\theta - \tau_{xy}(\cos^2\theta - \sin^2\theta)$$

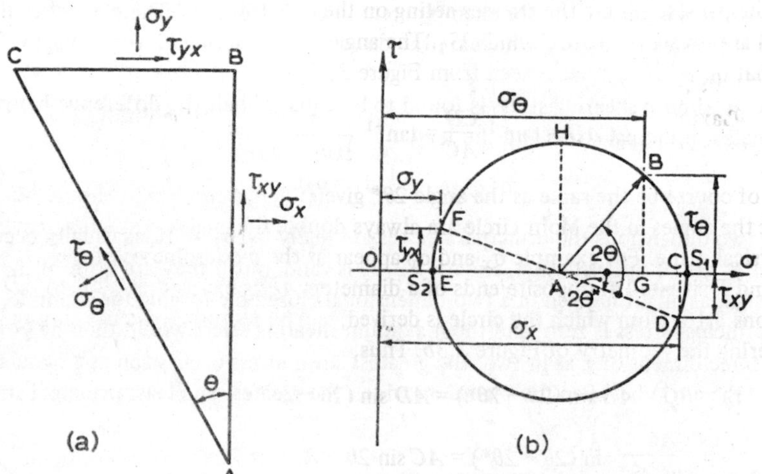

Figure 3.3

(*a*) The stress system acting on an elementary right-angled prism, in two-dimensional stress

(*b*) The Mohr circle corresponding to the above stress system

These equations may be rewritten

$$\sigma_\theta = \tfrac{1}{2}(\sigma_x + \sigma_y) + \tfrac{1}{2}(\sigma_x - \sigma_y) \cos 2\theta + \tau_{xy} \sin 2\theta \quad (3.13a)$$

$$\tau_\theta = \tfrac{1}{2}(\sigma_x - \sigma_y) \sin 2\theta - \tau_{xy} \cos 2\theta \quad (3.13b)$$

Previously the direct-stress and shear-stress equations were considered separately. They may be taken into consideration simultaneously, by squaring and adding the equations 3.13:

$$[\sigma_\theta - \tfrac{1}{2}(\sigma_x + \sigma_y)]^2 + \tau_\theta{}^2 = [\tfrac{1}{2}(\sigma_x - \sigma_y)]^2 + \tau_{xy}{}^2$$

This equation is of the form

$$(\sigma - A)^2 + \tau^2 = B^2 \quad (3.14)$$

which is the equation of a circle plotted with axes σ and τ, of radius

$$B = \sqrt{[\tfrac{1}{2}(\sigma_x - \sigma_y)]^2 + \tau_{xy}{}^2}$$

centred on the point $(A, 0)$ such that

$$A = \tfrac{1}{2}(\sigma_x + \sigma_y)$$

This is the Mohr circle, as shown in Figure 3.3*b*.

The plane on which σ_θ and τ_θ act is inclined at an angle θ, measured anticlockwise, to the plane on which σ_x and τ_{xy} act. To define these locations on the Mohr circle, it is necessary to adopt some convention about the sense of the shear stress. If clockwise shear is assumed to be positive, τ_{xy} has the negative value shown by $-CD$ in Figure 3.3*b*. The σ component of the coordinates of the point D is OC, equal to σ_x. Similarly $\tau_{yx} = +EF$, and $\sigma_y = OE$.

The principal stresses are the stresses acting on the planes where the shear stress is zero and are thus given by OS_1 and OS_2. The angle between the plane on which σ_x acts and that on which σ_1 acts is seen from Figure 3.3*b* to be

$$\angle CAD = \tan^{-1} \frac{CD}{AC} = \tan^{-1} \frac{\tau_{xy}}{\tfrac{1}{2}(\sigma_x - \sigma_y)}$$

This must of course be the same as the angle $2\theta^*$ given in equation 3.5. It should be noted that the angles in the Mohr circle are always double the angles in the diagram of the physical plane. For example σ_x and σ_y appear as the σ-coordinates of the points D and F situated at opposite ends of a diameter; AF is inclined at 180° to AD.

Equations 3.13, from which the circle is derived, can be reproduced quite simply by considering the geometry of Figure 3.3*b*. Thus,

$$\tau_\theta = BG = AB \sin(2\theta - 2\theta^*) = AD \sin(2\theta - 2\theta^*)$$

$$= \frac{AC}{\cos 2\theta^*} \sin(2\theta - 2\theta^*) = AC \sin 2\theta - AC \cos 2\theta \tan 2\theta^*$$

$$= AC \sin 2\theta - CD \cos 2\theta$$

$$= \tfrac{1}{2}(\sigma_x - \sigma_y) \sin 2\theta - \tau_{xy} \cos 2\theta$$

as in equation 3.13*b*.

The results obtained earlier in this chapter can be written down from inspection of the circle diagram:

1. It can immediately be seen that the principal stresses, defined as the direct stresses on planes across which the shearing stress is zero, are given by:

$$\left.\begin{aligned}\sigma_1 = OS_1 = A + B = \tfrac{1}{2}(\sigma_x + \sigma_y) + \tfrac{1}{2}\sqrt{(\sigma_x - \sigma_y)^2 + 4\tau_{xy}^2} \quad . \quad . \\ \sigma_2 = OS_2 = A - B = \tfrac{1}{2}(\sigma_x + \sigma_y) - \tfrac{1}{2}\sqrt{(\sigma_x - \sigma_y)^2 + 4\tau_{xy}^2} \quad . \quad .\end{aligned}\right\} \quad (3.15)$$

These are, of course, identical with equations 3.7.

2. It is also obvious, without further comment, that these are the maximum and minimum direct stresses (equation 3.5).

3. The plane across which the maximum shear stress acts is at an angle $2\theta = 90°$ to the plane on which the principal stress acts in the stress diagram. Consequently these planes intersect at $\theta = 45°$ in physical space (equation 3.9).

4. The maximum shearing stress

$$\tau_{max} = AH = B = \sqrt{\{\tfrac{1}{2}(\sigma_x - \sigma_y)\}^2 + \tau_{xy}^2},$$

as in equation 3.10. Also, since $AH = \tfrac{1}{2}S_1S_2 = \tfrac{1}{2}(OS_1 - OS_2)$ it is clear that the maximum shearing stress is equal to half the difference between the principal stresses (3.11).

The normal stress on the plane of maximum shear is given by $OA = \tfrac{1}{2}(OS_1 + OS_2)$, as found in Example 3.2 and equation 3.12. It is of course also equal to the constant $A = \tfrac{1}{2}(\sigma_x + \sigma_y)$, in equation 3.14.

The circle diagram is thus a powerful tool for obtaining these results, without any need to manipulate the equilibrium equations. Other important uses will be described in later chapters. It is quite easy to draw the appropriate circle for a given stress system, and this helps in visualising changes in stress in different directions. The construction is particularly simple if the principal stresses are known in magnitude and direction.

3.5.2 *Mohr's circle for two-dimensional stress, referred to principal axes*

The principal axes are the directions of the principal stresses, and therefore are normal to the principal planes. If these are known, it is convenient to choose the *X*- and *Y*-axes to coincide with them. Since they must be mutually perpendicular, it is in fact necessary to determine the direction of only one of them.

Then $\sigma_x = \sigma_1,\ \sigma_y = \sigma_2,\ \tau_{xy} = 0$ as in Figure 3.4*a*.

The equilibrium equations (3.13) simplify to

$$\begin{aligned}\sigma_\theta &= \tfrac{1}{2}(\sigma_1 + \sigma_2) + \tfrac{1}{2}(\sigma_1 - \sigma_2)\cos 2\theta \\ \tau_\theta &= \qquad\qquad\quad + \tfrac{1}{2}(\sigma_1 - \sigma_2)\sin 2\theta\end{aligned} \quad (3.16)$$

whence

$$\left\{\sigma_\theta - \frac{\sigma_1 + \sigma_2}{2}\right\}^2 + \tau_\theta^2 = \left\{\frac{\sigma_1 - \sigma_2}{2}\right\}^2 \quad (3.17)$$

The stress circle in the (σ, τ) plane can thus be constructed as in Figure 3.4*b* immediately, from a knowledge of σ_1 and σ_2 only. The stresses on a plane at any angl

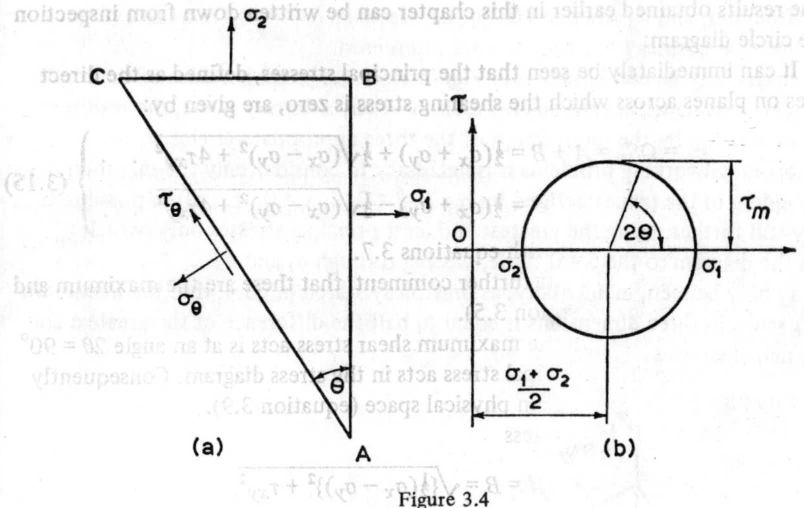

Figure 3.4

(*a*) The stress system referred to principal axes.
(*b*) The Mohr circle for this condition.

θ to the plane on which σ_1 acts, will be given by the coordinates of the extremity of the radius vector at an angle 2θ, measured in the same sense as θ, from the σ axis.

A very simple example may be chosen to illustrate the use of this presentation. Others will be found in the exercises. In the tensile test $\sigma_2 = \sigma_3 = 0$. The circle passes through σ_1 and the origin. The maximum shear stress may be read directly: $\tau_{max} = \frac{\sigma_1}{2}$, acting on a plane at 45° to the axis, as is well known.

3.5.3 *Mohr's circle for three-dimensional stress*

We shall encounter the three-dimensional stress system more frequently than the condition of plane stress, involving only two principal stresses. However, any arbitrary physical plane such as that shown in Figure 3.5 will have only one resultant direct stress and one resultant shearing stress acting upon it. Variations in σ and τ in three-dimensional physical space may therefore be plotted, as before, in two-dimensional stress space.

If the coordinate axes are chosen to coincide with the principal axes, the diagram may be developed by considering one physical plane at a time. For a plane parallel to the *Z*-axis, only the principal stresses σ_1 and σ_2 are involved, and the (σ, τ) diagram (Figure 3.5*a*) is the same as in Figure 3.4. It may be shown (see for example reference 3.4) that if the plane is inclined at a constant angle γ to the *Z*-axis, the centre of the circle remains unchanged, but the radius increases with γ from a minimum value $\frac{\sigma_1 - \sigma_2}{2}$ at $\gamma = 0$ to a maximum $\left(\sigma_3 - \frac{\sigma_1 + \sigma_2}{2}\right)$ at $\gamma = \frac{\pi}{2}$. The diagram for the stresses on a plane inclined at constant angle α to the *X*-axis is similar, but has extreme σ values equal to σ_2 and σ_3 for $\alpha = 0$ (Figure 3.5*b*). For the plane inclined to the *Y*-axis, which is parallel to the intermediate principal stress, the circle diagram passes through σ_1 and σ_3 when $\beta = 0$, but decreases in radius as β increases (Figure 3.5*c*). Direction cosines $l = \cos\alpha$, etc., are often used.

The circles may be combined, as in Figure 3.5*d*. All possible values of the resultant stresses must lie within the shaded areas. For any arbitrary plane, the resultant stress will be given by the intersection of the three appropriate circles.

In most metalworking problems it is necessary to consider only the major circles, corresponding to the planes defined by $\alpha = 0$, $\beta = 0$, or $\gamma = 0$. It is often possible to simplify still further, using the greatest and least principal stresses only, which reduces the diagram to the $\beta = 0$ circle, passing through σ_1 and σ_3.

It may now be seen, incidentally, as previously stated in §3.4, that the maximum shearing stress in three dimensions is equal to half the difference of the greatest and least principal stresses.

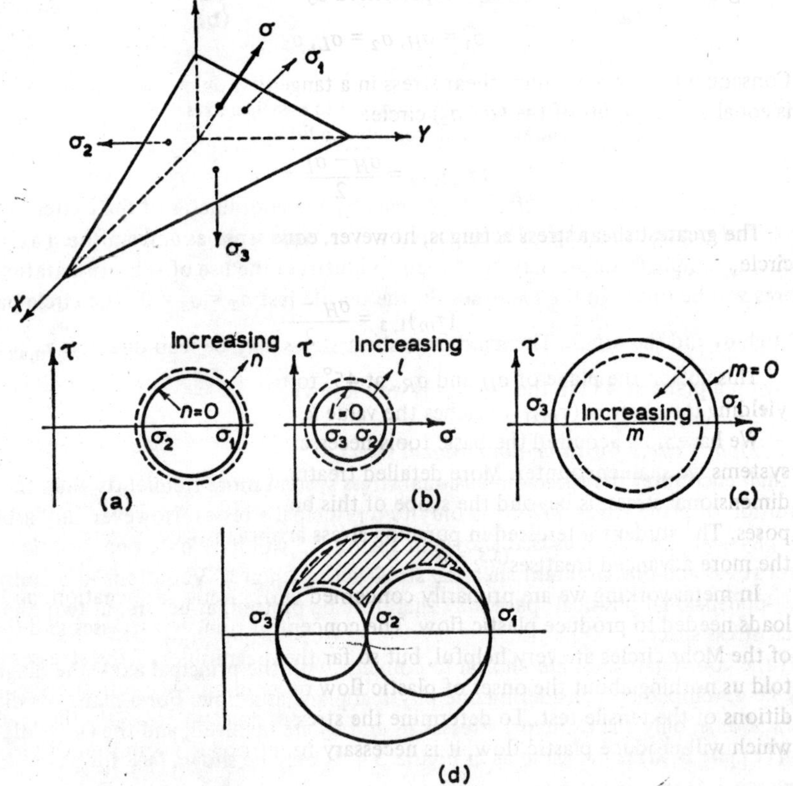

Figure

Figure 3.5 The stress acting on an arbitrary plane, and the Mohr circles for

(*a*) a plane parallel to the *Z*-axis
(*b*) a plane parallel to the *X*-axis
(*c*) a plane parallel to the *Y*-axis
(*d*) the two-dimensional representation of general three-dimensional stress

As a simple example, consider the stresses in a thin-walled cylinder of radius r and wall thickness t, subjected to an internal pressure p. The force tending to blow the ends off causes a longitudinal stress σ_L, so

$$\pi r^2 p = 2\pi r\, t\, \sigma_L\, ; \sigma_L = p \cdot \frac{r}{2t}$$

By considering the force equilibrium across a diametral plane, for an arbitrary length x, the circumferential or hoop stress σ_H can be found:

$$2rxp = 2tx\, \sigma_H ; \sigma_H = p \cdot \frac{r}{t}$$

The mean stress σ_R in the radial direction is $p/2$, which can be ignored in comparison with the others because r is assumed to be much greater than t. With reference to Figure 3.5(d), this situation is represented by

$$\sigma_1 = \sigma_H, \sigma_2 = \sigma_L, \sigma_3 \approx 0$$

Consequently the maximum shear stress in a tangential plane, containing σ_H and σ_L, is equal to the radius of the (σ_1, σ_2) circle:

$$(\tau_m)_{1,\,2} = \frac{\sigma_H - \sigma_L}{2} = \frac{\sigma_H}{4}$$

The greatest shear stress acting is, however, equal to the radius of the (σ_1, σ_3) circle,

$$(\tau_m)_{1,\,3} = \frac{\sigma_H - 0}{2}$$

This acts in the plane of σ_H and σ_R, at 45° to the wall surface. It could cause yielding long before $(\tau_m)_{1,2}$ reaches the value k.

We have now acquired the basic tools necessary to evaluate the stresses in the systems we shall encounter. More detailed treatment, particularly of general three-dimensional stress, is beyond the scope of this book and is not essential for our purposes. The student interested in pursuing stress analysis further is referred to one of the more advanced treatises[3.5, 3.6].

In metalworking we are primarily concerned with plastic deformation, and in the loads needed to produce plastic flow. The concept of principal stresses and the use of the Mohr circles are very helpful, but so far the magnitudes of the stresses have told us nothing about the onset of plastic flow or yielding, except in the special conditions of the tensile test. To determine the stresses and subsequently the loads, which will produce plastic flow, it is necessary to introduce a criterion of yielding.

3.6 Yield criteria

In uniaxial tension (or pure compression) the yield stress is, in principle, easily defined. If the metal has previously been deformed to any known extent, subsequent yielding will occur as soon as the stress reaches a magnitude Y, which may be read from the stress–strain curve at the appropriate strain. The stress–strain curve is in fact, as seen in Chapter 2, a yield stress–strain curve. With a real metal it is not quite so easy, since the onset of plastic flow is not sharply defined, but there is seldom serious doubt about the value.

Most stress systems are complex, and it is not feasible to perform separate yield tests for all possible variations, so we need to find some guide to the combinations of stresses which will produce plastic flow in an element of the body. Several yield criteria have been proposed, of which the most important are due to Tresca[3.7] (1864) and von Mises[3.8] (1913). These form the starting-point of useful metalworking, though in the older textbooks on Strength of Materials they were often referred to as Failure Criteria, because they indicate the cessation of the elastic behaviour. This limits the useful range of structural members which are not allowed to take up a permanent set, but elastic deformation is of secondary importance in metalworking.

3.6.1 *Tresca maximum shear-stress criterion*

Tresca's criterion suggests that plastic flow occurs when the maximum shear stress reaches a particular value. This appears reasonable, since it is known that plastic flow depends on slip, and is essentially a shear process. In fact, under many conditions this criterion appears to predict yielding with sufficient accuracy. Since the maximum shearing stress is equal to half the difference between the maximum and minimum principal stresses (Figure 3.5), the criterion may be written

$$\tau_{max} = \tfrac{1}{2}(\sigma_1 - \sigma_3) = \text{constant} \tag{3.18}$$

It thus implies that yielding is independent of the magnitude of the intermediate principal stress σ_2. Lode (1926) showed that this is not strictly true[3.9]

3.6.2 *von Mises maximum shear-strain-energy criterion*

In 1913 von Mises proposed a symmetrical quadratic condition:

$$(\sigma_1 - \sigma_2)^2 + (\sigma_2 - \sigma_3)^2 + (\sigma_3 - \sigma_1)^2 = \text{constant}$$

This includes all three principal stresses. Hencky (1924) interpreted this criterion[3.10] as meaning that plastic flow occurs when the shear-strain energy reaches a critical value.

$$\frac{1}{6G}\{(\sigma_1 - \sigma_2)^2 + (\sigma_2 - \sigma_3)^2 + (\sigma_3 - \sigma_1)^2\} = \text{constant} = A \tag{3.19}$$

Here G is the rigidity modulus; $G = \dfrac{E}{2(1 + \nu)}$

The shear-strain energy is the elastic energy of distortion, and the criterion implies that the elastic energy associated with change in volume has no influence on yielding. Bridgman[3.11] has confirmed experimentally that hydrostatic pressure has practically no effect on the stress system necessary to initiate plastic flow in metals. This should not be confused with the observation, also originally by Bridgman, that many materials which are normally brittle, such as marble, will flow plastically under high hydrostatic pressure[3.12]. The superposition of a uniform compressive stress may suppress tensile failure but will not significantly affect the yield stresses.

Experiments by Taylor and Quinney (1931) and many later workers have shown[3.13] that von Mises' shear-strain-energy criterion accords well with practical results, and is a closer approximation than the Tresca shear-stress criterion. It is now generally accepted, and will be used throughout this book.

3.6.3 *Relationship between tensile yield stress Y and shear yield stress k*

The yield criteria must obviously be applicable to any stress system. As the value of the constant A in equation 3.19 has not yet been specified, we are free to determine it from consideration of a simple tensile test. Within the gauge length of the tensile specimen, the stress system is a pure uniaxial tension, and yield is known to occur at the value Y.

Thus: $\sigma_1 = Y, \sigma_2 = 0, \sigma_3 = 0$, at the onset of plastic flow. Then the von Mises criterion gives

$$6GA = (\sigma_1 - \sigma_2)^2 + (\sigma_2 - \sigma_3)^2 + (\sigma_3 - \sigma_1)^2 = 2Y^2 \tag{3.20}$$

However, we might equally well have chosen to apply the criterion to a pure torsion test, in which yield occurs at a particular value of the applied shear stress, usually denoted by k. It can easily be shown that the principal stresses in a torsion test are equal in magnitude to the maximum shear stresses (the Mohr circle is centred on zero). Thus, at the onset of plastic flow in pure torsion: $\sigma_1 = +k, \sigma_2 = 0, \sigma_3 = -k$.

In this instance the von Mises criterion gives

$$6GA = k^2 + k^2 + 4k^2 \tag{3.21}$$

These values for A must be equal, since A is a constant and the criterion is known to be generally valid. This implies that, according to von Mises,

$$\underline{2Y^2 = 6k^2; \; 2k = \frac{2}{\sqrt{3}} Y = 1{\cdot}155Y} \tag{3.22}$$

It is convenient to express the result in terms of $2k$, rather than k, as will be seen in the next section.

If the Tresca criterion is applied to these two stress systems, a slightly different result is obtained. In pure torsion, yielding is predicted when

$$\tau_{max} = \tfrac{1}{2}(\sigma_1 - \sigma_3) = \tfrac{1}{2}(k + k) = k.$$

This follows also from the definition of k. In pure tension, however,

$$\tau_{max} = \tfrac{1}{2}(\sigma_1 - \sigma_3) = \frac{Y}{2}; \; \sigma_1 - \sigma_3 = Y \tag{3.23}$$

Thus, according to Tresca, $\qquad 2k = Y \qquad$ (3.24)

It may be noticed that when any two principal stresses are equal both criteria reduce to the same equation. Thus if $\sigma_2 = \sigma_3$, von Mises criterion gives, from equation 3.20,

$$(\sigma_1 - \sigma_3)^2 + (\sigma_3 - \sigma_1)^2 = 2Y^2$$

or

$$(\sigma_1 - \sigma_3)^2 = Y^2 \tag{3.25}$$

This is the same as the Tresca criterion, equation 3.23. The physical reason for this is that the stress system can be regarded as a combination of a uniaxial stress $(\sigma_1 - \sigma_2) = (\sigma_1 - \sigma_3)$ and a hydrostatic stress σ_2. Since the latter is known not to influence yielding, yield must occur at the uniaxial yield stress Y, and no criterion is required.

3.6.4 *Yield under plane-strain conditions*

It is also unnecessary to use a yield criterion for plane-strain conditions, where in fact yielding occurs due to pure shear, though this is not immediately obvious.

Plane strain is defined as a condition in which (*a*) the flow is everywhere parallel to a given plane, say the (x,y) plane, and (*b*) the motion is independent of z. Thus one principal strain-increment, say $d\epsilon_2$, is zero. It follows that if there is no volume change $d\epsilon_1 = -d\epsilon_3$, assuming no elastic deformation, that is assuming an incompressible rigid-plastic material. The deformation is thus pure shear-strain. It is assumed, as can be justified by an argument due to Hill[3.14], that pure shear strain is produced by pure shear stress. Yield consequently occurs in plane strain at the shear yield stress k. The Mohr circle for plane plastic strain, with no strain-hardening, therefore always has the radius $\tau_{max} = k$. There may be a superimposed hydrostatic stress σ_2 which will alter the values of σ_1 and σ_3 but will not influence yielding. This is usually a compressive stress in metalworking. Under these conditions, as in Figure 5.3 of Chapter 5,

$$\sigma_1 = \sigma_2 - k, \; \sigma_3 = \sigma_2 + k$$

Thus
$$\sigma_2 = \tfrac{1}{2}(\sigma_1 + \sigma_3) \tag{3.26}$$

and
$$\sigma_1 - \sigma_3 = 2k \tag{3.27}$$

The latter may be written

$$\sigma_1 - \sigma_3 = S,$$

which defines the yield stress in a plane-strain compression test, sometimes called the constrained yield stress. Using the von Mises relationship between k and Y (equation 3.22)

$$S = 2k = 1{\cdot}155Y$$

These equations are important in later developments of metalworking theory, since plane-strain conditions provide a simplification essential for slip-line field and most upper-bound techniques.

EXAMPLES

3.1 In a plane-stress system $\sigma_x = 750$ N/mm², $\sigma_y = 150$ N/mm², $\sigma_z = 0$ and $\tau_{xy} = 150$ N/mm². What are the magnitudes and directions of the principal stresses?

Solution. The magnitudes of the principal stresses are found from equation 3.7, which can itself be derived easily from a Mohr circle. This gives

$$\sigma_1 = \frac{900}{2} + \tfrac{1}{2}\sqrt{36 \times 10^4 + 9 \times 10^4} = 450 + 335$$

$$\sigma_1 = 785 \text{ N/mm}^2, \; \sigma^2 = 115 \text{ N/mm}^2 \; . \; (\sigma_3 = 0)$$

The angle between the principal plane and the plane on which σ_x acts is found from equation 3.5, giving

$$\tan 2\theta^* = 300/600; \ \theta^* = 13{\cdot}3°$$

The results are also obtainable rapidly by locating the points (σ_x, τ_{xy}) and (σ_y, τ_{yx}) on a (σ, τ) diagram. The Mohr circle passing through them is centred on (450, 0) and has a radius given by $R^2 = 150^2 + 300^2$; $R = 335$ N/mm².

3.2 Show that the normal stress on the planes of maximum shear is equal to the mean of the principal stresses
(*a*) from first principles
(*b*) by constructing the Mohr circle

Solution. (*a*) Following the procedure of §3.2 the stationary value of τ_θ is found by differentiating equation 3.2 w.r.t. θ and equating to zero. Thus $(\sigma_x - \sigma_y)\cos 2\theta = -2\tau \sin 2\theta$. A second differentiation gives $\dfrac{d^2\tau_\theta}{\theta^2} = -4(\tau_\theta)$, showing this value to be a maximum.

Substitution in equation 3.4 gives $\sigma_\theta = \frac{1}{2}(\sigma_x + \sigma_y)$ on the plane of maximum shear. This must be valid for any orientations of the axes X and Y, including the principal axes. Equation 3.7 also shows that $(\sigma_1 + \sigma_2) = (\sigma_x + \sigma_y)$.

(*b*) The answer is immediately seen from the symmetry of the Mohr circle.

3.3 A compressive stress of 150 N/mm² is applied between opposite faces of a 250 mm steel cube. Determine the normal forces on the other two pairs of faces that would prevent the cube from expanding by more than 0·050 mm.
For steel $E = 207$ kN/mm², $\nu = 0{\cdot}3$

Solution. If the load is applied along OZ, it is required to find the equal stresses σ_x and σ_y that produce equilibrium at the given strain; $\epsilon_x = \epsilon_y = 0{\cdot}050/250 = 2 \times 10^{-4}$.

$$\epsilon_x = \frac{\sigma_x}{E} - \frac{\nu\sigma_y}{E} - \frac{\nu\sigma_z}{E}$$

Thus

$$\sigma_x(1 - 0{\cdot}3) - 0{\cdot}3 \times (-150) = 2 \times 10^{-4} \times 207 \times 10^3$$

$$0{\cdot}7\sigma_x = 41{\cdot}4 - 45{\cdot}0$$

$$\sigma_x = 5{\cdot}14 \text{ N/mm}^2; \ F = 320 \text{ kN}$$

3.4 What are the directions and magnitudes of the maximum shear stresses in Example 3.1?

Solution. The results may be obtained as in Section 3.2, or directly from the Mohr circle. The plane on which the maximum shear stress acts is inclined at an angle θ^{**} to the plane on which σ_x acts:

$$\tan 2\theta^{**} = (750 - 150)/300; \ \theta^{**} = 31{\cdot}7°$$

This is also given by $2\theta^{**} = 90 - 2\theta^*$.
The maximum shear stress is equal to the radius of the circle:

$$R^2 = (300)^2 + (150)^2; \ \tau_m = 335 \text{ N/mm}^2$$

3.5 If $\sigma_x = 450$ N/mm², $\sigma_y = 450$ N/mm², $\sigma_z = 0$ and $\tau_{xy} = 150$ N/mm², what are the principal stresses and the maximum shear stresses?

Solution. The Mohr circle passing through $\sigma_x = 450$ and $\sigma_y = 450$ must be centred on $\sigma = 450$. Thus σ_x and σ_y are the direct stresses acting on the planes of maximum shear, and $\tau_m = \tau_{xy} = 150$ N/mm².
The principal stresses are 600 and 300 N/mm².
It should however be noted that in a real system the third principal stress, $\sigma_3 = 0$, ought to be included, so the maximum shear stress in a three-dimensional body would be $\frac{1}{2}(\sigma_1 - \sigma_3) = 300$ N/mm².

3.6 If the stress system of Example 3.1 just causes yielding, what is the uniaxial yield stress Y of the material, according to (*a*) the Tresca criterion (*b*) the von Mises criterion?

Solution. The principal stresses are found from equations 3.7

$$\sigma_1 = \frac{\sigma_x + \sigma_y}{2} + \tfrac{1}{2}\sqrt{(\sigma_x - \sigma_y)^2 + 4\tau_{xy}^2}$$

$$= \frac{750 + 150}{2} + \tfrac{1}{2}\sqrt{(750 - 150)^2 + 4(150)^2} = 450 + 335$$

$$= 785 \text{ N/mm}^2$$

$$\sigma_2 = 450 - 335 = 115 \text{ N/mm}^2$$

(They can also be found from the Mohr circle as in Example 3.1.)

In the hypothetical plane-stress system, there are no stresses acting out of this plane so

$$\tau_m = \frac{\sigma_1 - \sigma_2}{2} = 335 \text{ N/mm}^2$$

Thus according to Tresca, $Y = 2k = 2\tau_m = 670 \text{ N/mm}^2$

It is more realistic to recognise that although no stress is applied on the planes xz and yz, a shear stress will arise within the material on a plane normal to the xy plane, because $\sigma_3 = 0$.

Then, according to Tresca $Y = 2k = 2\tau_m = \sigma_1 - \sigma_3 = 785 - 0 = 785 \text{ N/mm}^2$ and, according to von Mises

$$(785 - 115)^2 + (115 - 0)^2 + (785 - 0)^2 = 2Y^2$$

$$Y = 734 \text{ N/mm}^2$$

3.7 If the same material is used as in Example 3.6, would the stress system of Example 3.5 cause plastic flow? Would the material yield if $\sigma_x = 450$, $\sigma_y = 450$, $\sigma_z = 450$ and $\tau_{xy} = 150$ N/mm²?

Solution. The maximum shear stress in example 3.5 (equal to the radius of the circle) is 300 N/mm², which is not sufficient to cause yielding at $k = 785/2 = 393$ N/mm² (Tresca), but using von Mises criterion, $(600 - 300)^2 + (300 - 0)^2 + (0 - 600)^2 = 735$ N/mm², which would just cause yielding, by shear in the plane containing x and z.

The addition of the stress $\sigma_z = 450$ would prevent yielding because the maximum shear stress would then be 150 N/mm² with a superimposed hydrostatic tension.

3.8 A compressive load of 4,000 KN is applied to a well-lubricated cube of metal of 80 mm side, and just causes yielding. What load would be required if the other sides were constrained by forces of 1,000 kN and 2,000 kN respectively?

Solution. The compressive load will produce the major principal stress $\sigma_1 = 625$ N/mm². Note that there is no friction and therefore no shear stress on this cube face. The yield stress Y is thus equal to 625 N/mm².

Assuming no friction on the other faces, $\sigma_2 = 312$ and $\sigma_3 = 156$ N/mm².

According to von Mises criterion, equation 3.20, at yielding

$$(\sigma_1 - 312)^2 + (312 - 156)^2 + (156 - \sigma_1)^2 = 2(625)^2$$

$$2\sigma_1^2 - 936\sigma_1 - 635234 = 0$$

whence $\sigma_1 = 844$ N/mm², ignoring the negative root.

The load required to cause yielding is thus increased to 5,400 kN.

3.9 What load is required to deform the cube of Example 3.8 if the constraining forces are both 1,500 kN? Show how this result can be deduced from the knowledge that yielding is independent of hydrostatic pressure.

Solution. The solution can be obtained in the same way as in Example 3.8, substituting $\sigma_2 = \sigma_3 = 234$ in equation 3.20, giving $2(\sigma_1 - 234)^2 = 2Y^2 = 2(625)^2$; $\sigma_1 = 859$ N/mm². The load is thus 5,498 kN.

The stress system can be considered as a uniaxial stress $(\sigma_1 - 234)$ together with a hydrostatic pressure 234 N/mm² (acting in all three principal stress directions). Yield thus occurs when $\sigma - 234 = Y$, giving the same answer.

3.10 The yield stress of a tensile specimen machined from a 500 mm wide, 5 mm thick copper strip is found to be 340 N/mm². The strip is further rolled, with an applied tension of 250 kN. What roll pressure would just cause deformation, ignoring any friction effects?

Solution. The relative dimensions show that deformation occurs under plane-strain conditions. Since $Y = 340$, $S = 2k = 391$ N/mm². (Equations 3.27 and 3.22.)

The applied tensile stress is +100 N/mm². If there is no friction, this is a principal stress. The Mohr stress circle for plane strain (Figure 3.5) extends from $\sigma_3 = +100$ to $\sigma_1 = -291$, since the diameter is $2k$. This can also be seen from equation 3.27, with due regard to sign

$$\sigma_1 - \sigma_3 = S;\quad \sigma_1 - 100 = -391.$$

The roll pressure is thus decreased by the applied tension to 291 N/mm².

(A more detailed method of evaluating roll pressures for various values of friction and applied tension will be described in Chapter 9, again using the condition of yielding in plane strain.)

REFERENCES

3.1 Hoffman, O. and Sachs, G., *Introduction to the Theory of Plasticity for Engineers*, McGraw-Hill, 1953, p. 7.

3.2 Johnson, W. and Mellor, P. B., *Plasticity for Mechanical Engineers*, Van Nostrand, 1962, p. 23.

3.3 Mohr, O., *Abhandlungen aus dem Gebiete der technischen Mechanik*, Ernst & Sohn, Berlin, 1914.

3.4 Hoffman, O. and Sachs, G., *op. cit.*, p. 12.

3.5 Timoshenko, S. and Goodier, J. N., *Theory of Elasticity*, 2nd ed., McGraw-Hill, 1951.

3.6 Ford, H., *Advanced Mechanics of Materials*, Longmans, 1963.

3.7 Tresca, H., *C. R. Acad. Sci., Paris*, 59, 754–756, 1864.

3.8 von Mises, R., 'Mechanik der festen Körper im plastisch-deformablen Zustand', *Nachr. Ges. Wiss. Göttingen, Math.-phys. Klasse*, 582–592, 1913.

3.9 Lode, W., 'Versuche über den Einfluss der mittleren Hauptspannung auf das Fliessen der Metalle Fe, Cu und Ni', *Zeits. Phys.*, 36, 913–939, 1926.

3.10 Hencky, H., 'Zur Theorie plastischer Deformationen und der hierdurch im Material hervorgerufenen Nachspannungen', *Zeits. angew. Math. Mech.*, 4, 323–334, 1924.

3.11 Bridgman, P. W., 'Effects of high hydrostatic pressure on the plastic properties of metals', *Rev. Mod. Phys.*, 17, 3–14, 1945.

3.12 Bridgman, P. W., *Studies in Large Plastic Flow and Fracture*, McGraw-Hill, 1952.

3.13 Taylor, G. I. and Quinney, H., 'The plastic distortion of metals', *Phil. Trans., Series A*, 230, 323–362, 1931.

3.14 Hill, R., *The Mathematical Theory of Plasticity*, Oxford Univ. Press, 1956, p. 36.

3.15 Pugh, H. Ll. D., (Ed.), *Mechanical Behaviour of metals under pressure*, Elsevier, Amsterdam, 1970.

4

Determination of Working Loads by Consideration of Work, and of Stress Distribution

4.1 Introduction

Metalworking theory has two main objectives: prediction of metal deformation, and of the loads required to produce this, but it can also help in predicting causes of fracture. Industrial practice is usually based on production of a required change in shape without introducing undesirable stresses and strains in the final material. This governs the type of process chosen, and the equipment is designed to provide and withstand the heavy loads involved. We shall not be concerned with design of equipment in this book, though some of the principles of selection will become apparent in Part II. The problem of predicting working loads and evaluating the effects of changes in process parameters for existing presses, rolling mills, and other metalworking machines, is however important in proper utilisation for maximum productivity. There are often simple empirical rules and formulae, which should not be ignored, but these are of most value when applied with an understanding of the particular process, obtained from more detailed study. In many instances, practical operations are too complex for full theoretical treatment, but prediction of working loads for the simpler basic processes of rolling, forging, extrusion, and drawing, can now be made with considerable accuracy. These processes are very widely used, so that the results also have direct economic significance.

In this chapter we shall discuss two methods of estimating metalworking loads by consideration of stresses. Two other important methods, which depend on metal flow patterns, will be described in the next chapter. These will be applied to individual processes in detail in Part II. It will then be seen that it is possible to influence performance considerably, within the limits imposed by maximum load, by suitably choosing working schedules, tool profiles, and lubrication on the basis of such calculations.

4.2 Load required to produce yielding in homogeneous deformation

The simplest way in which a metal can be deformed is exemplified by a tensile test (Chapter 2). Any element of the metal which is originally a small cube, becomes a parallelepiped after plastic deformation in simple tension. The whole specimen, apart from the ends, which are always ignored in tensile testing, is free to deform without restraint by any external body. It is a general principle that this homogeneous deformation requires less work, and consequently lower load than any other type of deformation. Calculation of the load for homogeneous deformation thus gives a lower limit for the load necessary in any other operation producing the same final change in cross-sectional area of the workpiece. For example, wire-drawing produces a shape change, elongation accompanied by reduction in diameter, which is essentially the same as that produced by a tensile test. However, the die used in wire-drawing introduces a frictional resistance and also some internal distortion in the wire, both of which increase the work required.

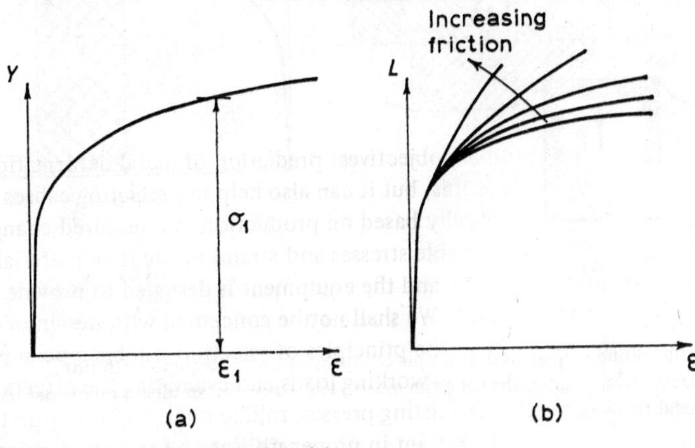

Figure 4.1

(*a*) A stress–strain curve showing the stress σ_1, required to compress a block of height h_0 to a strain $\epsilon_1 = \ln \frac{h_0}{h_1}$, without friction.

(*b*) Comparison of the loads required to compress a block with a good lubricant, and under dry conditions.

In some practical operations, the additional load arising from this frictional and mechanical constraint may be quite small. For example, the stress system and the deformation in axial compression of a small-diameter cylinder between well-lubricated platens (Chapter 2, §2.6) approximates to an ideal uniaxial system. In a tensile test, the load necessary to cause plastic flow in a specimen which has been deformed to a given strain is, of course, simply the product of the area and the yield stress Y corresponding to that strain. Thus the load P required for frictionless compression of a billet (Figure 4.1*a*) from a height h_0 to a height h_1 will increase during compression to a final value which is the product of the final area A_1 and the yield stress Y_1 at the strain ϵ_1 corresponding to $\ln(h_0/h_1)$:

$$P = A_1 Y_1 \tag{4.1}$$

Provided the lubrication is good, this value will be only slightly less than that measured experimentally. If, however, the friction at the platens is appreciable, as in dry compression (Figure 4.1*b*), the load may be considerably greater. This can be allowed for by the method of §4.4, and will be further considered in Chapter 10.

In well-lubricated rolling the friction is low ($\mu \sim 0{\cdot}05$) and the process can again be considered as essentially a compression, but the system is slightly more complex because the strain, and therefore the yield stress of the material, increases from entry to exit of the roll gap (Figure 4.2).

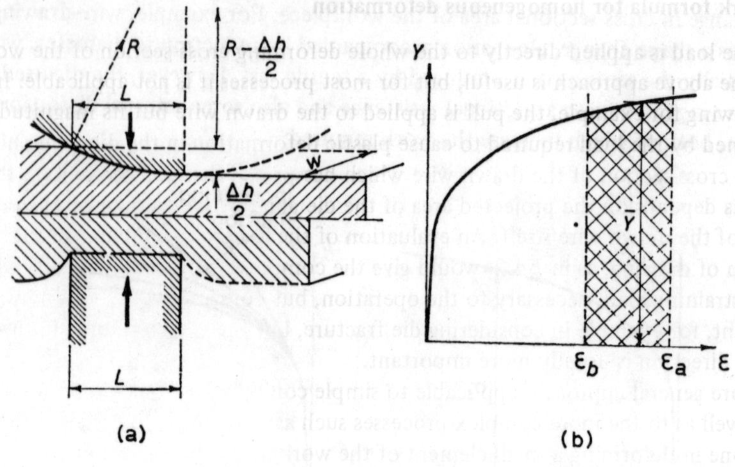

Figure 4.2

(*a*) **Strip-rolling, considered as simple compression with negligible friction.**
(*b*) **Stress–strain curve, showing the mean yield stress for an ideal process deforming metal from ϵ_b to ϵ_a:**

$$\bar{Y} = \frac{1}{\epsilon_a - \epsilon_b}\int_{\epsilon_b}^{\epsilon_a} Y \, d\epsilon.$$

The rolls are usually large enough for their curvature to be ignored. The variation in yield stress can be allowed for in detail (Chapter 9), but for many purposes it suffices to consider the mean yield stress $\bar{Y}$ as a first approximation. This gives a simple formula for the rolling load P

$$P = A\bar{Y} = Lw \, . \, \bar{Y}$$

Since $L^2 = R^2 - \left(R - \dfrac{\Delta h}{2}\right)^2$ and $R \gg \Delta h$, this may be written

$$P = w\bar{Y}\sqrt{R\Delta h} \tag{4.2}$$

Most cold-rolling involves wide strip, and the conditions approximate to plane strain (negligible lateral spread, $\epsilon_z = 0$). It is then necessary to use the yield stress in plane strain, S, which, as shown in Chapter 3 (§3.6.4), is equal to $1{\cdot}15Y$. It is usual to evaluate the load per unit width for wide-strip rolling.

$$P/w = \bar{S}\sqrt{R\Delta h} \tag{4.3}$$

As a rough guide, 20% may be allowed for friction, and the formula for the load P^*, including the frictional contribution,

$$P^*/w = 1{\cdot}2\bar{S}\sqrt{R\Delta h} \tag{4.4}$$

is very useful for rapid assessment of actual cold-rolling loads for wide strip.

4.3 Work formula for homogeneous deformation

When the load is applied directly to the whole deforming cross-section of the workpiece, the above approach is useful, but for most processes it is not applicable. In wire-drawing for example, the pull is applied to the drawn wire but its magnitude is determined by the load required to cause plastic deformation in the die zone, not over the cross-section of the drawn wire which has passed the die. The pull on the wire thus depends on the projected area of the die annulus rather than on the cross-section of the drawn wire itself. An evaluation of the load perpendicular to the direction of drawing, as in §4.2, would give the compressive force on the die, which is a constraining load, necessary to the operation, but doing no work. This may be significant, for example in considering die fracture, but the working force in the drawing direction is usually more important.

A more general approach, applicable to simple conditions of tension or compression as well as to the more complex processes such as drawing, is to consider the work done in deforming a small element of the workpiece, and then to integrate this over the whole deforming region.

Thus, for uniaxial tension the principal stresses at any point are:

$$\sigma_1 = Y, \ \sigma_2 = 0, \ \sigma_3 = 0,$$

where Y is the instantaneous yield stress at the strain ϵ corresponding to the appropriate cross-section area A and length l. The increment of work done in increasing the length of the specimen by δl beyond this strain is given by the product of force and displacement:

$$\delta W = (YA)\delta l \tag{4.5}$$

The increment of work, per unit volume V, is

$$\frac{\delta W}{V} = \frac{\delta W}{Al} = Y\frac{\delta l}{l}$$

We may assume no volume change, and integrate this expression between the original length l_0 and the final length l_1:

$$\frac{W}{V} = \int_{l_0}^{l_1} Y . \frac{dl}{l} = \int_{\epsilon_0}^{\epsilon_1} Y . d\epsilon \tag{4.6}$$

This gives the well-known result that the work done per unit volume in homogeneous deformation is equal to the area of the stress–strain curve, between the appropriate strain values.

This may be evaluated directly from the dimension change, assuming an average yield stress $\bar{Y}$.

$$\frac{W}{V} = \bar{Y}\int_{l_0}^{l_1} \frac{dl}{l} = \bar{Y} \ln \frac{l_1}{l_0} \tag{4.7}$$

Equation 4.7, often known as the work formula, gives a reasonable approximation for a metal which has been work-hardened before the tensile stretching, so that Y does not vary unduly in the process. It is less reliable for annealed material, where Y increases rapidly with strain, so it is then preferable to use equation 4.6, integrating the stress–strain curve graphically. This method can be applied to several practical operations.

4.3.1 *Work Formula for wire drawing*

The work done by the drawing force F_1 in moving from the starting position, adjacent to the die, to the full length l_1 of drawn wire, is given by

$$W_1 = F_1 l_1 \tag{4.8}$$

Assuming homogeneous deformation, the work done in deforming the wire in the die is, from equation 4.7

$$W = V\bar{Y}\ln\frac{l_1}{l_0}$$

In the absence of friction these will be equal:

$$F = \frac{V}{l_1}\bar{Y}\ln\frac{l_1}{l_0}$$

Since $V = l_0 A_0 = l_1 A_1$, this may be written

$$F = A_1\bar{Y}\ln\frac{l_1}{l_0} \tag{4.9a}$$

It is usual to consider reduction of area in wire-drawing, rather than increase in length, since change of area is the property required in practice. Using the constancy of volume again, equation 4.9a becomes

$$F = A_1\bar{Y}\ln\frac{A_0}{A_1} \tag{4.9b}$$

The reduction of area r is given by

$$r = \frac{A_0 - A_1}{A_0} = 1 - \frac{A_1}{A_0}$$

Thus

$$F = A_1\bar{Y}\ln\frac{1}{1-r} \tag{4.9c}$$

The drawing stress σ_1 is consequently

$$\sigma_1 = \frac{F}{A_1} = \bar{Y}\ln\frac{1}{1-r} \tag{4.10}$$

This result forms the basis of many wire-drawing calculations, corrections being applied to take account of the influences of mechanical and frictional constraints.

4.3.2 *Example of application of work formula for drawing: determination of maximum possible reduction of area in one pass*

Wire-drawing is limited eventually by tensile failure of the drawn wire. For such a heavy pass, the maximum tensile stress will be nearly equal to its yield stress, because of the severe strain-hardening which the wire has experienced. Thus at the limiting reduction

$$\sigma_1 = Y_1$$

The rate of strain-hardening will be small, so that the mean yield stress $\bar{Y}$ will also be nearly equal to Y_1. The maximum reduction r_m is thus given by the conditions

$$\frac{\sigma_1}{Y_1} = 1; \ \sigma_1 = \bar{Y} \ln \frac{1}{1 - r_m} = Y_1 \ln \frac{1}{1 - r_m} \tag{4.11}$$

Thus

$$\frac{1}{1 - r_m} = e^{1 \cdot 0} = 2 \cdot 7$$

$$r_m = 1 - \frac{1}{2 \cdot 7} = \underline{0 \cdot 63} \tag{4.12}$$

The maximum possible reduction with perfect lubrication would thus be about 63% or slightly more if the rate of strain-hardening is still appreciable. Internal distortion and friction reduce this limit, as will be seen in Chapter 6, but under favourable conditions wire may be drawn with reductions of area well over 50%.

4.3.3 *Extrusion of a bar*

The work formula can also be used to obtain a lower limit to extrusion pressure, assuming homogeneous deformation and zero friction. In extrusion, the force is applied to the original billet, of area A_0, not to the product whose area is A_1 as it was in the drawing example above. This force moves through a distance l_0 equal to the length of the billet and so does work

$$W_0 = (pA_0)l_0$$

This is equated to the work of homogeneous deformation

$$W = V\bar{Y} \ln \frac{l_1}{l_0} = V\bar{Y} \ln \frac{A_0}{A_1}$$

giving

$$p = \bar{Y} \ln \frac{A_0}{A_1} \tag{4.13}$$

We shall see later that this lower limit is a poor approximation for extrusion because the constraint factor is high for all useful extrusion ratios A_0/A_1. It is possible to calculate a maximum reduction of area, as for drawing, based on a limit $p = Y$, but this implies compressive yielding ahead of the die, not tensile failure. In fact this always happens in the early stage of practical extrusion, but is of no significance because the billet merely compresses to fill the container and is then restrained from further increase in diameter.

4.3.4 *Forging and rolling*

In forging with large flat platens, the area in contact with the workpiece increases continuously during the operation, so the differential form of the work formula is appropriate. The incremental work performed by the force (pA) at the instant when $A = A_x$ is

$$\mathrm{d}W_x = (pA)_x \cdot \mathrm{d}l,$$

and the work of homogeneous deformation is

$$\mathrm{d}W = VY_x \frac{\mathrm{d}l}{l_x} = A_x Y_x \mathrm{d}l$$

Thus $p_x = Y_x$, as in equation 4.1.

This is not a useful method since the same result can be obtained immediately, as in §4.2. The same is clearly true also of rolling (equation 4.2). However, equation 4.2 is quite commonly referred to as the work formula for rolling.

4.4 Allowance for frictional constraint by local stress-evaluation

In §4.3 it is shown that consideration of plastic work performed gives a lower limit for the working stress or load in drawing and extrusion, and in a rather trivial way for forging and rolling. All these industrially important processes normally involve friction at the tool face, since lubrication can never be perfect. In principle, in assessing the total work, it would be possible to allow for the work done against friction. This, however, would involve a knowledge of the forces acting on the tool. The stresses on the tool will not be the same over the whole face, because the frictional drag increases with distance against the direction of motion, so determination of the forces requires integration of the stresses acting on a small element. Having done this, we find that it is immediately possible to evaluate the working load without recourse to the work concept.

This method of stress consideration is quite generally applicable. It is simple to use and, unlike slip-line field solutions (Chapter 6), is equally suitable for round rod and wide strip. Under many conditions, however, it underestimates the load because it ignores the internal distortion of the workpiece. To illustrate the approach, we shall derive the drawing stress for strip drawing. The same technique will be applied, in Part II, to drawing and extrusion of round rod, forging, and strip-rolling.

4.4.1 *Drawing of wide, non-hardening, strip through wedgeshaped dies*[4.1]

Figure 4.3 shows a section of a wide strip being drawn through two dies of total included angle 2α. The original height of the strip, h_b, is assumed to be very much smaller than the width w. Then the conditions approximate closely to plane strain (negligible lateral strain and the same deformation on all planes parallel to the one shown). Because of friction, the stresses on the die increase from entry to exit. Figure 4.3 shows the stresses acting on an element of the strip at a distance x from the virtual apex of the dies.

If p is the pressure normal to the die surface, and μ is the coefficient of friction, we may determine the stress σ_x in the direction of drawing by consideration of the equilibrium of the forces acting on the element. There are three force components

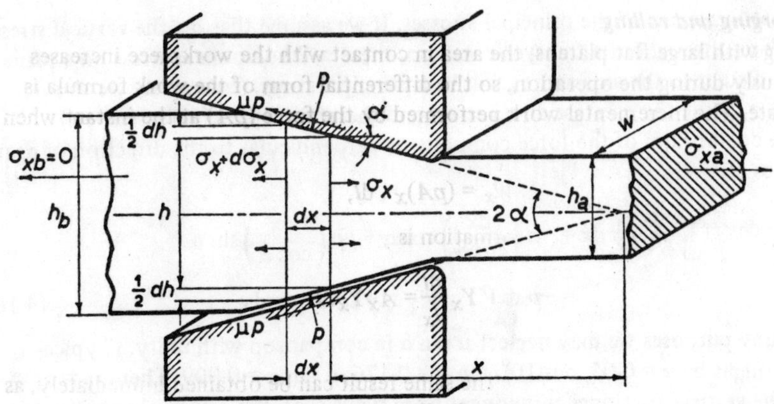

Figure 4.3. The stresses acting on an element in the deformation zone during drawing strip through dies of constant angle.

acting in the longitudinal direction, Ox:

due to the change in longitudinal stress (if any),

$$(\sigma_x + \mathrm{d}\sigma_x)(h + \mathrm{d}h)\, w - \sigma_x h w$$

due directly to the die pressure (on both dies),

$$2 \, . \, p \left(w \frac{\mathrm{d}x}{\cos \alpha} \right) \sin \alpha$$

due to the frictional stress (on both dies)

$$2\mu p \left(w \frac{\mathrm{d}x}{\cos \alpha} \right) \cos \alpha$$

For steady drawing conditions, these must be in equilibrium. Eliminating the constant width w and ignoring the product $\mathrm{d}\sigma_x \, . \, \mathrm{d}h$, this gives

$$\sigma_x \, \mathrm{d}h + h \, \mathrm{d}\sigma_x + 2p \, \mathrm{d}x \tan \alpha + 2\mu p \, \mathrm{d}x = 0 \tag{4.14}$$

h may be expressed geometrically in terms of x

$$h = 2x \tan \alpha; \quad \mathrm{d}h = 2\mathrm{d}x \tan \alpha$$

Equation 4.14 thus becomes

$$\sigma_x \, \mathrm{d}h + h \, \mathrm{d}\sigma_x + p \, \mathrm{d}h + \mu p \, \mathrm{d}h \cos \alpha = 0$$

$$h \, \mathrm{d}\sigma_x + [\sigma_x + p\,(1 + \mu \cos \alpha)] \, \mathrm{d}h = 0$$

It is convenient to introduce a parameter $B = \mu \cot \alpha$

$$h \, \mathrm{d}\sigma_x + [\sigma_x + p\,(1 + B)] \, \mathrm{d}h = 0 \tag{4.15}$$

The equation can then in principle be integrated if a relationship can be found between σ_x and p. This can be obtained from the yield criterion, if σ_x and p are expressed in terms of the principal stresses. If we assume that σ_y, the vertical stress, is uniform throughout a transverse section, there will be no shear stress on a plane perpendicular to the axis and σ_x will be a principal stress. (Intuitively we should expect σ_x to be the greatest direct stress and therefore a principal stress, σ_1.)

The equilibrium of the force components perpendicular to the direction of drawing gives

$$\sigma_y\,dx = -p\left(\frac{dx}{\cos\alpha}\right)\cos\alpha + \mu p\left(\frac{dx}{\cos\alpha}\right)\sin\alpha$$

$$= -p\,dx \qquad + \mu p \tan\alpha\,dx \tag{4.16}$$

For many purposes we may neglect $\mu\tan\alpha$ in comparison with unity. (Typical values might be $\mu = 0{\cdot}05$, $\alpha = 10^\circ$, $\tan\alpha = 0{\cdot}176$; $\mu\tan\alpha = 0{\cdot}009$. Then $\sigma_y = \sigma_3 \simeq -p$.

If the vertical frictional component were significant, the principal axes would be slightly rotated, and there would be shear on the plane perpendicular to the drawing direction. We shall see an example of this in connection with hot rolling later on, but for the moment the procedure is greatly simplified by assuming good lubrication conditions, giving

$$\sigma_3 = -p; \quad \sigma_1 = \sigma_x \tag{4.17}$$

Under plane-strain conditions, yield occurs in pure shear (§3.6.4) so $\sigma_1 - \sigma_3 = 2k = S$, which is independent of the choice of yield criterion. Thus

$$\sigma_x + p = S, \text{ or } p = S - \sigma_x \tag{4.18}$$

Substituting this value of p in equation 4.15, and rearranging slightly,

$$\frac{d\sigma_x}{\sigma_x + (S - \sigma_x)(1 + B)} = -\frac{dh}{h}$$

$$\frac{d\sigma_x}{B\sigma_x - S(1 + B)} = \frac{dh}{h} \tag{4.19}$$

This is the basic differential equation of wide-strip drawing, first given in this form in 1944 by Sachs, Lubahn and Tracy[4.2]. It applies equally to wedge dies or to curved dies. The parameter B has the value $\mu\cot\alpha$, where μ and α are the coefficient of friction and semi-die angle respectively, at the position x, and need not be constants independent of x. It can also be used for work-hardening metal, since S is the yield stress at the section, and a relationship between S and x can be deduced from the stress–strain curve.

To obtain the simplest solution, however, we may assume that the dies are straight, the friction is uniform, and the material does not work-harden. Thus, with α, μ and S constant, the equation can be integrated immediately:

$$\frac{1}{B}\ln\,[B\sigma_x - S(1 + B)] = \ln h + \text{constant } C.$$

or

$$B\sigma_x - S(1 + B) = C' h^B, \text{ where } C' = e^{CB}$$

The constant of integration can be found from the entry conditions, assuming no back tension at the die entry: $\sigma_x = \sigma_{xb} = 0, h = h_b$

Thus
$$-S(1+B) = C' h_b^B$$

$$\sigma_x = \frac{1}{B}\left[-S(1+B)\left(\frac{h}{h_b}\right)^B + S(1+B)\right] \tag{4.20}$$

It is often useful in metalworking calculations to use the dimensionless ratio of the working stress to the yield stress. This allows results to be plotted in a general way without reference to a particular metal, but the actual stresses can then readily be determined for any given metal whose yield stress is known. It also provides rapid comparison with the stress required for homogeneous deformation and so gives a measure of the efficiency of a process.

$$\frac{\sigma_x}{S} = \frac{1+B}{B}\left[1 - \left(\frac{h}{h_b}\right)^B\right] \tag{4.21}$$

This is the direct stress, in the direction of drawing, at any distance x from the virtual apex. Since yield occurs when

$$\frac{p}{S} = 1 - \frac{\sigma_x}{S} \tag{4.18}$$

the die pressure at any point may also be evaluated.

$$\left(\frac{p}{S}\right)_x = 1 - \frac{1+B}{B}\left[1 - \left(\frac{h}{h_b}\right)^B\right] \tag{4.22}$$

It will be seen that this varies with h and so with the position x. The actual drawing stress σ_{xa} exerted by the drawbench grips is the value of σ_x at the die exit, where $h = h_a$.

$$\text{Drawing stress} = \frac{\sigma_{xa}}{S} = \frac{1+B}{B}\left[1 - \left(\frac{h_a}{h_b}\right)^B\right] \tag{4.23}$$

4.4.2 *Example of application of drawing stress equation: determination of maximum reduction of area in one pass, allowing for friction*

As an instance of the use of this equation, we may estimate the maximum reduction in area per pass for some typical practical values: $\mu = 0{\cdot}05$, $\alpha = 15°$. Thus $\cot\alpha = 3{\cdot}73$, $B = \mu\cot\alpha = 0{\cdot}1865$, $\frac{1+B}{B} = 6{\cdot}34$. The limit to drawing will be set by tensile failure of the drawn bar ahead of the die. Since we have assumed no work-hardening, this will occur at the yield stress Y (not S since a drawn strip of finite width is free to contract laterally). Thus for maximum reduction r_m,

$$\frac{\sigma_{xa}}{S} = \frac{Y}{S} = \frac{1}{1{\cdot}15} = 6{\cdot}34\left[1 - \left(\frac{h_a}{h_b}\right)^{0{\cdot}1865}\right]$$

$$1 - \left(\frac{h_a}{h_b}\right)^{0{\cdot}1865} = 1 - 0{\cdot}1367 = 0{\cdot}8633; \quad \frac{h_a}{h_b} = 0{\cdot}455$$

$$r_m = 1 - \frac{h_a}{h_b} \simeq 55\% \tag{4.24}$$

The work formula (equation 4.11) gave $r_m = 63\%$ for round bar. If allowance is made for the constraint in plane-strain drawing, equation 4.11 becomes, for non-hardening metal

$$\frac{\sigma_1}{Y} = 1;\ \sigma_1 = S \ln \frac{1}{1 - r_m}$$

Then $$\ln \frac{1}{1 - r_m} = \frac{1}{1{\cdot}15};\ \frac{1}{1 - r_m} = e^{0{\cdot}866} \simeq 2{\cdot}38$$

$$r_m \simeq 58\% \qquad (4.25)$$

Thus, when the lubrication is sufficiently good to give $\mu = 0{\cdot}05$, the limiting reduction is little lower than the ideal value. Because of the advantage to be gained with work-hardening metal, with S_1 greater than $\bar{S}$, it has been possible to draw wide bars with 58% reduction of area in actual experiments[(4.3)].

4.5 Comparison of work formula and stress evaluation, for drawing

For perfectly-lubricated dies, the stress equation cannot be used in its integrated form (4.23), because $B = 0$. In the differential form, which is valid for all conditions

$$h\ \mathrm{d}\sigma_x + [\sigma_x + p(1 + \mu \cot \alpha)]\ \mathrm{d}h = 0 \qquad (4.15)$$

Thus for $\mu = 0$,

$$h\ \mathrm{d}\sigma_x + (\sigma_x + p)\ \mathrm{d}h = 0$$

Incorporating the yield relationship for plane strain (equation 4.18)

$$h\ \mathrm{d}\sigma_x + S\ \mathrm{d}h = 0$$

This may now be integrated:

$$\frac{\sigma_x}{S} = -\ln h + \text{const } K.$$

Using the boundary conditions at entry ($x = b$, $\sigma_{xb} = 0$, $h = h_b$) to obtain the value of K, this gives:

$$K = \ln h_b$$

$$\frac{\sigma_x}{S} = \ln \frac{h_b}{h};\ \frac{\sigma_{xa}}{S} = \ln \frac{h_b}{h_a} = \ln \frac{1}{1 - r} \qquad (4.26)$$

This corresponds to the evaluation from considerations of work done in plane-strain homogeneous deformation (compare equation 4.10 for round bar). When the die is perfectly lubricated, no additional work is done in overcoming friction and the same result is obtained by either method, taking the appropriate value for yield stress according to the geometry of wire (Y) or strip (S). The stress-analysis approach thus contains the work-formula information but extends it to include friction.

4.6 Allowance for work-hardening, in stress evaluations

As a first approximation, work-hardening can be included by using a mean value $\bar{S}$.

It can also be allowed for, more accurately, by deriving an analytical expression to fit the stress–strain curve, and using this together with the differential equation 4.19. An easier and better method, considered in Chapter 12, 1965 edition, is to simulate the stress–strain curve with an analogue computer, and then to solve equation 4.19 together with this data, using the computer. Once the instrument is set up, the results are very rapidly obtained graphically, for any chosen values of μ and α. This approach removes most of the tedium from the calculations, and may lead to more widespread use of the analytical method.

4.7 Validity of stress-evaluation approach

The analysis given above has been described in detail in terms of strip drawing. It is however, as we shall see in Part II, equally applicable to extrusion and rolling of strip, and to drawing and extrusion of round bars. The same type of approach can

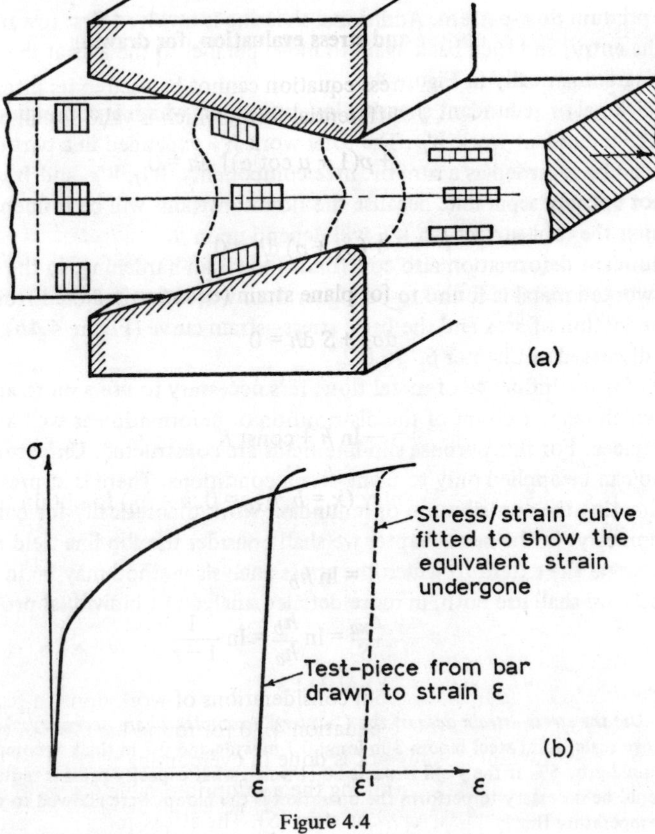

Figure 4.4

(*a*) Diagram showing the distortion of an element of strip, as it passes through a drawing die.

(*b*) A basic stress–strain curve with a superimposed tensile-test curve of drawn bar, showing that the yield stress has been increased by redundant work.

also be used for some forging operations. It therefore provides a valuable tool for estimation of working loads and examination of the influence of changes in process parameters, especially when used in its more sophisticated form which allows for work-hardening.

Under many conditions the predictions are in good accord with observed working loads, though it must be admitted that the coefficient of friction μ is often regarded as an adjustable parameter. However, within reasonable limits, values of μ may be estimated beforehand and the load calculations still appear reasonably accurate. We shall discuss the actual values of μ in Chapter 11.

Under some circumstances, however, it is found that the stress evaluation seriously underestimates working loads. The reason for this is that the total work done has three components, due to the homogeneous plastic deformation (W_H) the frictional resistance (W_F), and also to the internal distortion that may occur. For example in drawing through a wedge-shaped die, which we have just considered, the shape of the die constrains the metal to flow in a particular manner which usually differs from the optimum flow-pattern. Additional shearing is involved first towards the axis near the entry, and then back again to move parallel to the axis at the exit. This is shown diagrammatically in Figure 4.4.

This additional or redundant shearing involves work, which is consequently known as the redundant work W_R. The total work W_T expended in a particular deformation can be regarded as a sum of three components, W_H, W_F, and W_R, but these are not entirely separable, because the flow constraint will be influenced by the friction at the tool surface, so W_R will depend upon μ.

The redundant deformation also contributes to strain-hardening, so the yield stress of a worked metal is found to be higher than would be predicted from the measured reduction of area and the basic stress–strain curve (Figure 4.4*b*). This will be further discussed in Chapter 6, §6.6.2.

To allow for the influence of metal flow, it is necessary to use a more advanced approach which takes account of the distribution of deformation as well as of stress in the workpiece. For this purpose slip-line fields are constructed. Unfortunately this method can be applied only to plane-strain conditions. There is at present no way of estimating the contribution of redundant work theoretically for conditions of axial symmetry. In the next chapter we shall consider the slip-line field method, which will reveal the extent to which the stress analysis method may be in error. Subsequently we shall use both, in more detailed analysis of individual processes.

EXAMPLES

Use the stress–strain data of the Chapter 2 examples where necessary.

4.1 **To remove scale, a hot steel bloom 3 m long, 0·7 m wide and 0·7 m thick is compressed longitudinally by 5%. If the yield stress is 60 N/mm², what capacity press is required? What load would be necessary to perform the operation if the bloom were allowed to cool to room temperature first?**

Solution. **Assuming the volume to remain constant, the width and thickness each increase by $\sqrt{1{\cdot}05}$ during the compression. The force required is $(60 \times 700 \times 700 \times 1{\cdot}05) \approx 31 \times 10^6$ N.**

At room temperature $S_1 = 432$ N/mm² at $\epsilon_1 = 0{\cdot}052$ (for mild steel, Example 2.3*a*). Thus $Y \approx 380$ and the load would be about 200 MN. Note that S_1, not $\bar{S}$, is used.

It will be seen in Chapter 10 that there would be an appreciable increase in the hot-working load due to the high friction in an actual working operation.

4.2 What would be the approximate roll load necessary to reduce 2 m wide, 2·50 mm thick aluminium sheet to 2·00 mm thick in one pass at room temperature, using 350 mm diameter rolls?

Solution. The strain imparted is $\epsilon_1 = \ln 2{\cdot}50/2{\cdot}00 = 0{\cdot}223$. Approximating to the stress–strain curve of Example 2.3*b* by a straight line over this region, $\bar{S} = \frac{1}{2}(108 + 70) \approx 89$ N/mm². (It could be somewhat more accurately deduced from $\frac{1}{\epsilon_1}\int_0^{\epsilon_1} S\,d\epsilon$, but the difference is very small). The length of contact of the roll with the strip is given by equation 4.2, $L = \sqrt{R\Delta h} = 9{\cdot}35$ mm. The load per unit width (equation 4.3) is $P/w = \bar{S}L \approx 830$ N/mm.

Thus $P = 1{,}660$ kN. Allowing 20% for friction, $P \simeq 2{\cdot}0$ MN.

4.3 If the pass in Example 4.2 were increased from a draft 0·50 mm to 0·75 mm, what influence would there be on the roll load?

Solution. The strain imparted is $\epsilon_1 = \ln 2{\cdot}50/1{\cdot}75 = 0{\cdot}357$. A straight-line approximation to the stress–strain curve of Example 2.3*b* gives $\bar{S} = 90$ N/mm². (Note that this simple method becomes more accurate as ϵ_1 increases). From equation 4.2, $L = \sqrt{R\Delta h} = 11{\cdot}5$ mm. From equation 4.3, $P/w = \bar{S}L = 1{,}035$, $P = 2{\cdot}1$ MN. Allowing 20% for friction $P \approx 2{\cdot}5$ MN.

4.4 If the maximum roll capacity is 1 MN, suggest a suitable rolling schedule for the operation described in Example 4.2. Could a similar schedule be prepared if the material were mild steel instead of aluminium?

Solution. (a) The load permitted is about half that required to complete the operation in one pass. If the mean yield stress were unchanged, the final pass would, according to the relationship in equation 4.3 be about $\Delta h = (0{\cdot}5)^2 \times 0{\cdot}50 \approx 0{\cdot}13$ mm. The stress–strain curve suggests a mean yield stress of about 106 N/mm² for such a pass. The load would then be $P = 1{\cdot}2\, w\, \bar{S}\sqrt{R\Delta h}$. For $P = 10^6$ N, $w = 2{,}000$ mm, and $\bar{S} = 105$ N/mm²; $\sqrt{R\Delta h} = 3{\cdot}93$, $\Delta h = 0{\cdot}09$ mm. Choose the final pass to be 0·08 mm and calculate the load:

Pass no.	Δh mm	h_{N-1} mm	$\epsilon_N - \epsilon_{N-1}$	$\bar{S}$ kN/mm²	P kN	$\sqrt{R\Delta h}$
N	0·08	2·08	0·223 – 0·183	0·106	951	3·74
$N-1$	0·09	2·17	0·183 – 0·160	0·104	990	3·97
$N-2$	0·09	2·26	0·160 – 0·101	0·101	960	3·97
$N-3$	0·10	2·36	0·101 – 0·057	0·091	913	4·18
$N-4$	0·14	2·50	0·057 – 0	0·065	772	4·95

The draft for each pass is estimated on the basis of the preceding calculation.

Thus a suitable schedule for this operation would be five passes, respectively reducing the strip to 2·36, 2·26, 2·17, 2·08 and 2·00 mm. This would actually imply unwarranted confidence in the formula, to allow 1% safety factor! The first pass could in principle be heavier, but $\bar{S}$ is not well defined for the first pass after annealing.

(b) With mild steel the final pass would involve a mean yield stress $\bar{S} \approx 0{\cdot}6$ kN/mm². Equation 4.2 gives the final draft (assuming that the rolls do not deform – see § 9.4):

$$1{,}000 = 1{\cdot}2\,.\,2{,}000\,.\,0{\cdot}6\sqrt{175\Delta h};\quad \Delta h \approx 0{\cdot}0028 \text{ mm}$$

This mill is thus quite unsuitable for rolling steel with these dimensions.

4.5 (a) An annealed mild steel strip 2·50 mm thick is rolled to 2·00 mm in one pass and then to 1·50 mm. What is the mean yield stress in the second pass? What error would be involved in assuming the yield stress constant at its final value?

(b) If the strip is further rolled to 1·25 mm, what would be the mean yield stress for this pass, and by how much would it differ from the final yield stress?

Solution. (a) The strain imparted by the first pass is $\epsilon_{01} = \ln \dfrac{2{\cdot}50}{2{\cdot}00} = 0{\cdot}223$; the total of first and second is $\epsilon_{02} = \ln \dfrac{2{\cdot}50}{1{\cdot}50} = 0{\cdot}511$. From the stress–strain curve (Ex. 2.3) $S_2 = 0{\cdot}75$, $S_1 = 0{\cdot}61$. Thus the mean yield stress is 0·68 kN/mm² and the error in assuming the yield stress to be constant at S_2 would be about 10%.

(b) $\epsilon_{03} = \ln \dfrac{2{\cdot}50}{1{\cdot}25} = 0{\cdot}693$. $S_3 = 0{\cdot}80$ kN/mm²,

$\bar{S}_{23} = \frac{1}{2}(0{\cdot}75 + 0{\cdot}80) = 0{\cdot}78$ and the error in assuming

$\bar{S}_{23} = S_3$ is 0·2 kN/mm², about 3%.

4.6 A 2·50 mm diameter annealed brass wire is drawn to 2·15 mm diameter. Calculate the approximate minimum drawing load. What would this be if the wire had been annealed at 3·20 mm diameter and drawn to 2·50 mm before this pass?

Solution. (a) The strain imparted is $\epsilon_{01} = \ln A_0/A_1$ from equation 2.11. For this pass $\epsilon_{01} = \ln(2{\cdot}50)^2/(2{\cdot}15)^2 = \ln 1{\cdot}35 = 0{\cdot}301$. $\bar{S} = \frac{1}{2}(0{\cdot}58 + 0{\cdot}20) = 0{\cdot}39$ kN/mm². The minimum drawing load is that required to impart the deformation homogeneously. From equation 4.9 the drawing force is

$$F = A_1\,\bar{Y} \ln A_0/A_1 = A_1\,\bar{Y}\,\epsilon_{01} = 370 \text{ N}$$

(b) If the prior strain were $\epsilon_{01} = \ln(3{\cdot}20)^2/(2{\cdot}50)^2 = 0{\cdot}493$, the final strain would be $\epsilon_{02} = \ln(3{\cdot}20)^2/(2{\cdot}15)^2 = 0{\cdot}795$ (= 0·493 + 0·301). Then $\bar{S} = 0{\cdot}73$ kN/mm² and $F = 693$ N.

4.7 A 7·5 kW electric motor is geared to a draw-bench speed of 0·1 m/s. What would be the maximum size of round steel bar that could be drawn under ideal conditions in a pass reducing its diameter by 1·60 mm?

Solution. 7,500 watts = 7,500 Nm/s . Speed 0·1 m/s Drawbench pull $F = 75$ kN
Equation 4.9 gives $F = A_1\,\bar{Y} \ln A_0/A_1$.

The maximum size of bar that could be drawn would be expected to be the one for which 1·60 mm represented a light pass. Suppose that a reasonable value to choose for the yield stress is $\bar{S} = 450$ N/mm², say $Y = 400$.

From equation 4.9

$$75{\cdot}10^3 = \frac{\pi}{4} d_1^2 \,.\, 400 \,.\, \ln(d_1 + 1{\cdot}60)^2/d_1^2$$

But $$d_1 \gg 1{\cdot}6 \text{ so } \ln(d_1 + 1{\cdot}6)^2/d_1^2 \approx \ln(1 + 3{\cdot}2/d_1) \approx 3{\cdot}2/d_1$$

$$75{\cdot}10^3 \approx \frac{\pi}{4} d_1^2 \,.\, 400 \,.\, \frac{3{\cdot}2}{d_1};\ d_1 \approx 74{\cdot}6 \text{ mm}$$

This would involve a strain $\epsilon_{01} = \ln \dfrac{76{\cdot}2}{74{\cdot}6} = 0{\cdot}02$.

A better value for $\bar{Y}$ would thus be about 240 N, so d_1 would be correspondingly increased:

$$d_1 \approx 74{\cdot}6 \times 400/240 = 120 \text{ mm};\ \epsilon_{01} \approx 0{\cdot}01$$

The exact value chosen depends upon the effective mean yield stress for very light deformations, which cannot be found accurately from the data of Example 2.3. The lower yield point of mild steel in the tensile test is however 0·23 kN/mm² (Example 2.1), so $\bar{Y} \approx 240$ is reasonable, even for a very light pass.

The largest round steel bar that could be drawn under the specified conditions would thus be over 100 mm diameter, but in fact the deformation would be far from homogeneous (see chapter 6, §6.7).

Such 'sizing' passes are commonly used in the production of bright-drawn standard bars, and the reduction is often specified as 1·5 mm (formerly 1/16 inch) off the diameter, or for larger bars 3 mm.

4.8 A copper wire is annealed at 2·12 mm diameter. What is the smallest diameter to which it could theoretically be drawn in (a) one pass; (b) in three passes?

Solution. The maximum reduction possible under ideal conditions is given by equation 4.10, with the condition $\sigma_1 = Y_1 = \bar{Y}$, as in equation 4.12

(a) $\sigma_1 = \bar{Y} \ln A_0/A_1$; $A_0/A_1 = 2{\cdot}7, d_0/d_1 = 1{\cdot}65$ so $d_1 = 1{\cdot}28$ mm.

(b) The strain $\epsilon_{01} = \ln A_0/A_1 = 1{\cdot}0$ can be applied in each of the three passes, so the maximum possible in three passes is $\epsilon_{03} = 3{\cdot}0 = \ln A_0/A_3$; $A_0/A_3 = 20, d_0/d_3 = 4{\cdot}48, d_3 = 0{\cdot}47$ mm.

4.9 A nickel-silver ribbon 10 mm wide and 0·50 mm thick is drawn to 0·30 mm thick through dies of 20° included angle. The hardness of the drawn strip is found to be 165 V.P.N. Estimate the drawing load and the die pressure, assuming $\mu = 0{\cdot}08$. If all the work done appeared as heat, what would be the temperature rise in the strip?

Solution. (a) $h_0 = 0{\cdot}50$ mm, $h_1 = 0{\cdot}30$ mm, $A_1 = 0{\cdot}0707$ mm², $B = 0{\cdot}8 \cot 10° = 4{\cdot}54$. The drawing stress is found from equation 4.23

$$\frac{\sigma_x}{S} = \frac{1{\cdot}45}{0{\cdot}45}\,[1 - (0{\cdot}6)^{0{\cdot}45}] = 3{\cdot}22\,[1 - 0{\cdot}795] = 0{\cdot}66$$

The yield stress is found from the hardness, equation 2.25. 165 Vickers is approximately the same as 165 Brinell, both being expressed in kgf/mm². Thus $Y_1 \approx 165/0{\cdot}3 = 550$ N/mm², $\bar{S} \approx S_1 \approx 630$. The drawing stress is then 420 N/mm². A better value would be obtained if the hardness were measured before and after the pass, but high accuracy cannot be expected without determining yield stresses directly.

(b) The work done in drawing a length L metres of wire, equation 4.8, is

$$W = FL = 10\,.\,0{\cdot}30\,.\,420\,L \text{ Nm}$$

$$= 1{,}260\,L \text{ joules}$$

The volume of this length is $10 \times 0{\cdot}30 \times L \times 10^{-6}$ m³. Since the density of nickel silver is $8{\cdot}8 \times 10^3$ kg/m³, the weight of the strip is $26{\cdot}4 \times 10^{-3} L$ kg.

In the absence of accurate data, assume the specific heat to be 400 J/kg (Cu = 380, Ni = 443). Then

$$1{,}260\,L = 400\,.\,26\,.\,4\,.\,10^{-3} L\,(T_1 - T_0)$$

$$T_1 - T_0 \approx 120°\text{C}$$

4.10 An annealed steel strip 10 mm wide and 0·40 mm thick is drawn to 0·30 mm thick through dies of 24° included angle, at 2·0 m/s. What is the total work expended in drawing a 50 kg coil, assuming $\mu = 0{\cdot}10$? If the efficiency of the drawbench drive is 85%, what would be the input current to the driving motor on a 230-volt main? What would be the cost of electricity at 2·2p per unit (kWh)?

Solution. Equation 4.23 gives the drawing stress. With

$$\epsilon = \ln \frac{0{\cdot}40}{0{\cdot}30} = 0{\cdot}288,\quad B = 0{\cdot}1 \cot 12° = 0{\cdot}470$$

$$\frac{\sigma_1}{S} = \frac{1{\cdot}47}{0{\cdot}47}\,[1 - (0{\cdot}75)^{0{\cdot}47}] = 3{\cdot}13 \times (1 - 0{\cdot}873) = 0{\cdot}40$$

Using a straight-line approximation, $\bar{S} = \frac{1}{2}(660 + 350) = 500$ N/mm², $\sigma_1 = 200$ N/mm²; $A_1 = 3{\cdot}0$ mm², $F_1 = 600$ N.

Assuming the density of copper to be $7{\cdot}8 \times 10^3$ kg/m³, the length L_1 of the coil is given by

$$3{\cdot}0 \times 10^{-6} \times L_1 \times 7{\cdot}8 \times 10^3 = 50;\quad L_1 = 2{\cdot}14 \times 10^3 \text{ m}$$

The work expended in drawing the full length of the coil is $F_1 L_1 = 1{,}280$ kNm. Speed 2 m/s. Time to draw the coil $1{\cdot}07 \times 10^3$ s. Power required, 1,200 Nm/s = 1·2 kW. With 85% efficiency, input power = 1·41 kW. Input current for a single-phase motor at 230 V = 6·13 A. Total power consumed in drawing the coil

$$1{\cdot}41 \times \frac{1\,.\,07\,.\,10^3}{3{,}600} = 0{\cdot}42 \text{ kWh.}$$

The cost is thus 0·92p.

REFERENCES

4.1 Sachs, G., *Spanlose Formung der Metalle*, Springer, Berlin, 1934.

4.2 Sachs, G., Lubahn, J. D. and Tracy, D. P., 'Drawing of thin-walled tubing', *J. Appl. Mech*., 11, 199–210, 1944.

4.3 Lancaster, P. R. and Rowe, G. W., 'A comparison of boundary lubricants under light and heavy loads', *Wear*, 2, No. 6, 428–437, 1958.

5

Determination of Working Loads by Consideration of Metal Flow

5.1 Introduction

The analysis of surface stresses allows for the contribution of friction, which may easily amount to 20% or 30% of the total working load. The results are in quite good agreement with the loads measured in practical metalworking, provided that the coefficient of friction is known and that the internal distortion of the workpiece involves the least possible shearing, compatible with the change of external shape. The problem of determining appropriate coefficients of friction will be considered in Chapter 11. The practical loads may however be much greater than those predicted, due to constraint imposed by the necessity for the metal to flow in a particular way. In general, the more curved the shear lines, the higher the pressure required to make the metal flow. The work done in overcoming this excess constraint cannot be found from the overall change in shape, and is known as redundant work. We shall see (§5.6.6) that it increases the load in hardness testing, for example, by a factor of nearly 3, above that required for simple compression of a small cylinder. Factors of two or more are often encountered in extrusion.

Surface-stress analysis takes no account of redundant work, and there is in fact no general analytical theory. For the special condition where flow occurs entirely in one plane, with no deformation in the direction perpendicular to that plane (e.g. flow in the XY plane, with $\epsilon_z = 0$), a graphical type of solution can be obtained. This plane-strain condition is thus of considerable importance, though it does not often arise in practice. Where the practical geometry does produce approximately plane-strain deformation, as in small-tool forging of thin, wide strip (Figures 5.5 and and 5.6), there is very close agreement with predicted results. The results of plane-strain theory can however prove very helpful in understanding practical processes involving axial symmetry (wire drawing, bar extrusion etc), and more complex conditions. The graphical theory, known as slip-line field analysis, depends upon determination of the plastic flow pattern in the deforming workpiece.

Plastic flow in metals occurs predominantly by slip, on an atomic scale, along crystallographic planes. The details of the process of dislocation movement, which is the basic phenomenon of slip, have been extensively studied in recent years. Several comprehensive books are available for the reader interested in obtaining a deeper insight into the fundamental behaviour of deforming metals[5.1, 5.2]. For our present purpose, it is important to recognise only that plastic flow is essentially a shear process. In a real metal, slip occurs most easily on planes of close-packed atoms, and in the direction of the line of atoms which lies closest to the line of maximum shear stress. Metalworking, however, is concerned with metals containing very large numbers of crystals. It is thus, nearly always, reasonable to assume that a sufficient number will be favourably oriented for the general slip-direction to coincide with the direction of maximum shear stress.

For simplicity, it is usual to postulate a structureless, homogeneous, and isotropic material, in which slip always occurs precisely in the maximum shear-stress direction. This direction is consequently of importance in any consideration of metal flow. Once it is determined, the direction of plastic flow of the idealised material is known, and this gives a first approximation to the flow of a real metal. Allowance for anisotropy in real metals is very complex[5.3]; this is unfortunate because all metals, though highly polycrystalline, tend to become anisotropic during deformation The assumption of homogeneity is reasonably good, for most industrial metals; and the effects of structure are not usually important, except in metals with hexagonal crystal lattices (such as zinc, cadmium, magnesium and beryllium), which have only one active slip-plane in the crystal.

We shall confine our attention to the simple, idealised, material. It is also convenient to consider first a material which does not work-harden, so that the value of the yield stress remains constant throughout the deformation. This is a poor approximation for annealed metal, but the slope of the stress–strain curves is quite small for most metals after a moderate deformation. Making these assumptions, we can evaluate plane-strain working loads, including the redundant work factor, by the use of slip-line fields.

5.2 Deformation in simple compression

It has been shown in Chapter 3 that the directions of maximum shear are inclined at 45° to the directions of the principal stresses (Figure 3.4). If a cube is compressed across two opposite faces, it can easily be shown experimentally that it will fail by shear on diagonal planes. Figure 5.1 shows a cylinder of 10% aluminium bronze which has sheared along 45° cones, leaving a solid core in the form of a double cone, with smooth, bright sides. (The same pattern is found in the standard compression test for concrete.)

Figure 5.1 A tracing from a photograph of a 10% aluminium bronze cylinder which has been compressed to fracture. (*By courtesy of Dr. A. T. Male.*)

Metals which are more ductile start to deform in the same way, but the parts do not separate until the secondary tensile stresses have become much larger, after heavy deformation, which often obscures the pattern.

The general problem of three-dimensional deformation is still intractable, but for two-dimensional strain it is possible to predict working loads from consideration of slip-lines, which are lines showing the directions of maximum shear everywhere in the plastically-deforming body. We shall assume throughout this chapter that plane-strain conditions apply; namely that there is no strain in one principal direction so that $\epsilon_2 = 0$, and that the pattern of deformation is the same for all planes perpendicular to this direction.

If a wide strip is compressed between parallel flat-faced anvils of breadth b equal to the strip thickness h, the undeformed strip on each side of the anvils restricts the spread of the original width w. If w is greater than about $10b$, the lateral strain and the end effects may be ignored, so that plane-strain conditions apply. Figure 5.5 shows such a compression. Deformation starts by shearing along the diagonals, and the four blocks of metal I, II, III, IV start to move by sliding over one another as solid, undeformed bodies. This sliding coincides with the directions of maximum shearing stress, and it will be seen in §5.3 that deformation in plane strain is a pure shear process, in which yield occurs when $\tau_{max} = k$.

The direct stress necessary to produce a shear stress k in this example is easily found. The two principal stresses will act at 45° to the direction of maximum shear stress, and so will be horizontal and vertical in Figure 5.5. The horizontal stress acting at any point on the line AC must zero since there is no force acting on the rigid block II. The horizontal principal stress is thus $\sigma_3 = 0$. Since $k = \frac{1}{2}(\sigma_1 - \sigma_3)$, the vertical principal stress σ_1, acting on the surface AB of the anvil will be equal to $2k$. The indentation pressure is thus

$$q = \sigma_1 = 2k \tag{5.1}$$

This is the same as the yield stress $S = 2k$ which would be found if a tensile test could be conducted under conditions of plane strain. It should be recognised, however, that the normal tensile test of round bar, or even of wide strip, does not involve the constraint of plane strain, and so gives the value Y. If the von Mises criterion is used, $Y = 2k/1{\cdot}55$. The additional direct stress which is required to cause yielding in plane strain, arises from the externally imposed necessity for the metal to flow in such a way that no lateral contraction occurs in tension, and no lateral spread in compression.

5.3 Stress evaluation using slip-lines

The elementary calculation given above shows the way in which a knowledge of the directions of maximum shear can be used, together with the magnitude of the shear yield stress, to determine a working load. This is the basis of slip-line field solutions. The slip-line field diagram can be considered as a map, showing the directions of maximum shear at every point in the deforming body. Figure 5.5 shows, therefore, a very simple slip-line field. Since shear must always be accompanied by complementary shear of equal magnitude and opposite sense, to preserve rotational equilibrium

(§3.6.3), there will always be two mutually perpendicular directions of maximum shear at each point. A general map of these directions consequently consists of two sets of mutually orthogonal lines. In the simplest instance, these would all be straight and form a Cartesian network. They need not be straight however, and a network of straight radii and concentric circles, resembling a polar coordinate system, is frequently used. Both of these can be seen in Figure 5.7. Further developments utilise orthogonal networks in which the lines of both sets are curved (see Chapter 6, for example).

The spacing of the lines may be freely chosen. In straight regions it has no significance. Where the lines are curved, greater accuracy is obtained by drawing lines with small angular separation. Intervals of 5° usually give as high an accuracy as is useful, and 15° often suffices.

In the simple example of compression, given above, the stresses could be calculated directly in terms of the shear stress. In a more complex system, account must also be taken of a hydrostatic pressure which varies from point to point in the plastically deforming body. The significance of this can be appreciated by considering the Mohr circle, as follows.

Slip-line field solutions always refer to plane strain, and for this condition, as was stated in Chapter 3, the principal stresses for an incompressible, ideally plastic material are related:

$$\sigma_2 = \tfrac{1}{2}(\sigma_1 + \sigma_3) \tag{3.26}$$

This has been shown, experimentally, to be very nearly true for a real metal. The Mohr stress circle for plane strain conditions can thus be drawn with radius

$$\tau_{max} = \tfrac{1}{2}(\sigma_1 - \sigma_3) \tag{3.11a}$$

and centred on σ_2, in accordance with equation 3.26, as shown in Figure 5.2. Because the material is incompressible, plane-strain deformation with $d\epsilon_2 = 0$ implies that $d\epsilon_3 = -d\epsilon_1$ and the deformation must therefore be pure shear strain. It is assumed that this is produced by pure shear stress. Consequently yield will occur, as in any pure shear process, when the maximum shear stress reaches the value k. It is not necessary to use any yield criterion, because k is just the result of a pure-shear test. Thus the radius τ_{max} of the Mohr stress circle for plane plastic-strain is equal to k, and the principal stresses in Figure 5.2*b* are

$$\sigma_1 = \sigma_2 - k, \sigma_3 = \sigma_2 + k.$$

These can be expressed in terms of a hydrostatic pressure $-p$, equal to σ_2:

$$\sigma_1 = -p - k, \sigma_2 = -p, \sigma_3 = -p + k \tag{5.2}$$

It is known (Chapter 3, §3.6) that a hydrostatic pressure does not affect yielding. The complete stress system in plane strain is therefore pure shear with a superimposed hydrostatic pressure. (It is necessary to consider only the major stress circle, since we are interested only in the plane perpendicular to the direction of the intermediate principal stress σ_2.)

Thus in plane plastic strain, k is everywhere constant, for a non-hardening metal, but p may vary. The stress system at any point can be completely determined if we

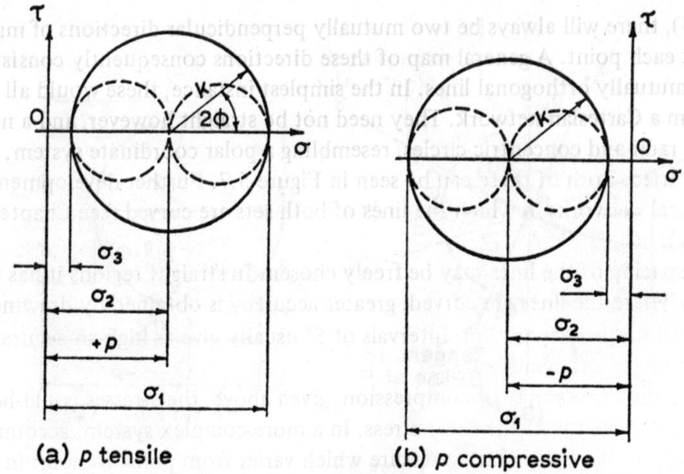

Figure 5.2 The Mohr stress circle for plane plastic strain, with maximum shear-stress k. The superimposed hydrostatic pressure is given by $p = \frac{1}{2}(\sigma_1 + \sigma_3)$.

can find the magnitude of p and the direction of k. The slip-lines show the directions of k at any point immediately. The changes in p can be deduced from the angular rotation of the slip-line between one point and another in the field, and the absolute magnitude of p is found by starting at some point on a boundary where p_0 is known from external conditions.

5.4 Determination of hydrostatic pressure from slip-line rotation. The Hencky equations

As we have seen (§5.3), at any point in the deforming body there will be two mutually-perpendicular directions of maximum shear stress, shown by the slip-lines. It is conventional to designate these lines α and β, such that, if they are regarded as a pair of right-handed axes of reference (like X and Y), the line of action of the *algebraically greatest* principal stress lies in the first (and third) quadrant. It should be noted that the α and β lines may be curved, and that the stress of largest numerical value need not be greatest algebraically. It quite often happens that the stress of greatest magnitude is compressive and therefore negative, while the algebraically greatest stress is in fact zero. (See, for example, §5.6.5.) Let us suppose that the α line, at some point P, is inclined at an angle ϕ, measured anticlockwise from the X-axis (Figure 5.3a).

The Mohr circle for the point P may be drawn (see Chapter 3, §3.5), assuming some value $-p$ for the hydrostatic compressive stress. If both principal stresses are compressive ($p > k$), σ_1 is given by the length OA. The stresses σ_x and σ_y may then be found from this circle. It must be remembered, as in Chapter 3, that the Mohr circle shows the planes on which stresses act. Direct stresses are perpendicular to their planes, while shear stresses are parallel. The plane [Y], on which the stress σ_x acts, is thus at an angle $\left(\frac{\pi}{4} + \phi\right)$ measured clockwise from the plane [1], on which

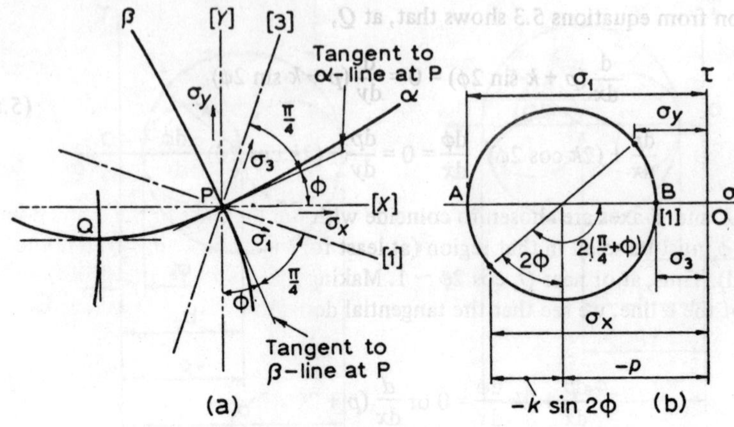

Figure 5.3

(*a*) The α- and β-slip-lines at points *P* and *Q* in the deforming body.
(*b*) The Mohr circle appropriate to the point *P*.

σ_3 acts. Since angular rotation 2θ in the Mohr circle is double the rotation θ in the physical diagram, the circle shows σ_x at $\left(\frac{\pi}{2}+2\phi\right)$ from the σ_3 direction, as in Figure 5.3*b*. The location of σ_y on the diagram can be found in the same way, and is on the opposite end of the diameter drawn from σ_x, since σ_x and σ_y are mutually perpendicular on the physical plane. The values of the stresses on the *X* and *Y*-planes can then be read directly from the coordinates of these points:

$$\begin{aligned} \sigma_x &= -p - k \sin 2\phi \\ \sigma_y &= -p + k \sin 2\phi \\ \tau_{xy} &= k \cos 2\phi \end{aligned} \tag{5.3}$$

It can be shown, by considering the equilibrium of a small element (Example 5.1), that the gradients of the direct stresses along and perpendicular to a slip-line are zero, at any point on the slip-line. Thus

$$\frac{d\sigma_\alpha}{d\alpha} = \frac{d\sigma_\beta}{d\beta} = 0 \tag{5.4}$$

In Figure 5.3, the slip-lines were supposed to be at some angle ϕ to the *X* and *Y*-axes. These axes were not however specified in relation to any external system, and can be freely chosen. We may therefore suppose them to be such that at some other point *Q* the *X* and *Y*-axes coincide with the tangents to the α and β slip-lines, respectively, as in Figure 5.3*a*.

Then, at the point *Q*, equation 5.4 may be written

$$\left(\frac{d\sigma_x}{dx}\right)_Q = \left(\frac{d\sigma_y}{dy}\right)_Q = 0 \tag{5.4a}$$

Substitution from equations 5.3 shows that, at Q,

$$\frac{d}{dx}(p + k \sin 2\phi) = 0 = \frac{d}{dy}(p - k \sin 2\phi)$$
$$\frac{dp}{dx} + (2k \cos 2\phi) \cdot \frac{d\phi}{dx} = 0 = \frac{dp}{dy} - (2k \cos 2\phi) \cdot \frac{d\phi}{dy} \tag{5.5}$$

But if the X and Y-axes are chosen to coincide with the tangents to the α and β-lines at Q, then ϕ must be small in that region (at least for the distance involved in the differential). Thus, at or near Q, $\cos 2\phi \sim 1$. Making this approximation for the equation of the α line, we see that the tangential derivative of $(p + 2k\phi)$ is zero at Q.

$$\frac{dp}{dx} + 2k\frac{d\phi}{dx} = 0 \text{ or } \frac{d}{dx} \cdot (p + 2k\phi) = 0$$

Consequently $p + 2k\phi$ = constant along the α-line at Q. The point Q was chosen quite arbitrarily, and the X and Y-axes were then specified. Thus they could equally well have been specified for any other point, so that the relationship is valid at any point on the α line:

$$p + 2k\phi = \text{constant}, C \tag{5.6}$$

This is, moreover, equally true if ϕ is measured from some other reference direction inclined at a constant angle ϕ' to the chosen axis OX, since that would merely alter the value of the constant C.

The corresponding variation with y gives the relationship along a β-line. These two conditions,

$$p + 2k\phi = \text{constant, along an } \alpha\text{-line} \tag{5.7a}$$

$$p - 2k\phi = \text{constant, along a } \beta\text{-line} \tag{5.7b}$$

were first applied to metal deformation by Hencky[5.4] in 1923, and are known as the Hencky slip-line equations. They had been formulated much earlier in the science of soil mechanics[5.5]. (The reader may find it helpful to associate α and + to remember these important equations correctly.)

The Hencky equations enable the hydrostatic pressure at any point in the deforming body to be determined, from the curvature of the slip-lines, provided that the value of the constant is known. This is found from equilibrium conditions at one of the boundaries.

5.5 Stresses and slip-lines at boundaries of the plastic body

These conditions, as with all slip-line field theory, apply in plane strain.

5.5.1 *Free Surface*

The plastic zone sometimes extends to the free surface beyond the confines of the tool. We shall see that this occurs, for example, in the vicinity of a punch indenting a large block. (Figure 5.7.)

At a free surface, there can be no normal stress, so $\sigma_3 = 0$, assuming σ_1 compressive. Since $\sigma_3 = -p + k$ (equation 5.2) the hydrostatic pressure must then be equal in magnitude to the shear stress. This determines the major principal stress:

$$p = k; \ \sigma_1 = -p - k = -2k \tag{5.8}$$

The directions of the slip-lines can also be found, from the condition that there can be no tangential stress at a free surface. The components of the shear stresses along the slip-lines, resolved parallel to the surface, must thus be equal, and consequently the slip-lines must meet a free surface at 45°. This follows also from the directions of maximum shear being at 45° to the principal stress directions.

It should be noted that there are two possible configurations, as shown in Figure 5.4*a*.

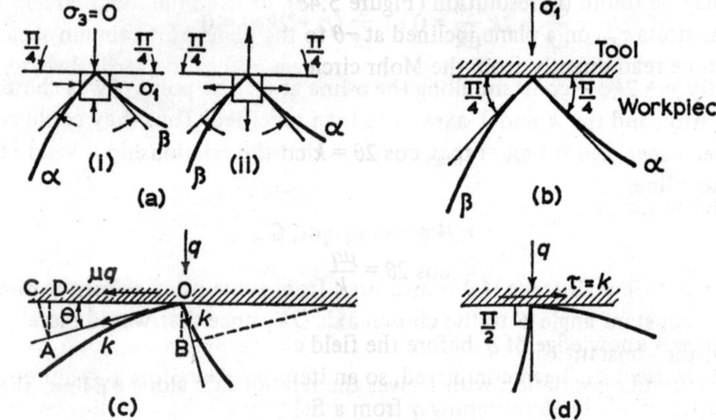

Figure 5.4

(*a*) Slip-lines at a free surface.

(i) major principal stress compressive

$$p = k, \sigma_1 = -2k$$

(ii) major principal stress tensile

$$p = -k, \sigma_1 = +2k$$

(*b*) Slip-lines at a frictionless interface.

$$\sigma_3 \neq 0, p \neq k$$

(*c*) Interface with Coulomb friction.

$$\theta = \tfrac{1}{2}\cos^{-1}\frac{\mu q}{k}$$

(*d*) Interface with sticking friction.

$$\theta = 0 \text{ or } \frac{\pi}{2}$$

The choice of α and β lines depends, as stated in §5.4, upon the convention that the direction of the algebraically greatest principal stress lies in the first quadrant between α and β.

5.5.2 *Frictionless interface*

Again, by definition, there can be no resultant shear parallel to the interface, so the slip-lines must meet the boundary at 45° (Figure 5.4*b*). There may be, and usually will be, a normal stress across the interface, so $\sigma_3 \neq 0$. The value of p is thus not usually equal to k.

5.5.3 *Interface with Coulomb friction* (μ = *constant*)

The resultant shear stress in an interface with Coulomb friction is equal to the product of normal stress q' and coefficient of friction μ. This is balanced by the resolved components of the forces on the slip-lines, which are inclined at some angle θ, respectively to the tangential and normal directions at the surface (Figure 5.4*c*). The value of θ may be found by resolution (Figure 5.4*c*), or from the Mohr circle.

The shear-stress τ_{xy} on a plane inclined at $-\theta$ to the plane of maximum shear (the slip-line) can be read directly from the Mohr circle, at -2θ to the vertical vector giving τ_{max}.

$$\tau_{xy} = \tau_{max} \cos 2\theta = k \cos 2\theta \tag{5.9}$$

Equating this to $\mu q'$:

$$\cos 2\theta = \frac{\mu q'}{k} \tag{5.10}$$

This implies a knowledge of q' before the field can be drawn. In fact q' cannot be found until the field has been completed, so an iterative procedure is required. The first approach is made by determining q from a field drawn for a frictionless interface, and assuming $q' \approx q$. This is in fact sufficiently accurate for the final solution in some instances, for example in strip-drawing.

5.5.4 *Perfectly rough interface*

If the friction is so high that there is no interfacial movement, the metal will yield beneath the interface, when the tangential stress reaches the value k, its yield stress in pure shear. The applied frictional stress cannot therefore be greater than k. So when the friction is very high a special condition arises, with

$$\tau_{xy} = k \tag{5.11}$$

independent of the normal stress. This is known as 'sticking friction', and the coefficient of friction is then meaningless. Equation 5.9 shows that, for any condition,

$$\tau_{xy} = k \cos 2\theta$$

Thus, for sticking friction,

$$\cos 2\theta = 1; \ \theta = 0 \tag{5.12}$$

Consequently, one slip-line meets the interface tangentially, the other normally, as in Figure 5.4*d*.

5.6 Application of the slip-line field to a static system. Plane-strain indentation with flat, frictionless, platens

We shall consider first the application of slip-line fields to a simple example of forging. A load is applied to the platen or punch; when it reaches a particular value, the metal will yield slightly, leaving a permanent indentation. The problem is to determine this load, ignoring any changes which would occur if the punch sank further in. In the subsequent section we shall consider the importance of velocity, when the system is not statically in equilibrium.

This example is of importance in two common test procedures; the Ford plane-strain compression test and, by analogy, the Brinell hardness test.

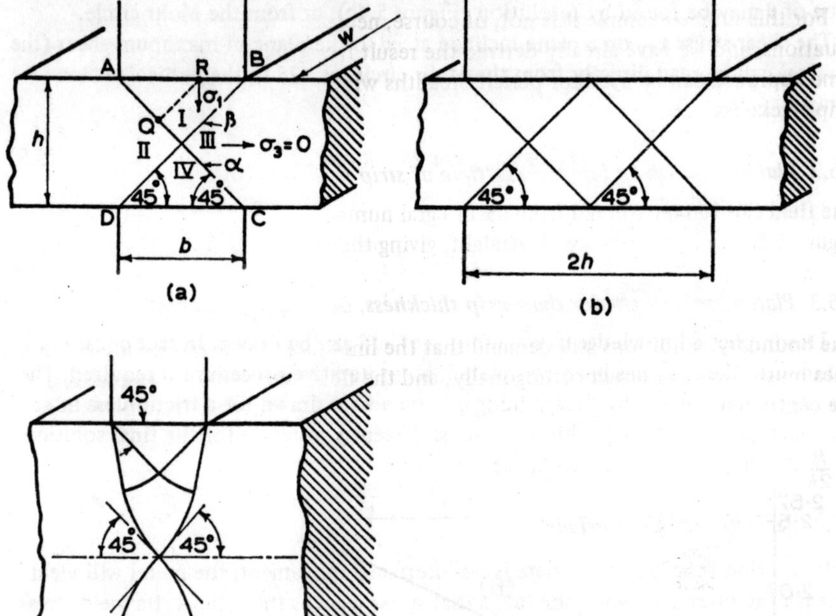

Figure 5.5 Plane-strain indentation. Simple slip-line fields for

(*a*) $b = h$
(*b*) $b = 2h$
(*c*) $b < h$

Figure 5.5 shows a thin, wide strip, indented between two flat, rectangular platens, with as little friction as possible. If the strip is wide enough, say 10 times b or h, whichever is the larger, the lateral spread may be neglected. The conditions then approach plane strain, and slip-line field theory can be applied.

The pressure σ_1 to produce yielding depends on the ratio of strip thickness to platen breadth.

5.6.1 *Strip thickness equal to platen breadth,* ($h = b$)

Because the interface is assumed to be frictionless, the slip-lines meet it as 45°. We have already seen that the diagonals are slip-lines (§5.3), so the simple network shown in Figure 5.5*a* fulfils the boundary conditions. At a point Q on the slip-line AC, the horizontal principal-stress σ_3 must be zero, so that $p = k$. The vertical stress σ_1 will then be $-2k$. Between Q and R, the β slip-line is straight, so $\phi = 0$ and, as we have seen, the Hencky equation ($p - 2k\phi$ = constant along a β-line) shows that the hydrostatic pressure remains constant along a straight slip-line. The hydrostatic pressure p at R is thus also equal to k, and $\sigma_1 = -2k$ at the platen surface.

For this simple example it is not, of course, necessary to use the slip-line field equations, and we have already derived the result from first principles (§5.2). The same approach can be used for platen breadths which are integral multiples of the strip thickness.

5.6.2 *Platen breadth an integral multiple of strip thickness* (b/h = 2, 3, 4 *etc.*)

The field can be constructed from an integral number of units corresponding to Figure 5.5*b*, and the lines are all straight, giving the same result: $\sigma_1 = -2k$.

5.6.3 *Platen breadth greater than strip thickness, but b/h not integral*

The boundary conditions still demand that the lines meet the interface at 45°. Slip-lines must always intersect orthogonally, and the field must be symmetrical about the centre line.

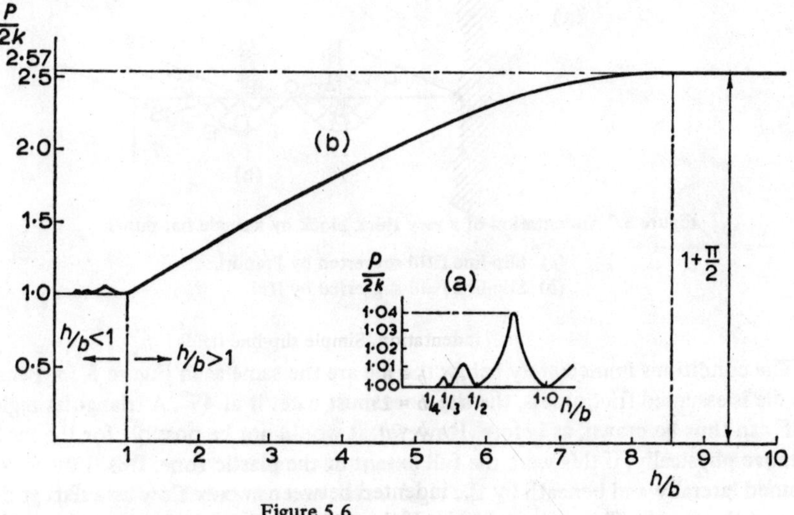

Figure 5.6

(*a*) Variation of p with the ratio h/b for values $b/h > 1$.
(*b*) Variation of p with the ratio h/b for values $h/b > 1$.

To fulfil these conditions, and to ensure that the field transforms steadily as the ratio h/b changes from 1 to $\frac{1}{2}$, the slip-lines must be curved. This means that the overall indentation pressure will be greater than when the simple shear pattern is possible. Green has provided a solution(5.6), which shows that the pressure distribution is then not uniform across the platen face, even though the friction is zero. The average pressure passes through maximum values between integral b/h ratios, but the greatest increase for $b/h > 1$ ($h/b < 1$) is only 4% (Figure 5.6).

5.6.4 *Strip thickness greater than platen breadth* (h/b *between* 1 *and* 10)

The field again involves curved slip-lines, and the pressure rises rapidly with decreasing relative platen breadth.

5.6.5 *Single-punch indentation of a semi-infinite block* ($h/b \sim \infty$).

(*a*) *Construction of the slip-line field.* When the block is very thick, the zones of plastic deformation do not extend completely across the block, and the problem becomes essentially that of single-sided indentation by one punch. Beyond a value $h/b \sim 10$, changes in geometry, by alteration of die breadth, have no further influence.

This may be considered in detail with reference to Figure 5.7.

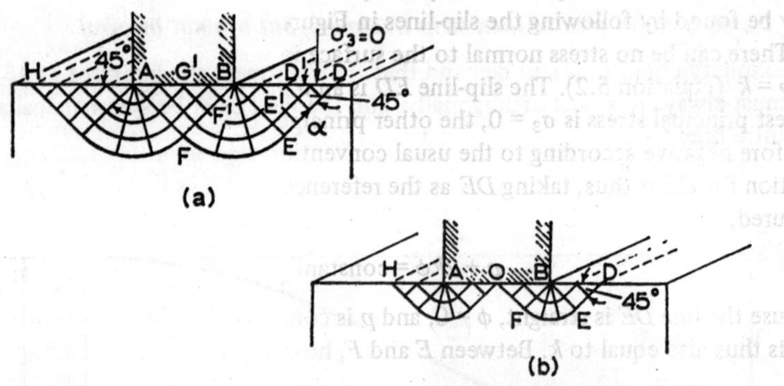

Figure 5.7 Indentation of a very thick block by a single flat punch.

(*a*) Slip-line field suggested by Prandtl.

(*b*) Slip-line field suggested by Hill.

The conditions immediately below the die are the same as in Figure 5.5*a*. Because the die is assumed frictionless, the slip-lines must meet it at 45°. A triangular region *ABF* can thus be drawn, as before. However, it would not be possible for the metal to move physically, if this were the full extent of the plastic zone. It is fully constrained laterally and beneath by the rigid metal, and can only flow upwards at the sides of the punch. This suggests that the plastic zone must be extended to the free surface along *AH* and *BD*.

The slip-lines must meet this surface also at 45°, and they must always be straight since the pressure cannot build up beneath a free surface, so another triangular network can be drawn, in the vicinity of B. The simplest way of connecting these two zones is to use a fan centred on B. It is useful to remember that fans can always be drawn round singularities, that is around points where there is no specified normal to the surface. The field so constructed at B defines the position of D, and the full solution is obtained by completing the left-hand side symmetrically.

Hill(5.7) has pointed out that this solution, due to Prandtl(5.8), would involve a considerable amount of distortion of the metal, before the plastic region had spread sufficiently to fill the postulated zone. Moreover, the velocity distribution in the triangle ABF is indeterminate. Prandtl assumed that the triangle formed a rigid nose or dead-zone, which moved downwards with the punch. This is found to occur when the friction is high, for example in hot piercing, and also represents fairly well the physical behaviour of annealed metal which strain-hardens rapidly. For the assumed conditions of low friction and constant yield stress, Hill proposed a solution in which the region beneath the indenter is divided into two parts (Figure 5.7*b*). Metal from the zone OBF flows along the slip-lines within $OFED$ to the free surface at the right. Metal from OA flows to the left. The flow of metal is therefore everywhere determinate. Both solutions lead to the same value for the indentation pressure, which is calculated using the Hencky equations, starting at a suitable point of known stress.

(*b*) *Stress determination from the slip-line field.* The pressure on the punch may easily be found by following the slip-lines in Figure 5.7*a*, starting at the free surface BD. There can be no stress normal to the surface in this region, so σ_3 is equal to zero and $p = k$ (equation 5.2). The slip-line ED is an α line, because the algebraically greatest principal stress is $\sigma_3 = 0$, the other principal stress being compressive and therefore negative according to the usual convention. The appropriate Hencky equation for DE is thus, taking DE as the reference direction from which ϕ is measured,

$$p + 2k\phi = \text{constant} = k \tag{5.13}$$

Because the line DE is straight, $\phi = 0$, and p is constant from D to E. The pressure p_E at E is thus also equal to k. Between E and F, however, the slip-line is curved, and the tangent rotates clockwise, until at F it makes an angle $\frac{\pi}{2}$ with DE. The angle ϕ, measured anticlockwise according to convention, therefore changes by $-\frac{\pi}{2}$ between E and F. Substituting this in equation 5.13 gives the pressure at F:

$$p_F + 2k\phi = p_F + 2k\left(-\frac{\pi}{2}\right) = \text{constant} = p_E = k.$$

or

$$p_F = k(1 + \pi)$$

It will be seen that as an α-line curves clockwise, the stress magnitude increases. This is a convenient way of selecting an α-line rapidly in any field. The converse is true of β-lines. The pressure remains unchanged along the straight lines FA and FB, so the pressure p_G at any point G' on the punch face is also equal to $k(1 + \pi)$. The major principal stress at G', acting in the vertical direction, is consequently given, according to equation 5.2, by

$$\sigma_1 = p_G + k = k(2 + \pi) = 2k\left(1 + \frac{\pi}{2}\right) \tag{5.14}$$

It is convenient to express the stress in terms of the dimensionless ratio $\sigma_1/2k$, where $2k$ is, as we have seen (equation 3.27), the yield stress in simple plane-strain compression, whether the maximum shear stress or maximum shear-strain energy criterion is used.

Slip-lines drawn anywhere in the plastic zone will be parallel to these lines and give the same result, so the indentation pressure P is uniform across AB, and is given by:

$$\frac{P}{2k} = 1 + \frac{\pi}{2} = 2{\cdot}57 \tag{5.15}$$

Thus the geometrical constraint on the flow increases the load required to cause yielding by a factor 2·57, apart from any frictional contribution.

This can be verified experimentally. The whole variation of yield pressure with width/thickness ratio (Figure 5.6) has been shown[5.9] to follow closely the predictions of slip-line field theory[5.6]. This provides, in fact, the most detailed correlation currently available. The striking cyclic variation of the pressure required to initiate incremental yielding was in fact observed independently, and was at first attributed to experimental errors. The theory deals only with small changes in deformation at any particular stage. If large changes are made, the geometry may change and the frictional conditions may also change, obscuring the pattern.

These results, like all slip-line field solutions, apply strictly to plane-strain conditions only. It should however be emphasised that the slip-line field solutions can also be used to suggest the type of deformation and stress variation to be expected, even when there is considerable departure from plane strain. Cylindrical strain, where there is axial symmetry, is an important example, which occurs in extrusion of round bars, wire-drawing and elsewhere. We shall see in Chapter 8 how the slip-line field solution has been used to suggest a formula into which two empirical constants can be inserted for estimation of rod-extrusion pressure[5.10]. The Brinell hardness test is an axially-symmetrical process resembling the indentation just considered. The true analogue would be a flat-ended cylindrical punch, but the results are only slightly modified when a ball is used, provided that the radius of the ball is large in comparison with that of the indentation.

5.6.6 *The Brinell Hardness Test*

In the Brinell test[5.11], a hard steel sphere is pressed into a softer metal under a given load. The diameter of the plastic indentation left, after removal of the sphere, is a measure of the hardness of the deformed metal. The Meyer hardness[5.12] of the indented metal is found by dividing the load by the projected area of the indentation, and is expressed in kilogrammes per square millimeter. The Brinell hardness is similarly obtained, but relates to the curved surface area, which for most practical conditions is only very slightly different. Tabor[5.13] has shown that the hardness H is closely related to the yield stress Y in homogeneous deformation, and for many metals

$$\frac{H}{Y} \approx 2{\cdot}8 - 2{\cdot}9 \tag{5.16}$$

This is a somewhat greater ratio than that due to the constraint in a comparable plane-strain indentation (equation 5.15), confirming the occurrence of such large factors in a common operation. The relationship (5.16) is very useful for rapid estimation of yield stress or degree of work-hardening of stock material, without destroying any of the metal. It can be used, for example, in approximate prediction of working loads.

5.7 Significance of velocity in slip-line field evaluations

The above sections relate to a simple forging or punching operation in which the punch is in static equilibrium. Even under these conditions we should, strictly speaking, consider the velocity as well as the direction of the metal flow. For example, Figures 5.7*a* and 5.7*b* show two different patterns which satisfy the stress conditions for a frictionless indenter.

In many metalworking processes, other than forging, the system is not static, but the metal flow rapidly attains a steady state which continues throughout the operation. For these, it is essential to verify that the chosen slip-line field conforms to the requirements of steady velocity, compatible with that of the rigid parts ahead of and behind the deformation zone.

A steady-state extrusion is chosen as an example in this section, but first it is necessary to derive an important pair of equations relating the velocity changes along slip-lines.

5.7.1 *Derivations of Geiringer's Velocity Equations*

The slip-lines are directions of maximum shear, and deformation along them occurs in pure shear. Elements lying along the slip-lines therefore distort but do not elongate or shorten. The superimposed hydrostatic pressure will not affect the dimensions of such an element, since the material is incompressible. Consequently there can be no change in velocity along a slip-line, arising from any stretch of the elements.

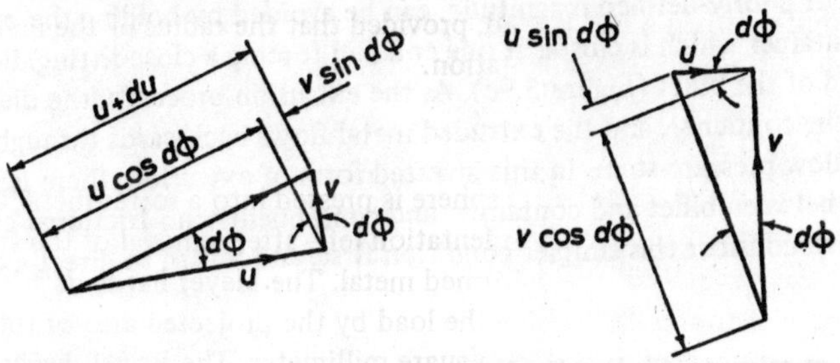

Figure 5.8 A diagram illustrating the change in velocity along a slip-line, due to change in direction.

There may, however, be a change in velocity due to change in direction of shear, which changes the velocity component perpendicular to the line. This may be seen from Figure 5.8.

The velocity component parallel to the α-line at Q is taken as u, and the velocity in the perpendicular direction of the β-line as v. Because the slip-line is curved at this point, the tangential velocity at the point Q', a distance ds further along the α line, will be $u + du$, comprised of $u \cos d\phi$ and $v \sin d\phi$. Thus,

$$u + du = u \cos d\phi + v \sin d\phi$$

$$du = v \, d\phi \tag{5.17a}$$

Considering the β-line, the component of the normal velocity will be negative;

$$v + dv = v \cos d\phi - u \sin d\phi$$

$$dv = -u \, d\phi \tag{5.17b}$$

These may be written in the form:

$$du - v \, d\phi = 0 \quad \text{along an } \alpha\text{-line} \tag{5.18a}$$

$$dv + u \, d\phi = 0 \quad \text{along a } \beta\text{-line} \tag{5.18b}$$

Equations 5.18 are known as the Geiringer equations[5.14], first proposed in 1930. It should be noted that the signs associated with the respective lines are opposite to those occurring in the Hencky stress equations 5.7. The Geiringer equations may be used to verify that the velocity predicted by the slip-line field is compatible with the imposed conditions.

5.8 Application of the slip-line field to steady-state motion: 50% inverted extrusion in plane strain, with unlubricated 180° die

Direct extrusion, which is more common than inverted extrusion, involves a container, closed at one end by the die, and at the other end by a close-fitting plunger (more usually a close-fitting pressure pad on the end of a mandrel), as in Figure 5.9*a*. The billet is squeezed by application of pressure to the plunger, causing metal to flow steadily out through the die. As extrusion proceeds, the plunger moves towards the die, and the billet slides forward against the inner wall of the container. This frictional contact, of poorly-defined magnitude, can be avoided by holding the billet stationary in a container which is closed at one end and forcing a close-fitting die against the free end of the billet (Figure 5.9*c*). As the extrusion proceeds, the die moves forward into the container, and the extruded metal flows backwards through the die and the hollow pressure-stem. In this inverted form of extrusion, there is no relative movement between billet and container, and consequently no frictional contribution. We shall consider this simpler condition first, and return to direct extrusion in Chapter 8.

5.8.1 *Construction of slip-line field*

It is convenient to assume that the die is at rest, and that the billet and container move towards it with unit velocity (Figure 5.9).

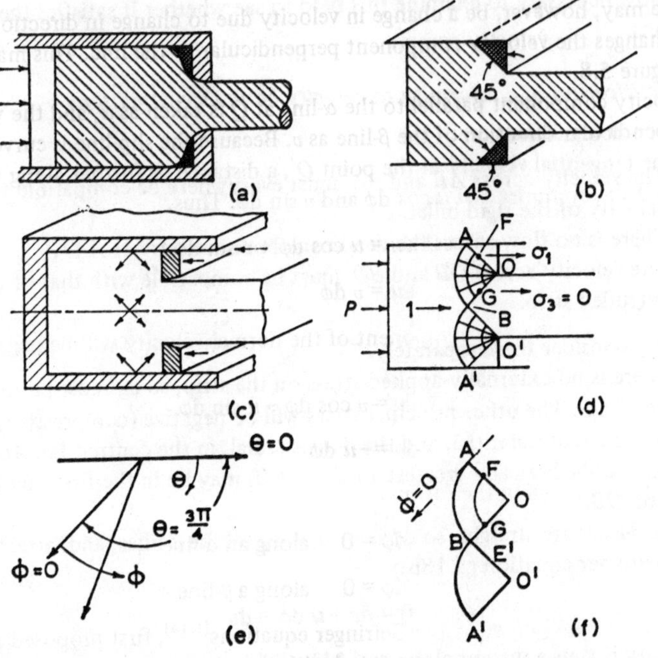

Figure 5.9

(*a*) The dead-metal zone in direct extrusion of a bar, and
(*b*) The idealised dead-metal boundary assumed.
(*c*) The slip-lines defined by boundary conditions in plane-strain inverted 50% extrusion.
(*d*) The complete slip-line field
(*e*), (*f*) The reference line for measurement of ϕ is chosen to coincide with the tangent the α line AB, at A.

From the boundary conditions for a frictionless interface (§5.5.2) we know that the slip-lines must meet the frictionless container wall at 45° (Figure 5.9*c*). The centre-line of the billet is an axis of symmetry, and can have no resultant shear component along it, so the slip lines must meet the centre-line also at 45°. To proceed further, we consider the flow pattern in a real extrusion with a 180° die. It is known that a 'dead-metal' zone forms in the corner between container and die (Figure 5.9*a*). Flow takes place by intense shear over the surface of this region. The billet skin flows into the interface, and in some circumstances, for example when the billet is coated with graphite so that welding is prevented, the dead zone may be physically separated from the end of the billet after the extrusion. The boundary is usually curved at each extremity against container and die orifice, but reasonably straight over most of its length. As an approximation, we may make the simple assumption that the dead zone is bounded by a straight line at 45° to the axis (Figure 5.9*b*). In fact, the exact position of this line does not greatly affect the result[5.15]. This assumption provides the first slip-line AO, which meets the wall at 45° as required by the boundary conditions. Recalling the example of §5.6.5, it can be seen that two radial fans centred on O and O' would complete the field (Figure 5.9*d*).

It is now necessary to examine this field to see whether it satisfies the velocity conditions.

5.8.2 *Verification of conformity to velocity boundary-conditions*

The imposed conditions are:

(*a*) The velocity across *AB* and *BA'* must everywhere be compatible with the velocity of the rigid billet.
(*b*) There is no flow across the dead-zone boundary *AO* or *A'O'*.
(*c*) The velocity across *OB* and *BO'* must be compatible with that of the rigid extruded strip.

We may consider these separately.

(*a*) There is no externally-applied stress on the strip, so σ_3 must be zero along the whole of *OB*. The other principal stress will be negative (compressive) so the α-lines must be circumferential, and the β-lines radial, in the centred-fan *AOB*, in order that the algebraically greatest stress $\sigma_3(=0)$ may lie in the first quadrant at any point *G*, on *OB*.

All the β-lines are straight, so $d\phi$ is zero along all the β-lines, and, according to the second Geiringer equation (5.18b),

$$0 = dv + u\ d\phi = dv$$

The velocity is thus constant along any β-line. This is a property of any straight slip-line.

Continuity of flow across *AB* at any point *D* gives $v^2 + u^2 = 1$, since the billet is assumed to have unit velocity. If a particular β-line is inclined to the extrusion direction (horizontal) at an angle θ (Figure 5.9*e*), the continuity condition determines the velocity on a β-line.

$$v = 1\ .\cos\theta \tag{5.19}$$

(*b*) We can find the velocity on a curved α-line, starting from *A*, with the knowledge that $u = 0$ at all points on *AO*. The direction $\phi = 0$ is still unspecified, and we may choose it to coincide with the tangent at *A* to the α-line *AB* (Figure 5.9*f*). This is in a direction $\theta = \dfrac{3\pi}{4}$, since θ is measured clockwise from the extrusion direction.

Thus $\theta + \phi = \dfrac{3\pi}{4}$.

The Geiringer equation 5.18a for any α-line is

$$du - v\ d\phi = 0$$

Combining this with the condition 5.19, and integrating,

$$\frac{du}{d\phi} = v = \cos\theta = \cos\left(\frac{3\pi}{4} - \phi\right)$$

$$u = -\sin\left(\frac{3\pi}{4} - \phi\right) + \text{constant.}$$

The constant may be found from the boundary condition, $u = 0$ when $\phi = 0$, thus determining the velocity on an α-line:

$$u = -\sin\left(\frac{3\pi}{4} - \phi\right) + \sin\frac{3\pi}{4} \tag{5.20}$$

(*c*) The velocity across the exit slip-line *OB* must be compatible with the rigid-body movement. The α slip-line has rotated through $\phi = \frac{\pi}{2}$, anticlockwise, between the point *F* on *AO*, and the point *G* on *OB*. The velocity u_G across *OB* at *G* is therefore given by

$$-u_G = \sin\left(\frac{3\pi}{4} - \frac{\pi}{2}\right) - \sin\frac{3\pi}{4}$$

$$= \frac{1}{\sqrt{2}} + \frac{1}{\sqrt{2}} = \sqrt{2} \tag{5.21}$$

The same result would be obtained for any point on *OB*. There is consequently a downward flow with a velocity of magnitude $\sqrt{2}$ across the whole boundary *OB*. It is necessary that the velocity across *OB* should be the same at all points, if the metal to the right of *OB* is rigid. There will be a corresponding flow outwards from the plastic zone $O'BA'$, with velocity $u_{G'} = \sqrt{2}$ across the boundary BO'. The net efflux velocity of metal from the plastic zone, across the whole exit boundary, will thus be the vector sum of these outward velocities. The net lateral velocity is zero, and the resultant axial velocity is

$$u_G \cos\frac{\pi}{4} + u_{G'} \cos\frac{\pi}{4} = 2 \,.\, \sqrt{2} \,.\, \frac{1}{\sqrt{2}} = 2$$

This is compatible with the motion of the rigid extruded-strip, which, for 50% reduction of area, must move twice as fast as the billet, whose velocity was assumed to be unity.

The same result would be obtained at any point on the straight line *OB*. It can however be seen that the velocity component v_G parallel to *OB*, according to equation 5.19 is

$$v_G = 1 \,.\, \cos\frac{\pi}{4} = \frac{1}{\sqrt{2}}$$

This is not compatible with the velocity component of the rigid metal parallel to *OB*, which is $2 \cos 45 = \sqrt{2}$. Such a jump or discontinuity in tangential velocity component is commonly found in slip-line field solutions. The physical significance is that a small element of the material on crossing such a slip-line abruptly changes its direction, undergoing a sudden shearing parallel to the discontinuity. This will be seen later in Figure 5.10, and is important in connection with upper-bound solutions as shown in §5.10. Because there can be no extension along a slip-line (equations 5.18), a velocity discontinuity is of constant magnitude along the whole of any given slip-line. In real metals, the discontinuity will be spread into a narrow band of severe shear.

Thus the velocities calculated from the slip-line field satisfy the boundary conditions. If they did not, it would be necessary to choose another slip-line field satisfying the stress conditions, and to test it also. This may be a tedious process, but fortunately there now exist many proven fields, one of which is usually suitable for a particular practical application. However, there is a simpler method of ensuring that a postulated slip-line field is compatible with velocity conditions. This is to incorporate the velocity vector diagram, or hodograph, in constructing the field. We shall discuss this in detail (§5.9) later. For the moment, having found that the chosen slip-line field for 50% extrusion is valid, it remains to calculate the stresses, in the same way as was done in evaluating the static indentation load.

5.8.3 *Stress-determination from the slip-line field*

It is convenient to start from the exit slip-line OB, since there is no externally applied force on the extruded strip. Thus, at any point G (Figure 5.9*d*) on OB, the principal stress in the extrusion direction is zero. Equation 5.2 then gives the magnitude of the hydrostatic pressure

$$\sigma_3 = 0; \;\; \sigma_3 = p - k; \;\; p = k,$$

as at a free surface. The appropriate Hencky equation for the β slip-line OB is

$$p - 2k\phi = \text{constant} \tag{5.7b}$$

Since OB is straight, ϕ does not vary between O and B, and the magnitude of p is constant and equal to k over the whole of OB. The equation 5.7a of the α-line starting at the point G is thus

$$p + 2k\phi = \text{constant} = p_G = k,$$

if ϕ is chosen to be zero in the direction OB. (There is of course no need to choose the same reference direction as was used for the velocity check. It is also possible to work in terms of $\Delta p + 2k\Delta\phi$, which avoids defining a zero direction.)

Between OB and OA, the tangent to the α-line rotates clockwise through an angle $\frac{\pi}{2}$, so at any point F on OA, $\phi = -\frac{\pi}{2}$;

$$p_F + 2k\left(-\frac{\pi}{2}\right) = k, \text{ or } p_F = k(1 + \pi)$$

The major principal stress σ_1 at F (acting now in the extrusion direction) is thus, using equation 5.2,

$$\sigma_1 = p + k = k(2 + \pi) \tag{5.22}$$

Because OA is straight, σ_1 is constant all along OA. This stress is transmitted through the dead zone and acts on the stationary die. The extrusion pressure P which must be applied to the end of the billet, via the container base, acts over the whole area, including that of the extruded bar. In 50% extrusion the area of the billet is twice that of the die, so

$$P = \frac{\sigma_1}{2} = k\left(1 + \frac{\pi}{2}\right)$$

or
$$\frac{P}{2k} = \frac{1}{2} + \frac{\pi}{4} = 1{\cdot}29 \tag{5.23}$$

The simple work formula for homogeneous deformation would predict (equation 4.7):

$$P = Y \ln\frac{A_1}{A_2} = Y \ln 2 = 0{\cdot}7Y = 0{\cdot}7 \times \frac{2k}{1{\cdot}15}$$

$$= 0{\cdot}605 \times 2k \tag{5.24}$$

The actual extrusion pressure is consequently about twice that which would be expected if no account were taken of the influence of constraint.

We shall consider extrusion more generally in Chapter 8. These simple examples have been chosen to show the method of applying slip-line fields in calculation of working loads, for static conditions and for steady-state motion.

5.8.4 *Slip-line fields for axi-symmetric deformation*

Provided that the basic assumptions of slip-line field theory are satisfied, the predictions are remarkably precise, and agree with detailed experimental observation of deformation patterns as well as forces. Unfortunately, very few practical processes meet the rigorous requirements, especially those of plane strain and non-hardening material. Strip-rolling approaches plane-strain conditions very closely, but it happens that redundant work is of little significance in cold rolling. Axisymmetric workpieces and products are much more common, for example in wire-drawing, tube-drawing and extrusion or forging of cylindrical billets.

In close-pass tube drawing however, the wall thickness is reduced but the diameter remains almost constant (see §7.1). The circumferential strain is thus negligible and there is a good approximation to plane-strain deformation. This provides an important clue to the solution of axi-symmetric metalworking problems. The body can be divided into annular elements, and those at large radius can be considered to deform in plane-strain. This assumption is not applicable to radial drawing, for example in cup drawing (§7.6) or in regions close to the axis, but it is nevertheless very useful[5.19].

Practical examinations of sections of incrementally-deformed cylindrical specimens, using dislocation etching or visioplasticity, show that the pattern of maximum shear-strain directions on a diametral section very closely resembles the slip-line field for an equivalent plane-strain section. This will be further discussed in §10.12.3. It is therefore postulated that a slip-line field can be directly applied to axial symmetry. The Hencky equations in their simple form are not valid, and the stress and velocity equations are not hyperbolic unless a heuristic assumption is made. The Haar and von Karman assumption is that the circumferential stress is equal to one of the principal stresses in the meridional planes. This has been widely discussed and is valid, for example, in the special conditions of expansion of a thick-walled sphere. For many purposes it is simpler to recognise that the Hencky equations for axial symmetry contain an additional term involving the reciprocal of the radius:

$$dp + 2k\, d\phi + (\sigma + p - k \cot \phi) \frac{dr}{r} = 0 \qquad (10.33)$$

If r is large in comparison with the depth of the annular element concerned, this term can simply be ignored, without introducing serious error. The value of p can thus be calculated, as for plane-strain. However, in converting to forces, it must be remembered that each stress acting on an element does so over a truncated conical surface generated by rotation of the element about the axis. Elements at large radius thus make the greatest contribution to the total force, and it is probably for this reason that the apparently unrealistic assumption of zero circumferential strain gives force values in good accordance with practical measurements. It is instructive to calculate the indentation pressure for a cylindrical indenter by this method, comparing the result with the hardness number relationship (equations 5.15 and 2.25) and noting the contribution made by inner and outer zones in the metal.

5.8.5 *Inclusion of strain-hardening in slip-line field theory*

To determine the force required to deform a metal in a particular operation, it is often sufficient to use the slip-line field for a non-hardening material and to apply the Hencky equations (5.7) with a mean value of the shear yield stress $\bar{k}$. This value is found using the concept of the work needed to deform the material homogeneously from strain ϵ_1 to ϵ_2, rather than the arithmetic average of k. Referring to equation 4.6 and Figure 4.2*b*,

$$2\bar{k} = \frac{1}{\epsilon_2 - \epsilon_1} \int_1^2 2k\, d\epsilon$$

where $2k$ is the yield stress measured, with correction for friction, in a plane-strain compression test. If the yield stress is measured in uniaxial tension or some other conditions, suitable conversions can be made using the generalised stress and strain (equations 13.16 and 13.17). For convenience, it is common to use an equation for the flow stress, such as

$$\sigma = B\epsilon^m, \qquad (2.14)$$

or

$$\sigma = \sigma_0 + B\epsilon^m \qquad (2.15)$$

The above procedure is usually accurate enough when pre-strained metal is worked, for example in the later stages of a multi-pass operation such as wire drawing. It is not however suitable for annealed metals, where there is a large change in yield stress through the process, and it also fails to take account of the modification of the slip-line field itself due to strain hardening.

Figure 5.10 compares a slip-line field determined experimentally, by a visioplasticity technique (§13.3) for an annealed specimen with the theoretical non-hardening solution[5.20].

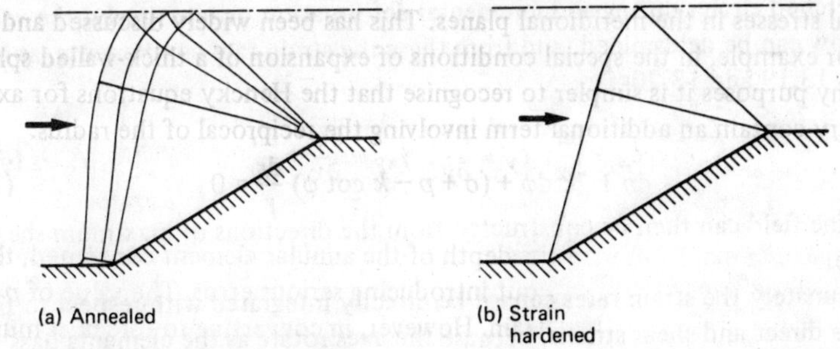

Figure 5.10 An experimental slip-line field for extrusion of annealed metal, compared with a theoretical field assuming no hardening(5.21).

The theoretical field assumes that velocity discontinuities exist on the entry and exit boundaries (see §§5.8.2 and 8.3), where the velocity component parallel to the slip-line changes suddenly in value from one side to the other of the line. This implies an infinite strain-rate, which is physically impossible, and in practice the zone of intense shear is spread out.

It has been shown(5.22) that the Hencky equations for strain-hardening material can be written*

$$\frac{\partial p}{\partial s_\alpha} + 2k\frac{\partial \phi}{\partial s_\alpha} - \frac{\partial k}{\partial s_\beta} = 0 \text{ along an } \alpha\text{-line} \tag{5.25}$$

$$\frac{\partial p}{\partial s_\beta} - 2k\frac{\partial \phi}{\partial s_\beta} - \frac{\partial k}{\partial s_\alpha} = 0 \text{ along a } \beta\text{-line}$$

where s_α and s_β are distances measured respectively along the α and β lines. Thus the hydrostatic pressure p depends on the change in yield stress k along a slip line, as well as on the rotation ϕ.

In general, k will vary with strain, strain rate, and temperature, so

$$\frac{\partial k}{\partial s_\alpha} = \frac{\partial k}{\partial \gamma}\cdot\frac{\partial \gamma}{\partial s_\alpha} + \frac{\partial k}{\partial \dot{\gamma}}\cdot\frac{\partial \dot{\gamma}}{\partial s_\alpha} + \frac{\partial k}{\partial T}\cdot\frac{\partial T}{\partial s_\alpha} \tag{5.26}$$

At room temperature and moderate speeds of deformation, the latter two terms can be neglected, leaving the Hencky equation for the α line as

$$\frac{\partial p}{\partial s_\alpha} + 2k\frac{\partial \phi}{\partial s_\alpha} - \frac{\partial k}{\partial \gamma}\cdot\frac{\partial \gamma}{\partial s_\beta} = 0 \tag{5.27}$$

General theoretical solutions are still not possible, but useful information can be obtained by using visioplasticity to obtain an initial slip-line field. As explained in §13.4.5, the directions of maximum shear strain can be traced from the distortion

* This arises because the equilibrium of forces along a curvilinear element of a slip line requires that $\frac{\partial p}{\partial \alpha} + \frac{\partial k}{\partial \beta} - \frac{2k}{r_\alpha} = 0$ along an α line and $\frac{1}{r_\alpha} = \frac{\partial \phi}{\partial \alpha}$.

of small etched circles during an incremental deformation. Alternatively, the velocities of flow can be determined, and from these the strain-rate components, as in equations 13.19 and 13.20:

$$\dot{\epsilon}_x = \frac{\partial u}{\partial x}; \; \dot{\epsilon}_y = \frac{\partial v}{\partial y}; \; \dot{\gamma}_{xy} = \frac{\partial v}{\partial x} + \frac{\partial u}{\partial y} \tag{5.28}$$

The slip-line field can then be constructed from the directions of maximum shear-strain rate.

Unfortunately the strain rates cannot be directly integrated with respect to time to give the direct and shear strains, because the axes rotate as the elements pass through the deformation zone. However the maximum shear-strain rate at any point can be determined, using the equation of a strain-rate circle (c.f. §13.2)

$$\left(\frac{\dot{\gamma}}{2}\right)^2 = \dot{\epsilon}_x^2 + \left(\frac{\dot{\gamma}_{xy}}{2}\right)^2 \tag{5.29}$$

which can be integrated along a stream line to give the generalised strain (see §13.24 and equation 13.30)

$$\bar{\epsilon} = \sqrt{\frac{1}{3}} \cdot \int_{\alpha} \dot{\gamma} \, dt$$

The generalised yield stress $\bar{\sigma}$/strain $\bar{\epsilon}$ curve for the material is then used to find $\bar{\sigma}$ and hence k, using the substitutions for a pure shear-stress system $\sigma_1 = +k$, $\sigma_2 = 0$, $\sigma_3 = -k$ in equation 13.16:

$$\begin{aligned}\bar{\sigma} &= \sqrt{\tfrac{1}{2}\{(\sigma_1 - \sigma_2)^2 + (\sigma_2 - \sigma_3)^2 + (\sigma_3 - \sigma_1)^2\}} \\ &= \sqrt{3}k\end{aligned}$$

(compare also §3.6.3).

Values for $\dfrac{\partial k}{\partial s_\alpha}$ and for $\dfrac{\partial k}{\partial s_\beta}$ can thus be found at each point and substituted into equation 5.27 for the calculation of hydrostatic pressure.

Alternatively, approximate values of the local strains can be evaluated from the major and minor axes of the ellipses produced from the circular grids.

A notable feature of equation 5.27 is that the strain-hardening factor may dominate so that the pressure calculated may change sign, becoming tensile. This can be important, and has been demonstrated, for example in machining.

5.8.6 *The influence of strain-rate and temperature*

For hot working, all the terms in equation 5.26 may be significant, but the variation of yield stress with temperature and with strain-rate usually dominates.

There is no direct way of incorporating these in slip-line field theory, though strain-rate by itself could be allowed for in the same way as strain hardening. It can be assumed that at a given temperature the flow-stress curve can be represented by an expression related to equation 2.14,

$$\sigma = B\,\epsilon^m\,\dot{\epsilon}^n \tag{5.30}$$

As these solutions are inevitably approximate and depend on point-to-point measurements from a model experiment, they eventually lead to the fully numerical techniques described in Chapter 13, but with suitable approximations the slip-line field approach provides valuable insight.

5.9 Velocity diagrams or hodographs

The procedure now usually adopted in construction of a slip-line field is to draw a hodograph. This is a diagram giving velocity vectors appropriate to any point in the slip-line field. From it, the compatibility of the chosen field with velocity boundary conditions can be seen directly.

The method may be illustrated with reference to the example of 50% inverted extrusion given above, though a simpler example is discussed in §§5.10.3 and 5.10.4.

In Figure 5.11*a*, any particle of metal to the left of the boundary *AB* is assumed to move in the extrusion direction with unit velocity. This is represented by a vector

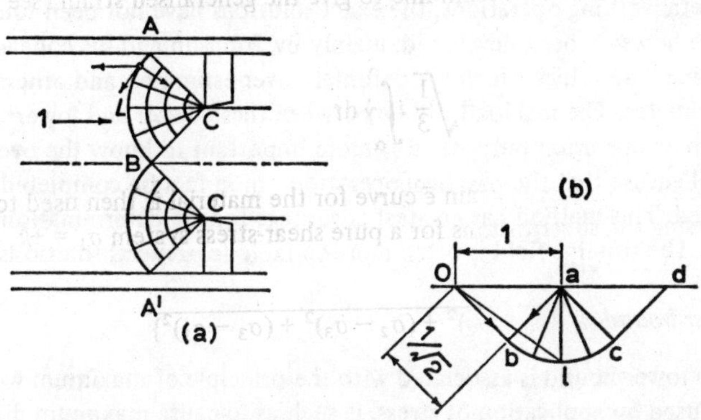

Figure 5.11 (*a*) The slip-line field as in Figure 5.9, and (*b*) the corresponding hodograph for 50% inverted extrusion.

Oa of unit length drawn from the origin in the hodograph, Figure 5.11*b*, parallel to the physical velocity. It is assumed that there is a velocity discontinuity along the boundary slip-line *AB*. The hodograph eventually shows this to be a valid assumption. A particle crossing *AB* in the vicinity of *A* will undergo a sudden shearing parallel to the tangent to *AB* at the point of crossing. For a particle crossing very near to *A*, this change in velocity will be represented by a vector drawn from the point *a* parallel to the tangent at *A*. After crossing the discontinuity, the element will be constrained to slide parallel to the assumed dead-metal boundary, so that its absolute velocity must be represented by a vector from *O* parallel to the line *AC*. This will intersect the vector from *a* in the point *b*.

Since the velocity discontinuity along *AB* has constant magnitude, equal to *ab* over the whole slip-line, a particle crossing *AB* at any other point will experience a velocity change parallel to the tangent at the point of crossing and equal in magnitude to *ab*. Thus at *L*, for example, the tangent is vertical and the velocity triangle is *Oal*.

In the vicinity of B, the tangent is at 45° to the centre line, so the discontinuity is ac and the absolute velocity within the plastic zone is Oc. The particle then traverses the plastic region and finally emerges from it by crossing the boundary slip-line BC, when it again experiences a sudden shear or velocity discontinuity parallel to BC, represented by a vector in the direction cd. The final velocity in the rigid metal to the right of BC must be in the extrusion direction, represented by the vector Oa produced. The intersection of cd and Od gives the final absolute velocity vector Od. It is clear from Figure 5.10*b* that $Od = 2Oa$, so that the final velocity has magnitude 2, showing that the chosen slip-line field is thus compatible with the velocity for 2:1 extrusion.

This method will be used in later chapters, since it is much simpler than the analytical method of §5.8.2.

5.10 Upper-bound and lower-bound techniques of load estimation

For many metalworking operations, the exact solutions have not been found. Methods have however been developed, mainly by Johnson and his colleagues[5.16], for determining load values which are definitely over-estimates, and others which are under-estimates. The real load will lie between these upper and lower bounds, but for design or operation purposes it is more important to know the over-estimate, since this will ensure that the practical operation can in fact be completed by the load calculated. This method has greatest advantage for the determination of particular solutions. The slip-line field is often more suitable for general solutions.

5.10.1 *Lower bound*

The idea of a lower bound is associated with the principle of maximum work. The distortion caused by application of stress, is such as to cause maximum dissipation of energy. Looked at in another way, the system tends to reach the state of minimum energy compatible with the equilibrium and yield conditions. Consequently, any other statically admissible stress system would produce an increment of work at most equal to that produced by the actual system, and probably less. Thus any system derived from *stress* equilibrium will be either just sufficient, or too little, to perform the operation. This gives a lower bound, but as we shall not need to use this, we shall not consider it in detail.

5.10.2 *Upper bound*

Consideration of upper bounds involves the conditions which have to be fulfilled by the *strain* increments in a fully plastic body, and does not concern itself with stress equilibrium.

The critical factor here is that the plastic volume should not change; that the material is incompressible. The principle of maximum work is used here also, but from the point of view of strain: an element deforms in such a way as to offer maximum resistance. If therefore we deduce the stress system from any assumed deformation, which must conform to the kinematic conditions, the value will be greater than, or equal to, that actually operating.

5.10.3 *Upper-Bound Theorem in Plane Strain*

The upper bound is the more valuable in metalworking, since it predicts a working load which is at least sufficient to perform the operation. In plane-strain problems, the solutions may be obtained entirely graphically. We shall consider first a simple deformation condition in plane strain, resembling a lathe cutting-operation. (Actually, the cutting operation is not typical of metalworking because the thickness $B'C'$ cos α is not defined by the tool geometry, and work-hardening plays an important part.)

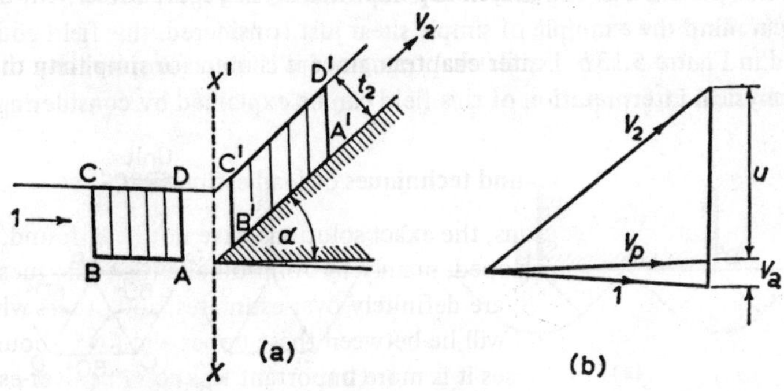

Figure 5.12

(*a*) Pure shear deformation on a single plane, as in an idealised machining operation.
(*b*) The associated velocity diagram or hodograph.

Suppose an element *ABCD* approaches a boundary (at the tool) *XX'* with unit velocity, and then becomes distorted into the form $A'B'C'D'$, moving with velocity v_2 at an angle α to the original direction. The hodograph is constructed as in §5.9, but is particularly simple (Figure 5.12*b*). The initial velocity has components v_a and v_p in the direction of the boundary and perpendicular to it. The velocity perpendicular to the boundary cannot be changed, because the material is incompressible, but the velocity along the boundary can be changed. In an idealised instance, this occurs instantaneously as in the velocity discontinuities already mentioned in §§5.8.2 and 5.9, but in practice the velocity would change over a small band, which only in the limit is a line. The material then has a new velocity *u*, whose magnitude is determined by the absolute value v_2, in the final direction completing the triangle.

The work done by the shear-stress on opposite sides of the block is equal to the product of force and distance.

$$\mathrm{d}W = F\,\mathrm{d}s = \tau A\,\mathrm{d}s$$

Thus
$$\frac{\mathrm{d}W}{\mathrm{d}t} = \tau A \frac{\mathrm{d}s}{\mathrm{d}t} = \tau A u \tag{5.31}$$

where *s* and *u* are measured along the straight boundary. Since plane strain is assumed, this may be evaluated for unit width.

$$A = s\,.\,1$$

Thus with $\tau = k$, $$\frac{\mathrm{d}W}{\mathrm{d}t} = k \,.\, u \,.\, s \tag{5.32}$$

If the boundary is not straight, $A \int \mathrm{d}s$, and $\frac{\mathrm{d}W}{\mathrm{d}t} = \int k \,.\, u \,.\, \mathrm{d}s$. Values of $\frac{\mathrm{d}W}{\mathrm{d}t}$ may be found graphically from this equation. The least value will be the best approximation.

5.10.4 *Application of upper-bound theory to plane-strain indentation*

We have already considered (§5.6.5) the frictionless indentation of a semi-infinite block by a flat punch. The slip-line field is reproduced in Figure 5.13*a*.

Bearing in mind the example of simple shear just considered, this field could be represented in Figure 5.13*b*. Equilateral triangles are chosen for simplicity of calculation. The physical interpretation of this field can be explained by considering the

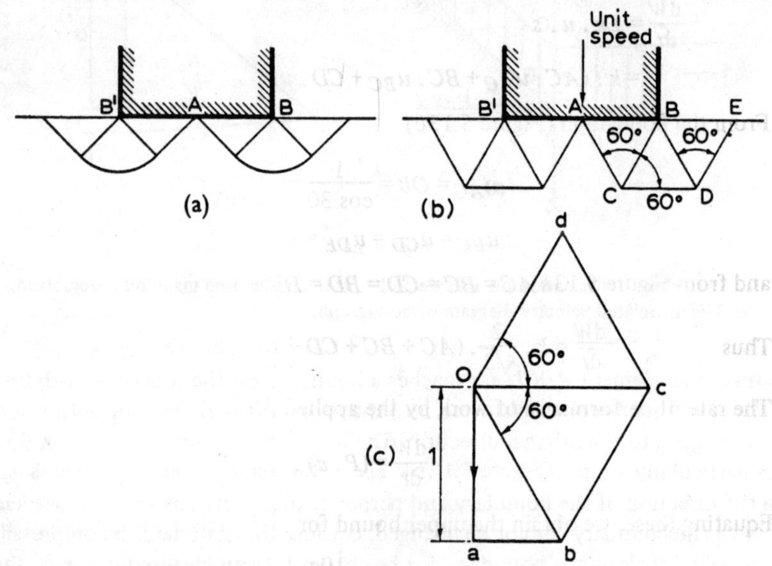

Figure 5.13

(*a*) Slip-line field for indentation of a large block.
(*b*) Approximate representation by straight shear lines.
(*c*) Hodograph for (*b*).

movement of any element in the plastic section *ABC*. The whole zone *ACDEBA* flows plastically and the metal below *ACDE* remains rigid. As the punch moves downwards with a velocity assumed equal to unity, any particle in the triangle *ABC* will move downwards with a vertical component of unity, but is constrained to slide parallel to the rigid boundary *AC* on which shear occurs. These velocities can be compounded, as in Figure 5.13*c*, showing that this produces a horizontal velocity component represented by the vector *ab*. As the particle crosses the boundary *BC*, its velocity is changed again by shear parallel to *CB* and it is constrained to move parallel to *CD*. The velocity triangle for this region *BCD* is thus represented by the original velocity Ob to the left of BC, the change in velocity bc parallel to CB, and the absolute velocity parallel to CD, which is thus equal to Oc.

A further change in velocity occurs on crossing BD, by addition of a shear velocity parallel to DB. The final absolute velocity in the triangle BDE is consequently given by Od, which is parallel to DE.

Figure 5.13*c* is the complete velocity diagram or hodograph for the right-hand part of the chosen plastic field. It should be noticed that the absolute velocity is always given by the appropriate vector from O. All particles in a particular zone of the field, bounded by shear lines, have the same velocity.

The velocity diagram is thus closely associated with the physical significance of the shear diagram. Having obtained the magnitudes of the velocities u along each boundary, we may deduce an upper bound to the load from equation 5.32:

$$\frac{dW}{dt} = \Sigma k \, . \, u \, . \, s$$

$$= k \, . \, (AC \, . \, u_{AC} + BC \, . \, u_{BC} + CD \, . \, u_{CD} + BD \, . \, u_{BD} + DE \, . \, u_{DE})$$

From the hodograph (Figure 5.13*c*)

$$u_{AC} = Ob = \frac{1}{\cos 30} = \frac{2}{\sqrt{3}}$$

$$u_{BC} = u_{CD} = u_{DE} = u_{BD} = u_{AC}$$

and from Figure 5.13*b* $AC = BC = CD = BD = DE = a$, $(B'B = 2a)$

Thus $$\frac{dW}{dt} = k \, . \, \frac{2}{\sqrt{3}} \, . \, (AC + BC + CD + BD + DE) = \frac{10a}{\sqrt{3}} \, . \, k \qquad (5.33)$$

The rate of performance of work by the applied pressure P is

$$\frac{dW}{dt} = (P \, . \, a) \, . \, 1 \qquad (5.34)$$

Equating these, we obtain the upper bound for the pressure to cause yielding:

$$P \, . \, a = \frac{10a}{\sqrt{3}} k.$$

or $$\frac{P}{2k} = 2{\cdot}89 \qquad (5.35)$$

It should be noted that this is not the only possible solution, since we have not considered stress boundary-conditions. If, for example, equal isosceles triangles with $\angle ACB = \angle CBD = \angle BDE = (\pi - 2\theta)$ had been chosen instead, the same procedure would yield the value:

$$\frac{P}{2k} = 2 \operatorname{cosec} 2\theta + \cot \theta \qquad (5.36)$$

This may readily be verified by drawing the hodograph. The lowest value of this pressure occurs for $\tan \theta = \sqrt{2}$ and is then

$$\frac{P}{2k} = 2\sqrt{2} = 2{\cdot}83 \qquad (5.37)$$

As we have seen, the lowest value of the upper bound is also the best approximation. This estimate may be compared with the accurate slip-line field result (equation 5.15), which is, of course, lower:

$$\frac{P}{2k} = 2{\cdot}57$$

It is not always possible to obtain a minimum analytically in this straight-forward manner, and there may be difficulties in deciding on the best estimate for the load. The method of obtaining at least one upper bound is however simple to use, and will be applied to other examples in later chapters. One other example may be mentioned here. It is not strictly within the scope of metalworking, but is a useful general concept, which gives an upper estimate of the collapse load for simple structures. It has been extensively used by Baker[5.17] and others (§5.10.6).

5.10.5 *Application of upper bounds to axial symmetry*[5.23, 5.24]

Kinematically – admissible velocity fields can be drawn for a diametral section of an axi-symmetric body. Most conveniently, it can be assumed that the field is spherical so that, in wire drawing for example, all the wire in the deforming zone is supposed to be moving towards the virtual apex of a conical die. Any element of the wire is assumed to move parallel to the axis until it encounters the spherical entry-boundary of the plastic zone and then to follow a radial path until it meets the second spherical boundary at the die exit. Reference to Figure 6.7 shows that this differs appreciably from the slip-line field configuration, which is much narrower on the axis than is assumed for the concentric spheres. The approximation improves for heavier reductions.

If the radii of the inlet and exit boundaries are respectively r_1 and r_2, the velocity v of an element flowing towards the apex, at an angle θ, is related to the initial velocity v_0 because volume is conserved. At some radius r,

$$v = v_0 \frac{r_1^2}{r^2} \cos\theta = v_f \frac{r_2^2}{r^2} \cos\theta \tag{5.38}$$

where v_f is the final velocity of the emerging wire. The velocity profile and hence the distortion of a transverse fibre can thus be predicted. Elements close to the axis will travel further along the axial direction in a given time than elements further out. This will produce a distortion similar to that shown in Figure 11.3.

Once the velocity field has been established, the corresponding strain-rate distribution can be deduced, as in §13.3. The rate of performing work is then calculated for the upper bound solution. There are three terms involved:

Throughout the deforming volume work is done at a rate

$$\dot{W}_H = \int_V \sigma\dot{\epsilon}\, \mathrm{d}V$$

due to the homogeneous deformation, as in equation 4.6.

On the surface S_D of a velocity discontinuity of magnitude u, there is redundant work

$$W_R = \int_{S_D} ku \, \mathrm{d}S$$

This resembles equation 5.32 for the simple upper bounds. Finally, work must be done in overcoming friction on the external faces S_E, on which the frictional stress can be assumed to be $\tau = mk$, and the velocity v_E

$$\dot{W}_F = \int_{S_E} mkv_E \, \mathrm{d}S$$

The upper-bound theorem states that the sum of these is greater than or equal to the rate of performing work by the external force in a real operation

$$\int \mathscr{P} v \mathrm{d}S \leqslant \int_V \sigma\epsilon \, \mathrm{d}V + \int_{S_D} ku \, \mathrm{d}S + \int_{S_E} mkv_E \, \mathrm{d}S \tag{5.39}$$

Comparison may be made with equations 13.57 and 13.88.

Many solutions of this type have been developed by Avitzur[5.23]. The best upper-bounds give results in good accordance with measured forces, but are less reliable for prediction of deformation.

The technique can however be further refined by using more realistic plastic-zone boundaries[5.24]. It can also be adapted for axi-symmetric polygonal bodies[5.25] and is particularly appropriate for square and hexagonal bar drawing, though the necessary computer programs are lengthy.

For more complex forms, upper-bound solutions can be obtained by dividing the workpiece into suitable blocks, as discussed in §13.5. At the expense of increased computation, strain-hardening and the presence of dead zones can be included with fair precision.

5.10.6 *The Plastic Hinge*[5.18]

If a simple beam is loaded at its centre, the bending moment increases steadily from zero at each end to a maximum at the centre. For a rigid-plastic metal no deflection would occur until a certain moment M_p caused the central region to deform plastically as a hinge. We suppose that, at the instant of collapse, the load moves downwards with unit speed and therefore does work at the rate $W \times 1$. This is entirely dissipated as heat, due to the plastic deformation of the hinge as the two rigid halves of the beam, each of length l, rotate about the hinge with angular velocity $1/l$. The total rate of energy dissipation at the hinge is thus $2M_p/l$. Equating these rates gives the value of M_p in terms of the load W which is found to cause collapse of a simple beam.

$$W \times 1 = 2M_p/l \tag{5.40}$$

In more complex structures, the positions of hinges are deduced from different postulated modes of deformation, and the collapse load is estimated, using the virtual work principle to obtain an upper bound for each. The lowest load, in terms of the moment M_p for a single hinge, will be the best approximation for the whole structure. Design on these lines can save appreciable weight in structural members.

EXAMPLES

5.1 Show, by considering the equilibrium of a small element in the plastic zone, that $\frac{\partial\sigma_x}{\partial x}+\frac{\partial\tau_{xy}}{\partial y}=0$ and $\frac{\partial\sigma_y}{\partial y}+\frac{\partial\tau_{xy}}{\partial x}=0$. Hence, using the fact that k is constant along a slip-line, show that $\frac{\partial\sigma_y}{\partial\alpha}=0=\frac{\partial\sigma_\beta}{\partial\beta}$

Solution. Sketch an elementary cube $\delta x, \delta y, \delta z$ at the origin of the OXYZ axes. Resolve the forces along OX and equate to zero for equilibrium:

$$\left(\sigma_x+\frac{\partial\sigma_x}{\partial x}.\delta x\right)\delta y\delta z-\sigma_x\delta y\delta z+\left(\tau_{yx}+\frac{\partial\tau_{yx}}{\partial y}.\delta y\right)\delta x\delta z-\tau_{yx}.\delta x\delta z=0$$

$$\left(\frac{\partial\sigma_x}{\partial x}+\frac{\partial\tau_{yx}}{\partial y}\right).\delta x\delta y\delta z=0$$

Similarly,
$$\frac{\partial\sigma_y}{\partial y}+\frac{\partial\tau_{yx}}{\partial x}=0$$

At the point Q (Figure 5.3) where the tangents to the slip-lines coincide with OX and OY axes, $\frac{\partial\tau_{xy}}{\partial x}=0$, because $\tau_{xy}=k$ which is a constant. Thus at $Q, \frac{\partial\sigma_y}{\partial y}=\frac{\partial\sigma_x}{\partial x}=0$ and since Q is any point at which the axes coincide with the slip-lines, this may be written generally

$$\frac{\partial\sigma_\alpha}{\partial\alpha}=0=\frac{\partial\sigma_\beta}{\partial\beta}$$

5.2 The yield stress of an ideal rigid-plastic material in plane-strain compression is 600 N/mm². If a block of this material is subjected to a hydrostatic compression of 450 N/mm², what principal stresses would just cause plastic flow? What is the magnitude of the shear stress on a plane inclined at 20° to one principal plane?

Solution. $S = 2k = 600$ N/mm². Draw the Mohr circle with centre (−450, 0) and radius 300. The principal stresses are given by the ends of the diameter along the σ axis, namely $\sigma_1 = -750$ N/mm² and $\sigma_3 = -150$ N/mm². The shear stress $\tau_\theta = \pm\ 300 \sin 40° = \pm 192$ N/mm².

5.3 Show by consideration of two pairs of α and β lines, using the Hencky equations, that the angle between two slip-lines of one family, where they are cut by a slip-line of the other family is constant along their length.

Solution. Consider two pairs of slip-lines. The α-lines AP and BQ are cut orthogonally by two β-lines. The Hencky equation 5.7 shows that along the α-line QB, $p_Q + 2k\phi_Q = p_B + 2k\phi_B$; and along the β-line BA, $p_B - 2k\phi_B = p_A - 2k\phi_A$. Thus

$$p_Q-p_A=(p_Q-p_B)+(p_B-p_A)=2k(\phi_B-\phi_Q)+2k(\phi_B-\phi_A)$$

Similarly $p_Q - p_A = 2k(\phi_Q - \phi_P) + 2k(\phi_A - \phi_P)$ since the general rotation by $\pi/2$ does not influence the relationship. These must be the same since the pressure can have only one value at a given point, so

$$2\phi_B-\phi_Q-\phi_A=\phi_Q+\phi_A-2\phi_P$$

$$\phi_Q-\phi_P=\phi_A-\phi_B$$

This is known as Hencky's first theorem.

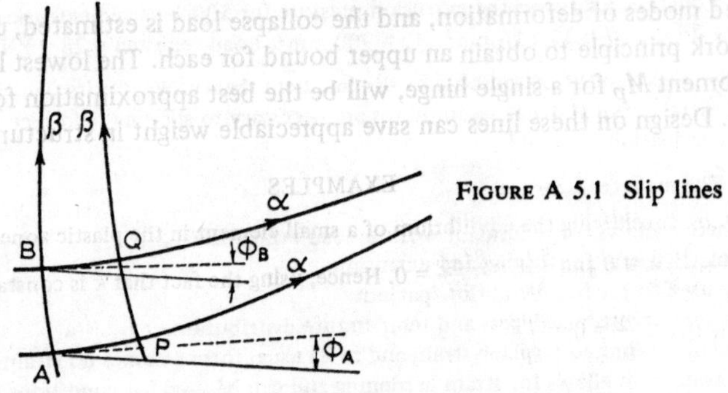

FIGURE A 5.1 Slip lines

5.4 The coefficient of friction for lubrication of mild steel is found to be $\mu = 0{\cdot}04$ for soap and $\mu = 0{\cdot}11$ for an oil. At what angles would the slip-lines be inclined to the tool/workpiece interface when using each of these lubricants?

Solution. Equation 5.10 gives the angle of inclination of the slip-lines to an interface with Coulomb friction.

$$2\theta = \cos^{-1} \mu q'/k$$

This cannot however be evaluated unless q' is known. A first approximation is made by supposing that $q' \approx q$, which is found by drawing the slip-line field for a frictionless interface. An example is given in Chapter 6, Example 6.7. In a simple instance such as plane-strain compression of a thin strip, $q = 2k$.

Thus for $\mu = 0{\cdot}04$	$2\theta = \cos^{-1} 0{\cdot}08$	$\theta = 43°$ (and 47)
$\mu = 0{\cdot}11$	$2\theta = \cos^{-1} 0{\cdot}22$	$\theta = 39°$ (and 51)

5.5 Sketch the distribution of hydrostatic pressure 10 mm beneath a long 75 mm wide, flat indenter which is loaded so that plastic flow just occurs in a very thick block.

Solution. Two solutions are possible, according to the choice of slip-line field suggested by Prandtl (Figure 5.7*a*) or by Hill (Figure 5.7*b*). The hydrostatic pressure distribution is found by drawing the chosen field to scale and ruling a horizontal line 10 mm below the surface.

From Figure 5.7*a* p/k is constant at the value 1·0 throughout the triangular region *BDE*, because $\sigma_3 = 0$ and $p = k$ over *BD*, and the slip-lines are straight. In the fan from *BE* to *BF* the pressure rises by increments $2k \times 0{\cdot}262$ as the horizontal line intercepts successive radii drawn at 15° intervals. The pressure reaches $p/k = 1 + \pi$ at the intersection with *BF*, and thereafter remains constant over the central region until *AF* is reached. It then decreases in a symmetrical manner.

From Figure 5.7*b* the pattern is similar, but the deformation zone does not extend so far outwards from the indenter. In the central region between *F* and *F'* it is reasonable to expect that the pressure would remain constant, and Hill's full solution in fact postulates a continuation of the straight slip-lines over this region, as in the Prandtl solution.

Both solutions give a pressure $p = 2{\cdot}57k$ immediately beneath the edges of the indenter, and the pressure varies almost linearly with distance inwards and outwards from these points.

5.6 An experimental inverted extrusion with tellurium lead is planned to reduce a 75 mm wide × 12 mm thick billet to 75 mm wide × 6 mm thick, using a square die. What capacity press will be needed? Assume that $S = 25$ N/mm². What effect does the length of the billet have on the pressure?

Solution. Equation 5.23 gives the extrusion pressure for 50% inverted extrusion $p = k \times (1 + \pi/2) = \frac{1}{2} \times 25 \times 2{\cdot}57 = 32{\cdot}1$ N/mm². $A_0 = 900$ mm². Load required = 28·9 kN.

5.7 Write a short account of the purpose of slip-line field theory. What is its significance in practical metalworking? What assumptions are necessary to slip-line field theory but not to the stress evaluation method?

Solution. The main features are:

(*a*) s.l.f. theory allows for redundant work in load evaluation.
(*b*) Particularly useful for forging and extrusion.
(*c*) Can be used to predict metal flow pattern.
(*d*) Can be used to estimate stress and temperature distribution.
(*e*) s.l.f. theory is limited to plane strain and in its usual form assumes no strain hardening. Stress evaluation allows for strain hardening and can be used for conditions of axial symmetry.
(*f*) Plane-strain analogies, for which s.l.f. solutions are known, may be useful, for example in wire drawing and more complex processes.

5.8 What are the Hencky equations and the Geiringer equations, and when are they used in slip-line field theory? Describe in general terms the sequence of drawing a slip-line field and estimating the metalworking load from it.

Solution. This problem is an exercise in recapitulating the basic features of s.l.f. solutions, as described in §§5.4–5.7.

5.9 Sketch a possible upper-bound solution for 50% inverted plane-strain extrusion and estimate the extrusion pressure in terms of $2k$.

Solution. (*a*) Consider the simple solution in Figure A5.2*a*.

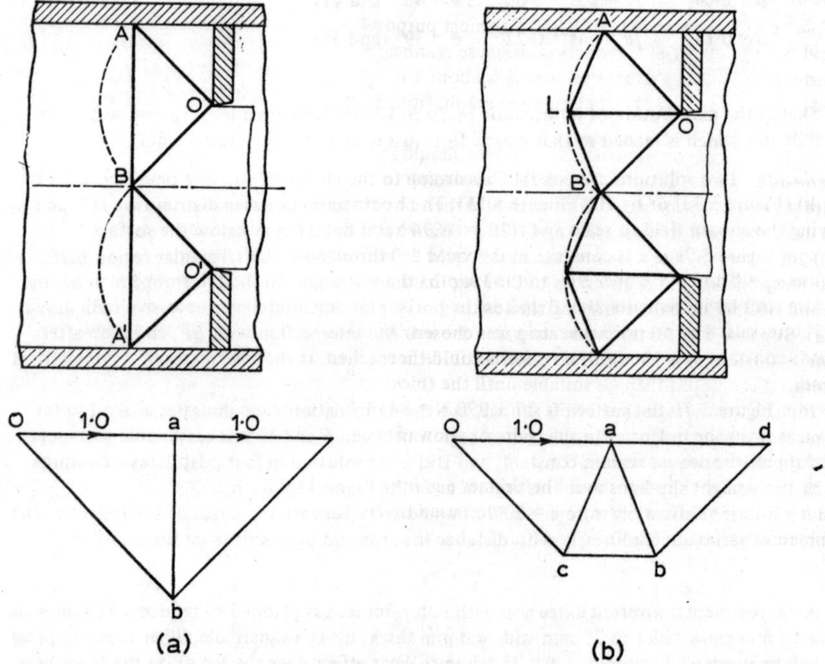

FIGURE A 5.2

From equation 5.32, $\frac{dW}{dt} = kus = k(u_{AB}AB + u_{AO}AO + u_{BO}BO)$. Taking the lengths from the field and the hodograph, for the upper half

$$\frac{dW}{dt} = k(1 \times 2 + \sqrt{2} \times \sqrt{2} + \sqrt{2} \times \sqrt{2}) = 6k$$

But

$$\frac{dW}{dt} = P \times AB \times 1{\cdot}0 = 2P; \frac{P}{2k} = 1{\cdot}5$$

This, like all upper-bound solutions, is greater than the extrusion pressure calculated from the slip-line field (equation 5.23):

$$P/2k = 1{\cdot}29$$

(*b*) With the closer approximation to the s.l.f. shown in Figure A5.2*b*

$$\frac{dW}{dt} = k(u_{AL}AL + u_{AO}AO + u_{LO}LO + u_{LB}LB + u_{BO}BO)$$

By calculating or measuring the lengths

$$\frac{dW}{dt} = k(0{\cdot}77 \times 1{\cdot}08 + 1{\cdot}0 \times 1{\cdot}41 + 0{\cdot}59 \times 1{\cdot}41 + 0{\cdot}77 \times 1{\cdot}08 + 1{\cdot}0 \times 1{\cdot}41)$$

$$= 5{\cdot}31k$$

$$P/2k = 1{\cdot}33$$

5.10 Suggest suitable dimensions for a plane-strain compression test apparatus to be used on 6 mm thick 0·2%C steel, assuming that the stress-strain curve resembles that of Example 2.3 but that the yield stress may be about 10% higher. A 500 kN press is available.

Solution. Maximum load 500 kN. For most purposes the compression test is not required to extend beyond about 80% reduction, because reannealing would be necessary before further deformation. Thus the greater value of ϵ is about 1·6 and the corresponding yield stress about 1·0 kN/mm², found by extrapolating the graph, from Example 2.3 to 0·92 kN/mm² at $\epsilon = 1{\cdot}6$ and adding 10%.

The plane-strain compression test platens should have a breadth b between 2 and 4 times the specimen thickness (Chapter 2 §2.6.2). This suggests a platen breadth b = 12 mm at the start. To ensure plane strain, the strip width w should then be $5b$, namely 60 mm. These platens can be used until the yield stress reaches 500/720 = 0·694 kN/mm², which permits a strain of only $\epsilon \approx 0{\cdot}26$. Either (*a*) the width requirement could be relaxed to say $4b$, permitting a yield stress up to 868 kN/mm² and a strain up to 0·63; or (*b*) the stock material could be reduced in thickness and reannealed before starting the test.

(*a*) Suppose that 50 mm wide strip was chosen. At $\epsilon = 0{\cdot}63$ the thickness would be given by $\ln 6/h_1 = 0{\cdot}63$, $h_1 = 3{\cdot}19$ mm. The platens could then be replaced by a second set with 7 mm breadth. These would then be suitable until the thickness was reduced to 7/4 = 1·7 mm, a strain of 1·26. The load would thus be 50 × 7 × 990 N = 346 kN, within the capacity of the machine. The platens could then be changed again, or allowance could be made for friction in the remainder of the test, using equation 10.5.

(*b*) Preferably the strip should be thinner e.g. h = 5 mm, with b = 10 mm, w = 50 mm, used from h = 5 to h = 2·5; and b = 5, w = 50 mm used thereafter. (At $\epsilon = 1{\cdot}6$ $\bar{S} = 1{\cdot}0$ kN/mm², the final load would be 250 kN).

REFERENCES

5.1 Cottrell, A. H., *Theoretical Structural Metallurgy*, Edward Arnold, 1948.

5.2 Barrett, C. S., *Structure of Metals*, McGraw-Hill, 1952.

5.3 Bourne, L. and Hill, R., 'On correlation of directional properties of rolled sheet in tension and cupping tests', *Phil. Mag.*, **41**, 671–681, 1950.

5.4 Hencky, H., 'Über einige statisch bestimmte Fälle des Gleichgewichts in plastischen Körpern', *Zeits. angew. Math. Mech.*, **3**, 241–251, 1923.

5.5 Kötter, F., 'Die Bestimmung des Drucks an gekrümmten Gleitflächen, eine Aufgabe aus der Lehre vom Erddruck'. *Berlin. Akad. Berichte*, pp. 229–233, 1903.

5.6 Green, A. P., 'A theoretical investigation of compression of ductile material between smooth flat dies', *Phil. Mag.*, **42**, 900–918, 1951.

5.7 Hill, R., *The Mathematical Theory of Plasticity*, Oxford Univ. Press, p. 255.

5.8 Prandtl, L., 'Über die Härte plastischer Körper', *Nachr. Ges. Wiss. Göttingen*, pp. 74–85, 1920.

5.9 Watts, A. B. and Ford, H., 'An experimental investigation of the yielding of strip between smooth dies', *Proc. Inst. Mech. Engrs.*, **B1**, 448–453, 1952.

5.10 Johnson, W., 'The pressure for cold extrusion of lubricated rod through square dies', *J. Inst. Met.*, **85**, 403–408, 1957.

5.11 Brinell, J. A., 'Methods of testing steel', *Cong. Int. Méthodes d'Essai, Paris*, Vol. 2, 1900.

5.12 Meyer, E., 'Untersuchungen über Härteprüfung und Härte', *Zeits. Ver. Deutsch. Ing.*, **52**, 645–654, 1908.

5.13 Tabor, D., *The Hardness of Metals*, Oxford Univ. Press, 1951.

5.14 Geiringer, H., 'Beitrag zum vollständigen ebenen Plastizitätsproblem', *Proc. 3rd Int. Cong. Appl. Mech. Stockholm*, Vol. 2, pp. 185–190, 1930.

5.15 Hill, R., Discussion of a paper by E. Siebel, 'Application to shaping processes of Hencky's laws of equilibrium', *J. Iron and Steel Inst.*, **156**, 513–517, 1947.

5.16 Johnson, W. and Kudo, H., *The Mechanics of Metal Extrusion*, Manchester Univ. Press, 1962.

5.17 Baker, J. F., *The Steel Skeleton*, Vol. I: *Elastic Behaviour and Design*; Baker, J. F., Horne, M. R. and Heyman, J., *The Steel Skeleton*, Vol. II: *Plastic Behaviour and Design*, Cambridge Univ. Press, 1954–1956.

5.18 Green, A. P., 'The plastic yielding of notched bars due to bending', *Q. J. Mech. Appl. Maths.*, **6**, 223–239, 1953.

5.19 Johnson, R. W., and Rowe, G. W., 'Redundant work in drawing cylindrical stock', *J. Inst. Metals*, **96**, 97–105, 1968.

5.20 Farmer, L. E. and Oxley, P. L. B., 'Slip-line field for plane-strain extrusion of strain-hardening material', *J. Mech. Phys. Solids*, **19**, 369–388, 1971.

5.21 Oxley, P. L. B., 'Allowing for strain rate effects in the analysis of metal working processes', *Inst. Phys. Conf. Ser. No. 21*, 359–381, 1974.

5.22 Christopherson, D. G., Oxley, P. L. B., and Palmer, W. B., 'Orthogonal cutting of a work-hardening material', *Engineering*, **186**, 113–115, 1958.

5.23 Avitzur, B., *Metal forming processes and analysis,* McGraw-Hill, New York, 1968.

5.24 Johnson, W., and Mellor, P.B., *Engineering plasticity*, van Nostrand Reinhold, London, Ch. 16, 1973.

5.25 Shield, R. T. and Drucker, D. C., 'The application of limit analysis to punch indentation problems', *Trans. A.S.M.E. J. App. Mech.*, **20**, 453–460, 1953.

5.26 Juneja, B. L. and Prakash, R., 'An analysis for drawing and extrusion of polygonal sections', *Int. J. Mach. Tool Des. Res.*, **15**, 1–18, 1975.

Part 2: Examination of Processes

6

Drawing of Round Bars and Flat Strip

6.1 Introduction

Large quantities of rods, tubes, wires and special sections are finished by cold drawing. Large bars, up to 150 mm (6 in.) diameter or sometimes more, are frequently given a light sizing pass, reducing the diameter by 1·5 mm ($\frac{1}{16}$ in.), to improve the surface finish and dimensional tolerance. Light reductions are also made in the production of bright-drawn angles, channels and strips. These sizing passes are usually regarded as relatively easy to perform, and theory has little contribution to make, except perhaps in predicting the limits below which the stock may deform ahead of the die, forming a damaging 'bulge' (§6.5.4). Many of the smaller sizes of round bar are, however, drawn to very much greater reductions of area, up to 50% reduction per pass, and wires may be reduced by 90% in sequential passes from the annealed state before they are re-annealed. Some wires, which may finish at 0·025 mm (0·001 in.) diameter or even less, are drawn through a very large number of dies before reaching the final size, and may be re-annealed several times as drawing proceeds.

The large sizes are drawn on heavy benches, after having been pointed so that they may be passed through the die and gripped by the jaws of a pulling 'dog'. The dog is driven by a hook engaging in an endless chain, and travels along a straight track, which may be up to 15 m (50 ft) long for bars of say 25 mm (1 in.) diameter. Speeds up to about 0·5 m/s (100 ft/min) are common for ferrous materials, but higher speeds are used for copper alloys. Many modern drawbenches are driven hydraulically, which gives much smoother starting action. As the diameter is reduced, it becomes uneconomic to handle these short lengths, and at about 10 mm ($\frac{1}{2}$ in.) diameter it is preferable to use a 'bull-block' in place of the straight bench so that long uninterrupted coils of wire may be produced. This is a large capstan, to which the pointed end of a coil is attached. As the capstan rotates, wire is drawn steadily through the die. Very long coils can be drawn in this way, and by special techniques it is possible to weld coils together for continuous production. Linear speeds of about 2·5 m/s (500 ft/min) may be used. Below about 3 mm ($\frac{1}{8}$ in.) diameter, much higher speeds are possible with moderate power, and even this system cannot handle sufficiently long coils. Mass production of the smaller sizes of wire usually involves several dies in tandem. Six or more may be mounted in a single machine. Power-driven capstans are provided between each die and the one ahead of it, and the wire is wrapped round the capstan to give enough frictional drag to pull the wire through the die. The speeds of the capstans are carefully adjusted throughout the train, to match, with about 1% slip, the increasing speed of the wire as it is elongated and reduced in cross-sectional area. After the last capstan, the wire is collected on a take-up spool. The reduction at each die is determined by the size of the die and the preceding one, and can be altered only by changing a die. It is therefore necessary to ensure that the dies do not wear beyond a certain oversize limit. Tungsten-carbide dies are often used for hard wires, and diamond is the usual choice for very fine gauges. Very large quantities of wire can be processed rapdily, final speeds often reaching 20 m/s (3 to 4,000 ft/min).

Much more time is taken in annealing. Batch annealing of coils may take 1 to 2 hours for each; and strand annealing, in which several wires are individually pulled through a long furnace, usually proceeds at a few inches per minute. High-temperature annealing for very short times has been tried successfully on an experimental basis, but requires careful control. Thus, when it is necessary to impart very large total reductions in area, it is important to reduce the number of annealing processes, and their associated descaling treatments, as far as possible by increasing the reduction of area given between one and the next.

Theory can predict the maximum reduction of area possible in a single pass. The limit is set by tensile failure of the drawn wire at or near its yield stress. Under ideal conditions, the drawing stress σ would be simply that necessary to deform the wire homogeneously by the required amount.

$$\sigma = Y \ln \frac{1}{1-r} \tag{6.3}$$

In drawing round bar, the limiting reduction r_m would be given by

$$\sigma = Y; \ln \frac{1}{1-r_m} = 1, r_m = 63\%$$

There is, however, always a frictional contribution which increases the drawing stress to σ':

$$\frac{\sigma'}{Y} = \frac{1+B}{B}[1-(1-r)^B] \tag{6.23}$$

For typical values of die angle α and coefficient of friction μ, the parameter $B = \mu \cot \alpha$ is about 0·19, and the limiting reduction r'_m is reduced to about 60%.

In addition to the frictional term, work is done internally in shearing the metal, first one way and then the other, as it passes through the die. This redundant work can be allowed for accurately in frictionless strip-drawing, where the strain is two dimensional, giving a drawing stress t:

$$\frac{t}{2k} = f\left(\frac{c}{d}\right) \ln \frac{1}{1-r} \tag{6.38}$$

The correction factor is a function of geometrical parameters c and d, corresponding to the strip thickness and platen breadth in compression. A similar relationship is found experimentally to be valid for a range of common drawing conditions with round bar, for which no general theory exists, though partial theories have been suggested:

$$\frac{\sigma}{Y} = \phi(\alpha, r) \ln \frac{1}{1-r} \tag{6.50}$$

This factor is of most importance for high die angles and light reductions, and does not appreciably diminish the maximum reduction possible with the low die angles commonly used.

To allow a reasonable margin, and particularly to avoid the danger of metallic contact and adherence of metal to the die face, the reductions taken in practice are considerably less than the absolute maximum, usually 35% to 45% per pass, and the redundant work factor may then be significant. The redundant shearing increases the strain-hardening, and the effect is additive; so that a number of light passes, to a given overall reduction of area, hardens the stock significantly more than one or two heavy passes to the same size. Multiple passes should be designed with this in mind.

Die design is also influenced. Friction and redundant work both increase the drawing load, but redundant work is greatest for steep die angles, which introduce most distortion, and the resolved component of the frictional drag is greatest for low die angles. There is consequently an optimum die angle which gives the lowest drawing force for a given reduction. This value varies with reduction, because the distortion and the die length vary, but for strip drawing the optimum is about $\alpha = 15°$, and for wire drawing, $\alpha = 6°$. The effects of variation in die angle can be predicted with good accuracy for straight-sided dies. The low die angles are also to be recommended because the throughput of lubricant is improved, reducing the danger of metallic transfer or 'pickup', which is the most serious problem in drawing.

Drawing-dies may also have a curved profile, and there is a formal resemblance between the differential equation for drawing and that for rolling:

Drawing: $$h\,d\sigma_x + \sigma_x\,dh + p(1 + \mu \cot \alpha)\,dh = 0 \tag{4.15}$$

Rolling: $$d(h\sigma_x) = -p_r(1 \pm \mu \cot \alpha)\,dh \tag{9.12}$$

The friction acts in only one direction over the die face during drawing, opposing the entry of metal into the die, and it is always necessary to apply a front tension. Some back tension will also be present in multiple-die tandem drawing, as in tandem rolling. There is, in fact, a process known as roller-drawing which is a hybrid process employing front tension with undriven rolls. A small roll assembly, with appropriately shaped rolls, is mounted on a drawbench, and used for example in forming strip for turbine blades. Other composite processes will be discussed in connection with tube-drawing.

6.2 Elementary assessment of drawing force: homogeneous-deformation contribution

The least possible work necessary to produce a given reduction of area is that required for homogeneous deformation. For unit volume of metal, this is equal to the integral of the stress–strain curve between the appropriate strain values (see Chapter 4, §4.3).

$$\frac{W}{V} = \int_{\epsilon_1}^{\epsilon_2} Y \, d\epsilon \tag{6.1}$$

If the metal has been strain-hardened before the operation, for example by prior drawing from a length l_0 to a length l_1, it is reasonably accurate to assume a mean value $\overline{Y}$ for most purposes (see also §6.3.2); though this is rather unreliable for annealed materials. For second and subsequent passes, equation 6.1 may be written

$$W = V\overline{Y} \int_{l_1}^{l_2} \frac{dl}{l} = V\overline{Y} \ln \frac{l_2}{l_1} \tag{6.2}$$

As shown in Chapter 4, §4.3.1, this may be equated to the work done by the drawing force $F = \sigma_2 A_2$ in travelling from the die to the full length l_2 of the drawn stock, assuming no friction or other energy-dissipating process, leading to

$$\sigma_2 = \overline{Y} \int_{\epsilon_1}^{\epsilon_2} d\epsilon = \overline{Y} \ln \frac{l_2}{l_1} = \overline{Y} \ln \frac{A_1}{A_2} = \overline{Y} \ln \frac{1}{1-r} \tag{6.3}$$

Even when friction and redundant work are significant, this remains the contribution of the homogeneous deformation to the drawing stress. It will reappear in several formulae.

6.3 Determination of plane-strain drawing load from local stress-evaluation

6.3.1 *Drawing of wide, flat strip with wedge-shaped dies (B constant, S constant)*

The basic equation for plane-strain drawing was derived in Chapter 4, §4.4.1, from consideration of the horizontal equilibrium of the forces acting on an element of deforming metal:

$$h \, d\sigma_x + \sigma_x \, dh + p(1 + \mu \cot \alpha) \, dh = 0 \tag{6.4}$$

For practical purposes the drawing stress σ_x and the die pressure p are principal stresses and so are related by the condition of yielding in plane strain (equation 3.27)

$$\sigma_x + p = S \tag{6.5}$$

Eliminating p and writing $B = \mu \cot \alpha$, equation 6.4 becomes

$$\frac{d\sigma_x}{B\sigma_x - S(1+B)} = \frac{dh}{h} \tag{6.6}$$

This can be integrated directly if B and S are both constant. This gives the drawing stress σ_{xa} at the exit, assuming no back-pull.

$$\frac{\sigma_{xa}}{S} = \frac{1+B}{B}\left[1-\left(\frac{h_a}{h_b}\right)^B\right] = \frac{1+B}{B}\left[1-(1-r)^B\right] \tag{6.7}$$

This may be compared with equation 6.23*b* for rod drawing.

6.3.2 *Drawing of strain-hardening strip with wedge-shaped dies*

The assumption of constant yield stress is reasonably reliable for heavily-worked strips, because the metal strain-hardens slowly and steadily. The value of yield stress chosen is the mean value derived from the area of the stress–strain curve between the appropriate ordinates, which, as we have seen in §6.2, represents the work done in homogeneous deformation. This is more realistic than a yield stress taken at the mean strain ordinate, but will usually not be very different in value. This method is less reliable for first passes after annealing, because of the relatively high initial rate of strain-hardening, which decreases as the metal is deformed. For such passes, or for any passes on metal which strain-hardens rapidly, it is preferable to incorporate the stress-strain relationship in the initial equations. This is most conveniently done with the aid of a small analogue computer, but an analytical approach is possible.

In terms of strain, $d\epsilon = -dh/h$, the horizontal equilibrium equation, 6.4,

$$h\, d\sigma_x + [\sigma_x + (S-\sigma_x)(1+B)]\, dh = 0$$

can be written

$$\frac{d\sigma_x}{d\epsilon} + B\sigma_x = S(1+B) \tag{6.8}$$

If this equation is multiplied throughout by an integrating factor $e^{B\epsilon}$, it may be integrated between the limits ϵ_b at entry and ϵ_a at exit:

$$\int_{\epsilon_b}^{\epsilon_a} e^{B\epsilon}\frac{d\sigma_x}{d\epsilon}\, d\epsilon + \int_{\epsilon_b}^{\epsilon_a} B\sigma_x e^{B\epsilon}\, d\epsilon = \int_{\epsilon_b}^{\epsilon_a} S(1+B)\, e^{B\epsilon}\, d\epsilon$$

$$\left[e^{B\epsilon}\sigma_x\right]_{\epsilon_b}^{\epsilon_a} - \int_{\epsilon_b}^{\epsilon_a} \sigma_x \,.\, B\, e^{B\epsilon}\, d\epsilon + \int_{\epsilon_b}^{\epsilon_a} B\sigma_x e^{B\epsilon}\, d\epsilon = \int_{\epsilon_b}^{\epsilon_a} S(1+B)\, e^{B\epsilon}\, d\epsilon$$

$$\sigma_{xa}\, e^{B\epsilon_a} - \sigma_{xb}\, e^{B\epsilon_b} = (1+B)\int_{\epsilon_b}^{\epsilon_a} S\, e^{B\epsilon}\, d\epsilon \tag{6.9}$$

The right-hand side of this equation may be evaluated for different assumed stress–strain relationships.

(*a*) *No strain-hardening, S = constant*

When the yield stress remains constant, *S* may be brought outside the integrand, giving

$$\sigma_{xa}\, e^{B\epsilon_a} - \sigma_{xb}\, e^{B\epsilon_b} = (1+B)\, S \,.\, \frac{e^{B\epsilon_a} - e^{B\epsilon_b}}{B}$$

If there is no back-pull the longitudinal stress is zero at the die entry:

$$\sigma_{xb} = 0,$$

$$\frac{\sigma_{xa}}{S} = \frac{1+B}{B}\left[1 - \frac{e^{B\epsilon_b}}{e^{B\epsilon_a}}\right]$$

Since $\epsilon_a - \epsilon_b = \ln h_b/h_a$ this is equivalent to

$$\frac{\sigma_{xa}}{S} = \frac{1+B}{B}\left[1 - \left(\frac{h_a}{h_b}\right)^B\right]$$

This accords, as is to be expected, with equation 6.7, which was obtained by ignoring strain-hardening throughout.

(b) Linear strain-hardening

For many metals, the strain-hardening is approximately linear after passing the knee of the stress–strain curve. It can then be assumed that

$$S = S_b + c\epsilon \tag{6.10}$$

Equation 6.9 then becomes

$$\sigma_{xa}\, e^{B\epsilon_a} - \sigma_{xb}\, e^{B\epsilon_b} = (1+B)\int_{\epsilon_b}^{\epsilon_a} (S_b + c\epsilon)\, e^{B\epsilon}\, d\epsilon$$

$$\sigma_{xb}\, e^{B\epsilon_a} = (1+B)\left[S_b\frac{e^{B\epsilon}}{B} + c\epsilon\frac{e^{B\epsilon}}{B} - \int c\frac{e^{B\epsilon}}{B}d\epsilon\right]_{\epsilon_b}^{\epsilon_a}$$

$$= (1+B)\left[\frac{S_b}{B}e^{B\epsilon} + \frac{c\epsilon\, e^{B\epsilon}}{B} - \frac{c\, e^{B\epsilon}}{B^2}\right]_{\epsilon_b}^{\epsilon_a}$$

This may be simplified if ϵ_b is taken as zero. Logarithmic strains are of course additive (Chapter 2, §2.2.2), so this does not restrict the result. Then

$$\sigma_{xa}\, e^{B\epsilon_a} = \frac{1+B}{B}\left[(S_b + c\epsilon)\, e^{B\epsilon} - \frac{c\, e^{B\epsilon}}{B}\right]_0^{\epsilon_a}$$

$$= \frac{1+B}{B}\left[(S_b + c\epsilon_a)\, e^{B\epsilon_a} - S_b - \frac{c\, e^{B\epsilon_a}}{B} + \frac{c}{B}\right]$$

Since $\quad S_a = S_b + c\epsilon_a$

$$\sigma_{xa} = \frac{1+B}{B}\left[\left(S_a - \frac{c}{B}\right) - \left(S_b - \frac{c}{B}\right)e^{-B\epsilon_a}\right] \tag{6.11}$$

(c) Exponential strain-hardening

A closer approximation to the strain-hardening curve for annealed metal is given by an exponential formula (Chapter 2, §2.4).

Thus for annealed metal we could assume

$$S = a + b\,e^{c\epsilon} \tag{6.12}$$

The constants a, b, and c are chosen to give the best approximation to the experimental results. This equation for S can be substituted into the integrand of equation 6.9, but the solution is cumbersome. It is recommended that an approximate chord is drawn through the stress–strain curve so that the linear-hardening solution can be applied, or, preferably, that a computer is used, as for the general solution.

(*d*) *General strain-hardening characteristic*

The value of S can be found numerically in terms of the strain ϵ at any point of a given stress–strain curve, whatever its shape. The integration of equation 6.9 can thus be performed numerically, using a digital computer. Once the programme has been prepared, solutions can be obtained rapidly for any chosen values of $B = \mu \cot \alpha$

It is much easier to use an analogue computer. This is first set up to generate sufficient straight lines to approximate closely to the actual stress–strain curve; four or five lines usually suffice. It then solves equation 6.8 with this data, and presents the result graphically for any chosen values of B. Quite simple, and relatively inexpensive, analogue computers give ample accuracy for practical purposes. This method can also be used for solution of equation 6.8 for cylindrical dies.

6.3.3 *Drawing of strain-hardening strip with cylindrical dies*

Most profiles may be represented approximately either by a wedge die, as already considered, or by a circular arc. For the latter, solution proceeds in the same way as in solving the rolling equation (Chapter 9, §9.3.1). Variation of S and α can be considered simultaneously.

The stress required for drawing strip through cylindrical dies is most easily calculated by solving the equation for die-pressure first. The equation of horizontal equilibrium (6.4) is

$$\mathrm{d}(\sigma_x h) + p(1 + \mu \cot \alpha)\,\mathrm{d}h = 0$$

Combining this with the condition of yielding, equation 6.5, and eliminating σ_x instead of p:

$$\mathrm{d}(hS - hp) + p(1 + \mu \cot \alpha)\,\mathrm{d}h = 0 \tag{6.13}$$

The radius of the die is assumed constant and equal to R (Figure 6.1), and the equation can then be expressed in terms of the polar coordinates (R, α), by substituting for $\mathrm{d}h$:

$$\mathrm{d}h = 2(R\,\mathrm{d}\alpha) \sin \alpha$$

$$\mathrm{d}(hS - hp) = -2Rp \sin \alpha\,(1 + \mu \cot \alpha)\,\mathrm{d}\alpha \tag{6.14}$$

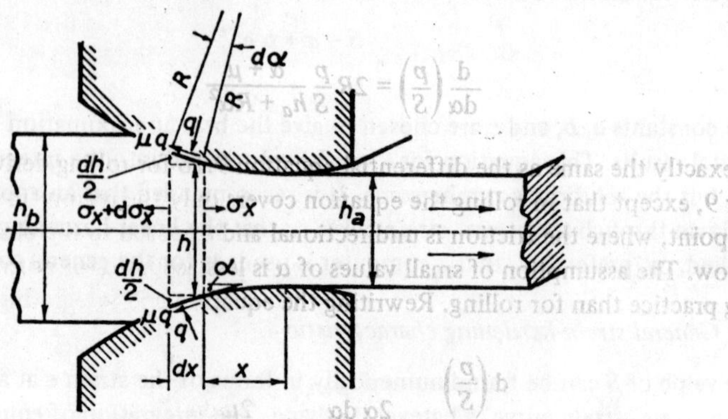

Figure 6.1 A diagram of strip-drawing through dies of circular profile.

In terms of the dimensionless ratio p/S,

$$\frac{\mathrm{d}}{\mathrm{d}\alpha}\left[hS\left(1-\frac{p}{S}\right)\right] = -2Rp\,(\sin\alpha + \mu\cos\alpha)$$

$$hS\frac{\mathrm{d}}{\mathrm{d}\alpha}\left(1-\frac{p}{S}\right) + \left(1-\frac{p}{S}\right)\frac{\mathrm{d}}{\mathrm{d}\alpha}(hS) = -2Rp\,(\sin\alpha + \mu\cos\alpha) \qquad (6.15)$$

Under most circumstances, the variation in die-pressure p through the deformation zone will exceed the variation in yield stress S, and the variation in the product hS will be less still, since S increases as h decreases. The term $\left(1-\frac{p}{S}\right)\frac{\mathrm{d}}{\mathrm{d}\alpha}(hS)$ can usually be neglected in comparison with $hS\frac{\mathrm{d}}{\mathrm{d}\alpha}\left(1-\frac{p}{S}\right)$, as first suggested by Bland and Ford[6.1]. This approximation is not valid when the rate of strain-hardening is high, as it often is for the first pass on annealed strip, nor when a high back-tension is applied, since the latter reduces the variation in p over the arc of contact (see Chapter 9, §9.3.3). For second and subsequent passes, under most drawing conditions, the approximation is sufficiently accurate. Equation 6.15 may then be written:

$$h\frac{\mathrm{d}}{\mathrm{d}\alpha}\left(\frac{p}{S}\right) = 2R\frac{p}{S}(\sin\alpha + \mu\cos\alpha)$$

If the angle of contact α is small, further approximations may be made:

$$\sin\alpha \approx \alpha, \quad \cos\alpha \approx 1 - \frac{\alpha^2}{2} \approx 1$$

$$h = h_a + 2R(1-\cos\alpha) \approx h_a + 2R\frac{\alpha^2}{2}$$

Then,

$$\frac{d}{d\alpha}\left(\frac{p}{S}\right) = 2R\frac{p}{S}\frac{\alpha+\mu}{h_a + R\alpha^2} \tag{6.16}$$

This is exactly the same as the differential equation 9.16 for rolling derived in Chapter 9, except that in rolling the equation covers only the region from exit to neutral point, where the friction is undirectional and opposed to the direction of metal flow. The assumption of small values of α is less accurate for normal strip-drawing practice than for rolling. Rewriting the equation and integrating,

$$\frac{d\left(\frac{p}{S}\right)}{\frac{p}{S}} = \frac{2\alpha\, d\alpha}{h_a/R + \alpha^2} + \frac{2\mu\, d\alpha}{h_a/R + \alpha^2}$$

$$\ln\left(\frac{p}{S}\right) = \ln\,(h_a/R + \alpha^2) + \frac{2\mu}{\sqrt{h_a/R}}\,.\,\tan^{-1}\frac{\alpha}{\sqrt{h_a/R}} + \text{constant}$$

It is convenient to introduce the symbol H,

$$H = 2\sqrt{R/h_a}\,.\tan^{-1}(\sqrt{R/h_a}\,.\,\alpha).$$

So

$$\ln\left(\frac{p}{S}\right) = \ln\left(\frac{h}{R}\right) + \mu H + \text{constant},$$

$$\frac{p}{S} = c\,\frac{h}{R}\,e^{\mu H} \tag{6.17}$$

Assuming that there is no longitudinal stress at the entry, $\sigma_{xb} = 0$ at $\alpha = \alpha_b$. Consequently $p_b = S_b - \sigma_{xb} = S_b$, and

$$1 = c\,\frac{h_b}{R}\,e^{\mu H_b};\; H_b = 2\sqrt{R/h_a}\,.\,\tan^{-1}(\sqrt{R/h_a}\,.\,\alpha_b)$$

Equation 6.17 can thus be written:

$$\frac{p}{S} = \frac{h}{h_b}\,e^{\mu(H - H_b)}$$

The longitudinal stress σ_x is found by combining this with the condition of yielding, equation 6.5:

$$\frac{\sigma_x}{S} = 1 - \frac{h}{h_b}\,e^{\mu(H - H_b)}$$

The drawing stress is the longitudinal stress σ_{xa} at the die exit:

$$\frac{\sigma_{xa}}{S} = 1 - \frac{h_a}{h_b}\,e^{\mu(H_a - H_b)} \tag{6.18}$$

6.4 Determination of drawing-load for cylindrical rod, from local stress evaluation

The method of stress evaluation can be applied to conditions of axial symmetry in the same way as to plane-strain drawing.

6.4.1 *Cylindrical-rod drawing, with a conical die.* (α, μ, Y constant)

The equilibrium of a small element in the working zone is again considered. Figure 6.2 shows the stresses acting on a thin frustum, at a distance x from the virtual apex of the conical die.

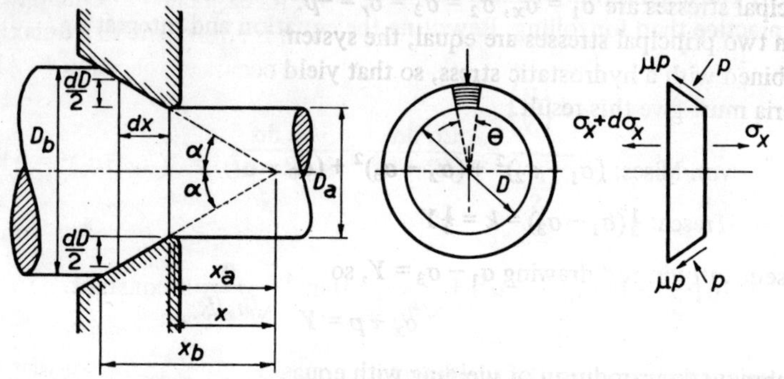

Figure 6.2 A diagram showing the stresses acting on an element of rod drawn through a conical die.

There are three force components acting in the axial direction:

due to the change in longitudinal stress,

$$(\sigma_x + \mathrm{d}\sigma_x)\frac{\pi}{4}(D + \mathrm{d}D)^2 - \sigma_x\frac{\pi}{4}D^2;$$

due directly to the die pressure on the circumference,

$$p\left(\pi D \cdot \frac{\mathrm{d}x}{\cos\alpha}\right)\sin\alpha;$$

due to the frictional drag at the circumference,

$$\mu p\left(\pi D\,\frac{\mathrm{d}x}{\cos\alpha}\right)\cos\alpha.$$

Under steady drawing conditions, these must be in equilibrium. Thus, ignoring the products of infinitesimals,

$$\frac{\sigma_x D\,\mathrm{d}D}{2} + \frac{D^2\,\mathrm{d}\sigma_x}{4} + p\,D\,\mathrm{d}x\tan\alpha + \mu p\,D\,\mathrm{d}x = 0$$

Since $\mathrm{d}D = 2\mathrm{d}x\tan\alpha$, this may be written

$$2\sigma_x\,\mathrm{d}D + D\,\mathrm{d}\sigma_x + 2p\,\mathrm{d}D + 2\mu p\,\mathrm{d}D\cot\alpha = 0$$

$$D\,\mathrm{d}\sigma_x + 2[\sigma_x + p(1 + \mu\cot\alpha)]\,\mathrm{d}D = 0 \qquad (6.19)$$

Radial equilibrium gives

$$\sigma_r(\pi D\,\mathrm{d}x) = -p\left(\pi D\frac{\mathrm{d}x}{\cos\alpha}\right)\cos\alpha + \mu p\left(\pi D\frac{\mathrm{d}x}{\cos\alpha}\right)\sin\alpha$$

$$\sigma_r = -p(1-\mu\tan\alpha)$$

It is usually permissible to ignore $\mu\tan\alpha$ in comparison with unity. Typical values are $\mu = 0{\cdot}05$, $\alpha = 6^\circ$, $\mu\tan\alpha = 0{\cdot}005$. Then the state of stress is cylindrical, and the principal stresses are $\sigma_1 = \sigma_x$, $\sigma_2 = \sigma_3 = \sigma_r = -p$. As explained in Chapter 3, §3.6.3, when two principal stresses are equal, the system is equivalent to a uniaxial stress combined with a hydrostatic stress, so that yield occurs at the value Y. All yield criteria must give this result.

von Mises: $(\sigma_1-\sigma_2)^2 + (\sigma_2-\sigma_3)^2 + (\sigma_3-\sigma_1)^2 = 2(\sigma_1-\sigma_3)^2 = 2Y^2$

Tresca: $\frac{1}{2}(\sigma_1-\sigma_3) = k = \frac{1}{2}Y$

Consequently in rod drawing $\sigma_1 - \sigma_3 = Y$, so

$$\sigma_x + p = Y \tag{6.20}$$

Combining this condition of yielding with equation 6.19, and again writing $B = \mu\cot\alpha$,

$$\frac{\mathrm{d}\sigma_x}{B\sigma_x - Y(1+B)} = 2\frac{\mathrm{d}D}{D} \tag{6.21}$$

This basic differential equation may be integrated directly if B and Y are constant. (Compare with equation 6.6 for strip-drawing.)

$$\frac{1}{B}\ln\left[B\sigma_x - Y(1+B)\right] = 2\ln D + \text{constant}$$

$$B\sigma_x - Y(1+B) = cD^{2B}$$

The constant of integration is found from the assumption that there is no longitudinal stress at the entry, as in single-hole drawing with no back-pull.

$$\sigma_x = \sigma_{xb} = 0,\ D = D_b$$

$$c = -Y(1+B)/D_b^{\,2B}$$

Thus

$$\frac{\sigma_x}{Y} = \frac{1+B}{B}\left[1-\left(\frac{D}{D_b}\right)^{2B}\right] \tag{6.22}$$

The drawing stress σ_{xa} is given by

$$\frac{\sigma_{xa}}{Y} = \frac{1+B}{B}\left[1-\left(\frac{D_a}{D_b}\right)^{2B}\right] \tag{6.23a}$$

This resembles the equation 6.7 for strip drawing, and the resemblance is still closer in terms of the reduction of area,

$$r = \frac{\pi}{4}(D_b^2 - D_a^2)\Big/\frac{\pi}{4}D_b^2;\ r = 1 - \left(\frac{D_a}{D_b}\right)^2$$

$$\frac{\sigma_{xa}}{Y} = \frac{1+B}{B}[1-(1-r)^B] \tag{6.23b}$$

This is the same as equation 6.7, except that the appropriate yield stress, Y for rod and S for strip, must be used.

6.4.2 *Frictionless drawing of cylindrical rod.* (Y *constant*)

If the coefficient of friction is zero, the parameter B is also zero, and equation 6.23 cannot be used. It is then necessary to revert to the differential equation 6.19, which becomes, with the condition $\mu = 0$,

$$D\,d\sigma_x + 2(\sigma_x + p)\,dD = 0$$

Eliminating p by use of the condition of yielding, equation 6.20

$$D\,d\sigma_x + 2Y\,dD = 0$$

This may be integrated if Y is constant:

$$\frac{\sigma_x}{Y} = -2\ln D + \text{constant}$$

At entry, $\sigma_x = \sigma_{xb} = 0,\ D = D_b$, so

$$\frac{\sigma_x}{Y} = \ln\left(\frac{D_b}{D}\right)^2$$

The drawing stress is thus

$$\frac{\sigma_{xa}}{Y} = \ln\left(\frac{D_b}{D_a}\right)^2 = \ln\frac{1}{1-r} \tag{6.24}$$

This result was also obtained in §6.2, equation 6.3, as the contribution of homogeneous plastic-deformation. This is to be expected, since the stress-evaluation method takes no account of redundant work but does allow for the homogeneous deformation and the friction. It may be noticed that equation 6.24 is independent of the die angle. Indeed, the dies might equally well be curved, provided that there is perfect lubrication.

6.4.3 *Allowance for strain-hardening in rod-drawing*

The normal procedure is to assume a mean value of the yield stress as in §6.3.2 based on the area of the stress-strain curve, but a more sophisticated approach, using an analogue or digital computer gives more reliable results for annealed metal, or for any conditions in which the strain-hardening is significantly large.

6.4.4. *Maximum reduction of area per pass, in rod-drawing*

A limit is set in cold-drawing by tensile failure of the drawn section. If there is no strain-hardening, this will occur at a stress equal to the yield stress of the drawn bar. In practice, a bar drawn to the maximum reduction in one pass will be so heavily strain-hardened that its maximum tensile stress (Chapter 2, §2.1.4) will be very nearly equal to its yield stress, so this assumption is a valid one. The limiting reduction is thus obtained using equation 6.23b.

$$\frac{\sigma_{xa}}{Y} = 1 = \frac{1+B}{B}[1-(1-r)^B]$$

For example, using the values selected for a similar calculation with strip (Chapter 4, §4.4.2)

$$\mu = 0{\cdot}05,\ \alpha = 15°,\ B = \mu \cot \alpha = 0{\cdot}1865$$

$$1 = \frac{1{\cdot}187}{0{\cdot}1865}[1-(1-r)^B]$$

$$(1-r)^{0{\cdot}1865} = 1 - 0{\cdot}158;\ 1 - r = 0{\cdot}398$$

$$r \approx 60\% \tag{6.25}$$

This is a rather heavier reduction than was found for the limit in strip-drawing of non-hardening material, 55% (equation 4.24) because of the elastic constraint increasing the effective yield stress during drawing in plane strain (equation 3.27).

The limit for frictionless rod drawing is found from equation 6.24.

$$\frac{\sigma_{xa}}{Y} = 1 = \ln\frac{1}{1-r}$$

$$\frac{1}{1-r} = e^{1{\cdot}0} = 2{\cdot}72;\ 1 - r = 0{\cdot}368$$

$$r \approx 63\%$$

Thus the friction found in typical drawing operations does not very greatly influence the theoretical maximum reduction of area per pass. Because the metal strain-hardens, the yield stress of the drawn bar will exceed the mean value of the yield stress, which increases the possible reduction slightly. On the other hand, the contribution of redundant work increases the drawing stress but it is a small factor for normal die angles at these heavy reductions. The practical limit is in the vicinity of 60%, but passes are usually restricted to appreciably smaller reductions, often 35–45%, because of the deterioration in lubrication in very heavy passes and the consequent danger of pickup of drawn metal on the die.

6.5 Slip-line field solution for plane-strain frictionless drawing, with wedge-shaped dies (α constant)

The stress-evaluation method gives values of the drawing stress and the die pressure which include the effects of friction. Under good lubrication conditions, this may amount to 10 to 20%. For example, with $\alpha = 15°$, $\mu = 0{\cdot}05$ as before, the drawing stress for 40% reduction is

$$\frac{\sigma_{xa}}{S} = \frac{1+B}{B}[1-(1-r)^B] \qquad (6.7)$$

$$= \frac{1{\cdot}187}{0{\cdot}1865}[1-(0{\cdot}6)^{0{\cdot}1865}]$$

$$= 6{\cdot}34(1-0{\cdot}908) = \underline{0{\cdot}584}$$

This may be compared with the frictionless drawing stress for the same pass

$$\frac{\sigma_x}{S} = \ln\frac{1}{1-r} = \ln\frac{1}{0{\cdot}6} = \underline{0{\cdot}513}$$

When heavy reductions of area are made using low-angle dies, these equations are reasonably accurate, but for lighter reductions, particularly with steeper dies, the experimental drawing stresses are appreciably higher than the predicted results, as shown in Figure 6.3.

This is due to the redundant work performed in redundant shearing of an element of the strip, as explained in Chapter 4, §4.7. The contribution of redundant work to the drawing stress or the die pressure can be found accurately by use of slip-line fields. The elements of slip-line field theory have been discussed in Chapter 5.

Construction of the network can be facilitated by the use of Prager's stress-plane cycloid technique[6.3], but this will not be necessary for the comprehension of problems considered in this chapter. For many purposes likely to be encountered in practice, slip-line fields have already been constructed and verified. It is recommended that one of these should be selected, even if it does not exactly represent the actual conditions. Construction of a slip-line field for an unsolved problem is a trial-and-error process which requires experience and intuition, and can take a long time. Failing any known field which approximates to the operation, the upper-bound technique is a quicker approach and may give sufficient accuracy.

A velocity solution, or *hodograph*, is completed at the same time as the slip-line field, starting with the given velocity boundary-conditions. If the slip-line field is a valid one, this velocity solution must be compatible with all the boundary conditions affecting the plastic zone. If this is not so, the original choice of field must be modified. In addition, it is necessary for the plastic work to be positive everywhere, and the associated stress distribution in the rigid metal adjacent to the plastic zone should not show any inconsistencies with the yield criterion. It can, however, usually be assumed with reasonable confidence that the latter conditions will be satisfied if the problem approximates to one of the solutions which have now become standard forms, but the velocity conditions should not be ignored.

6.5.1 *Simple slip-line field solutions for frictionless strip-drawing,* $[r = 2\sin\alpha/(1+2\sin\alpha)]$

The reason for choosing this particular reduction will become apparent when the field has been drawn. To construct the field, we consider first the boundary condi-

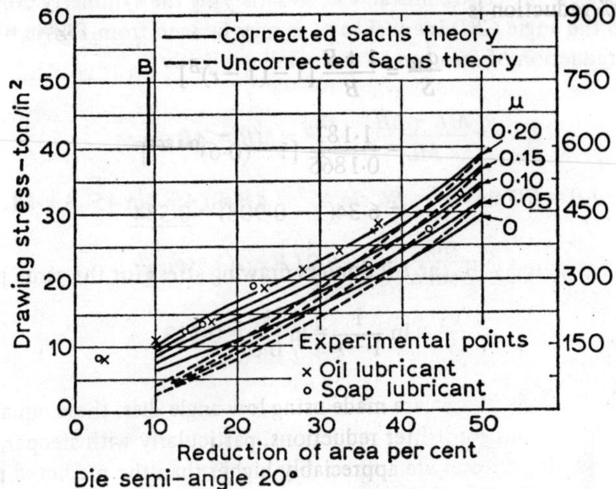

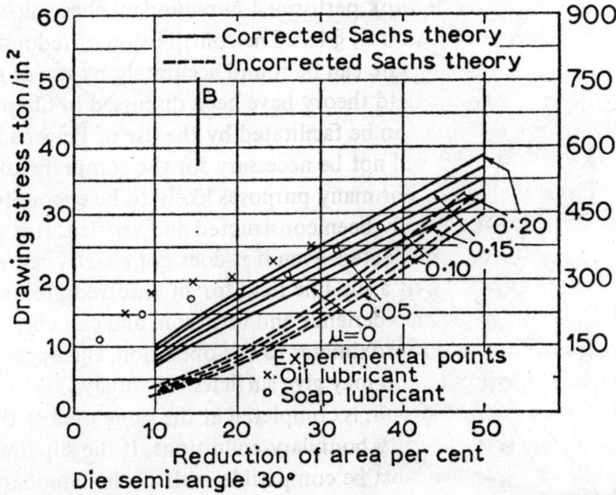

Figure 6.3 The variation of drawing stress with reduction of area for wide steel strips drawn through 20° and 30° dies[6.2]. The line *B* represents the predicted onset of bulging. (150 N/mm² ≃ 10 tons/in²)

tions of stress. It is assumed that there is no friction at the die face, so the slip lines must meet the die face at 45°, to ensure that their resolved components exactly cancel along the interface (see also Chapter 5, §5.5). This suggests that the field starts as in Figure 6.4*a*.

It is clear that when a moderate reduction of area is made, the zone of plastic deformation must reach right across the strip. As was seen in considering punch indentation (Chapter 5, §5.6.5), the field is extended by constructing radial fans centred on the two singularities *A* and *B*. This procedure is quite generally applicable, but the simplest field is obtained if the fan centred on *B* can be drawn so that the line DB meets the centre-line at 45°, so satisfying the symmetry condition. This occurs when the angle CBD is equal to α, as may be seen from Figure 6.4*c*. The corresponding reduction of area can be evaluated explicitly,

$$r = \frac{ML - LB}{ML}, \quad \frac{rh_b}{2} = MB = AB \sin \alpha$$

$$LB = BD \sin 45° = BC \sin 45° = AB \sin 45° \sin 45° = \tfrac{1}{2}AB.$$

$$\therefore \quad \frac{rh_b}{2 \sin \alpha} = AB = 2LB = 2(1 - r)\frac{h_b}{2}.$$

$$\therefore \quad r = \frac{2 \sin \alpha}{1 + 2 \sin \alpha}$$

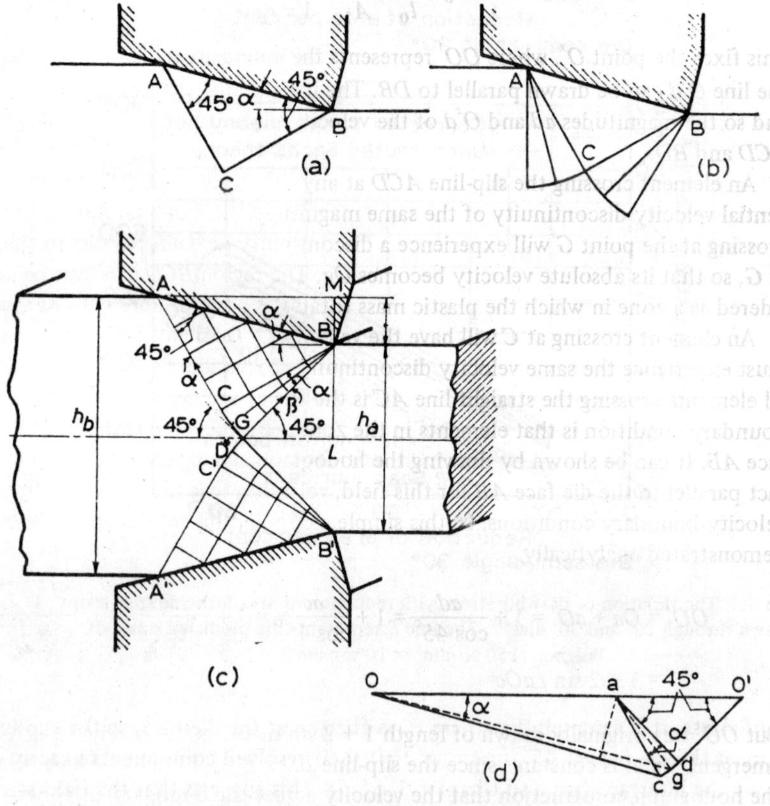

Figure 6.4 Stages in the construction of a slip-line field for strip drawing with reduction of area $r = 2 \sin \alpha/(1 + 2 \sin \alpha)$

(*a*) Slip-lines meeting the frictionless die at 45°.
(*b*) Extension by radial fans centred on *A* and *B*.
(*c*) Complete field for this special reduction.
(*d*) The hodograph.

The validity of this field is checked by constructing the hodograph (Figure 6.4*d*). Before entering the deformation zone, the billet is assumed to have an axial velocity equal to unity, represented on the hodograph by a horizontal line *Oa* of unit length, measured from *O* in the direction of drawing.

An element of metal crossing the slip-line *ACD* in the vicinity of *D* experiences a sudden change in velocity, produced by a discontinuity in the component of its velocity parallel to the slip-line at the point of crossing. Thus, immediately adjacent to *D* the discontinuity is parallel to the tangent at *D*, that is, at 45° to the axis. This is represented on the hodograph by a line drawn from *a* in the direction *ad*. When the element has passed through the plastic zone, it emerges across the slip line *BD*, experiencing a second velocity discontinuity parallel to *DB*, which brings its absolute velocity back to the longitudinal direction. The final velocity is defined by the reduction of area, since the volume remains constant.

$$v = \frac{l_1}{l_0} = \frac{A_0}{A_1} = \frac{1}{1-r}$$

This fixes the point O', where OO' represents the emergent velocity of the strip, and the line $O'd$ can be drawn parallel to *DB*. The intersection determines the point *d*, and so the magnitudes *ad* and $O'd$ of the velocity discontinuities on the slip-lines *ACD* and *BD*.

An element crossing the slip-line *ACD* at any other point must experience a tangential velocity discontinuity of the same magnitude, *ad*. For example, the element crossing at the point *G* will experience a discontinuity $ag = ad$, parallel to the tangent at *G*, so that its absolute velocity becomes *Og*. The region *BCD* can thus be considered as a zone in which the plastic mass rotates as a whole about the centre *B*.

An element crossing at *C* will have the velocity *Oc*, and any element crossing *AC* must experience the same velocity discontinuity *ac*, parallel to *AC*. The velocity of all elements crossing the straight line *AC* is the same, and is given by *Oc*. The physical boundary condition is that elements in the zone *ABC* must move parallel to the die face *AB*. It can be shown by drawing the hodograph accurately that the line *Oc* is in fact parallel to the die face *AB* for this field, verifying that the field conforms to the velocity boundary conditions. In this simple example, this result can easily be demonstrated analytically.

$$OO' = Oa + aO' = 1 + \frac{ad}{\cos 45^\circ} = 1 + \frac{ac}{\cos 45^\circ} = 1 + \frac{1 \,.\, \sin \angle aOc}{\cos 45^\circ} \cdot \frac{1}{\cos 45^\circ}$$

$$= 1 + 2 \sin \angle aOc \tag{6.26}$$

But OO' was originally drawn of length $1 + 2 \sin \alpha$, so $\angle aOc = \alpha$. The velocity of the emergent metal is constant, since the slip-line *BD* is straight, and we have shown in the hodograph construction that the velocity across the boundary slip-line *ACD* is everywhere compatible with unit initial velocity, so all the velocity boundary conditions are satisfied. Strictly speaking, it is necessary also to verify that the rate of plastic working is everywhere positive within the deformation zone, and that the stress required in the rigid regions is admissible. There are methods available for these verifications, but it may reasonably be assumed for general purposes that, if the velocity conditions are satisfied and the field resembles one of the many proven fields now available, the field is a valid one. Errors can arise when these checks are ignored, for example in predicting the dead zone in extrusion, but the risk is small.

The stress solution for this field is particularly simple for extrusion (Chapter 8, §8.3.1), because there is then no applied stress on the emergent strip, so that $\sigma_3 = 0$, on the straight line *BD*, which provides a simple starting-point. The procedure is the same for drawing, except that the principal stress σ_3 acting on the exit slip-line is not zero, but is equal to the tensile drawing stress $+ t_x$. The boundary condition is thus

$$\sigma_3 = +t_x; \; -p_D + k = t_x \tag{6.27}$$

The hydrostatic pressure $-p_D$ can no longer be determined immediately, but it can be eliminated by considering the die pressure $-q$, which is derived in terms of $-p_D$, the pressure at *D* or at any point on the straight slip-line *BD*. The latter is a β slip-line because σ_3 is algebraically greater than the compressive stress σ_1. The angle $\angle CBD$ is equal to α, so the Hencky equation for clockwise rotation of the α-line between *D* and *C* gives the pressure p_C at any point on the straight slip-line *BC*

$$p_C + 2k(-\alpha) = \text{constant} = p_D$$

$$p_C = p_D + 2k\alpha$$

The slip-line is straight from *C* to *A*, so the pressure p_A at the die face is equal to p_C. The principal stress σ_1 acting on the die face (which we may call $-q$, the die pressure) is thus

$$\sigma_1 = -q = -(p_C + k) = -(p_D + k + 2k\alpha) \tag{6.28}$$

The lines in the zone *ACB* are parallel so the pressure is uniform over *AB*.

The longitudinal component of the force acting on both dies, assuming unit width, is

$$Q_L = 2(qAB) \sin \alpha$$

Under steady drawing conditions this must be equal to the drawing force *T*. It is convenient to express both in terms of the reduction of area:

$$r = \frac{h_b - h_a}{h_b}; \; rh_b = h_b - h_a = 2AB \sin \alpha$$

$$T = t_x h_a = t_x(1 - r)\, h_b$$

$$Q = 2qAB \sin \alpha = qrh_b$$

Thus

$$t_x = q\,\frac{r}{1-r}; \; t_x = (t_x + q)r \tag{6.29}$$

From equations 6.27 and 6.28,

$$p_D = k - t_x = q - k(1 + 2\alpha)$$

$$q + t_x = 2k(1 + \alpha)$$

Combining this with equation 6.29, for the chosen reduction of area,

$$\left.\begin{aligned} \frac{t_x}{2k} &= r(1+\alpha) = \frac{2(1+\alpha)\sin\alpha}{1+2\sin\alpha} \\ \frac{q}{2k} &= (1+\alpha) - \frac{t_x}{2k} = \frac{1+\alpha}{1+2\sin\alpha} \end{aligned}\right\} \quad (6.30)$$

These explicit values can be obtained for only one reduction, such that BD is straight. For 15° semi-angle, this reduction is 34%. Then

$$\frac{t_x}{2k} = r(1+\alpha) = 0{\cdot}34\,(1 + 0{\cdot}262) = 0{\cdot}429$$

The stress required to produce homogeneous deformation to 34% is given by equation 6.3, making allowance for the increased yield stress in plane strain,

$$\frac{\sigma}{2k} = \ln\frac{1}{1-r} = \ln\frac{1}{0{\cdot}66} = 0{\cdot}415$$

The contribution of redundant work is thus small, and can be neglected. If lower die angles are used the redundant work contribution at 34% reduction is even less. It is therefore of greater interest to consider the more complex slip-line field solution for lower reductions or higher die angles, where the redundant work is appreciable.

When the reduction of area is less than $2 \sin\alpha/(1 + 2\sin\alpha)$, it is necessary to extend the field further, from the fans centred on A and B.

6.5.2 *The slip-line field for reductions less than* $2\sin\alpha/(1 + 2\sin\alpha)$. *Construction for extending slip-line fields from radial fans*

The field may conveniently be extended by a graphical method[6.4]. The basic arcs are divided into suitable equal intervals by radii from A and B, as in Figure 6.5. It is usual to choose 5° intervals, which give good accuracy, but the explanation can be given more simply in terms of a 10° net.

The slip-line BC_1 must rotate through 10° in passing from C_1 to the point 6. This is easily shown from the Hencky equations and is known as Henckys' 1st Theorem (see Example 5.3). The slip-line AC'_1 must similarly rotate through 10° between C'_1 and the point 6. The position of 6 is thus found accurately by drawing a line $C_1 6$ from C_1, at half this angle (5°) to the radius BC_1, to represent the chord of the arc $C_1 6$. A chord is similarly drawn from C'_1 at 5° to AC'_1. The intersection of these lines gives the position of 6. The next mesh is started with a line from C_2 at 5° to BC_2, showing the direction to 5. This intersects the chord starting at 6 in a direction at 10° to $C'_1 6$, because each chord at 6 will be at 5° to the tangent, giving the position of 5. The field is thus built up in unit cells. Finally, smooth orthogonal curves are drawn through the points of intersection. The accuracy is obviously improved if a

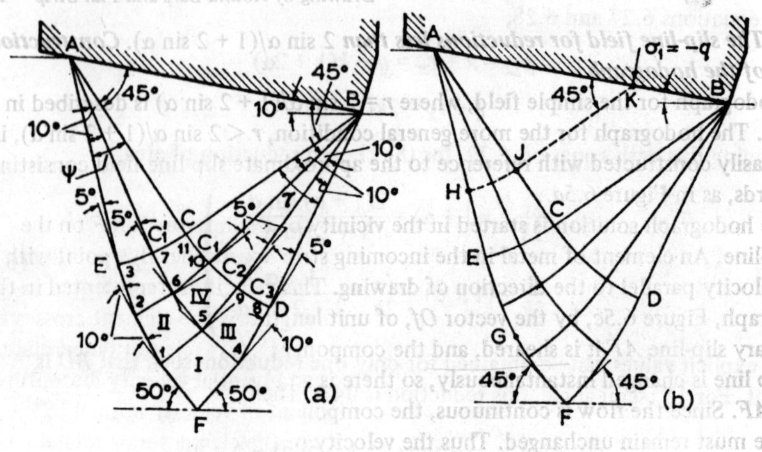

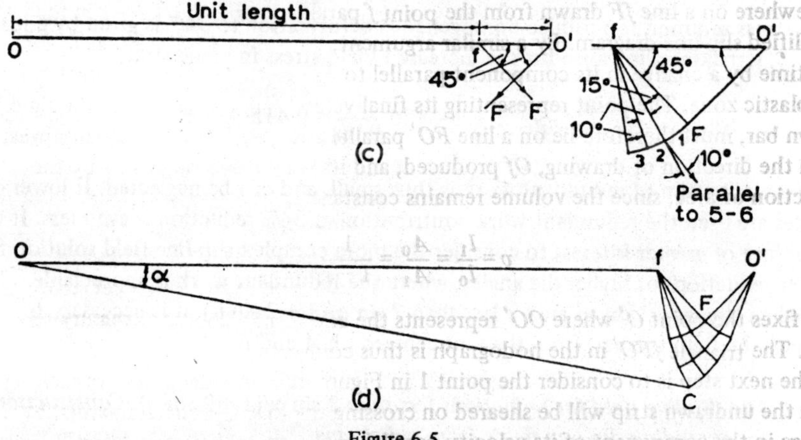

Figure 6.5

(*a*) A 10° network constructed by the graphical chord method, starting with the radial fans centred on *A* and *B*.
(*b*) The final smoothed slip-line field. (*c*), (*d*) The hodograph.

network with smaller intervals is chosen initially. This type of field, based on equal or unequal arcs, is frequently encountered in slip-line fields for metalworking problems.

In the drawing problem, the extent of the field based on the arcs *CD* and *CE* is determined by the condition that the slip lines must intersect the centre-line at 45°, so that there is no resultant shear stress on the axis of symmetry. This condition fixes the angles ζ and ψ, and the field is then complete. It remains to check the compatibility of this solution with the velocity boundary conditions. This may be done analytically, but it is easier to construct the hodograph, following the procedure of Chapter 5, §5.9, by considering the directions of motion of an element of the strip as it passes through the deformation zone depicted by the slip-line field.

6.5.3 *The slip-line field for reductions less than* 2 sin $\alpha/(1 + 2 \sin \alpha)$. *Construction of the hodograph*

The hodograph for the simple field, where $r = 2 \sin \alpha/(1 + 2 \sin \alpha)$ is described in §6.5.1. The hodograph for the more general condition, $r < 2 \sin \alpha/(1 + 2 \sin \alpha)$, is most easily constructed with reference to the approximate slip-line field consisting of chords, as in Figure 6.5*a*.

The hodograph solution is started in the vicinity of the intersection *F* on the centre-line. An element of metal in the incoming strip approaches this point with unit velocity parallel to the direction of drawing. This velocity is represented in the hodograph, Figure 6.5*c*, by the vector *Of*, of unit length. As the element crosses the boundary slip-line *AF* it is sheared, and the component of its velocity tangential to the slip line is changed instantaneously, so there is a tangential velocity discontinuity along *AF*. Since the flow is continuous, the component of velocity normal to the slip-line must remain unchanged. Thus the velocity of the element immediately after crossing the slip line near *F* must be represented in the hodograph by a point lying somewhere on a line *fF* drawn from the point *f* parallel to the chord *F*–1 in the simplified slip-line diagram. By a similar argument, the velocity will change again, this time by a change in its component parallel to *FD*, as the element emerges from the plastic zone. The point representing its final velocity, equal to that of the rigid drawn bar, must therefore lie on a line *FO*′ parallel to *F*–4. The final velocity must be in the direction of drawing, *Of* produced, and its magnitude is defined by the reduction of area, since the volume remains constant:

$$v = \frac{l_1}{l_0} = \frac{A_0}{A_1} = \frac{1}{1 - r}$$

This fixes the point *O*′ where *OO*′ represents the emergent velocity of the drawn strip. The triangle *fFO*′ in the hodograph is thus completed.

The next step is to consider the point 1 in Figure 6.5*a*. An element approaching from the undrawn strip will be sheared on crossing the line 1–2, accompanied by a change in the component of its velocity parallel to the chord between 1 and 2. This is represented in the approximate hodograph by the line *f*–1 drawn from the point *f*, parallel to 1–2. The magnitude of the velocity change must be the same as that represented by *fF*, since it occurs on the same slip-line *AF*, and the tangential velocity discontinuity must remain constant along a slip-line. The length *f*–1 is thus equal to the length *f*–*F*. This fixes the point 1 in the hodograph. As the element leaves this mesh II of the network, it passes into the mesh I, where it must have the velocity represented by *O*–*F*. It will be recalled that the postulated approximate slip-field consists of lines of shear between which the 'blocks' are rigid. The velocity of any one 'block' represented by a mesh of the slip-line network, is therefore the same at all points enclosed by the bounding lines. Consequently the element passing from II to I across the chord 1–5 will suffer a change in velocity parallel to 1–5, to bring its velocity to that of mesh I. The line 1*F* must of course close the triangle since the velocity of mesh I represented by *OF* cannot have two different values.

The same procedure is followed for an element approaching the point 2 and crossing the line 2–3. This adds the triangle *f*12 to the hodograph, building up a fan of equal radial segments, centred on *f*. It should be noticed that the number of segments is equal to the number of intersections along AF, and thus equal to the number of segments in the fan centred on B in the slip-line field, not the fan centred on A.

A similar section of the hodograph is built up round O' in the same way, by considering the efflux of elements across BF. The number of segments in the fan centred on O' in the hodograph is equal to that in the radial slip-line fan centred on A.

The next point to consider is 5, where an element of metal from the mesh II crosses the line 5–6 into IV. The velocity in IV must be determined in the hodograph by a line from 1 drawn parallel to 5–6. On crossing the slip line 5–9 into mesh III, its velocity must revert to that represented in the hodograph by O–4. The velocity change across 5–9 in the slip-line field is thus represented by a line in the hodograph drawn from 4 parallel to 5–9, completing the next mesh of the hodograph.

In this way, the hodograph is built up, mesh by mesh. When it is completed, the intersections are joined by smooth curves instead of the chords, except of course where the radial lines such as fF are known to be straight. As in the straight-line technique for constructing approximate slip-line fields, the accuracy of the final field depends on the size of angular interval chosen.

It will be seen that the slip-line field and the hodograph (Figure 6.5*a* and *d*) bear a strong resemblance to each other. The hodograph is in fact always orthogonal to the slip-line field. The purpose of drawing the hodograph is to determine the conformity of the slip-line field solution to the velocity boundary conditions. In this example, the exit velocity was used in constructing the field. If the solution is correct, the velocity represented by the vector OC should be parallel to the die face, since any element in the zone ABC must move parallel to the die face.

6.5.4 *The slip-line field for reductions less than* $2 \sin \alpha/(1 + 2 \sin \alpha)$. *Stress-determination from the slip-line field*

When it has been shown that the chosen field is compatible with the boundary conditions of both stress and velocity, the stresses in the plastic zone may be evaluated using the Hencky equations, as in §6.5.1 and in the simple examples of Chapter 5.

The slip-line field shown in Figure 6.5*b* is more complex than that of Figure 6.4. Neither entry nor exit boundary is straight, so the hydrostatic pressure varies along both boundaries of the plastic zone. The analytical solution is reasonably straightforward but rather tedious, and a better method for use in actual calculations is given in the §6.5.5. The analysis does, however, illustrate a common procedure in general plasticity theory.

The hydrostatic pressure p_F at the intersection can be found from the condition that there is no back tension, or that the back tension has a known, finite, value. In the absence of back-tension, the average longitudinal stress acting over the rearward boundary must be equal to zero for steady drawing.

The stress at any point on AEF may be found from the Hencky equation in terms of the compressive stress p_F at F, on the centre-line. AEF is an α-line because the longitudinal stress σ_3, though zero, is algebraically greater than σ_1 which is compressive. The appropriate Hencky equation is

$$p + 2k\phi = \text{constant} \tag{5.7a}$$

At some point G on AEF, the slip-line has rotated clockwise by an angle ϕ' from its direction at F. In the Hencky equations, as in the Mohr circle, angles are measured anticlockwise, so the hydrostatic pressure at G is

$$p_G + 2k(-\phi') = p_F; \; p_G = p_F + 2k\phi' \tag{6.31}$$

Because the slip-line is curved, the direction of the principal stress will rotate continuously. The shear stress k on the slip line contributes a component $(k\ \mathrm{d}s)\cos\left(\frac{\pi}{4}+\phi'\right)$ to the longitudinal stress. The stress normal to the slip-line is equal to p_G and contributes a component $(p_G\ \mathrm{d}s)\sin\left(\frac{\pi}{4}+\phi'\right)$ in the opposite direction. Thus the total force acting over the boundary AEF, which must be equal to zero in the absence of back tension, is

$$\int_0^h \sigma_x dy = \int p_G\ \mathrm{d}s \sin\left(\frac{\pi}{4}+\phi'\right) - k\int \mathrm{d}s \cos\left(\frac{\pi}{4}+\phi'\right) = 0 \tag{6.32}$$

$$\int (p_F + 2k\phi') \sin\left(\frac{\pi}{4}+\phi'\right) \mathrm{d}s - k\int \cos\left(\frac{\pi}{4}+\phi'\right) \mathrm{d}s = 0$$

A value is thus found for the pressure p_F at the starting point, and from this the absolute value of the pressure at any point on AEF. The slip-line AEF is straight along the portion AE, so the pressure remains uniform in this region. An element of metal at a point H between A and E thus experiences a pressure p_E. Between H and J the β slip-line rotates by an angle ψ anticlockwise, so the pressure increases, according to the Hencky equation for a β-line (5.7b).

$$p_J - 2k\psi = p_E$$

This pressure is uniform along the straight slip-line AC. The slip-line JK parallel to BC is also straight, so this pressure also acts on the die face AB. The principal stress σ_1, acting on the die, will be normal to the die because the slip-lines meet the frictionless interface at 45°, and is thus equal to the die pressure $-q$:

$$-q = (\sigma_1)_K = p_K + k = p_J + k = p_E + 2k\psi + k$$

The angle through which the α slip-line AEF rotates clockwise between F and E is equal to ζ, the angle of the fan centred on B, because the network is orthogonal, so

$$p_E + 2k(-\zeta) = p_F$$

Thus
$$q = p_F + 2k\zeta + 2k\psi + k \tag{6.33}$$

The angles ζ and ψ must satisfy the relationship

$$\zeta - \psi = \alpha \tag{6.34}$$

because the slip-lines intersect the centre-line at 45°. This may be seen by considering the rotation of the slip-lines. The tangent to FB at F makes an angle 45° with the axis. The tangent to EB at E thus makes an angle $\left(\frac{\pi}{4}-\zeta\right)$ to the axis. The tangent to CB at C makes an angle $\left(\frac{\pi}{4}-\zeta+\psi\right)$, but CB is a straight line at 45° to the die and so makes an angle $\left(\frac{\pi}{4}-\alpha\right)$ with the axis. Thus $\frac{\pi}{4}-\zeta+\psi=\frac{\pi}{4}-\alpha$, confirming the relationship 6.34.

This solution may be checked for a simple example, where $\psi = 0$. Then $\zeta = \alpha$ and

$$q = p_F + 2k\alpha + k \tag{6.35}$$

The condition $\psi = 0$ occurs in fact at the reduction $r = 2 \sin \alpha/(1 + 2 \sin \alpha)$. Equation 6.35 gives the same value as was found in §6.5.1, by considering this reduction explicitly (equation 6.28). It is also the same value as is found for frictionless extrusion with this reduction (Chapter 8, equation 8.14), but in the extrusion example, p_F was found to be equal to k. The die pressure q is not always uniform, as is shown for heavy reductions in extrusion (§8.3.3).

Values of p_F and q were first obtained by Hill and Tupper(6.5). The drawing stress is found by equating the drawing force to the longitudinal component of the die pressure, as in §6.5.1

$$t_x = q\frac{r}{r-1} \tag{6.29}$$

For example, if $r = 0{\cdot}1$ and $\alpha = 15°$ the predicted drawing stress is

$$\frac{t_x}{2k} = 0{\cdot}21$$

This may be compared with the homogeneous-deformation stress,

$$\frac{\sigma}{2k} = \ln 1{\cdot}11 = 0{\cdot}104$$

This is actually a larger correction than would be encountered in practical drawing conditions, because die angles steeper than 15° are unfavourable also from the point of view of lubrication. Reductions of area appreciably less than 10%, with 15° dies, are insufficient to cause plastic deformation across the whole section. The dies then act separately, and the slip-line field is more like that for indentation of a very thick block (Chapter 5, §5.6.5). The constraint is then very large, the die pressure rises, and the metal tends to be displaced laterally, causing a bulge in the surface before the strip enters the die, rather like the pile-up of metal round an indentation. The process can also be likened to machining with a very high, negative, rake angle. This bulging occurs when the die pressure reaches a value

$$q = 2k\left(1+\frac{\pi}{2}-\alpha\right) \text{ at } r = \alpha\left(0{\cdot}23+\frac{\alpha}{9}\right) \tag{6.36}$$

The slip-line field described above is not valid beyond the bulging limit. Bulging is observed experimentally at about the predicted stage, but it is detrimental to the lubrication and to the strip itself, and drawing should not be performed under these conditions.

Between these limits of bulging, on the one hand, and negligible redundant work on the other, the theory gives remarkably good agreement with experiments conducted under the best available conditions of lubrication. The calculation given above is reasonably straightforward and has been cited at length to demonstrate the procedure, but it is rather tedious. It can be circumvented by using results obtained by Hill and Green.

6.5.5 *Redundant-work factor in terms of geometrical parameters*

Hill and Green[6.6] developed the analogy between plane-strain compression, or forging, and plane-strain drawing. We have seen that the pressure required for plane-strain compression between parallel platens depends on the ratio of strip-thickness h to platen-breadth b (Chapter 5, §5.6).

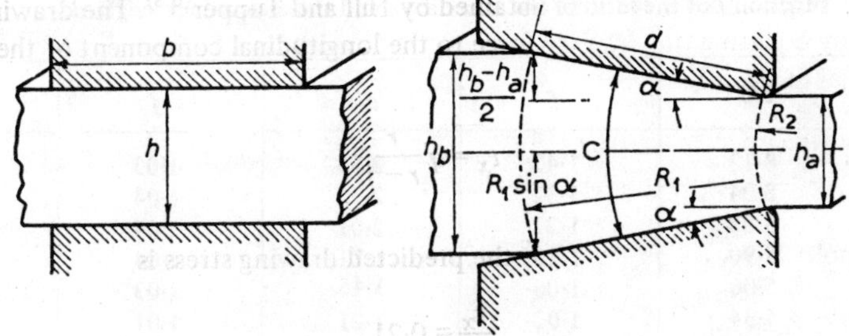

Figure 6.6 The geometrical analogy between plane-strain compression and plane-strain drawing.

Figure 6.6 shows how these parameters correspond to the dimensions c and d when inclined platens are used. The ratio c/d is easily found:

$$c = \frac{R_1 + R_2}{2} \,.\, 2\alpha, \; d = R_1 - R_2; \; \frac{c}{d} = \frac{R_1 + R_2}{R_1 - R_2} \,.\, \alpha$$

$$r = \frac{R_1 \sin\alpha - R_2 \sin\alpha}{R_1 \sin\alpha}; \; R_1 r = R_1 - R_2$$

$$\frac{c}{d} = \frac{2-r}{r} \,.\, \alpha \tag{6.37}$$

Hill and Green showed that the redundant-work correction in drawing is a function of c/d. Thus, in the absence of friction, the drawing stress is given by

$$\frac{t_x}{2k} = f\left(\frac{c}{d}\right)\sigma,$$

where σ is the stress for homogeneous deformation, so

$$\frac{t_x}{2k} = f\left(\frac{c}{d}\right) \ln \frac{1}{1-r} \tag{6.38}$$

Table 6.1 gives the values of this function from $c/d = 0{\cdot}5$ to $8{\cdot}72$. For values of c/d less than 0·5, there is no redundant work contribution. The value 8·72 corresponds to indentation by separate dies, which is the bulge limit corresponding to the maximum value of f, equal to $f\left(\frac{c}{d}\right) = 1 + \frac{\pi}{2} = 2{\cdot}57$, as in single flat-punch indentation of a semi-infinite block (equation 5.15). It is found that if c/d is less than 1·3, the function f lies between 1·0 and 1·04, so for these values the redundant work factor is negligible for most purposes. We have already seen this to be so for 30% reduction with 15° dies. With 5° dies, the redundant work can be neglected for any reduction exceeding about 15%, that is, virtually any useful reduction.

Table 6.1
The redundant-work factor for frictionless strip drawing, which is a function f of the ratio of the mean arc c and the die-contact length d.

f	c/d	f	c/d	f	c/d
2·57	8·72	1·43	2·75	1·03	0·77
2·48	8·04	1·32	2·35	1·04	0·72
2·31	6·88	1·22	2·01	1·04	0·71
2·14	5·90	1·13	1·71	1·04	0·69
1·98	5·06	1·06	1·45	1·03	0·65
1·83	4·34	1·02	1·21	1·01	0·58
1·69	3·73	1·00	1·00	1·00	0·50
1·55	3·20	1·01	0·86		

(*Courtesy of A. P. Green*[6.7] *and the Institution of Mechanical Engineers*)

6.6 Determination of plane-strain drawing stress, allowing for friction, redundant work and strain-hardening

6.6.1 *Slip-line field solution including friction*

The slip-line field shown in Figure 6.7 is very similar to that constructed for frictionless drawing in Figure 6.5. The starting condition is however modified. Instead of the slip-lines meeting the die face at 45°, their angles are determined by the condition that the resolved components of the shear forces must equal the frictional drag. This gives the angle θ between one slip-line and the die face:

$$\cos 2\theta = \frac{\mu q'}{k} \qquad (5.10)$$

where μ is the coefficient of friction, and q' the die-pressure, the superscript being used to indicate evaluation with finite friction. It is necessary to assume a value of q' before the field can be started, so the solution involves trial and error. It is convenient to assume as a starting condition that friction does not affect the die pressure, so that $q' = q$.

The pressure is again uniform over the die face, because the slip lines AC and BC are straight. For other reductions, where the mesh does not fit exactly, q' is not uniform. The average longitudinal component of the die-force is given by

$$2\bar{q}'AB \sin\alpha + 2\mu\bar{q}'AB \cos\alpha = 2\bar{q}' \cdot \frac{rh_b}{2\sin\alpha}(\sin\alpha + \mu\cos\alpha)$$

$$= \bar{q}'rh_b(1 + \mu\cot\alpha)$$

This may be equated to the drawing force T'

$$T' = t_x' \cdot h_a \cdot 1 = t'(1-r)\,h_b$$

giving

$$t_x' = \bar{q}'\,\frac{r}{1-r}(1 + \mu\cot\alpha) \tag{6.39}$$

This may be compared with equation 6.29 for frictionless drawing.

$$t_x = \bar{q} \cdot \frac{r}{1-r} \tag{6.29}$$

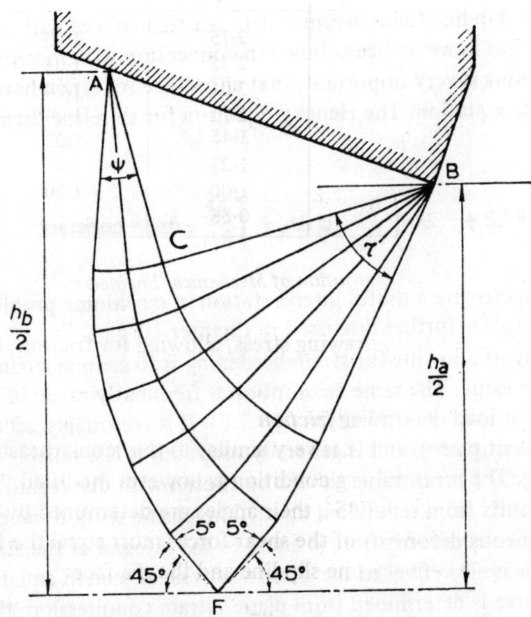

Figure 6.7 The slip-line field for strip-drawing through wedge-shaped dies with finite friction.

The full solution for q' has not been obtained, but it is found experimentally that the die pressure is not appreciably influenced by friction, and the first assumption that $\bar{q}' = \bar{q}$, is sufficiently accurate at least for well-lubricated drawing.

Then $$\left[\frac{t'}{t}\right]_x = 1 + \mu\cot\alpha \tag{6.40}$$

This provides a correction factor, to allow for the frictional contribution to the drawing stress. Thus, combining equations 6.38 and 6.40, the full solution, with friction and redundant work, is

$$\frac{t_x}{2k} = (1 + \mu \cot \alpha) f\left(\frac{c}{d}\right) \ln \frac{1}{1-r} \tag{6.41}$$

In fact, because the frictional stress contributes to yielding, the die pressure is reduced by friction, and the assumption that $\overline{q}' = \overline{q}$ is not strictly valid for heavy reductions and high friction. A more accurate form of equation 6.40 was obtained by Green and Hill[(6.6)]:

$$\left[\frac{t'}{t}\right]_x = (1 + \mu \cot \alpha) - \mu\,(0{\cdot}2 + 0{\cdot}08r \cot^2 \alpha) \tag{6.40a}$$

Their calculations are accurate for die angles between 5° and 15°, and for coefficients of friction less than 0·1, but should not be used outside this range.

6.6.2 *Allowance for strain-hardening. Zero friction*

The basic theory of slip-line fields assumes rigid-plastic material with a constant yield stress, $2k$. It has however been shown in connection with machining[(6.8)], where strain-hardening is very important, that allowance for strain-hardening can be made in the slip-line equation. The Hencky equation for an α-line then takes the form

$$p + 2k\,\phi - 2k \int \frac{\partial k}{\partial s_\alpha} \cdot \phi\, \mathrm{d}s_\alpha + \int \frac{\partial k}{\partial s_\beta} \cdot \mathrm{d}s_\alpha = \text{constant} \tag{6.42}$$

This method appears to give a useful interpretation in machining problems, but it is not easy to apply. This is further discussed in Chapter 10 §12.

The simplest way of allowing for strain-hardening is to assume a constant yield stress at the average value. The same assumption is frequently made in the stress-evaluation method of load determination (§6.3.2). It is reasonably accurate for second and subsequent passes, and is generally used, but is less reliable for the first pass after annealing. The mean value should be chosen with due regard to the strain-hardening which results from redundant shear. The plastic work done per unit volume, in homogeneous deformation, is equal to the integral of the stress–strain curve $\int Y \mathrm{d}\epsilon$, and this is also equal to the drawing stress, as seen in equation 6.3. If the stress–strain curve is determined from plane–strain compression, the corresponding relationship giving t_H, the drawing stress for wide strip which strain-hardens, is

$$t_H = \int_0^{\epsilon_2} 2k\, \mathrm{d}\epsilon \tag{6.43}$$

This defines the effective strain ϵ_2 for a given drawing pass. It will in general be greater than ϵ_a, the strain given by $\ln \dfrac{h_b}{h_a}$, because the drawing stress t_H is increased by the redundant-work contribution. The material has in fact been sheared more than would be predicted from the change in dimensions, and a practical measurement of the stress–strain curve of the drawn metal shows the yield stress to be greater than that corresponding to ϵ_a (see Figure 4.4*b*). Equations 6.43 and 6.38 may be compared:

$$t_H = \int_0^{\epsilon_2} 2k \, d\epsilon \tag{6.43}$$

$$\frac{t}{2k} = f\left(\frac{c}{d}\right)\int_0^{\epsilon_a} d\epsilon = f\left(\frac{c}{d}\right) \ln \frac{1}{1-r} \tag{6.38}$$

It is assumed that the mean effective strain imparted by a given drawing operation is the same whatever the strain-hardening characteristics of the metal. This has been shown to be valid in wire-drawing(6.9), except for very small reductions of area, and implies that

$$\epsilon_2 = f\left(\frac{c}{d}\right)\epsilon_a \tag{6.44}$$

The redundant-work factor can be determined experimentally by recording the stress–strain curve for the drawn metal and superimposing it on the original curve, as in Figure 4.4*b*, so that $f\left(\frac{c}{d}\right)$ can be read off directly as the ratio ϵ_2/ϵ_a. This method has been used in several investigations but it is not easy to fit the curves accurately.

If the integration of the stress–strain curve is taken between the apparent strain limits 0 and ϵ_a, it is necessary to define an increased value of $2k$, as the mean equivalent yield stress, which will give the same area as the integral of the real stress–strain curve between 0 and ϵ_2. This method does not allow for possible modification of the slip-line field by strain-hardening, but it appears from experiments that this can be neglected without appreciable error.

6.6.3 *Allowance for friction, redundant work, and strain-hardening simultaneously*

There is no rigorous theory allowing for friction, redundant work and strain-hardening simultaneously, but a useful method has been suggested by Green(6.7). It predicts drawing loads in good agreement with experimental results(6.2), though the prediction of die forces appears to be less reliable. Green calculated the correction factor for friction, assuming non-hardening material, by a stress-evaluation method(6.10) giving σ_x'/σ_x, and from a slip-line field solution(6.5) giving t_x'/t_x. These ratios were found to be the same, within 1%, over a practical range of values of α, μ and r, showing that, though the stress analysis ignores redundant work, it can be used to provide a friction correction. It is then assumed that this remains true when the material work-hardens, so that $\sigma_H'/\sigma_H = t_H'/t_H$. The drawing stress t_H' including redundant work, friction and work hardening is then given by

$$t_H' = \frac{\sigma_H'}{\sigma_H} \cdot t_H \qquad (6.45)$$

The value of t_H is found from equation 6.43, assuming that the mean equivalent strain is independent of the strain-hardening characteristics, and of the magnitude of the friction coefficient.

$$t_H' = \frac{\sigma_H'}{\sigma_H} \cdot \int_0^{\epsilon_2} 2k \, d\epsilon \qquad (6.46)$$

or, since $\quad \epsilon_2 = f\left(\frac{c}{d}\right)\epsilon_a, \; t_H' = f\left(\frac{c}{d}\right)\sigma_H'$

The ratio σ_H'/σ_H can easily be found with the aid of a computer. The basic differential equation derived from stress evaluation, including friction, is given by equation 6.6.

$$\frac{d\sigma_x'}{d\epsilon} + B\sigma_x' = (1 + B)\, 2k \qquad (6.47a)$$

In the absence of friction:

$$\frac{d\sigma_x}{d\epsilon} = 2k \qquad (6.47b)$$

The relationship between k and ϵ is found from the stress–strain curve, and the ratio σ_x'/σ_x can be obtained by solving these equations with the computer for any given stress–strain curve.

An experimental check of the predicted drawing stresses t_H', for a wide range of reductions of area and die angles, showed good correlation. The calculations involved are all simple, though most expeditiously performed by a small computer. It appears that the necessary coefficients of friction can be determined independently, provided that care is taken to ensure similarity of lubrication conditions. The theory is thus suitable for practical application, but strip-drawing is not a common industrial procedure. Green has suggested that, with small modifications, it could be applied to drawing of thin-walled tubes, which is industrially important. (See Chapter 7.)

6.7 Determination of drawing stress for round bar, allowing for friction, redundant work and strain-hardening

There is no rigorous theory to allow for redundant work in drawing round bar, but the stress-evaluation method allows for friction and strain-hardening. Because low die angles are used industrially, and heavy reductions are commonly taken, the redundant work factor is often small in practical wire-drawing, but it cannot always be neglected. It is particularly significant in multiple light passes, since the additional hardening is cumulative.

If the material does not strain-harden, or if a mean yield stress can be used, the effect of friction on the drawing stress is found from equation 6.23.

$$\frac{\sigma'_{xa}}{Y} = \frac{1+B}{B}[1-(1-r)^B] \quad (6.23)$$

A simpler way of allowing for friction, as in §6.6.1, is to resolve the die force parallel to the axis, and to equate this component to the drawing-force,

$$\sigma' = q' \frac{r}{1-r}(1 + \mu \cot \alpha) \qquad \text{with friction}$$

$$\sigma = q \frac{r}{1-r} \qquad \text{with no friction}$$

It is then assumed that the mean die pressure is not appreciably influenced by the friction, so that $\bar{q} \approx \bar{q}'$, and a correction factor for the drawing stress is obtained as in equation 6.40.

$$\frac{\sigma'}{\sigma} = 1 + \mu \cot \alpha$$

The drawing stress in homogeneous frictionless deformation is given by

$$\sigma = \int Y \, d\epsilon = \bar{Y} \ln \frac{1}{1-r} \quad (6.3)$$

It is found experimentally that the variation of die pressure with friction can be neglected, for small values of $B = \mu \cot \alpha$, so the drawing stress can be expressed in the form:

$$\sigma' = (1 + \mu \cot \alpha)\, \bar{Y} \ln \frac{1}{1-r} \quad (6.48)$$

To allow for the increased strain resulting from redundant deformation, a mean effective strain ϵ_2 may be defined by analogy with the plane-strain condition.

$$\epsilon_2 = \phi \,.\, \epsilon_b \quad (6.49)$$

The redundant-work factor can be accurately calculated in terms of the geometrical ratio c/d in plane strain, but this is not possible for axial symmetry. It has however been shown experimentally[(6.11)] that ϕ is a function of the corresponding geometrical ratio, of the mean cross-sectional area A_m to the area of contact with the die surface $(A_b - A_a)/\sin \alpha$. Using the same argument as in §6.6.1, allowance can be made for the influence of redundant work by multiplying the drawing stress by this factor ϕ. The symbol t_x is again used for the drawing stress when corrected for redundant work, with a superscript when friction is also included.

$$t_x' = (1 + \mu \cot \alpha)\, \phi\, \bar{Y} \ln \frac{1}{1-r} \quad (6.50)$$

This has been shown to be valid[(6.11)] for common wire-drawing practice. The correction factor can also be applied to the stress-evaluation equation 6.23.

$$t_x' = \phi\, \bar{Y} \frac{1+B}{B}[1-(1-r)^B] \quad (6.51)$$

The latter has the advantage that the redundant work factor can be incorporated in the solution of the differential equation for a strongly work-hardening metal, as in §6.6.3. The numerical values of ϕ for non-hardening metal, determined by R. W. Johnson(6.12) using equation 6.51 agree closely with those of Wistreich(6.11) using equation 6.50. It is found that for a range of metals and lubricants there is an approximate relationship

$$\phi = 0{\cdot}87 + \left(\frac{1-r}{r}\right) \sin \alpha \tag{6.52}$$

Some results are shown in Figure 6.8, from which it can be seen that the redundant work is greater in rod drawing than strip-drawing.

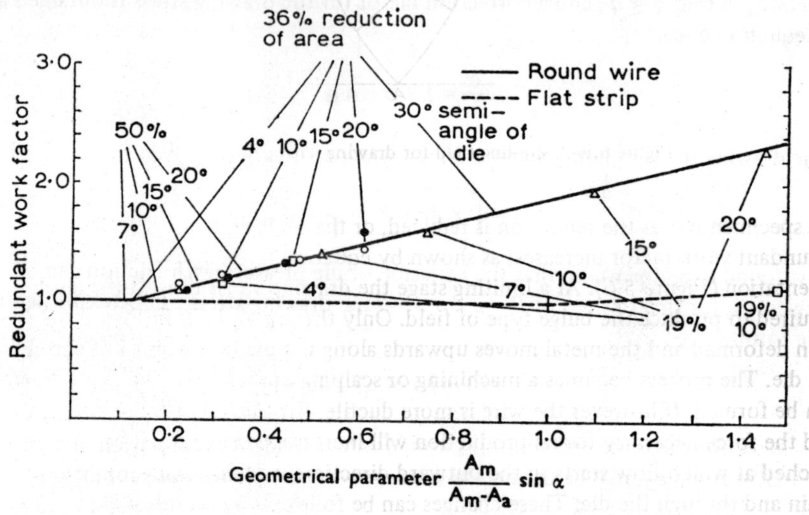

Figure 6.8 The redundant work factors ϕ and f for drawing, in axial-symmetry and plane-strain respectively. (Various wire materials and lubricants6·12.)

It should be noted that ϕ is defined in equation 6.49 in terms of strain, and in equation 6.51 in terms of stress. These values are identical for non-hardening material, but experimental measurements should not be transferred from one equation to the other for metals that strain-harden strongly.

A formula similar to equation 6.50 has been obtained theoretically by Shield(6.13) from an analysis of the internal distortion in a wire as it flows through a long conical channel (§6.11.2).

6.8 Bulge formation

Figure 6.9 shows a slip-line field in which the wire bulges before it enters the die. There is in fact a direct analogy with the transition in an indentation problem, from a field extending across the specimen as in Figure 5.5*c*, to a localised deformation zone for a thicker specimen, as in Figure 5.7.

In drawing of strip or wire with a heavy reduction, the field extends throughout

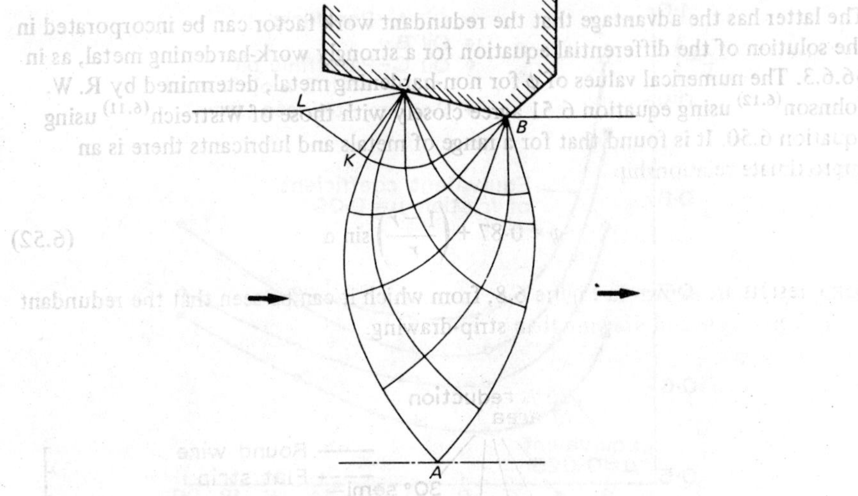

Figure 6.9 A slip-line field for drawing with a bulge[(6.16)].

the specimen but as the reduction is reduced, or the semi-angle is decreased, the redundant work factor increases, as shown by equation 6.52, again analogously to indentation (Figure 5.6). At a limiting stage the drawing stress rises to the level required to produce the bulge type of field. Only the region *LKB* in Figure 6.9 is then deformed and the metal moves upwards along the die face instead of through the die. The process becomes a machining or scalping operation, and chips of metal can be formed. If however the wire is more ductile, the pile-up or bulge will grow and the force necessary for its production will increase, until an equilibrium is reached at which flow starts in the outward direction and then continues down again and through the die. These changes can be followed by sectioning the partly-drawn wire, and can be predicted accurately for plane–strain drawing and also fairly closely for cylindrical wire.

It is interesting that the chip-formation field can be related directly to negative-rake metal cutting and to grinding processes.

6.9 Optimum die angles

There is an optimum die angle for a given reduction of area and coefficient of friction, which gives least drawing stress, because the contributions of redundant work and friction increase and decrease respectively with increasing die angle. The redundant work factor is, however, larger for round-bar drawing so the optimum die angles are lower than for strip-drawing. Some experimental measurements with mild steel are shown in Figure 6.9. Similar results are obtained with other metals.

It is also possible to design dies of special profiles, sometimes known as *sigmoidal dies*, that will produce desired conditions, such as constant longitudinal strain, streamline flow or minimal redundant work. Practical factors including wear and the

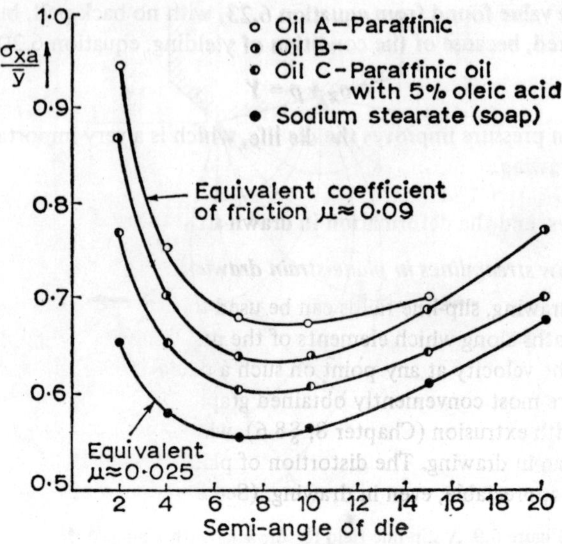

Figure 6.10 Drawing stresses for various die angles and coefficients of friction, showing the optimum angles for mild steel. Reduction of area 36%. (R. W. Johnson[6.12].)

presence of an intentional parallel land, as well as normal variations in ingoing wire dimensions tend to obscure the advantages of these tools. They are further discussed in the context of extrusion (§8.8.3) and tube rolling (§ 7.7.2).

6.10 Tandem drawing

Most wire-drawing involves passage through several dies in series, especially with copper wire, which is very ductile. To avoid excessive tension on the drawn wire, it is usual to wind one or two turns round a capstan between each pair of dies. The capstans are driven at carefully regulated speeds, so that there is a frictional drag at the capstan surface, which pulls the wire through the die, and the capstan also provides a small back-tension on the wire entering the next die.

In the solution of equation 6.21, it was assumed that there was no back-pull. If the back-tension has a value σ_b, the constant of integration is altered:

$$B\sigma_x - Y(1+B) = cD^{2B}$$

$$\sigma_x = \sigma_b \text{ when } D = D_b$$

$$B\sigma_b - Y(1+B) = cD_b{}^{2B};$$

$$\frac{\sigma_x}{Y} - \frac{1+B}{B} = \left[\frac{\sigma_b}{Y} - \frac{1+B}{B}\right]\left[\frac{D}{D_b}\right]^{2B}$$

The drawing stress σ_{xa} is given by

$$\frac{\sigma_{xa}}{Y} = \frac{1+B}{B}\left[1 - \left(\frac{D_a}{D_b}\right)^{2B}\right] + \frac{\sigma_b}{Y}\left(\frac{D_a}{D_b}\right)^{2B} \tag{6.53}$$

This exceeds the value found from equation 6.23, with no back-pull, but the die pressure is lowered, because of the condition of yielding, equation 6.20

$$\sigma_x + p = Y$$

This reduction in pressure improves the die life, which is a very important factor in practical wire-drawing.

6.11 Streamlines, and the deformation in drawn strip and rod

6.11.1 *Metal-flow streamlines in plane-strain drawing*

In plane-strain drawing, slip-line fields can be used together with the hodographs to determine the paths along which elements of the metal move through the deformation zone, and the velocity at any point on such a path.

The results are most conveniently obtained graphically, by the method described in connection with extrusion (Chapter 8, §8.6), where there is much greater inhomogeneity than in drawing. The distortion of planes originally perpendicular to the axis may be appreciable, even in drawing. (See Example 6.10.)

6.11.2 *Metal-flow streamlines in round-bar drawing*

Shield(6.13) has analysed the metal flow in a long convergent conical channel. The solution is rather complex but can be outlined briefly. The direct stresses are: σ_R radially towards the virtual apex of the conical channel, and σ_ϕ and σ_θ on the tangential plane inclined at ϕ to the axis and rotated by θ round the axis. There will also be a shear stress τ, dependent on ϕ only. It is assumed that $\sigma_\theta = \sigma_\phi$, and the yield condition is of the form

$$\sigma_R - \sigma_\phi = G(\tau)$$

On the axis $\tau = 0$, but at the die face it has some value, related to the friction, which may be described as a fraction m of the maximum shear stress k ($0 \leqslant m \geqslant 1$). The difference between σ_R and σ_ϕ increases as m increases, and so also as ϕ increases. The strains ϵ_R and ϵ_ϕ are similarly related, for an isotropic material, and from this the radial velocity u_R can be evaluated as a function of ϕ for an incompressible material. The results are in reasonable agreement with practical observations for well-lubricated long dies. Shield's work has been extended by Naylor(6.14) and by Wells(6.15). An interesting feature of the theory is that a maximum tensile stress is predicted on the axis, which may even exceed the yield stress. 'Cuppy' wire, in which the central core fractures inside a sound skin, is occasionally observed.

The upper-bound technique of Chapter 8, §8.7, can also be used for axisymmetrical drawing, and approximate streamlines can be constructed from the solution, as in §8.7.4, assuming zero friction. Sticking friction is not encountered in good drawing practice.

6.11.3 *Ductility of drawn wire*

In many wire-drawing operations it is important that the product should be ductile, whether for redrawing or for subsequent manipulation. It has been suggested that the effective strain produced in a given pass can be estimated from equations 6.49

and 6.52, or a revised form of the latter, averaged for various metals and lubricants:

$$\phi = 0{\cdot}88 + 0{\cdot}78\frac{D_1 + D_2}{D_1 - D_2} \cdot \frac{1 - \cos\alpha}{2\sin\alpha} \tag{6.54}$$

The ductility can be expressed as percentage reduction of area, or as percentage elongation to fracture, in a tensile test on homogeneously-deformed wire. Such data are available, for example for copper, or can readily be produced.

A convenient practical method of predicting ductility for any given pass is to use a nomogram solution of the equation,[6.28] together with a suitable ductility scale for the given metal placed against the equivalent reduction scale.

6.12 Lubrication in practical wire drawing

Some specialised techniques are available for manufacturing wires, as discussed in §6.13, and rods are often made by rolling, but the great majority of precision bars and wires are drawn through simple dies. Traditionally the lubrication in wire drawing is either dry, in which soap powder is picked up by the wire from a box ahead of the die; or wet, in which the wire and die are submerged in or flooded by oil. Generally speaking the dry lubricants, often fortified by lime or by a conversion coating on the wire, allow the heaviest reductions of area without pickup, while the wet lubricants impart better surface and provide cooling for high-speed drawing. Independent cooling of the dies, and more significantly of the wire, is nevertheless often important in multi-stand machines.

The lubrication regime is usually a mixture of hydrodynamic, in which thick velocity-dependent films tend to give good protection from metallic contact with the die; and boundary, in which chemical reaction products provide protection from pickup but allow close contact and burnishing of the wire, giving a very smooth surface. Hydrostatic effects are often important on a small scale, when pools of lubricant are trapped in surface depressions. Extreme pressure compounds may also be added to provide additional protection.

Whatever lubrication method is used, the initial preparation of the wire is of great importance. All oxide-scale particles must be removed, by mechanical or chemical processes, and care should be taken to avoid any stray grit or dirt being picked up. Filtration is essential for all oil-circulating systems, especially when fine wire is drawn. Ultrasonic cleaning is beneficial for the finest wires, since a small amount of metallic debris is always generated during industrial drawing and this must not be allowed to clog the dies.

Undoubtedly the most important aspect of lubrication in wire drawing is the reliable avoidance of pickup, especially for ferrous materials. In general, the smoother the wire surface, the greater the danger of pickup. Wear is next in importance.

6.12.1 *Steel wire*

Because of the dangers of pickup, dry drawing is often used, with a protective initial coating. The oldest method used a *sull coating*, which is a soft hydrated oxide, together with dry lime or borax. This is still effective, when further lubricated with soap, but it is not very controllable. For critical operations, phosphating is superior, but it is more expensive. Zinc phosphate is the most common ingredient. In warm acid solution the following reactions are believed to occur:

$$Fe^{+++} + 2H_2PO_4 \rightarrow Fe(H_2PO_4)_2$$

$$3Fe^{++} + 3Zn^{++} + 4PO_4^{---} \rightarrow Zn_3(PO_4)_2 + Fe_3(PO_4)_2$$

Other ions such as Mn^{++++} may be present, and activators such as nitrates, nitrites or chlorates are incorporated to accelerate the attack. Temperature is important in determining the reaction rate, and pH must be controlled to avoid direct acidic corrosion. Commercial solutions are carefully prescribed.

Phosphate conversion-coatings by themselves will protect from pickup but give high friction and heavy wear. It is always necessary to use an additional lubricant, though some phosphating baths contain a lubricating ingredient. Dry calcium stearate powder is often used. The slightly-rough coated wire picks this up from a box just ahead of the die and draws the powder forward into the die mouth, where it is strongly compacted and passes through the die as a coherent thin film. The process is critical, and the soap must not be allowed to coagulate so that the wire passes through uncoated, by channelling a stable hole through the powder.

Water-soluble sodium stearate avoids this problem, but involves an additional separate sequence of washing, dipping in the soap solution, and drying. Wet soap solution is much less effective than the dry coating deposited from it. This technique is more widely used for drawing tubes, where the bore lubrication is crucial and the unit cost of the workpiece is higher.

For less severe wire-drawing operations, and for higher speeds, borax or lime coatings are used as carriers for the lubricant. Large bars, 50–100 mm in diameter are usually drawn slowly with only light sizing passes. Grease or tallow, applied at the die, gives adequate lubrication.

Wet drawing is used for moderate passes at high speeds. Soap-fat emulsions are commonly used, with a typical composition:

potassium stearate 35%, tallow 25%, mineral oil 8%, stearic acid 2%, water 30%.

Since the stability of emulsions can cause trouble, and an inversion to water-in-oil emulsion may lead to corrosion of the machine, many operators prefer compounded neat oils. These are mineral oils with fatty acid, sulphonated, chlorinated or sulphurised additives in various proportions. It is believed that these oils give better finish than the emulsions, but there is the usual balance between high surface reflectivity with thin lubricant films, and matt surfaces with thicker films giving greater protection. The more viscous oils give less shiny surfaces.

Bright-drawing to size, of angles and other sections, also uses oils. Stainless steels are more prone to pickup on the dies than other steels, for reasons that are not fully explained but may depend upon the thin natural oxide film being easily ruptured, and on the high strain-hardening rate of stainless steel. Oils containing chlorinated wax, or straight chlorinated oils containing a high weight proportion of chlorine are effective. The major constituent is a hydrocarbon containing 35–40% Cl, and care must always be taken to control the production of corrosive HCl in moist conditions.

Phosphate conversion coatings cannot be produced on the unreactive surface of stainless steels, but *oxalate coatings* with similar carrier properties can be deposited. These are effective when used with calcium or sodium soap, but are more common in tube drawing, because of their expense. Lime or common salt (NaCl) coatings are cheaper but less effective than oxalates.

6.12.2 *Aluminium wire*

Aluminium, like stainless steel, has a brittle and thin oxide film. It tends to pick up on a steel tool once a layer of aluminium has been deposited locally, but can be drawn much more easily than stainless steel.

Aluminium and aluminium-alloy rods and bars are usually drawn with grease or with soaps, without using a carrier or chemical surface pretreatment.

Compounded mineral oils are commonly used for drawing wires. They may contain about 10% of fats or fatty oils, together with sulphurised or preferably chlorinated compounds. Water-base lubricants are not usually suitable because of the white stains that may be produced, causing surface deterioration.

6.12.3 *Copper and copper alloys*

Copper is particularly easy to draw, and gives very little trouble from pickup, though worn steel dies normally show a distinct copper colouration. Fine copper debris is generated even with slightly roughened dies, and should be filtered out of the lubricant.

Grease or tallow is used for large sizes of bar. Soap-fat emulsions are the most common lubricants for copper wire drawing, in much lower concentrations than are used for steel, typically 3% soap fat in water, but oil emulsions are preferred for fine wires. These fluids have good cooling properties, which become important at the very high speeds possible for copper wire drawing. At lower speeds, compounded oils give good results and excellent surface finish.

α-brass and bronzes can also be drawn effectively with high-dilution soap-fat emulsions or compounded oils, but α-β brasses are less ductile (see §12.2.2).

Staining is a problem with all copper alloys. Sulphur should be avoided entirely, and care should be taken to reduce as far as possible the amount of free fatty acid present, especially by removing lubricant residues from drawn wire before storage.

Water-base lubricants are being increasingly used.

6.12.4 *Other alloys*

Nickel alloys are prone to pick up on steel and carbide tools, in much the same way as stainless steel (18 Ni, 8 Cr, Fe) and should be similarly treated. Oxalate carrier-coatings with supplementary soap lubrication are most effective.

Titanium presents still more serious pickup problems, and is the worst of all metals for galling. Again oxalate and soap may be used. A cheaper pretreatment is light oxidation at about 700°C followed by a dip in a lime suspension and lubrication with soap. It is possible to form a complex fluoride-phosphate conversion ccating and this has been used effectively. Cost is less important for titanium because of the high cost of the material and its relatively low scrap value.

Chlorinated wax can be used under light conditions. Polymer coatings, of methacrylate or other resins, have been used. These give good protection.

Another approach is to use dispersion-hardening aluminium bronze dies. These avoid the pickup problem, but wear rapidly.

Tantalum can also be drawn with aluminium bronze dies, because relatively small quantities are required. Beeswax is a suitable lubricant.

Molybdenum and Tungsten are drawn at high temperatures using a graphited lubricant applied by dip or spray. The coated wires are sometimes baked before being drawn, but this additional operation is not always necessary.

6.13 Recent developments in wire manufacture(6.27)

6.13.1 *Theoretical contributions*

Both slip-line field and upper-bound solutions can be used to show when defects are likely to occur in drawn wire. Bulging ahead of the die may cause surface damage, while the hydrostatic tension on the axis can produce central bursts. For drawing-force calculations and pass design, simplified forms of the stress analysis equations are available with corrections for redundant work.

Analytical methods based on velocity and strain distributions allow die profiles to be designed to give streamline flow(6.17) or constancy of strain rate(6.18), or other specified features. Such dies can also reduce the occurrence of tearing at the surface of the wire. In greater detail, computerised stress-analysis methods, upper bounds, finite-element techniques and the weighted-residuals procedures, described in Chapter 13, have all been applied to wire drawing. These can provide fine-scale stress and temperature distribution patterns(6.19).

A significant question arises from some of this analysis about the interpretation of frictional stress at the tool face. It appears that neither of the usual assumptions, of constant friction coefficient or constant shear stress, is fully satisfactory, though for force calculations these are adequate.

Extensions of the upper-bound solutions allow forces to be calculated for square and hexagonal wires(6.20), and also for composite wires.

6.13.2 *Improvements in conventional processes*

Automation increased in the wire-drawing industry some years ago, but is reaching a steady level, with more emphasis being placed on versatility of equipment including for example, the possibility of switching production from copper to aluminium wire.

Very high speed drawing, up to 30 m/s, has been shown to be possible for copper and mild steel on an experimental basis(6.21). It may be limited not by lubrication problems as originally supposed, but by coil size and the time required to weld coils end to end.

At conventional speeds the main emphasis is on the further development and introduction of in-line processes for descaling, pickling and lubricating wires. Annealing is an important area receiving attention. Fluidised beds and other heating systems can be applied for wire speeds approaching draw-bench speeds. Non-destructive testing has considerably improved, and in-line test facilities have been developed.

Metallurgically, it is possible to increase the strength of steel wires by thermo-mechanical treatment, based on drawing in the appropriate temperature range. Warm drawing is of interest, particularly for stainless steel, but there is still a lack of suitable lubricants for temperatures of about 300°C. Superplastic alloys can be drawn in the form of wire. It is even possible to draw these in a die-less process by heating a narrow zone inductively into the superplastic range and applying a tensile force. The wire then thins out in much the same way as in silica or glass fibre-drawing.

Improved surface finish has resulted from laser drilling of diamond dies and electrolytic piercing of tungsten carbide dies. An interesting development in die treatment is boronising of tungsten carbide, by heating the die in a boron-rich atmosphere to convert the surface carbide, and more significantly the cobalt, to borides. This is claimed to reduce wear and chemical attack by the lubricant(6.22).

The most important improvements in lubrication relate to cooling of the wire, rather than the die, and better filtration to remove wire debris. Ultrasonic cleaning helps greatly with very fine wires. It is also claimed that suitable radial or axial vibration of the die improves lubrication as well as probably softening the wire by raising its temperature(6.23). Few changes are apparent in lubricants, but a method of depositing lubricant films electrochemically just ahead of the die has been demonstrated on a laboratory scale(6.27). Rotating and roller dies have been recommended for reduction of pickup.

6.13.3 *Newer processes*

Hydrostatic extrusion of wire is now established and is particularly suitable for brittle materials and composites(6.24). By applying a front tension, an efficient and well-controlled extrusion-drawing process is available.

An entirely new process of continuous forming of small-diameter wire from much heavier gauge material has been introduced recently(6.25). The stock is fed round a wheel and driven by friction into a converging chamber, from which it can escape only through a small circular or other suitably-shaped orifice. The pressure necessary to force the metal through this hole is quite easily produced by the friction, and very large nominal reductions in area can be made, in one pass, using very simple equipment.

Another novel technique, also proposed by Green(6.26), is known as helical extrusion, though it more closely resembles machining. A conical punch is driven slowly into the end of a copper billet to form a short tube with an annular face. This is then steadily deformed by a tool, somewhat like a negative rake cutting tool, that rotates about the axis of the billet and slowly advances in a helical path. Instead of forming a waste swarf, the metal is trapped in a small chamber, from which it escapes through an orifice as a wire. The equivalent reduction in cross section from the billet to the wire is enormous, perhaps from 150 mm diameter to 1 mm. The forces are low, partly because only a small volume is deformed at a given instant and partly because the severe local heating reduces the yield stress.

Direct production of rod by rolling of continuously-cast bars is now considered to be commercially viable, as it has long been for aluminium strip.

EXAMPLES

Use the stress–strain data of the Chapter 2 examples where appropriate. When it is necessary to assume constant yield stress, use suitable mean values.

6.1 Draw two graphs showing the redundant work factors for strip-drawing and for wire-drawing, in terms of appropriate geometrical parameters. Re-draw these graphs to show the redundant work factors as functions of die angle for various reductions of area between 10% and 60%.

Solution. This is a purely computational exercise based on Table 6.1 and equation 6.52. It provides graphical results of use in drawing calculations.

6.2 What are the optimum die angles for strip drawing and wire drawing with 30% reduction of area if $\mu = 0{\cdot}04$? How is the power requirement affected by departure from the optimum? (Ignore the influence of strain-hardening.)

Solution. For both strip drawing (equation 6.7) and wire drawing (equation 6.23) the ratio of drawing stress to mean yield stress ($\bar{S}$ and $\bar{Y}$ respectively) is given by

$$\frac{1+B}{B}\,[1-(1-r)^B]$$

The values obtained must be multiplied by the redundant work factors f and ϕ given by Table 6.1 and equation 6.52. $\phi = 0{\cdot}87 + A_2/M = 0{\cdot}87 + \dfrac{1-r}{r}\sin\alpha$. Equation 6.37 gives

$$\frac{c}{d} = \frac{2-r}{r}\,.\,\alpha$$

Thus:

$\alpha =$	2°	5°	14°
σ_{xa}/S or $\sigma_{xa}/Y =$	0·64	0·46	0·40
$A_2/M =$	0·081	0·204	0·561
$\phi =$	(0·95)	1·07	1·43
$c/d =$	–	–	1·38
$f =$	1·0	1·0	1·04
$\phi\,.\,\sigma_{xa}/Y =$	0·64	0·50	0·57
$f\,.\,\sigma_{xa}/S =$	0·64	0·46	0·42

The optimum angles giving least drawing force for the specified reduction of area and coefficient of friction are respectively about 5° for wire and 14° for sheet. The drawing stress is little affected by a small increase in angle, but rises more steeply with decreasing angle.

6.3 Calculate the drawing load for 40% reduction of area of 25 mm x 6 mm annealed mild steel strip using 12 mm radius dies, and compare this with the load using straight-tapered dies (*a*) of the same entry angle (*b*) of the same mean angle. Assume $\mu = 0{\cdot}1$. The contribution of redundant work may be neglected.

Solution. The appropriate stress equation is equation 6.18

$$\frac{\sigma_{xa}}{S} = 1 - \frac{h_a}{h_b}\,e^{\mu(H_a - H_b)};\; H_b = 2\sqrt{\frac{R}{h_a}}\tan^{-1}\sqrt{\frac{R}{h_a}}\,\alpha_b$$

(For drawing problems it can be assumed that $R' = R$).

$$h_b = 6,\; h_a = 3{\cdot}6 \text{ mm}.$$

From the geometry of the system $\tan \alpha_b = 0{\cdot}484$, $\alpha = 25{\cdot}8° = 0{\cdot}45$ rad. (Using an approximation because $R \gg \Delta h/2$, $\tan \alpha_b = \sqrt{\Delta h/(R - \Delta h/2)} = 0{\cdot}5$, $\alpha_b = 26{\cdot}5°$.

Then
$$H_b = 2 \times 1{\cdot}89 \tan^{-1} 0{\cdot}85 = 2{\cdot}66, \quad H_a = 0$$

$$\frac{\sigma_{xa}}{S} = 1 - 0{\cdot}6\, e^{-0{\cdot}266} = 0{\cdot}54$$

For a straight-tapered die, equation 4.23 gives

$$\frac{\sigma_{xa}}{S} = \frac{1+B}{B}\left[1 - \left(\frac{h_a}{h_b}\right)^B\right]$$

If $\alpha = \alpha_b$, as found above, $B = 0{\cdot}1/0{\cdot}484 = 0{\cdot}207$

$$\frac{\sigma_{xa}}{S} = 5{\cdot}83\,(1 - 0{\cdot}90) = 0{\cdot}58; \quad \sigma_{xa} \approx 0{\cdot}34, \quad P = 31 \text{ kN}$$

If $\alpha = \frac{1}{2}\alpha_b$, $B = 0{\cdot}1/0{\cdot}229 = 0{\cdot}436$

$$\frac{\sigma_{xa}}{S} = 3{\cdot}29\,(1 - 0{\cdot}8) = 0{\cdot}66; \quad \sigma_{xa} = 0{\cdot}39, \quad P = 35 \text{ kN}$$

6.4 **Compare the force required to draw 25 mm × 6 mm copper strip to 45% reduction of area with that required for an equal reduction on round bar of the same cross-sectional area, using *(a)* 12° included-angle dies *(b)* 30° included-angle dies, if $\mu = 0{\cdot}07$. Redundant work may be neglected.**

Solution. For strip, equation 4.23 gives

$$\frac{\sigma_{xa}}{S} = \frac{1+B}{B}\left[1 - \left(\frac{h_a}{h_b}\right)^B\right]; \quad B = 0{\cdot}07 \cot 6° \text{ or } 0{\cdot}07 \cot 15° = 0{\cdot}666 \text{ or } 0{\cdot}261$$

$$\alpha = 6°: \frac{\sigma_{xa}}{S} = 2{\cdot}50\,(1 - 0{\cdot}67) = 0{\cdot}825$$

$$\alpha = 15°: \frac{\sigma_{xa}}{S} = 4{\cdot}83\,(1 - 0{\cdot}855) = 0{\cdot}703$$

For round bar, equation 6.23 gives

$$\frac{\sigma_{xa}}{Y} = \frac{1+B}{B}\left[1 - \left(\frac{D_a}{D_b}\right)^{2B}\right] = \frac{1+B}{B}\left[1 - (1-r)^B\right]$$

Thus the r.h.s. is the same as in equation 4.23, but the yield stress used is Y, so the force required to draw the round bar is 15% less than that required for the strip.

From the data of Example 2.3, $\bar{S} \approx 0{\cdot}30$ kN/mm² so with $\alpha = 6°$

$$F_{\text{strip}} = 0{\cdot}825 \times 150 \times 0{\cdot}45 \times 0{\cdot}30 = 16{\cdot}7 \text{ kN}$$

$$F_{\text{bar}} = 0{\cdot}825 \times 150 \times 0{\cdot}45 \times 0{\cdot}26 = 14{\cdot}5 \text{ kN}$$

6.5 **Annealed brass wire is to be drawn from 6 mm diameter to 1·5 mm with an intermediate annealing at about 3 mm. Suggest a suitable drawing schedule for a wire-drawing machine with a maximum pull of 4·5 kN. The dies may be ground to any angle, but the best available lubricant gives $\mu = 0{\cdot}04$.**

Solution. The schedule is most easily prepared in an approximate way using equation 6.3, $F_n = A_n \bar{Y} \ln A_{n-1}/A_n$, and allowing about 20% for friction, equation 6.48, e.g.

$$(1 + 0{\cdot}04 \cot 10°)$$

The reductions of area after each annealing are $\epsilon' = \ln\left(\frac{6\cdot0}{3\cdot0}\right)^2 = 1\cdot39$ and $\epsilon'' = \ln\left(\frac{3\cdot0}{1\cdot5}\right)^2 = 1\cdot39$; $r' = r'' = 0\cdot75$.

(*a*) From Ex. 2.2, at $\epsilon = 1\cdot39$, $S_n = 0\cdot77$, $Y_n = 0\cdot67$ kN/mm²

The drawing force that would fracture the final wire is given by $F = A_n Y_n \approx 1\cdot77 \times 0\cdot67 = 1\cdot19$ kN. The limit is thus set by fracture rather than by capacity of the machine.

The maximum possible reduction, allowing 20% friction, is given by

$$1 = \frac{\sigma_x}{Y} = 1\cdot2 \ln A_{n-1}/A_n;\; A_{n-1}/A_n = 2\cdot30,\; r_m = 0\cdot56$$

Two equal passes of $\epsilon = 0\cdot69$ ($r = 50\%$) would complete the second stage. The intermediate area would be given by $\ln A_{n-1}/A_n = 0\cdot69$; $A_{n-1}/A_n = 1\cdot99$, $A_{n-1} = 3\cdot53$ mm²; $\bar{Y}_n \approx 0\cdot65$, $\bar{Y}_{n-1} \approx 0\cdot52$. Thus

$$F_n = 1\cdot2 \times 1\cdot77 \times 0\cdot65 \times 0\cdot69 = 0\cdot95 \text{ kN}$$

$$F_{n-1} = 1\cdot2 \times 3\cdot53 \times 0\cdot52 \times 0\cdot69 = 1\cdot52 \text{ kN}$$

(*b*) The drawing force that would fracture the wire after drawing to $\epsilon' = 1\cdot39$ would be $F = A_{n-2}\, Y_{n-2} \approx 7\cdot07 \times 0\cdot67 = 4\cdot74$ kN, so the limit is set by the drawbench capacity, 4·5 kN.

The maximum permissible pass before the interanneal is thus given, making a rough prior estimate of $\bar{Y}$ for the pass, by

$$F_{n-2} = 4\cdot5 \approx 1\cdot2 \times 7\cdot07 \times 0\cdot65 \times \epsilon;\; \epsilon = 0\cdot81,\; r = 0\cdot55$$

The wire would be taken from $\epsilon = 0\cdot58$ to $\epsilon = 1\cdot39$ in this pass. The preceding pass would be similarly limited:

$$\epsilon = \ln A_{n-3}/A_{n-2} = 0\cdot81, \text{ so } A_{n-3} = 15\cdot8 \text{ mm}^2$$

Suppose $\bar{Y} = 0\cdot55$

$$F_{n-3} = 4\cdot5 \approx 1\cdot2 \times 15\cdot8 \times 0\cdot55 \times \epsilon;\; \epsilon_{n-3} = 0\cdot43,\; r = 0\cdot35$$

This pass takes the wire from $\epsilon = 0\cdot15$ to $\epsilon = 0\cdot58$. The next preceding pass is calculated in the same way:

$$\epsilon = \ln A_{n-4}/A_{n-3} = 0\cdot15 \text{ so } A_{n-4} = 18\cdot4 \text{ mm}^2$$

Suppose $\bar{Y} = 0\cdot35$

$$F_{n-4} = 4\cdot5 \approx 1\cdot2 \times 18\cdot4 \times 0\cdot35 \times \epsilon;\; \epsilon_{n-4} = 0\cdot58,\; r = 0\cdot44$$

This would in fact take wire of a greater diameter than that given, so these three passes would be adequate, totalling 1·82. A suitable schedule could be slightly easier, say

Pass 1	$\epsilon = 0$ to $\epsilon = 0\cdot38$
2	$\epsilon = 0\cdot38$ to $\epsilon = 0\cdot76$
3	$\epsilon = 0\cdot76$ to $\epsilon = 1\cdot39$

The corresponding diameters given by $\epsilon = \ln (d_n/d_{n-1})^2$ are

$$d_1 = 4\cdot96$$
$$d_2 = 4\cdot10$$
$$d_3 = 3\cdot00 \text{ mm}$$

These passes are close to the maximum, and a better schedule for practical purposes would allow a greater margin of safety, so that four passes before annealing at $d = 3$ mm would be preferable. Greater accuracy in load calculation could be obtained by using equation 6.23 and allowing for redundant work if necessary.

6.6 Determine the total equivalent strain in copper wire drawn from the annealed condition at 3 mm diameter to 0·075 mm using a die of 20° included angle, with (*a*) the theoretical maximum reduction in each pass, (*b*) 19% reduction of area in each pass.

Solution. The total nominal strain imparted to the wire is $\epsilon = \ln \dfrac{3{\cdot}0}{0{\cdot}075} = 3{\cdot}68$ ($r = 0{\cdot}98$). The theoretical maximum in each pass is $\epsilon = 1$. With 20° included angle the redundant work at this reduction of area is negligible (Figure 6.8 or Equation 6.52. Even in the fourth pass ($\epsilon = 0{\cdot}68$, $r = 0{\cdot}50$), $\phi \approx 1{\cdot}04$, so the total strain is equal to the nominal strain.

With 19% passes, $\epsilon = 0{\cdot}21, \phi = 1{\cdot}61; \epsilon' = 0{\cdot}338$, ($r' = 0{\cdot}29$). There will be 17 such passes ($\epsilon = 3{\cdot}57$) and one pass $\epsilon = 0{\cdot}11$, ($r = 0{\cdot}104$). For the latter, $\phi = 2{\cdot}37$ and the equivalent strain is 0·259. The total equivalent strain would be $\epsilon_T = 17 \times 0{\cdot}338 + 0{\cdot}259 = 6{\cdot}01$, ($r_T = 0{\cdot}997$).

6.7 Draw the slip-line field and the hodograph for 20% reduction of area of non-hardening wide strip with dies of semi-angle 20°, assuming $\mu = 0{\cdot}10$. What redundant strain is involved? (It is necessary to evaluate q by drawing the field for $\mu = 0$ first).

Solution. The field is constructed as in §6.5.2, starting with the 45° triangle shown in Figure 6.5, and the radial fans centred on A and B. It will be found that the field almost matches the required reduction if there are three 10° segments centred on A and five on B. Using the straight-line approximation, without drawing the smooth curves for the true field, the contributions of the sections on the entry slip-line can be summed. Thus, for the segment starting on the centre line, $\phi = 50°$, $\cos\phi = 0{\cdot}643$, $\sin\phi = 0{\cdot}766$ and the length $s = 0{\cdot}78$ (measured in mm, taking $\frac{1}{2}h_b = 125$ mm and $\frac{1}{2}h_a = 100$ mm as suitable diagram dimensions). The pressure increment to the first intersection is $2k\Delta\phi = 2k(0{\cdot}175)$ so the mean pressure on this section is $p_x + 0{\cdot}175k$, if p_x is the pressure on the centre line. Thus $\bar{p}s \sin\phi = 0{\cdot}596p_x + 0{\cdot}104k$, and $ks\cos\phi = 0{\cdot}501k$.

The total rearward horizontal force component (equation 6.32) is $\Sigma \bar{p}s\sin\phi + ks\cos\phi$. Equating this to zero (no backpull) gives $-(5{\cdot}09p_x + 6{\cdot}25k) + 1{\cdot}37k = 0$, $p_x = -0{\cdot}96k$. The pressure at the apex C of the 45° triangle is $p_c = p_x + 2{\cdot}80k = -1{\cdot}84k$, which is also the hydrostatic pressure at the die face. The major principal stress at the die face is thus $\sigma_1 = q = -(p + k) = 2{\cdot}84k$. Thus $q/2k = 1{\cdot}42$.

The same procedure is followed when $\mu = 0{\cdot}10$, except that the initial triangle is set at an angle given by $\cos 2\theta = \mu q/k$, taking $q = 2{\cdot}84k$ as found above. To meet the condition for 45° intersection at the centre-line, suitable values for ψ and ζ are 25° and 55° respectively. Equating the rearward horizontal force component to zero gives $p_x = -0{\cdot}75k$ and thence $q = 2{\cdot}88k$. It will be seen that this differs little from the value for frictionless drawing.

6.8 If the ductility of homogeneously-deformed brass can be expressed in the form

$$e1\% = 42\exp(-7{\cdot}5\epsilon)$$

determine the ductility of a 25 mm × 3 mm brass strip drawn to 25 mm × 2·4 mm using dies of 20° semi-angle, ignoring the influence of friction. Compare this with the ductility when drawn to the same reduction of area under ideal conditions.

Solution. For this pass $r_{01} = 0{\cdot}20$, $\epsilon_{01} = \ln\dfrac{3{\cdot}0}{2{\cdot}4} = 0{\cdot}223$, but account must be taken of the redundant deformation. From equation 6.37, $\dfrac{c}{d} = \dfrac{2-r}{r}\cdot\alpha = 3{\cdot}14$, so using Table 6.1, $f\left(\dfrac{c}{d}\right) = 1{\cdot}53$ and the effective strain is $\epsilon' = f\epsilon_{01} = 0{\cdot}341$ ($r = 0{\cdot}29$). Substituting this in the given ductility equation, (e1%) = 42 exp(−2·56) = 42/12·9 = 3·3%. Under ideal conditions (e1%) = 42 exp (−1·67) = 7·9%. (Actually the exponential equation given ceases to represent the ductility beyond about $r = 0{\cdot}3$, $\epsilon = 0{\cdot}36$, and the percentage elongation then remains almost independent of prior strain over an appreciable range.)

6.9 Estimate the average temperature rise of the strip in Example 6.8, assuming the frictional heating to be equal to the heat loss by convection.

Solution. (*a*) The work done per unit volume of strip, if deformed with zero redundant work, is given by

$$\frac{W}{V} = \int_0^{\epsilon_1} 2k\mathrm{d}\epsilon \approx \overline{2k}\epsilon_1$$

From the data of Example 2.2 the mean flow stress in a pass $\epsilon = 0$ to $\epsilon = 0{\cdot}223$ is $2k \approx 0{\cdot}38$ kN/mm²

$$\frac{W}{V} = 0.085\ V\ \mathrm{kN/mm^2}$$

The density of brass is $8{\cdot}38 \times 10^{-3}$ g/mm³, so for a mass m

$$W = \frac{0{\cdot}085}{8{\cdot}38 \times 10^{-3}} m\ \mathrm{kN/g} = 10{\cdot}1\ m\ \mathrm{J/g}$$

The specific heat of brass is 0·376 J/g so

$$10{\cdot}1\ m = m \times 0{\cdot}376 \times (T_1 - T_0);\ T_1 - T_0 = 26°\mathrm{C}.$$

(*b*) Allowing for redundant work $W/V = \overline{Y}\epsilon'$. The temperature rise is thus increased in ratio ϵ_1/ϵ_1 to 39°C.

6.10 Draw the metal-flow streamlines for wide non-hardening strip drawn through dies of 15° semi-angle with 34% reduction of area, neglecting friction effects. (Read Chapter VIII, §8.6 first.)

Solution. The slip-line field is of the simple type shown in Figure 6.4. The x component of the velocity at C is 1·18, at D is 1·26 and the mean value between C and D is 12·25, as found from the hodograph. The exit velocity is 1·515 ($=1/1 - r$). The reciprocals of these velocities are plotted against distance along the x-direction, assuming unit initial velocity and constant velocities through each section, for different stream lines. The areas beneath these sections are then plotted progressively to show $t = \int_0^x \frac{\mathrm{d}x}{V_x}$. Lines of constant t can be drawn arbitrarily to find the position x reached by any element after time t. The resulting points would represent the distortion of a grid drawn on the ingoing strip edge. The pattern resembles that of Figure 8.5 but the distortion is less severe.

REFERENCES

6.1 Bland, D. R. and Ford, H., 'The calculation of roll force and torque in cold strip-rolling with tensions', *Proc. Inst. Mech. Engrs*, 159, 144–153, 1948.

6.2 Lancaster, P. R. and Rowe, G. W., 'Experimental study of the influence of lubrication upon cold drawing', *Proc. Inst. Mech. Engrs.*, 178, 69–89, 1963.

6.3 Prager, W. and Hodge, P. G., *The Theory of Perfectly Plastic Solids*, John Wiley, 1951.

6.4 Johnson, W. and Mellor, P. B., *Plasticity for Mechanical Engineers*, Van Nostrand, 1962, p. 271.

6.5 Hill, R. and Tupper, S. J., 'A new theory of plastic deformation in wire drawing', *J. Iron and Steel Inst.*, 159, 353–359, 1948.

6.6 Green, A. P. and Hill, R., 'Calculations on the influence of friction and die geometry in sheet drawing', *J. Mech. Phys. Solids*, 1, 31–36, 1953.

6.7 Green, A. P., 'Plane strain theories of drawing', *Proc. Inst. Mech. Engrs.*, 174, 847–864, 1960.

6.8 Christopherson, D. G., Oxley, P. L. B. and Palmer, W. B., 'Orthogonal cutting of work-hardening material', *Engineering*, 186, 113–115, 1958; 'Stress relations along slip lines for a material with variable flow stress', *Int. J. Mech. Sci.*, 8, 63–64, 1966.

6.9 Wistreich, J. G., 'The fundamentals of wire-drawing', *Metall. Rev.*, 3, 97–142, 1958.
6.10 Sachs, G., Lubahn, J. D. and Tracy, D. P., 'Drawing of thin-walled tubing', *J. Appl. Mech.*, 11, 199–210, 1944.
6.11 Wistreich, J. G., 'Investigation of the mechanics of wire-drawing', *Proc. Inst. Mech. Engrs.*, 169, 654–665, 1955.
6.12 Johnson, R. W., *Ph.D. Dissertation*, Birmingham University, 1965.
6.13 Shield, R. T., 'Plastic flow in a converging conical channel', *J. Mech. Phys. Solids*, 3, 246–258, 1955.
6.14 Naylor, H., *Ph. D. Dissertation*, Leeds University, 1957.
6.15 Wells, J., *M.Sc. Dissertation*, Leeds University, 1962.
6.16 Johnson, R. W. and Rowe, G. W., 'Bulge formation in strip drawing with light reductions in area', *Proc. Inst. Mech. Engrs.*, 182, 521–526, 1968.
6.17 Richmond, O. and Morrison, H. L., 'Streamlined wire-drawing dies of minimum length', *J. Mech. Phys. Solids*, 15, 195–203, 1967.
6.18 Blazynski, T. Z., 'Theoretical methods of designing tools for metal-forming processes', *Metal Form.*, 34, 143–150, 1967.
6.19 Avitzur, B., 'Analysis of wire drawing and extrusion through conical dies of large cone angle', *Trans. A.S.M.E.*, (B) 86, 305–316, 1964.
6.20 Juneja, B. L. and Prakash, R., 'An analysis for drawing and extrusion of polygonal sections', *Int. J. Mach. Tool Des. Res.*, 15, 1–18, 1975.
6.21 Lancaster, P. R., 'A study of lubrication in the high-speed drawing process', *Wire Ind.*, 39, 294–297, 1972.
6.22 Linial, A. V., 'New process-boronizing', *Ind. Heat*, 41, 28–31, 1974.
6.23 Sansome, D. H., 'Recent developments in oscillatory metal working', *Engg.*, 213, 243–247, 1973.
6.24 Pugh, H. Ll. D. (Ed.), '*Mechanical behaviour of materials under pressure,* Elsevier, Amsterdam, 1970.
6.25 Green, D., 'Continuous extrusion-forming of wire sections', *J. Inst. Metals*, 100, 295–300, 1972.
6.26 Green, D., 'Making copper wire and sections by helical extrusion', *Metall. Engg. Q.*, 13, 18–25, 1973.
6.27 Rowe, G. W., 'Recent advances in wire manufacture', *Int. Metall. Rev.*, 1976.
6.28 Johnson, R. W. and Rowe, G. W. "Predicting the ductility of drawn wire" *Wire Ind.* (Jan), 61–62, 1967.

7

Tube Making and Deep Drawing

7.1 Introduction

Tubes are used for many purposes, and can be produced in a variety of ways. Direct casting is often used for very large pipes, but this does not come within the scope of the present book. Neither does the fabrication of tubes from single- or multiple-layers of strip, which are bent into circular form by rollers and subsequently welded. Such welded tubes are however sometimes redrawn to non-standard sizes, and can then be considered from the point of view of metalworking theory in the same way as seamless tubes.

Most seamless tubes are made by elongation of pierced or bored billets, and pass through many successive operations. In the early stages, it is economic to impart large reductions of cross-sectional area to hot billets by extrusion or cross-rolling, or by pushing on a mandrel through stationary or roller dies. For example, billets of 150 mm (6 in.) diameter may be extruded with as much as 50:1 reduction of area into hollows, as they are called at this stage, which may be 6 m (20 ft) long with 5 mm (0·2 in.) wall thickness. Thick-walled tubes are conveniently produced by cross-rolling or push-bench reduction, and are often used in the hot-finished condition. Most hollows are however redrawn one or many times at room temperature, some to wall-thicknesses of only a few thousandths of an inch. Hypodermic needles of stainless steel may be only one hundredth of an inch in outside diameter. On the other hand, large storage cylinders 0·5 m (2 ft) or more in diameter, may be drawn on large hydraulic benches.

There are four major cold-drawing processes for production of tubes, as shown diagrammatically in Figure 7.1.

As explained in §7.2.4, the heaviest reductions may be taken with a moving mandrel, because the friction at the inner surface carries some of the drawing load. This process also introduces little danger of pickup from the bore of the tube, which is a serious problem in plug-drawing, because there is only a small relative motion between mandrel and tube. On the other hand, mandrels are expensive, difficult to handle, and must be removed by a reeling process. Plugs are comparatively small and can be made of wear-resisting materials. They introduce a second friction surface which is difficult to lubricate, especially because their thermal isolation at the end of a long plug-bar allows the temperature to rise quite rapidly to a value close to that of the rubbing surface. The limiting reduction in area by plug-drawing is more often set by pickup than by stress considerations alone, except for tubes of copper alloys which can be drawn close to the theoretical limits because they are much less prone to pickup. Floating plugs, which are profiled to match the internal surface of the deforming tube and are held in position by the equilibrium of normal force and frictional drag, operate under less severe conditions and permit heavy reductions with ferrous tubing.

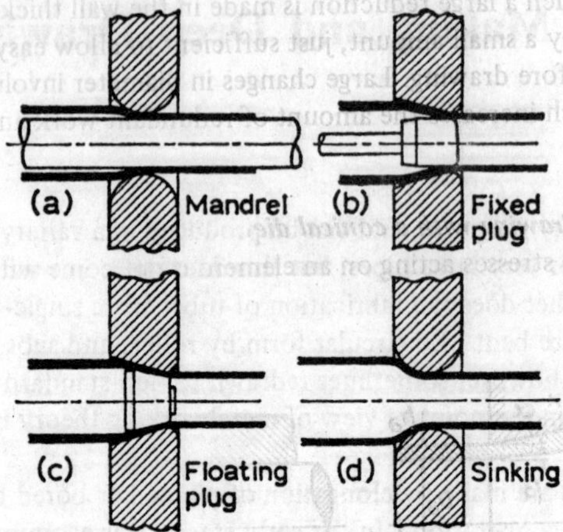

Figure 7.1 Tube elongation by drawing with internal support by (*a*) mandrel, (*b*) plug, (*c*) floating plug; and without internal support by sinking (*d*).

In these three processes the major part of the deformation is reduction of wall thickness, but it is also possible for the diameter to be reduced, accompanied usually by a small increase in wall thickness, by sinking the tube without internal support. The reflex-bending involved may however cause internal cracking, so the process is not widely used; but most tube-drawing with a mandrel or plug involves some reduction in internal diameter, if only to allow initial clearance in the bore. There is consequently a proportion of sinking for all except the hypothetical close-pass, in which the tool fits the undeformed tube bore exactly.

Simple stress evaluation, analogous to that for wide-strip drawing, permits reasonably accurate calculation of drawing stress and maximum reduction for many practical schedules approximating to close-pass conditions. The hoop strain in such passes in negligible, so that plane-strain theory can be applied, but no slip-line field has been proposed which will take account of the sinking part of the deformation, which probably provides the greater part of the redundant work. It is however possible to make semi-empirical allowance for the contribution of the redundant work to the drawing stress, on the basis of measurements of yield stress and hence of the effective strain in drawn tubes.

Deep-drawing is similar in some respects to tube sinking but is more complex because of the severe shape change even in simple cupforming. The major mechanical feature of interest in deep drawing is the change in local thickness which occurs at various positions, and this is best studied empirically. Anisotropy may be very important in practical deep-drawing but this is beyond the scope of this book.

7.2 Determination by stress evaluation of the load for close-pass drawing of thin-walled tube

It is convenient at first to suppose that the diameter of the tube remains constant, and that the wall thickness alone is changed during drawing. There is then no hoop strain, and plane-strain conditions can be assumed. This is nearly true of many industrial passes, in which a large reduction is made in the wall thickness while the diameter is decreased by a small amount, just sufficient to allow easy insertion of the plug or mandrel before drawing. Large changes in diameter involve a high proportion of sinking which increases the amount of redundant work and reduces the efficiency.

7.2.1 *Close-pass plug-drawing with a conical die*

Figure 7.2(*a*) shows the stresses acting on an element of tube between the die and the plug[7.1].

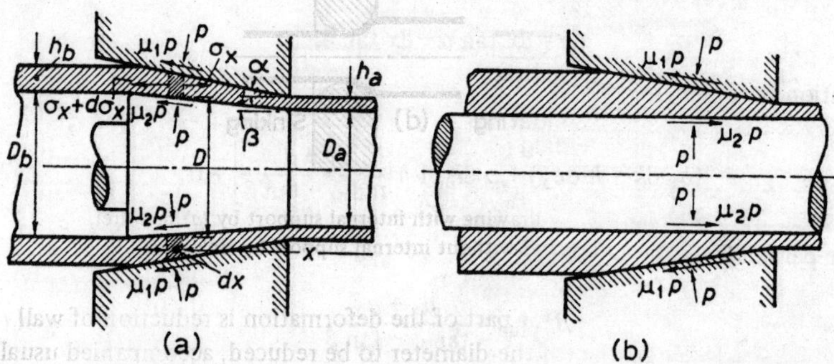

Figure 7.2 The stresses acting on an element of thin-walled tube in close-pass drawing (*a*) over a slightly-tapered plug; (*b*) with a moving mandrel.

If the wall thickness h is small in comparison with the mean tube diameter D, which remains nearly constant, the determination of drawing stress proceeds as for plane strain strip-drawing (Chapter 4, §4.4.1), except that the strip width w is replaced by the tube circumference. Assuming the die pressure p to be equal to the plug pressure, the axial components of the forces acting on the element are:

due to the longitudinal stress,

$$(\sigma_x + d\sigma_x)(h + dh)\pi D - \sigma_x h\pi D = (\sigma_x\, dh + h\, d\sigma_x)\pi D;$$

due to the die pressure

$$\int_0^{2\pi} p\left(\frac{dx}{\cos\alpha}\,\frac{D}{2}\,d\theta\right)\sin\alpha = p\pi D\tan\alpha\, dx;$$

due to the plug pressure,

$$-\int_0^{2\pi} p\left(\frac{dx}{\cos\beta}\,\frac{D}{2}\,d\theta\right)\sin\beta = -p\pi D\tan\beta\, dx;$$

due to the die friction,

$$\int_0^{2\pi} \mu_1 p\left(\frac{dx}{\cos\alpha}\,\frac{D}{2}\,d\theta\right)\cos\alpha = \mu_1 p\pi D\, dx;$$

due to the plug friction,

$$\int_0^{2\pi} \mu_2 p \left(\frac{dx}{\cos \beta} \frac{D}{2} \, d\theta \right) \cos \beta = \mu_2 p \pi D \, dx$$

Under steady drawing conditions, these must be in equilibrium. Thus

$$(\sigma_x \, dh + h \, d\sigma_x) \pi D + p\pi D (\tan \alpha - \tan \beta) \, dx + p\pi D (\mu_1 + \mu_2) \, dx = 0$$

Since the net wall-thickness change is

$$dh = dx \tan \alpha - dx \tan \beta,$$

the equation becomes

$$(\sigma_x \, dh + h \, d\sigma_x) + p \, dh \left[1 + \frac{\mu_1 + \mu_2}{\tan \alpha - \tan \beta} \right] = 0 \qquad (7.1)$$

With the parameter

$$B^*_{\text{plug}} = \frac{\mu_1 + \mu_2}{\tan \alpha - \tan \beta} \qquad (7.2)$$

this may be written

$$h \, d\sigma_x + [\sigma_x + p(1 + B^*)] \, dh = 0 \qquad (7.3)$$

which is of the same form as equation 4.15 for strip drawing, appropriate allowance being made for the value of the parameter B for strip and B^* for tube. As in Chapter 4, consideration of radial equilibrium suggests that the frictional contribution to the die pressure is small, and that the principal stresses can be taken as

$$\sigma_1 = \sigma_x, \;\; \sigma_2 = -p$$

(This also justifies the assumption that die and plug pressures are equal.) Since in a close pass there is no change in diameter, these will be related by the condition of yielding in plane strain, equation 3.27:

$$\sigma_1 - \sigma_2 = S = 1{\cdot}15Y; \;\; \sigma_x + p = S$$

By substitution for p in equation 7.3,

$$\frac{d\sigma_x}{\sigma_x + (S - \sigma_x)(1 + B^*)} = -\frac{dh}{h}$$

$$\frac{d\sigma_x}{B^* \sigma_x - S(1 + B^*)} = \frac{dh}{h} \qquad (7.4)$$

This equation is valid for any values of B^* and S, but the simplest solution is obtained if μ and S are constant or average values, and the die and plug have straight sides, so that α and β are constant. (In practical drawing the plug is usually cylindrical, so

$\beta = 0$). Then, integrating directly,

$$\frac{1}{B^*} \ln [B^*\sigma_x - S(1 + B^*)] = \ln h + \text{constant}$$

$$B^*\sigma_x - S(1 + B^*) = ch^{B^*} \tag{7.5}$$

The constant of integration may be found from the entry conditions. Assuming no back-pull, $h = h_b$, $\sigma_x = \sigma_{xb} = 0$, and

$$c = -S(1 + B^*)h_b^{-B^*}$$

Equation 7.5 can thus be written

$$\frac{\sigma_x}{S} = \frac{1 + B^*}{B^*}\left[1 - \left(\frac{h}{h_b}\right)^{B^*}\right] \tag{7.6}$$

The drawing stress is the axial stress σ_{xa} at the exit, where $h = h_a$.

$$\frac{\sigma_{xa}}{S} = \frac{1 + B^*}{B^*}\left[1 - \left(\frac{h_a}{h_b}\right)^{B^*}\right] \tag{7.7}$$

The die or plug pressure is obtained from the condition $\frac{p}{S} = 1 - \frac{\sigma_x}{S}$.

7.2.2 *Close-pass mandrel-drawing with a conical die*

In plug-drawing, the frictional drag acts in the backward direction on both inside and outside of the tube. When a mandrel is drawn forward with the tube, however, the relative motion on the inside is reversed because the tube elongates while the mandrel remains undeformed. The direction of the frictional force between mandrel and tube is therefore opposite to that between tube and die, as shown in Figure 7.2*b*. The stress equation and its solution are exactly the same as for plug-drawing except that the parameter B^* is changed to

$$B^*_{\text{mandrel}} = \frac{\mu_1 - \mu_2}{\tan\alpha - \tan\beta} \tag{7.8}$$

Thus if $\mu_1 = \mu_2$, as is often true, B^* takes the value zero. Under these circumstances the above integration cannot be used. The differential equation 7.3 with the value $B^* = 0$ becomes simply

$$h\, d\sigma_x + (\sigma_x + p)\, dh = 0$$

$$h\, d\sigma_x + S\, dh = 0$$

This may be directly integrated with the boundary condition $\sigma_{xb} = 0$ at $h = h_b$, giving

$$\frac{\sigma_{xa}}{S} = \ln\frac{h_b}{h_a} = \ln\frac{1}{1 - r} \tag{7.9}$$

This is the familiar expression for homogeneous deformation.

It is however possible for the friction coefficient μ_2 on the mandrel to exceed μ_1 on the die, and B^* is then negative. The analysis is otherwise unaltered, so the drawing stress for mandrel drawing can be less than that for frictionless drawing to the same reduction of area.

7.2.3 *Plug-drawing with circular-profile dies*

One common type of wear during tube-drawing, particularly with relatively hard metals, is the formation of an annular depression or ring at the entry position. Dies of circular profile, symmetrical about the central plane, were once widely used as 'double-entry' dies which could be reversed in the die holder for further drawing after the ring had formed. This was not of course possible if wear or pickup occurred in the throat region, and the large entry angle of small-radius profiles tended to impair lubrication and to introduce inhomogeneity in the product. Straight-taper dies of small angle give better drawing and have generally superseded the double-entry die. Single-entry dies with circular profiles of larger radius are, however, quite satisfactory.

The analytical approach for dies of curved profile is essentially the same for tube as for strip, which was discussed in Chapter 6, §6.3.3. The basic differential equation is the same as that derived above (§7.2.1) in connection with conical dies

$$d(h\sigma_x) + p\,dh\left[1 + \frac{\mu_1 + \mu_2}{\tan\alpha - \tan\beta}\right] = 0 \tag{7.1}$$

If the plug is parallel-sided, as is usual, $\beta = 0$. It is convenient to solve this equation in terms of p/S, using the yield condition $\sigma_x + p = S$. Thus

$$d(hS - hp) + p[1 + (\mu_1 + \mu_2)\cot\alpha]\,dh = 0 \tag{7.10}$$

This is the same as equation 6.13, except that the die friction μ has been replaced by the sum of die and plug friction, $(\mu_1 + \mu_2)$. The solution is thus obtained by making this substitution in the final equations 6.17 and 6.18 for strip-drawing through circular-profile dies:

$$\frac{p}{S} = \frac{h}{h_b}\,e^{(\mu_1 + \mu_2)(H - H_b)}$$

$$\frac{\sigma_{xa}}{S} = 1 - \frac{h_a}{h_b}\,e^{(\mu_1 + \mu_2)(H_a - H_b)} \tag{7.11}$$

where $$H_b = 2\sqrt{\frac{R}{h_a}}\tan^{-1}\left(\sqrt{\frac{R}{h_a}}\,.\,\alpha_b\right)$$

It is commonly assumed that $\mu_1 = \mu_2$. Then

$$\frac{\sigma_{xa}}{S} = 1 - \frac{h_a}{h_b}\,e^{2\mu(H_a - H_b)}$$

7.2.4 *Maximum reductions of area in tube drawing*

As in drawing of strip (Chapter 4, §4.4.2) or round rod (Chapter 6, §6.4.4) the greatest possible reduction of area per pass is limited by tensile fracture of the drawn metal, which occurs at the yield stress if there is no appreciable strain-hardening.

Plug-drawing: Because the tube is unsupported after leaving the plug, it can contract circumferentially under the influence of the axial tension. The tensile yield stress eventually limiting the pass is therefore Y, not S, though plane-strain conditions are maintained in the actual working zone. Assuming failure to occur as soon as the yield stress Y in the drawn tube is reached, the maximum reduction in one pass may be found from equation 7.7:

$$\frac{Y}{S} = \frac{1 + B^*}{B^*}\left[1 - \left(\frac{h_a}{h_b}\right)^{B^*}\right]_{max}$$

With typical values ($\mu_1 = \mu_2 = 0{\cdot}05, \alpha = 15°, \beta = 0$)

$$B^* = 2\mu \cot \alpha = 0{\cdot}373, \quad (1 + B^*)/B^* = 3{\cdot}68$$

$$\left(\frac{h_a}{h_b}\right)_{max} = 0{\cdot}489; \; r_{max} = 1 - \frac{h_a}{h_b} = 0{\cdot}51 \tag{7.12}$$

This is lower than 0·55, the limit for comparable strip-drawing (equation 4.24) because of the additional frictional drag on the parallel plug. Both limits will be increased slightly if the metal work-hardens, because the final yield stress Y_a will then exceed the mean tensile yield stress, $\overline{Y}(=\overline{S}/1{\cdot}15)$.

Strip-drawing can be considered as a special instance of the more general theory, in which $\mu_1 = \mu_2$ and $\beta = -\alpha$. Then

$$B^* = \frac{\mu_1 + \mu_2}{\tan \alpha + \tan \beta} = \frac{2\mu}{2 \tan \alpha} = B.$$

Mandrel-drawing. In mandrel drawing, if the coefficient of friction is the same on the inside and outside of the tube, B^* is zero, and the drawing stress is the same as for frictionless strip-drawing (equation 7.9). The limiting reduction is however greater in tube drawing on a moving mandrel, because the tube is restrained from contracting circumferentially both during passage through the die and after the deformation has been completed. Plane-strain conditions therefore operate throughout and the limit is given by

$$\ln \frac{1}{1 - r_m} = \frac{S}{S}; \; r_m \approx 0{\cdot}63$$

This is of course the same as that for frictionless drawing of round rod, where the yield stress is Y throughout (equation 4.11), but it is greater than 58%, the limit for frictionless strip drawing (equation 4.25).

It is actually possible to draw to greater reduction still with higher friction on the mandrel, since some of the load is carried by the mandrel. This is best utilised in tandem drawing.

7.3 Tandem drawing of tubes on a mandrel

If two dies are mounted on the same bench, with a small separation between them, a tube may be drawn sequentially through them on a mandrel passing through both. Greater reductions in area are then possible in one pass than can be obtained with a single die. The reason is that some of the drawing stress is provided directly by tension in the tube wall, but a proportion is derived from the frictional drag between the drawn tube and the mandrel. This portion depends on the area of contact and increases with the length of drawn tube ahead of the second die, and so with the separation of the two dies. If the separation is large enough, the frictional drag may be sufficient to permit a reduction in area r_2 at the second die equal to that in the first, r_1; no direct contribution of tension in the tube being necessary. The total reduction of area r_{12} of a tube drawn from area A_c to A_b in one die and then to A_a in the second is found in either of the usual ways:

$$r_1 = \frac{A_b - A_a}{A_b}, \quad r_2 = \frac{A_c - A_b}{A_c}, \quad r_{12} = \frac{A_c - A_a}{A_c}$$

$$r_{12} = 1 - \frac{A_a}{A_c} = 1 - (1 - r_1)(1 - r_2) = r_1 + r_2 - r_1 r_2 \tag{7.13}$$

Alternatively, $$\epsilon_{02} = \epsilon_{01} + \epsilon_{12} = \ln\frac{1}{1 - r_1} + \ln\frac{1}{1 - r_2}$$

Thus for frictionless drawing the limits are $r_1 = r_2 = 0{\cdot}63$ and $r_{12} = 0{\cdot}86$.

$$(\epsilon_{01} = 0{\cdot}994, \ \epsilon_{02} = 2\epsilon_{01} = 1{\cdot}988)$$

With practical separations the limit is reduced, but it has been shown in a comprehensive experimental study by Sachs and Espey[(7.2)] that reductions of area of 75% are possible in one pass with two dies in tandem.

This principle can be extended and is particularly useful in hot working, where μ is naturally high, and S does not increase. Five or six dies may be used, for example in hot-forming large gas-storage cylinders. The dies are lubricated with a graphitic material and the mandrel is unlubricated. Since these tubes have a closed end, it is convenient to push against this end with the mandrel, no direct tensile force being applied. The drawbench thus becomes a push-bench.

The friction on the outside of the tube may be reduced still further by replacing each die by three equally-spaced contoured rollers, with their axes parallel to three tangents mutually inclined at 120°. Each die has its rollers in a different orientation from the others so that a truly circular tube is produced. As many as 12 or even 20 dies may be used with a single very long mandrel, and very heavy reductions of area are possible on such a roller push-bench.

7.4 Tube sinking

The sinking process, of diameter reduction with no internal support for the tube, is not often used with metals of moderate or poor ductility because of the danger of cracking in the tube bore during the reverse bend at the exit. It is however important as part of a normal drawing pass, since it is usually desirable to reduce the diameter of the tube as well as its wall thickness. Some diameter reduction is always necessary because clearance must be provided for the insertion of the plug or mandrel. All practical drawing passes thus include a proportion of sinking before the bore of the tube contacts the inner tool. The differential equation derived by Sachs and Baldwin[(7.3)] for this process is similar to that for wire-drawing (equation 6.21) except that for the more complex stress system in tube sinking a modified yield stress $Y' =$

mY is used as a simplification of the von Mises yield criterion. The best value for the constant multiplier m is 1·10, so that

$$\sigma_1 - \sigma_3 = 1{\cdot}10Y \tag{7.14}$$

The appropriate equilibrium equation, assuming no wall-thickness change, is

$$\frac{d\sigma_\alpha}{\sigma_\alpha B - 1{\cdot}1Y(1+B)} = \frac{dr}{r} \tag{7.15}$$

The stress σ_α is the direct stress acting on a section perpendicular to the conical surface. At the exit this becomes the drawing stress σ_{xa}, given by the solution of equation 7.15 with the boundary condition $\sigma_{xb} = 0$, in exactly the same way as the solution for wire drawing (§6.4.1)

$$\frac{\sigma_{xa}}{1{\cdot}1Y} = \frac{1+B}{B}\left[1 - \left(\frac{D_a}{D_b}\right)^B\right] \tag{7.16}$$

The ratio D_a/D_b is raised to the power B, not $2B$ as in wire-drawing because it represents an area ratio, as explained in §6.4.1, and the cross-sectional area in tube sinking is approximately πDh, not $\pi/4 \,.\, D^2$ as in rod-drawing. Thus for given values of die angle, coefficient of friction and reduction of area, the drawing stress for tube sinking is the same as that for rod-drawing, except for the 10% increase due to the greater yield stress in sinking. This has been verified experimentally by Sachs and Baldwin[7.3].

Under most practical conditions, the wall thickness is slightly increased by sinking, but this does not seriously affect the calculated stress.

7.5 Redundant work in tube drawing

There is no general slip-line field for prediction of redundant work in tube-drawing. Green[7.4] has suggested that it should be possible to apply a redundant-work correction to the stress evaluations for close-pass drawing, in a manner similar to that described for strip-drawing in Chapter 6, §6.6.3. It is however probable that a high proportion of the redundant work is attributable to the sinking which precedes drawing in most practical passes, and this has yet to be studied in detail.

A semi-empirical approach has been taken by Blazynski and Cole[7.5]. The effective strain is estimated by comparing data from a plane-strain compression test on a sample cut from the drawn tube, with the test data obtained on annealed metal. As in Figure 4.4, the curves can be fitted to show the redundant work factor. The mean yield stress for the pass, including the redundant strain, is then used in computing the frictionless drawing stress (equation 7.9). The contributions of homogeneous deformation and redundant work to the total load are thus determined and the frictional contribution may be found by subtraction from the measured load. In the experiments quoted, the frictional contribution was also determined independently by measuring the mean die pressure, using a calibrated die. Good agreement was found, when an allowance was made for the departure from plane-strain conditions during the sinking part of the operation. Selecting one example, at 40% reduction of area the friction was found to contribute about 20% of the total drawing load, and the redundant deformation about 8%, in a pass involving 6% sinking.

7.6 Deep-drawing and pressing

Deep-drawing is a complex operation which in its simplest form, often used for test purposes, can be represented by the formation of a cylindrical cup from a thin disc. A circular punch forces the sheet metal through a die orifice with a small radial clearance. To avoid crinkling, the residual flange is held in contact with the flat face of the die throughout the operation. The passage of metal through the die resembles mandrel push-bench forming, but there is no deliberate reduction in wall thickness and the prior part of the operation is an exaggerated type of sinking. The process is governed by equations similar to those for tube sinking, but the outer radius b of the blank decreases continuously throughout the pass. At any instant, the radial stress in the disc at radius x is given, for frictionless drawing, by

$$\frac{\sigma}{1{\cdot}1Y} = \ln\frac{b}{x} \tag{7.17}$$

and the tangential stress is consequently

$$\frac{\sigma_\theta}{1{\cdot}1Y} = 1 - \ln\frac{b}{x}$$

These are the homogeneous deformation results, with allowance for the yield conditions. A more important feature than the drawing load is the local strain in the sheet, which may lead to local necking and finally to fracture. Frictionless radial drawing tends to thicken the sheet, while the bending and sliding over the die profile and the punch head tends to thin it. The most serious thinning arises from the stretching over the punch head and particularly between the punch head and the die. To reduce the thinning as far as possible it is desirable to maintain high friction on the punch, as in mandrel drawing, but low friction everywhere else. Swift[7.6] has made extensive studies of the effects of the many external variables in deep drawing, and Jevons[7.7] has examined the relevant properties of materials for deep-drawing.

7.7 Tube production by rolling and extrusion

If a cylindrical billet is cross-rolled by two large-diameter barrel-shaped rolls whose axes are inclined to its axis, a tension is produced at the centre of the billet. In a Mannesmann-type piercer[7.8] this tension is increased until a cavity appears in the core. A plug is inserted to smooth out the bore of this hole, and the wall thickness of the hollow so made is then reduced by compressive rolling action between plug and rolls.

Three rolls of special design, all inclined to the tube axis, are used in an Assel elongator[7.9]

7.7.1 *Pilgering and cold reducing*

A widely-used method for hot-reduction of both diameter and wall thickness is known as the Pilger process. The hollow bloom is carried on a central rod and fed into the gap between two transverse rolls. Each roll has a circumferential groove of

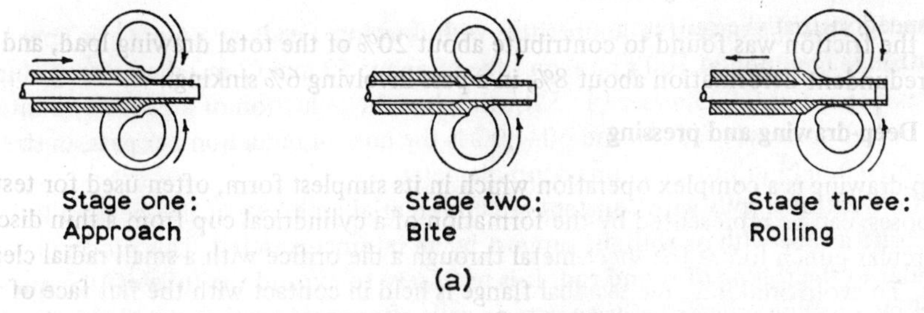

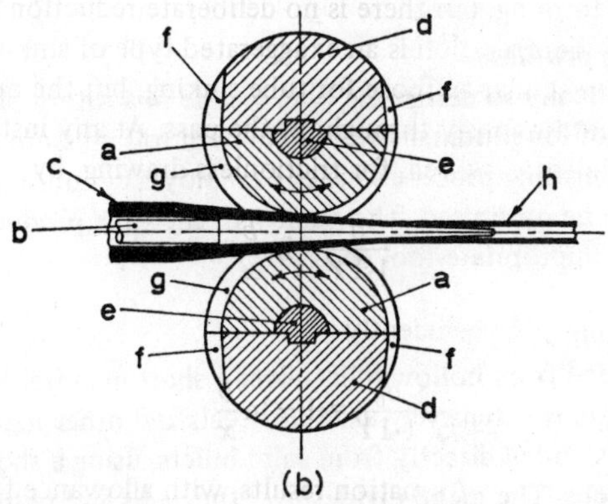

Figure 7.3

(*a*) A diagram of the Pilger process for hot tubes.
(*b*) A diagrammatic view of the cold-reducing process.

(*Courtesy of Tube Investments Ltd. and Mannesmann-Meer AG.*)

decreasing depth, which rolls the bloom on to the central rod (Figure 7.3*a*). The rolls rotate against the movement of the bloom and are partly cut away, so that the bloom can be advanced a little after each stroke. It is also rotated through 90° each time to maintain uniform walls. The name derives from the German pilgern, to go on a pilgrimage, because of the fancied resemblance to a pilgrim's advance, two steps forward one step back.

Cold-reducing[7.10] resembles Pilgering in some respects but is a high-precision process used to impart heavy reductions of wall thickness at room temperature, especially for expensive metals such as stainless steel which are prone to pickup trouble in drawing. There is much less interfacial sliding than in drawing, and lubrication is relatively simple.

Two rolls are used with a tapered semi-circular groove in each, as in Figure 7.3*. The rolls rotate in the direction of elongation of the tube and the whole roll assembly reciprocates along the tube as the rolls rotate. The wall thickness is thus rolled down on the fixed tapered mandrel much more steadily than in the hot Pilger process. Between each cycle the tube is again rotated by 90°. The large reciprocating mass requires careful balancing, but modern designs have increased the operating speed so that the output is competitive with drawing. Feeds of 5 mm (¼ in.) per stroke are common at 100 r.p.m. on tube of about 50 mm (2 in.) diameter, giving an exit speed of 0·025 m/sec (4–5 ft/min), but the reduction of area is not limited by pickup or tensile failure, and 60–80% is the normal reduction per pass, in sizes from 10 mm (½ in.) to 200 mm (8 in.) diameter.

Conventional rolls with constant circumferential grooves can also be used to elongate a tube with or without support by an internal mandrel. This process is very similar to the rolling of round rod. It is necessary to expand the tube off the mandrel by a reeling process between convex rolls afterwards, so that the mandrel can be extracted easily.

7.7.2 *Roll and plug profiles*

It is possible theoretically to design the plug profile in a rotary piercing operation to produce constancy of longitudinal strain, minimum roll pressure or other desired effects(7.11). The elongating processes usually employ a cylindrical mandrel, and the roll profile can then be optimised. The quality of the tube produced can be significantly improved by appropriate tool design(7.12).

7.7.3 *Tube extrusion*

Tubes can be extruded from hollow billets over a short mandrel held from the pressure pad. This method is extensively used for steels and other hard alloys. Lead and aluminium can be extruded directly from solid billets, using a short plug supported in the die by thin webs. The metal divides over the webs and reunites by pressure welding as it passes through the die beyond them. This technique can also be applied to copper and copper alloys but care must be taken to avoid entraining oxides of these metals. Many hollow sections, as well as simple tubes, can be produced by such bridge-die extrusion. Because the plug is rigidly held, the wall thickness and eccentricity tolerances can be maintained within much smaller limits than for conventional extrusion. The strength of the bridge at working temperature is a major limitation.

The pressure required for extrusion of tubes can be estimated with sufficient accuracy by ignoring the detailed disturbance by the internal tool and calculating the pressure for extrusion of bar to the same area ratio. More accurate solutions can be obtained by application of upper-bound technique, as explained in Chapter 8.

7.8 Lubrication in practical tube making

The main lubrication problems in tube making appear in the drawing processes, especially fixed-plug drawing(7.13).

Hot rolling and piercing of tubes require adequate cooling of the tools but as this is completed in a short time-cycle, using water sprays, thermal crazing often occurs. This makes any form of lubrication very difficult. Graphite is added to the piercing-plug coolant, but otherwise little attempt is made to provide good lubrication. Pilger mills and tube elongators also usually rely on water cooling alone. Roller push-benches, using sets of shaped rolls whose axes are transverse to the tube, are lubricated with graphite or, for stainless steel, with glass. Care must be taken to achieve the correct balance of lubrication on the mandrel; too high friction leads to sticking and problems of removing the mandrel, but too low friction fails to provide the necessary drawing force, which cannot be sustained by the tube end only, when several die stands are involved, as explained for tandem drawing in §7.3. The large push benches with simple tandem dies also require high mandrel friction, usually obtained with plain water sprays, and low die friction, produced by swabbing with graphited grease.

Hot and cold extrusion of tubes follows extrusion lubrication practice generally and is considered in Chapter 8.

Cold drawing uses the same techniques as the severer forms of wire drawing. Lubrication in sinking is relatively easy, and few problems arise in lubricating the outside of tubes if good wire-drawing practice is adopted. *Mandrel drawing* (§7.2.2) permits heavy reductions, and a fairly high mandrel friction is advantageous (equation 7.8). The main purpose of the internal tube lubricant is then to avoid pickup and to facilitate the subsequent removal of the mandrel. The tubes are often reeled to expand them off the mandrel. Floating plugs position themselves according to the balance of forces and usually encourage a fairly thick film of lubricant to form. Heavy reductions are thus possible without a limitation from the lubricant, though stability may be a problem. Fixed plugs, in contrast, are difficult to lubricate. They are inaccessible and the geometric conditions do not usually favour hydrodynamic lubrication. The plug temperature tends to rise towards the interfacial temperature during long draws, because the heat conduction down a long bar is inadequate. Attempts have been made to pass coolant down hollow plug bars, to cool the plug and also to provide a better chance of hydrodynamic lubrication. Unfortunately this increases the stress in the plug bar because of its reduced cross-section, increasing the risk of chatter.

It is possible to improve the external lubricant film thickness by drawing through two dies, the first of which acts as a sizing and sealing die to allow pressure to build up in a chamber joining it to the second die, in which the real reduction occurs. The pressure may be applied by an external pump, or preferably by hydrodynamic action at suitable drawing speeds. Dried soap coatings can be used. However the major problem remains, namely the lubrication of the plug. Tandem plug arrangements have been proposed, but with little, if any, acceptance in practice.

7.8.1 *Steel tube drawing*

For carbon steels phosphate conversion coatings with a sodium stearate lubricant are widely used. Chlorinated or even boundary-lubricating oils can be used with advantage on lighter draws. Handling between the various coating operations is then avoided and high-speed on-line lubrication is possible.

Stainless steels require oxalate coating with soap, or for light work, chlorinated oil. Ultrasonic vibration of the plug is believed to enhance lubrication and is sometimes used industrially. Various other lubricants have been proposed, notably polymeric coatings, which were advocated in about 1960, but these are now usually considered uneconomic. For very thin-walled tubes, oxalate coating or pickling processes cannot be used because of the danger of pinholes developing, so mechanical roughening of the surfaces is employed to provide enhanced lubricant throughout. Cold-reducing, lubricated with chlorinated oil, has become economically competitive in recent years.

7.8.2 *Other alloys*

Copper tubes are much easier to lubricate than steel tubes. As in wire drawing, compounded oils can be used for small tubes drawn at high speed. Grease or dried soap is preferred for large-diameter stock. Staining, especially by sulphur, must be avoided.

Aluminium tubes also present little trouble. Compounded oils, soap or waxes may be used.

Titanium and nickel alloys follow the pattern for stainless steels. Oxalate coatings with soap give good performance, possibly improved by fluoride/phosphate with soap for titanium. There is always a danger of pickup, and at one time polymer lubricants were favoured because they practically eliminate pickup. Cold reducing, which imposęs less demands than drawing on the lubricant, has always been competitive for titanium tube and has recently been improved with better-balanced high-speed machines.

7.9 Lubrication in deep-drawing and pressing

The severity of lubrication conditions in sheet forming varies from hydrostatic *bulge forming*, which requires no lubrication, to *ironing*, which closely resembles tube drawing. The lubricants used for mandrel-drawing of tubes are generally applicable to ironing, but additional problems arise because of the very thin walls that may be required. Small defects, increasing the force only slightly, may lead to fracture.

Stretch forming introduces little danger of pickup, but it is essential to provide uniform low friction over the large areas. For small specimens, polyethylene sheet is very effective and is widely used in drawability testing, but in industrial sizes it tends to tear. Boundary lubricants or lamellar solids may be used. An interesting technique introduced some years ago involves chilling the forming tool to about −2°C, and spraying a thin film of water onto it. The ice so produced is melted in contact with the deforming sheet but cannot escape, so it provides a very uniform low-friction layer. More recently a technique of supplying oil through fine capillary orifices to the surface of the tool has been developed. This improves substantially on the performance of a polyethylene film. The oil seeps out onto the surface under moderate pressure and is then trapped, providing a low friction layer.

Ultrasonic vibration has also been used to improve the lubrication in stretch forming[7.16, 7.17].

Deep drawing is of intermediate severity, and it is important to recognise that the lubrication conditions differ considerably in different regions. The sheet is restrained by a blank holder, whose pressure adjustment is critical. If the pressure is too high there is a danger of lubricant breakdown, with scoring or, in extreme cases, pickup. If it is too low, the sheet will buckle as it draws radially inwards. At the correct setting, soap, oil or a graphite film is adequate. As the sheet enters the die it is stretched over the die mouth profile, but as it progresses further the pressure on both sides increases and the main body of the cup or other product is made by a mandrel-drawing type of process. In addition, the end of the punch is important. If this is rounded, there is significant stretch forming over the punch nose, while if it is flat the bottom of the product will remain at the original thickness. There is thus considerable scope for optimising lubrication conditions. In common with mandrel tube drawing, a high punch friction with a low die friction gives the lowest force.

The performance in a deep-drawing test is often assessed by the limiting drawing ratio (LDR), which relates to the greatest diameter of blank that can be drawn completely into a cup of given diameter. Low clamping pressure, low blankholder friction and die friction but high punch friction contribute to a large LDR. Differential or selective lubrication is therefore important, and the punch may be left unlubricated. Surface roughness also affects the results, and the punch is sometimes deliberately roughened, while the die is highly polished. Slight roughening of the blank can also be used. This somewhat increases the friction but increases the amount of lubricant carried into the tool interface, thereby reducing the danger of pickup. For comparison of drawability of different alloys and the effect of different mechanical and thermal pretreatments a standardised cupping test proposed by Swift is frequently used[7.13]. A common reference oil (Esso TSD 996) is widely used for comparative purposes, though polyethylene gives a deeper draw and is also often quoted, particularly when a large stretching is involved, using a rounded punch.

Vibrational energy, when suitably applied, can also increase the possible depth of draw, especially with sheets of alloys otherwise difficult to lubricate[7.16].

In industrial practice the shapes are seldom simple, and skilful application of lubricant is important. Local areas are lubricated while others are left dry, or are deliberately roughened, to distribute the flow of metal. *Draw beads* are widely used to restrain the sheet, for example along the sides of a rectangular pressing. These are raised bars of approximately circular or rectangular profile protruding from the die and matching suitable cavities in the blankholder. The sheet is forced to pass through the convoluted channel so formed, and is severely restrained. Scoring and pickup often occur on drawbeads but the damaged material may be confined to the residual flash.

Bending, spinning and *hydraulic forming* (rubber punching) can be lubricated easily. Explosive, electromagnetic and electrospark (hydrospark) bulging techniques are better employed without lubricant, to avoid local pools being trapped and causing blemishes.

Almost all major lubricants have been used at some time in some sheet forming operation, including vanishing oils, whose main characteristic is to disappear by evaporation after the process.

7.9.1 *Sheet steel forming*

For shallow pressings, mineral oils are usually adequate. Removal of the lubricant after the pressing is an important criterion and for this reason water-base fluids, mainly soap solutions and emulsions, are preferred. Soap-fat emulsions or compounded oils are used when wear is an important consideration. Thicker films of emulsion can be provided by incorporating fillers such as lime or talc to produce pastes.

On more severe operations, chlorinated oils may be used, and when appreciable ironing is involved, phosphate and soap gives the deepest draws, though removal is then more expensive. Lamellar solids are also effective but difficult to remove without clogging filters. Soap and fat compounds have been widely used.

Stainless steels are more likely to cause trouble through lubricant breakdown. For light work, soap-fat emulsions can be used, but these are often pigmented with graphite or MoS_2. Mineral oil is satisfactory for spinning and similar operations, but chlorinated oils or fortified emulsions give better protection from pickup. Dry soap films are widely used, with or without an oxalate carrier. Thin polymeric coatings give good protection during forming but are expensive to apply and to remove.

For hot punching it is possible to use glass as an effective lubricant.

7.9.2 *Copper alloys*

Copper is easy to form and gives very little scoring or adhesion trouble. It is however important to avoid staining, particularly by sulphur compounds. Brasses and bronzes require more-protective lubricants. Soaps are probably the best lubricants for copper alloys. For stamping and other light work mineral oil, water-base solutions or emulsions can be used. Chlorinated or fatty additives may be incorporated, but not sulphurised E.P. compounds. Soap-fat emulsions are suitable for pressings. Deep drawing is usually performed with chlorinated oils or dried soap films.

7.9.3 *Other alloys*

Pickup can cause trouble with aluminium. Mineral oil with a boundary additive is preferred but polyethylene films can be used. These are particularly effective in small components and can be improved by embossing the polyethylene to entrap a subsidiary oil lubricant. A straight mineral oil suffices for stamping. Water-base emulsions may be used under moderately severe conditions and again have the advantage of ease of removal, which is particularly important because of the danger of staining aluminium. They are suitable for stretch forming.

Pigmented soap-fat emulsions are the main lubricants for deep drawing aluminium, but dry wax films are quite widely used.

Titanium is usually hot formed at 500°C–800°C, with a bentonite grease and graphite. Pickup presents serious problems at room temperature. Methacrylate resins give the best protection. If these are inconvenient or expensive, the titanium surface may be lightly oxidised, or preferably fluoride-phosphate coated, and lubricated with sulphonated oil or a MoS_2 resin.

Nickel alloys also usually require a polymer coating or an oxalate with an E.P. lubricant.

The refractory metals tungsten and molybdenum can be warm formed with graphite or with basalt. For cold forming molybdenum, aluminium-bronze dies are used with castor oil as the lubricant.

7.10 Recent developments in tube manufacture

7.10.1 *Theoretical contributions*

As in many other areas of metalworking, there is a strong interest in linking the theory of tube-making processes with practice, especially in the design of tools and the prediction of the final properties and possible defects in the tubes.

Visioplasticity methods, sometimes including computer-aided solutions, are very helpful. Deformation patterns can be obtained and correlated with theoretical analysis. Improved profiles can be predicted for extrusion and for rotary processes of cross-roll piercing and elongation[7.11].

Upper-bound solutions are available for extrusion and for drawing processes, including floating-plug drawing. These too can be used to optimise pass and profile design, as for wire drawing[7.12].

Longitudinal tube rolling involves little redundant work and can conveniently be studied by stress-analysis methods.

7.10.2 *Improvements in conventional processes*

Quality is a very important feature of tubes, especially those made of expensive materials or destined for high-pressure uses. Defects, if any, are most likely to occur on the interior surface, where they are difficult to see. Some of the most important developments in tube manufacture have consequently been those providing automatic on-line inspection at high speeds by eddy current, ultrasonic or inductive methods. Floating plugs are more widely used than before, partly because they provide a more consistently satisfactory bore than fixed plugs, and partly because longer lengths can be processed. The main problem in the use of floating plugs for stainless and carbon steel is now the coiling of stiff tubes. Cold pilgering (cold reducing) has also increased in popularity. It provides high surface quality and freedom from pickup even with titanium and stainless steel, and the newer high-speed, balanced machines compete well economically with drawbenches.

For hot rolling, three-roll elongators with improved roll profiles are available and are sometimes used to feed a long sequence of roller dies. There is considerable interest in the warm working of stainless steel, since the yield stress can be reduced by over 30%, and a comparable or better improvement in drawability obtained, by operating at 300°C.

Savings in annealing costs can be very significant even in conventional drawing, if proper attention is paid to pass sequences and to annealing cycles.

7.10.3 *Newer processes*[7.15]

Combined rotary piercing and elongating, and sometimes repiercing, can give better bore surfaces than the usual piercing operation. This is important because it is recognised that defects introduced at this stage cannot be rectified, and usually become worse in subsequent processing[7.11].

The production of thin-walled tubing has always presented special problems, but various methods can now be used. Hydrostatic extrusion can produce very thin-walled tube to high tolerances in titanium and other alloys, and can also be used for alloys of relatively low ductility that cannot otherwise be extruded[7.14]. Roll forming, which resembles sheet spinning, offers another possibility, while explosive and implosive forming can be used for special purposes, especially for the production of short lengths of bimetallic tubing.

Longer lengths of clad tubes can be drawn coaxially or by drawing a helically-wound pair of strips, one of which carries a groove near the edge to form a mechanical lock. Helical welding is also possible.

Electromagnetic, electrohydraulic and hydraulic techniques can all be used to deform tubes in various ways, to make crimped joints, bulges or constrictions.

Lubricants remain little changed in general classification, though accelerated conversion coatings have been introduced. Several studies of the application of ultrasonic vibration suggest that die forces can be reduced and that better lubrication and tool life result from suitable vibration of the die or the plug in tube drawing. This appears to be particularly relevant in the ironing of thin-walled cups.

EXAMPLES

7.1 Compare the drawing loads for plug and mandrel drawing of annealed mild steel tube from 50 mm internal diameter and 2·5 mm wall thickness to 49 mm I.D. x 1·8 mm. Assume that the included angle of the die is $2\alpha = 30°$ and that $\mu = 0{\cdot}10$.

Solution. Area $A_0 = \pi\bar{D}_0 h_0 \approx \pi D_0 h_0 = 392$ mm², $A_1 = 277$, $r = 0{\cdot}293$. Using the stress equation 7.7

$$\frac{\sigma_{xa}}{S} = \frac{1+B^*}{B^*}\left[1-\left(\frac{A_1}{A_0}\right)^{B^*}\right]$$

(*a*) $B^*_{\text{plug}} = (\mu_1 + \mu_2)/(\tan\alpha - \tan\beta) = 0{\cdot}2 \cot 15° = 0{\cdot}746$

$$\frac{\sigma_{xa}}{S} = \frac{1{\cdot}746}{0{\cdot}746}\left[1-\left(\frac{277}{392}\right)^{0.746}\right] = 2{\cdot}34\,(1-0{\cdot}772)$$

$$= 0{\cdot}534$$

(If the ratio h_1/h_0 were used, the value would be 2·34 (1 – 0·782) = 0·51)

(*b*) $B^*_{\text{mandrel}} = (\mu_1 - \mu_2)/(\tan\alpha - \tan\beta) = 0$

$$\frac{\sigma_{xa}}{S} = \ln\frac{A_0}{A_1} = 0{\cdot}35$$

The load for plug drawing is thus about 50% greater than for mandrel drawing.

7.2 Suggest a tandem die arrangement to allow tubes to be drawn with 65% reduction of area in one pass.

Solution. $r_{02} = 0{\cdot}65$, $\epsilon_{02} = \ln\dfrac{1}{1-r} = 1{\cdot}05$. Two passes $\epsilon_{01} = 0{\cdot}55$ and $\epsilon_{12} = 0{\cdot}50$ ($r_{01} \doteq 0{\cdot}42$, $r_{12} = 0{\cdot}39$) could be used. For most drawing operations it is preferable to take a heavier reduction in the first pass, e.g. $\epsilon_{01} = 0{\cdot}60$ and $\epsilon_{12} = 0{\cdot}45$, ($r_{01} = 0{\cdot}45$ and $r_{12} = 0{\cdot}36$), because of work-hardening, but for tandem drawing this would involve a greater separation of the dies, and the order should be reversed, i.e. 36% in the first die and 45% in the second.

7.3 Estimate the draw-bench pull for plug drawing of stainless steel from 45mm ID x 1·60 mm to 43 mm x 1·25 mm with circular-profile dies of 50 mm radius, assuming $\mu = 0{\cdot}08$ and $S = 0{\cdot}60$ kN/mm². Would it be necessary to make allowance for redundant work? To what extent is the result influenced by the coefficient of friction?

Solution. Use equation 7.11, assuming that the internal diameter remains unchanged. The angle of contact is given by $\cos\alpha_b = (R - \Delta h)/R$,

$$\alpha_b = \cos^{-1} 0{\cdot}993 = 0{\cdot}118 \text{ rad}$$

$$H_b = 2\sqrt{\frac{50}{1{\cdot}25}}\tan^{-1}\left(\sqrt{\frac{50}{1{\cdot}25}} \times 0{\cdot}118\right) = 12{\cdot}65 \tan^{-1} 0{\cdot}746$$

$$= 8{\cdot}11$$

Since $\alpha_a = 0$, $H_a = 0$. Assuming $\mu_1 = \mu_2 = 0{\cdot}08$

$$\frac{\sigma_{xa}}{S} = 1 - \frac{h_a}{h_b}e^{2\mu(H_a - H_b)} = 1 - \frac{43}{45}e^{-1.30} = 0{\cdot}74$$

Since $A_a = 169\ \text{mm}^2$ and $\overline{S} = 0{\cdot}60, P = 75\ \text{kN}$

If $\mu = 0{\cdot}05, \dfrac{\sigma_{xa}}{S} = 1 - 0{\cdot}955\ e^{-0{\cdot}811} = 0{\cdot}58$

$\mu = 0{\cdot}10, \dfrac{\sigma_{xa}}{S} = 1 - 0{\cdot}955\ e^{-1{\cdot}62} = 0{\cdot}81$

Comparison with the results quoted in reference 7.5 suggests that the contribution of redundant work could be neglected in this pass. This can also be seen by analogy with strip drawing (equation 6.37 and Table 6.1):

$$r = \frac{1{\cdot}6 - 1{\cdot}25}{1{\cdot}6} = 0{\cdot}22;\ \frac{c}{d} = \frac{2 - 0{\cdot}22}{0{\cdot}22}\ .\ 0{\cdot}118 = 0{\cdot}95;\ f\left(\frac{c}{d}\right) \approx 1$$

7.4 What is the theoretical maximum reduction of area that can be taken in one close pass with 25 mm I.D. x 1 mm mild steel tube on a fixed plug with (*a*) $\alpha = 15°$ straight-taper and (*b*) 50 mm radius-profile dies, if $\mu = 0{\cdot}05$?

Solution. (*a*) Use equation 7.7. $B^* = 2\mu \cot \alpha = 0{\cdot}373$. For maximum pass $\sigma_{xa} = Y$, so

$$\frac{\sigma_{xa}}{S} = 0{\cdot}866 = \frac{1{\cdot}373}{0{\cdot}373}\left[1 - \left(\frac{h_a}{h_b}\right)^{0{\cdot}373}\right]$$

$$\left(\frac{h_a}{h_b}\right)^{0{\cdot}373}_{\max} = 0{\cdot}765;\ \frac{h_a}{h_b} = 0{\cdot}49$$

Maximum reduction of area = 51% assuming no hardening.

(*b*) For curved dies it is necessary to approximate or to use an iterative method. Thus, suppose the maximum reduction is about 50%, i.e. 25 x 1 mm to 25 x 0·5 mm, $\epsilon = 0{\cdot}7$

$$\alpha_b = \cos^{-1}\frac{R - \Delta h}{R} = \cos^{-1} 0{\cdot}99 = 0{\cdot}141\ \text{rad}$$

$$H_b = 2\sqrt{\frac{50}{0{\cdot}5}}\tan^{-1}\left(\sqrt{\frac{50}{0{\cdot}5}} \times 0{\cdot}14\right) = 20\tan^{-1} 1{\cdot}4 = 19{\cdot}0$$

$$\frac{\sigma_{xa}}{S} = 1 - 0{\cdot}5\ e^{-1{\cdot}9} = 0{\cdot}925$$

The practical limit is set, as in (*a*), by fracture at $\sigma_{xa} = Y_a = 0{\cdot}866\ S_a$, so the maximum reduction of area would be less than 50% if there were no hardening. However, the mean value $\overline{S}$ between $\epsilon_b = 0$ and $\epsilon_a = 0{\cdot}7$ is about 0·65 kN/mm², (see Example 2.3*a*) whereas S_a at $\epsilon = 0{\cdot}7$ is about 0·80 kN/mm². Thus the drawing stress for a 50% pass would be

$$\sigma_x = 0{\cdot}93\ \overline{S} = 0{\cdot}60\ \text{kN/mm}^2$$

and the fracture stress

$$(\sigma_x)_{\max} = 0{\cdot}866\ \overline{S}_a = 0{\cdot}69\ \text{kN/mm}^2$$

This would permit a pass somewhat greater than 50%.

It is not usually justifiable to attempt greater accuracy, in view of the uncertainty of μ.

7.5 What is the percentage contribution of friction to the drawing stress at 40% reduction of area using two lubricants giving $\mu = 0{\cdot}05$ and $\mu = 0{\cdot}10$, with $\alpha = 15°$ die and a fixed parallel plug?

Solution. Use equation 7.7.

(*a*) $B = 2 \times 0{\cdot}05 \cot 15° = 0{\cdot}373, h_a/h_b = 0{\cdot}6$

$$\frac{\sigma_{xa}}{S} = \frac{1{\cdot}373}{0{\cdot}373}[1-(0{\cdot}6)^{0{\cdot}373}] = 3{\cdot}68(1-0{\cdot}826) = 0{\cdot}641$$

(*b*) $B = 0{\cdot}746$

$$\frac{\sigma_{xa}}{S} = \frac{1{\cdot}746}{0{\cdot}746}[1-(0{\cdot}6)^{0{\cdot}746}] = 2{\cdot}34(1-0{\cdot}685) = 0{\cdot}737$$

With no friction

$$\frac{\sigma_{xa}}{S} = \epsilon = 0{\cdot}51$$

The frictional contributions are respectively 26% and 45%.

7.6 Sketch four methods used for reducing the diameter and wall thickness of cold tubes. What are their relative advantages and disadvantages?

Solution. See §7.1 for the essential features.

7.7 Estimate the force required for sinking a 100 mm O.D. x 12 mm mild steel tube to 90 mm O.D., using a die with semi-angle 15°, and a lubricant giving $\mu = 0{\cdot}05$.

Solution. Use equation 7.16.

$$B = 0{\cdot}05 \cot 15° = 0{\cdot}187$$

$$\frac{\sigma_{xa}}{1{\cdot}1\overline{Y}} = \frac{1{\cdot}187}{0{\cdot}187}\left[1-\left(\frac{90}{100}\right)^{0{\cdot}187}\right] = 6{\cdot}35(1-0{\cdot}980) = 0{\cdot}124$$

$$\epsilon = \ln \overline{D}_b/\overline{D}_a = 0{\cdot}112;\ \overline{S} = 0{\cdot}40\ \text{kN/mm}^2$$

$$1{\cdot}1\overline{Y} = 0{\cdot}38\ \text{kN/mm}^2$$

$$\sigma_{xa} = 0{\cdot}047\ \text{kN/mm}^2,$$

$A_a = \pi \times 78 \times 12 = 2{,}940\ \text{mm}^2$, ignoring any thickening or thinning that may occur. Drawing force ≈ 140 kN.

7.8 Suggest a plug-drawing schedule for reducing 50 mm O.D. x 6 mm mild steel tube to 12 mm O.D. x 0·9 mm, and calculate the size of equipment required, assuming reasonable values of the necessary parameters. The strain should not exceed 80% between annealing processes.

Solution. Initial cross-sectional area = $\pi \times 44 \times 6 = 829\ \text{mm}^2$. Final cross-sectional area = $\pi \times 11{\cdot}1 \times 0{\cdot}9 = 31{\cdot}4\ \text{mm}^2$. Total strain $\epsilon = \ln 26{\cdot}4 = 3{\cdot}27$.

Suppose $\mu = 0{\cdot}05$, $\alpha = 15°$, $B^* = 0{\cdot}373$. From equation 7.7, the maximum reduction in a close pass is given by

$$\frac{\sigma_{xa}}{S} \approx \frac{Y}{S} = 0{\cdot}866 = \frac{1{\cdot}373}{0{\cdot}373}\left[1-\left(\frac{A_a}{A_b}\right)^{0{\cdot}373}\right]$$

$$\left(\frac{A_a}{A_b}\right)^{0{\cdot}373} = 0{\cdot}765;\ \frac{A_a}{A_b} = 0{\cdot}488$$

$$r_{max} \approx 0{\cdot}51\ \epsilon_{max} = \ln\frac{A_b}{A_a} = 0{\cdot}718$$

The maximum strain permitted before annealing is given by $r = 0{\cdot}8$, $\epsilon_A = 1{\cdot}61$, so the total strain is just too great to allow only one inter-anneal, according to the rough criterion of 80% per anneal. However, the cost of an additional annealing would be considerable, so an attempt might be made to use say $\epsilon_A^* = 1{\cdot}7$ before the first annealing and $\epsilon_A^{**} = 1{\cdot}6$ for the second stage. Each stage would then have three passes of about $\epsilon = 0{\cdot}56$ ($r = 0{\cdot}43$).

In detail the total thickness strain is $\epsilon'_{06} = \ln\dfrac{6}{0{\cdot}9} = 1{\cdot}90$, and the diametral strain is $\epsilon''_{06} = \ln\dfrac{50}{12} = 1{\cdot}43$, ($\epsilon'_{06} + \epsilon''_{06} = 3{\cdot}33$). Equally divided, this implies that in each pass $\epsilon' = 0{\cdot}316$, $\epsilon'' = 0{\cdot}238$, $\epsilon = 0{\cdot}554$, $r = 0{\cdot}42$.

Thus $h_0 = 6{\cdot}0$ mm, $\epsilon'_{01} = 0{\cdot}316$, $h_0/h_1 = 1{\cdot}372$; $h_1 = 4{\cdot}37$ mm and similarly $h_2 = 3{\cdot}18$, $h_3 = 2{\cdot}32$. The diameters are given by $D_0 = 50$, $\epsilon''_{01} = 0{\cdot}238$ $D_0/D_1 = 1{\cdot}268$;

$$D_1 = 39{\cdot}4,\ D_2 = 31{\cdot}1,\ D_3 = 24{\cdot}5$$

After the interanneal, $h_4 = 1{\cdot}69$, $h_5 = 1{\cdot}23$, $h_6 = 0{\cdot}90$

$$D_4 = 19{\cdot}3,\ D_5 = 15{\cdot}2,\ D_6 = 12{\cdot}0$$

The die diameters would usually be adjusted to standard values, e.g.

50, 40, 30, 25; 20, 15, 12 mm

The drawing stress for $\epsilon = 0{\cdot}554$, corresponding to $A_a/A_b = 0{\cdot}575$, is given approximately by assuming the whole reduction to be close pass, as in equation 7.7:

$$\frac{\sigma_{xa}}{S} = 3{\cdot}68[1 - (0{\cdot}575)^{0.373}] = 0{\cdot}686$$

This will overestimate, because the diameter reductions involve a contribution from sinking, as in example 7.7.

The area A_1, after the first pass, is $\pi \times 35 \times 4{\cdot}4 = 484$ mm² and $\bar{S}$ for this pass is $0{\cdot}59$ kN/mm², so the drawing force is $F_1 = 195$ kN. The force F_6 for the final pass is $\approx 0{\cdot}686 \times 0{\cdot}90 \times 31{\cdot}4 = 19$ kN.

7.9 What is the maximum size of stainless steel tube that can be drawn by 25% reduction of area in a close pass on a drawbench with 200 kN pull? Assume $S \approx 0{\cdot}75$ kN/mm².

Solution. Use equation 7.7 for plug drawing. Assume $B^* = 0{\cdot}373$

$$\frac{\sigma_{xa}}{S} = 3{\cdot}68\left[1 - \left(\frac{A_a}{A_b}\right)^{0.373}\right]$$

Now $\bar{S} = 0{\cdot}75$ kN/mm² and $\sigma_{xa} = 200/A_a$ kN/mm²

$$1 - \left(\frac{A_a}{A_b}\right)^{0.373} = \frac{200}{3{\cdot}68 \times 0{\cdot}75\, A_a} = \frac{72{\cdot}5}{A_a}$$

But $\dfrac{A_a}{A_b} = 1 - r = 0{\cdot}75$

$$1 - 0{\cdot}898 = 72{\cdot}5/A_a;\ A_a = 647 \text{ mm}^2.$$

If $B = 0$, as with a mandrel,

$$\frac{\sigma_{xa}}{S} = \epsilon = 0{\cdot}287;\ \sigma_{xa} = 0{\cdot}215 \text{ kN/mm}^2$$

$$A_a = 200/0{\cdot}215 = 930 \text{ mm}^2$$

Thus $\pi Dh = 930$, $Dh = 296$, e.g. $D = 50$, $h = 6$ mm.

7.10 Approximately what power is required for a drawbench driven by a geared motor of overall efficiency 85%, to be used in drawing annealed copper tube from 40 mm O.D. x 3 mm to 30 mm O.D. x 2·4 mm with a conical die of 24° included angle and a suitable floating plug, at a speed of 1·6 m/sec?

Solution. With a floating plug the condition of stability is seen from Figure 7.2 to be:

$$p \pi D \tan\beta \, dx = \mu_2 p \pi D \, dx; \; \tan\beta = \mu_2$$

If $\mu = 0{\cdot}04$, $\beta = 2^\circ 18'$; $2\alpha = 24^\circ$

$$A_b = \pi \times 37 \times 3 = 349 \text{ mm}^2, \; A_a = \pi \times 27{\cdot}6 \times 2{\cdot}4 = 208 \text{ mm}^2$$

$$\epsilon = \ln 349/208 = 0{\cdot}52, \; \bar{S} \approx 0{\cdot}28 \text{ kN/mm}^2$$

$$B^* = \frac{0{\cdot}08}{\tan\alpha - \tan\beta} = 0{\cdot}47$$

Using equation 7.7

$$\frac{\sigma_{xa}}{S} = \frac{1 + B^*}{B^*}\left[1 - \left(\frac{A_a}{A_b}\right)^{B^*}\right]$$

$$= \frac{1{\cdot}47}{0{\cdot}47}\left[1 - (0{\cdot}595)^{0{\cdot}47}\right]$$

$$= 0{\cdot}675$$

$$F = 0{\cdot}675 \times 0{\cdot}28 \times 208 = 39{\cdot}3 \text{ kN}$$

$$\text{Work done per second} = 39{\cdot}3 \times 1{\cdot}6 \text{ kN m/sec}$$

$$= 62{\cdot}9 \text{ k J/s} = 63 \text{ kW}$$

$$(= 84 \text{ hp})$$

Allowing 85% efficiency ≈ 75 kW

(Homogeneous work would involve $\sigma_{xa}/S = 0{\cdot}45$)

In practice, floating plugs are preferably of curved profile and are used with curved dies.

REFERENCES

7.1 Sachs, G., Lubahn, J. D. and Tracy, D. R., 'Drawing of thin-walled tubing', *J. Appl. Mech.*, **11**, 199–210, 1944.

7.2 Sachs, G. and Espey, G., Experimentation on tube drawing with a moving mandrel', *J. Appl. Mech.*, (ASME) **14**, 81–87, 1947; 'Effect of spacing between dies in tandem drawing of tubular parts', *Trans. ASME*, 69, 139–143, 1947.

7.3 Sachs, G. and Baldwin, W. M., 'Stress analysis of tube sinking', *Trans. ASME*, **68**, 655–662, 1946.

7.4 Green, A. P., 'Plane-strain theories of drawing', *Proc. Inst. Mech. Engrs.*, **174**, 847–864, 1960.

7.5 Blazynski, T. Z. and Cole, I. M., 'An investigation of the plug-drawing process', *Proc. Inst. Mech. Engrs.*, **174**, 797–804, 1960.

7.6 Chung, S. Y. and Swift, H. W., 'Cup-drawing from a flat blank', *Proc. Inst. Mech. Engrs.* **165**, 199–223, 1951.

7.7 Jevons, J. D., *The Metallurgy of Deep Drawing and Pressing*, 2nd Ed., Chapman & Hall, 1945.

7.8 Mooshake, R. and Hillmer, H., 'Die technische und betriebswirtschaftliche Entwicklung des Mannesmannrohr-Walzverfahrens', *Stahl und Eisen*, **53**, 465–488, 1933.

7.9 Snee, C. E., 'Use of modern Assel mill in production of seamless tubing', *Iron and Steel Engr.*, **33**, 124–133, 1956.

7.10 Polushkin, E. P., 'Effect of cold work on microstructure of low carbon steel tubes', *Trans. Amer. Soc. Metals*, **22**, 635–656, 1934.

7.11 Blazynski, T. Z., 'Theoretical methods of designing tools for metal-forming processes', *Metal Form.*, 34, 143–150, 1967.

7.12 Avitzur, B., *Metal forming processes and analysis*, McGraw Hill, New York, 1968.

7.13 Schey, J. A. (Ed.), *Metal deformation processes: friction and lubrication*, Marcel Dekker, New York, 1970.

7.14 Pugh, H. Ll. D. (Ed.), *Mechanical behaviour of materials under pressure*, Elsevier, Amsterdam, 1970.

7.15 Blazynski, T. Z., 'Recent advances in tube production', *Int. Metall. Rev.*, 1976.

7.16 Sansome, D. H., 'Recent developments in oscillatory metal working', *Engg.*, 213, 243–247, 1973.

7.17 Jones, J. B., 'Draw ironing, flare and flange forming with ultrasonic assist', *Metal Prog.*, 93, 103–107, 1968.

8
Extrusion

8.1 Introduction

Extrusion is a relatively young metalworking process. Commercial extrusion of lead pipes started early in the nineteenth century, and it was not until near the end of that century that it was possible to extrude even brass. This was largely because the heavy and sustained pressures required were not available. The lack was eventually overcome by heating the billets to high temperatures, to reduce their yield stress, but this introduced a further problem of making billet containers for extrusion presses, which would withstand the severe combination of high temperature and high pressure. A history of extrusion and a full account of modern practice, are given by Pearson and Parkins[8.1] in their book *The Extrusion of Metals.*

Steel extrusion was not commercially practicable until about 1930, but it has advanced very rapidly, particularly with the introduction of molten-glass lubrication by Sejournet[8.2]. This a considerable improvement on lubrication with graphite, giving lower extrusion pressures, better die life, and the possibility of using longer billets. Most carbon and stainless steels are now extruded at about 1,200°C, with glass as the lubricant though other materials are beginning to compete with glass. Developments in furnaces include induction heaters and hydrogen-atmosphere furnaces for molybdenum extrusion at temperatures up to 1,800°C. Hydraulic extrusion-presses with a thrust of 200 MN (20,000 tons) are in use, though about 20 MN (2,000 tons) is the most common capacity. Billets are usually small, in comparison with forging blooms, perhaps 0·15 m (6 in.) diameter and 0·5 m (18 in.) long, though aluminium billets up to 0·75 m (30 in.) diameter and 1·8 m (70 in.) long can be extruded in a very large press. Very heavy reductions in area are however possible, and the rate of production is high. The harder alloys are usually restricted to about 20:1 area-ratio, but 60:1 or even 100:1 may be used on easily-extruded metals like aluminium. Modern extrusion presses can be run fully automatically from a tape-controller, and may produce sixty complete extrusions per hour. Very complex sections can be produced, especially in brass and aluminium, and, because it is essentially compressive, extrusion can be used for working relatively brittle alloys.

There has in recent years been a revival of interest in cold extrusion, which is now possible even with steels, using an impact technique. There may be very large differences in section in the final product, and the whole slug can be used; for example making a tube with a heavy flange, in one operation. The mechanical properties of cold extrusions are very good.

At present the theory is applicable only to the simpler processes. Stress evaluation is of little direct value in predicting extrusion loads, because the reductions of area are very large necessitating large die angles, often 180°, which introduce considerable inhomogeneity in the deformation. There is appreciable redundant work in all practical extrusion. Plane-strain solutions apply to only a small proportion of real

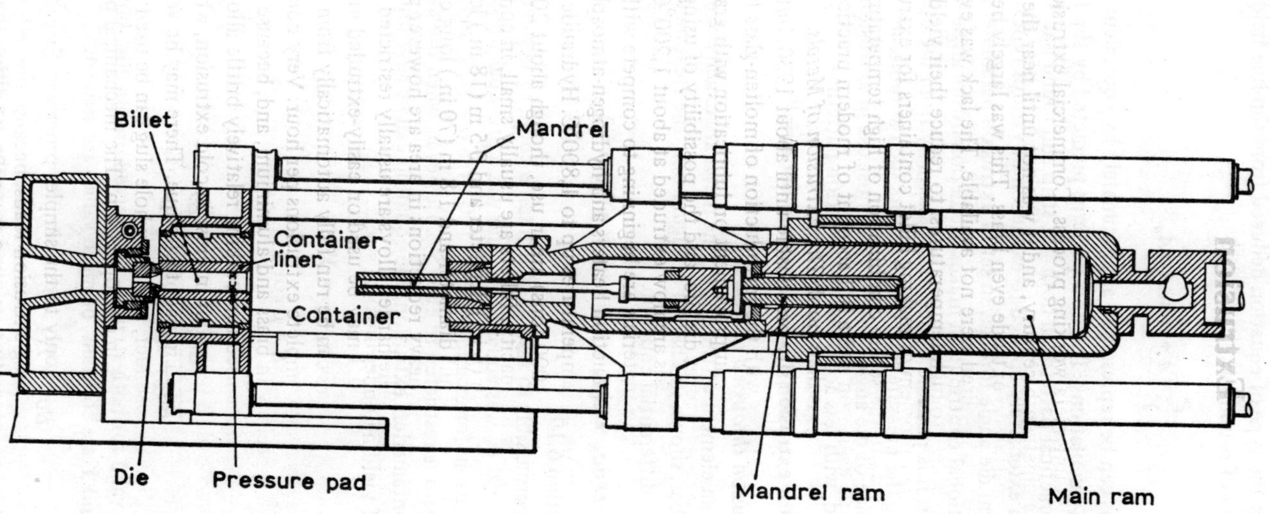

Figure 8.1 A diagram of a modern 1,000-tonnes press for tube extrusion.

(*Courtesy of Fielding & Platt Ltd.*)

extrusion problems, but the type of equation deduced from slip-line fields can be applied also to extrusion of round bar.

$$\frac{P}{Y} = a + b \ln A_b/A_a \tag{8.25}$$

Many extruded sections can be represented approximately by equivalent flat strips or round bars, for load calculations. The flow patterns predicted by slip-line fields also closely resemble practical patterns in both flat strip and round bar. Upper-bound theory has been extensively applied to extrusion of various types. As in drawing, there is an optimum die angle for a given set of extrusion parameters, but the uncertainty in the frictional contribution makes prediction of the angle less accurate for extrusion. However, it is possible to depart considerably from the optimum without seriously increasing the pressure, and other factors such as lubricant retention may dominate the choice of die angle. Theory is of most use in extrusion for predicting working loads, and, with less accuracy, deformation patterns.

8.2 Stress evaluation for extrusion of round bar and flat strip[8.3]

8.2.1 *Round-bar extrusion through a conical die*

The stresses acting on an element of the deforming metal are exactly the same as in the drawing of round bar, shown in Chapter 6, Figure 6.2. The treatment follows that of §6.4, giving the differential equation

$$D\, d\sigma_x + 2[\sigma_x + p(1 + \mu \cot \alpha)]\, dD = 0 \tag{8.1}$$

corresponding to equation 6.19.

In drawing, it was permissible to assume that $\mu \tan \alpha$ could be neglected in comparison with unity, so that σ_y was equal to $-p$, and was a principal stress. This is not generally true for extrusion; for example if $2\alpha = 120°$, $\tan \alpha = 1{\cdot}73$, and μ must be less than 0·03 if the error is restricted to even 5%. However, when the assumption is valid, as in open or containerless extrusion, the equation can be simplified as before using the condition of yielding under uniaxial stress,

$$\sigma_x + p = Y \tag{8.2}$$

Using the notation

$$B = \mu \cot \alpha \tag{8.3}$$

this leads to

$$\frac{d\sigma_x}{B\sigma_x - Y(1 + B)} = 2\frac{dD}{D} \tag{8.4}$$

This basic equation is valid for both straight- and curved-profile dies, but it is simpler to integrate when B and Y are constant, or mean values are assumed.

$$B\sigma_x - Y(1 + B) = c'D^{2B}$$

In extrusion the constant of integration is found from the condition that there is no longitudinal stress at the die exit

$$\frac{\sigma_x}{Y} = \frac{1 + B}{B}\left[1 - \left(\frac{D}{D_a}\right)^{2B}\right] \tag{8.5}$$

The extrusion pressure is the pressure σ_{xb} acting on the end of the ingoing billet

$$\frac{\sigma_{xb}}{Y} = \frac{1+B}{B}\left[1 - \left(\frac{D_b}{D_a}\right)^{2B}\right] \tag{8.6}$$

This may be contrasted with equation 6.23a for rod-drawing

$$\frac{\sigma_{xa}}{Y} = \frac{1+B}{B}\left[1 - \left(\frac{D_a}{D_b}\right)^{2B}\right] \tag{6.23a}$$

The difference in entry and exit conditions inverts the diameter ratio.

Equation 8.6 predicts a limit to extrusion when plastic deformation occurs before the metal enters the die, with $\sigma_{xb} = Y$. The greatest value the limit can take is, of course, that predicted for perfect lubrication, but when $\mu = 0$ equation 8.6 cannot be used, and it is necessary to revert to the differential form, equation 8.4, which becomes

$$D\,\mathrm{d}\sigma_x + 2Y\,\mathrm{d}D = 0$$

With the boundary condition $\sigma_{xa} = 0$

$$\frac{\sigma_{xb}}{Y} = \ln\left(\frac{D_b}{D_a}\right)^2 = \ln R \tag{8.7}$$

The limiting extrusion ratio R is thus

$$\ln R = 1, \quad R = 2{\cdot}7 \tag{8.8}$$

which is the same as $r = 63\%$ reduction of area, the limit in drawing. This is a low value, and in all practical extrusions the billet does deform by upsetting within the container, before the extrusion starts. It is, of course, restrained from further radial expansion, by the container walls, and extrusion ratios of 20:1 or even 100:1 can be used. There is then a very high pressure between the billet and the container, so that it is necessary to provide good lubrication there, as well as on the billet-face entering the die.

8.2.2 *Allowance for container friction*

The same type of analysis can be applied to estimation of the influence of container friction.

A transverse slice, as shown in Figure 8.2*a*, is subjected to axial and radial compression, and to a frictional force acting along the container walls. Under steady conditions, equilibrium in the axial direction requires that

$$(\sigma_x + \mathrm{d}\sigma_x)\frac{\pi}{4}(D_b + \mathrm{d}D_b)^2 - \sigma_x\frac{\pi}{4}D_b{}^2 = \mu\sigma_r\pi D_b\,\mathrm{d}x.$$

The container walls are usually parallel, so

$$D_b\,\mathrm{d}\sigma_x = 4\mu\sigma_r\,\mathrm{d}x$$

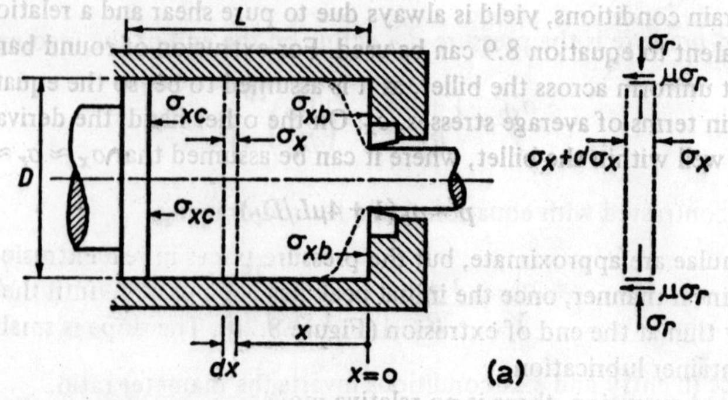

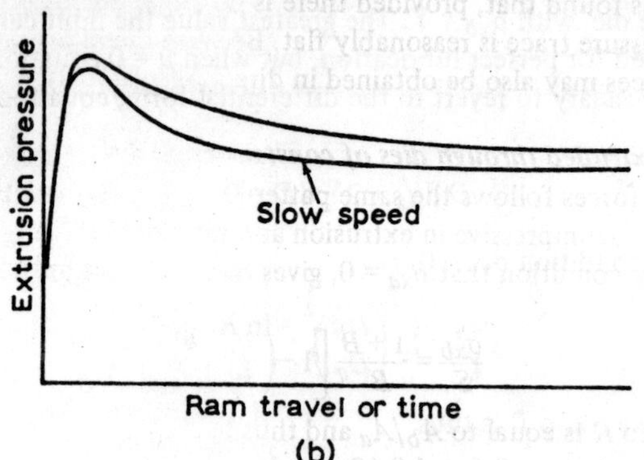

Figure 8.2
(*a*) The stresses acting on a slice of billet metal in the container.
(*b*) Experimental pressure-diagrams.

Assuming that $\sigma_x = \sigma_1$ and $\sigma_z = \sigma_r = \sigma_3$, yielding is due to uniaxial stress, and

$$\sigma_x - \sigma_r = Y \tag{8.9}$$

$$\frac{d\sigma_x}{\sigma_x - Y} = \frac{4\mu}{D_b}\,dx$$

This may be integrated between the limits $x = 0$ at the die, where $\sigma_x = \sigma_{xb}$; and $x = L$ at the end of the billet, where $\sigma_x = \sigma_{xc}$.

$$\frac{\sigma_{xc} - Y}{\sigma_{xb} - Y} = e^{\frac{4\mu L}{D_b}}$$

$$\sigma_{xc} = (\sigma_{xb} - Y)\,e^{\frac{4\mu L}{D_b}} + Y \tag{8.10a}$$

Because μ is usually small, it is often permissible to expand the exponential and to use the linear approximation:

$$(\sigma_{xc} - Y) = (\sigma_{xb} - Y)(1 + 4\mu L/D_b) \tag{8.10b}$$

In plane-strain conditions, yield is always due to pure shear and a relationship $\sigma_x - \sigma_z = 2k$ equivalent to equation 8.9 can be used. For extrusion of round bar, the axial stress is not uniform across the billet, as it is assumed to be, so the equation should be written in terms of average stresses, σ_x. On the other hand, the derivation is likely to be valid well within the billet, where it can be assumed that $\sigma_x \approx \sigma_r \approx -p$, giving

$$p = p_b(1 + 4\mu L/D_b) \qquad (8.11)$$

These formulae are approximate, but the pressure traces in real extrusions do fall in a roughly linear manner, once the initial peak has been passed, until the billet becomes very thin at the end of extrusion (Figure 8.2*b*). The slope is small when there is good container lubrication.

In inverted extrusion, there is no relative movement between the billet and the container, and it is found that, provided there is no recrystallisation during the extrusion, the pressure trace is reasonably flat. Because the billet cools during slow extrusion, flat traces may also be obtained in direct extrusion as in Figure 8.2*b*.

8.2.3 *Flat strip extruded through dies of constant angle*[8.3]

The resolution of forces follows the same pattern as for strip drawing (equation 4.19) but σ_x is, of course, compressive in extrusion and tensile in drawing. Integration with the boundary condition that $\sigma_{xa} = 0$, gives the extrusion pressure

$$\frac{\sigma_{xb}}{S} = \frac{1+B}{B}\left[1 - \left(\frac{h_b}{h_a}\right)^B\right] \qquad (8.12)$$

The extrusion ratio R is equal to A_b/A_a and thus to h_b/h_a for flat strip and $(D_b/D_a)^2$ for round bar, so equations 8.6 and 8.12 may be rewritten:

$$\frac{\sigma_{xb}}{Y} = \frac{1+B}{B}[1 - R^B] \quad \text{for round bar}$$

$$\frac{\sigma_{xb}}{S} = \frac{1+B}{B}[1 - R^B] \quad \text{for flat strip}$$

8.2.4 *Flat strip extruded through cylindrical dies*

The stress evaluation can be completed in the same way as for drawing and rolling, but the result is seldom needed.

8.2.5 *Limitations of stress evaluation for extrusion*

We have already seen that the neglect of $\mu \tan \alpha$ in deriving the yield criterion in equation 8.2 is less valid for extrusion than for drawing, because both α and μ are usually greater in extrusion.

There is a much more serious limitation, in neglecting redundant work. One of the great attractions of extrusion in production is that very large reductions of cross-sectional area are possible in a single pass. This is economically favourable, and also often improves the metallurgical quality of the material. Short dies, with high die angle, are desirable to reduce the physical size of the die pellet, which is often made of an expensive tungsten-steel; to reduce the radial component of the pressure, which may lead to bursting; to reduce the size of the discard; and to reduce the frictional drag. High die angles, however, introduce more redundant work, as was seen in considering drawing. The contributions of redundant work and friction to the axial stress are most important with high and low die angles respectively. The balance between them leads, as in drawing, to an optimum angle requiring least performance of work in the operation. As the reduction in area increases, the value of the optimum angle also increases, from between 7 and 15° semi-angle for drawing, to about 60° for a typical extrusion. It is more usual in extrusion to refer to the total included angle, and 120° is often quoted as the optimum angle for extrusion dies. It is, however, quite common to use flat, 180°, dies. The redundant-work contribution for 50% frictionless extrusion with 180° dies has been evaluated in Chapter 5, §5.8, where it was shown that the pressure was nearly double that required for unconstrained deformation by the same ratio. As in drawing, the redundant work factor decreases for heavier reductions of area, but with 180° dies the contribution is still appreciable even at 20:1 extrusion ratio (95% reduction). It is therefore very important to take account of redundant work in extrusion.

8.3 Slip-line field solutions for strip extrusion through tapered dies(8.4, 8.5)

8.3.1 *Frictionless extrusion* $r = \dfrac{2 \sin \alpha}{1 + 2 \sin \alpha}$

This particular reduction of area gives a very simple slip-line field, which has been described, together with the hodograph, in connection with strip-drawing. Chapter 6, §6.5.1 should be read before proceeding with this Section. However, this reduction is not an important condition for extrusion, because even with $\alpha = 90°$ the reduction is only $\frac{2}{3}$, but the simplicity of the solution makes it a useful introduction.

The slip-line field is identical with that constructed for strip-drawing in Figure 6.4. The minor principal stress σ_3 acting on the exit slip-line *BD* is zero for extrusion. Thus p_D can be found directly; equation 5.2 gives

$$\sigma_3 = -p_D + k; \; p_D = k \tag{8.13}$$

The α slip-line rotates clockwise through an angle α between *D* and *C* so the Hencky equation 5.7 is

$$p_C + 2k(-\alpha) = p_D; \; p_C = k(1 + 2\alpha)$$

From *C* to *A* the slip-line is straight, as are all lines in the zone *ACB*, so the pressure is the same at *A* and between *A* and *B*

$$p_A = p_C = k(1 + 2\alpha)$$

The major principal stress σ_1 in a direction normal to the die face is consequently uniform and equal to the die pressure $-q$.

$$\sigma_1 = -q = -(p_A + k) = -2k(1 + \alpha) \tag{8.14}$$

This equation may be compared with equation 6.28 for drawing. The component of the force on the die, resolved parallel to the extrusion direction, is $q(w \,.\, 2AB \sin \alpha)$, and the uniform pressure P applied to the billet by the pressure pad is given by

$$Ph_b w = qw \,.\, 2AB \sin\alpha = qh_b w \,.\, r$$

$$P = qr = -\sigma_1 r = 2k(1+\alpha)\,.\,r$$

$$\frac{P}{2k} = (1+\alpha)\,.\,\frac{2\sin\alpha}{1+2\sin\alpha} \tag{8.15}$$

The magnitude of the extrusion pressure can be compared with that required for homogeneous deformation, for common die angles.

$$\underline{\underline{2\alpha = 120^\circ}}, \quad r = \frac{\sqrt{3}}{1+\sqrt{3}} = 0{\cdot}63$$

$$\frac{P}{2k} = 0{\cdot}63\left(1+\frac{\pi}{3}\right) = 1{\cdot}3; \quad \left(\frac{P}{2k}\right)_H = \ln\frac{1}{1-r} = 1{\cdot}02 \tag{8.16}$$

$$\underline{\underline{2\alpha = 180^\circ}}, \quad r = \frac{2}{3}$$

$$\frac{P}{2k} = \frac{2}{3}\left(1+\frac{\pi}{2}\right) = 1{\cdot}71; \quad \left(\frac{P}{2k}\right)_H = \ln\frac{1}{1-r} = 1{\cdot}10 \tag{8.17}$$

8.3.2 *Frictionless extrusion;* r less than $\dfrac{2\sin\alpha}{1+2\sin\alpha}$

These smaller reductions of area are of much more importance in drawing than in extrusion, but the slip-line fields and hodographs are the same for both. They have been described in Chapter 6, §6.5.

8.3.3 *Frictionless extrusion;* r greater than $\dfrac{2\sin\alpha}{1+2\sin\alpha}$

The slip-line field starts as before with the right-angled triangle ABC and a radial fan centred on B (Figure 8.3*a*). For heavy reductions the point A lies on the die face, at a position determined by $BA = \sqrt{2}BC = \sqrt{2}BD$. To extend the field by the approximate chord method, with 10° intervals, a chord is drawn from C_1 at 5° to BC_1 to cut the axis at D_1. Two chords are then drawn, one from C_2 at 5° to BC_2, the other at 50° to the central plane at D_1. These intersect at C_3. The position of C_4 is found from the intersection of the chords CC_4 at 5° to BC and C_3C_4 at 10° to C_3D_1. The line C_4A_1 is a straight slip-line because CA is straight, and it is tangential to the slip-line D_1C_4. It is thus at 5° to the chord C_3C_4. Similarly the chord A_1E_1 is at 5° to C_4A_1. So the field is built up, as shown in Figure 8.3*a*. The hodograph is given in Figure 8.3*b*. BD is again a β-line.

The stress solution from the slip-line field starts with the boundary condition of zero external force on the extruded strip. At any point H on BD,

$$\sigma_3 = 0, \quad p = k$$

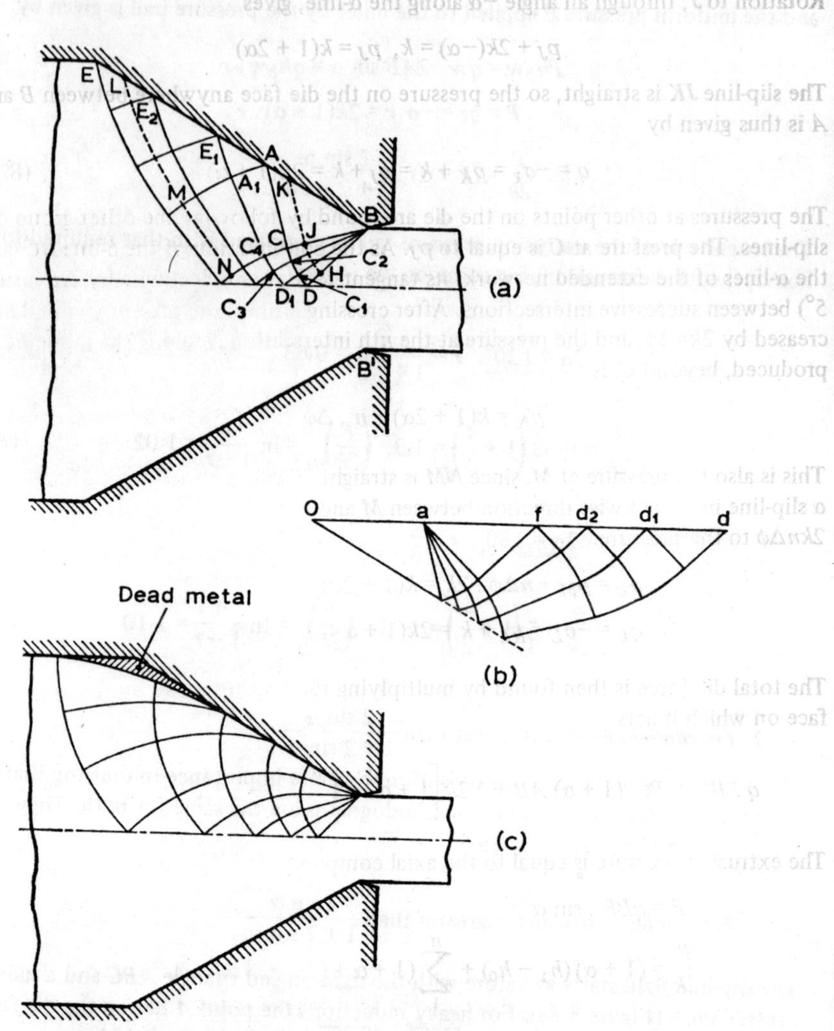

Figure 8.3

(*a*) The slip-line field for extrusion through frictionless tapered dies with a reduction of area greater than $\dfrac{2 \sin \alpha}{1 + 2 \sin \alpha}$

(*b*) The corresponding hodograph.

(*c*) The slip-line field with sticking friction at both die and container.

Rotation to J, through an angle $-\alpha$ along the α-line, gives

$$p_J + 2k(-\alpha) = k, \; p_J = k(1 + 2\alpha)$$

The slip-line JK is straight, so the pressure on the die face anywhere between B and A is thus given by

$$q = -\sigma_1 = p_K + k = p_J + k = 2k(1 + \alpha) \tag{8.18}$$

The pressures at other points on the die are found by following the other group of slip-lines. The pressure at C is equal to p_J. As the continuation of the β-line BC cuts the α-lines of the extended network, its tangent rotates anticlockwise by $\Delta\phi$ (usually 5°) between successive intersections. After crossing n lines, the pressure will be increased by $2kn\Delta\phi$, and the pressure at the nth intersection N, along the β-line BC produced, beyond C, is

$$p_N = k(1 + 2\alpha) + n \,.\, \Delta\phi \,.\, 2k$$

This is also the pressure at M, since NM is straight. There is a further rotation of the α slip-line in a clockwise direction between M and L, which again adds a component $2kn\Delta\phi$ to the pressure:

$$p_L = p_M + n\Delta\phi \,.\, 2k = k(1 + 2\alpha) + 2n \,.\, \Delta\phi \,.\, 2k$$

$$q_L = -\sigma_L = p_L + k = 2k(1 + \alpha + 2n \,.\, \Delta\phi)$$

The total die force is then found by multiplying the local pressure by the area of die face on which it acts

$$q \,.\, BE = 2k \,.\, (1 + \alpha)\, AB + \Sigma 2k \left[1 + \alpha + 2\left(\frac{n + \overline{n - 1}}{2}\right)\Delta\phi\right]\left[\frac{h_n - h_{n-1}}{\sin\alpha}\right]$$

The extrusion pressure is equal to the axial component of this force ($h_n = 1$).

$$P = qBE \,.\, \sin\alpha$$

$$\frac{P}{2k} = (1 + \alpha)(h_1 - h_0) + \sum_{1}^{n}(1 + \alpha + (2n - 1)\Delta\phi)(h_n - h_{n-1}) \tag{8.19}$$

The value of the extrusion pressure can be found quite simply by drawing the slip-line field accurately to scale. The field can also be applied to drawing to heavy reductions of area.

8.3.4 *Extrusion with friction at the die and the container*[8.5]

If there is friction at the die, the slip-lines will meet the die face at an angle θ given by

$$\cos 2\theta = \mu q/k$$

For light reductions of area, the field is the same as for strip-drawing (Figure 6.7); for heavy reductions it resembles Figure 8.3, except that the slip lines do not meet the die face at 45°.

When there is friction at both die and container, the slip-lines meet the container at an angle θ', which may of course differ from θ. Figure 8.3*c* shows the slip-line field for extrusion with sticking-friction at the container wall. The slip-lines then meet the wall tangentially, and shear occurs across a zone of dead metal, which remains in the corner of the container, but takes no part in the deformation. Such dead-metal regions are observed in practice when the friction is high, particularly with square dies. For a given reduction there is a particular die angle below which the dead-metal interface would lie on the die face itself. Beyond this, no dead zone would be expected. Similarly, for a given die angle, the dead-zone boundary retreats into the corner as the reduction of area is increased. As long as there is a full dead zone, the results are the same for wedge dies as for square dies, and the die angle is immaterial.

8.4 Slip-line fields for extrusion through square dies

Figure 8.4 shows some slip-line fields for extrusion through a square die. Figure 8.4*a* relates to frictionless conditions at both die and container. It is actually the limiting state of the field shown in Chapter 6, Figure 6.4, with $r = 2 \sin \alpha/(1 + 2 \sin \alpha) = \frac{2}{3}$ and $\alpha = 90°$. This is a smaller reduction than is normally used in extrusion. When the extrusion ratio is greater, the field must be extended as in Figure 8.4*b*, to give the limiting condition of the field shown in Figure 8.3. There is no dead-metal zone in either of these fields.

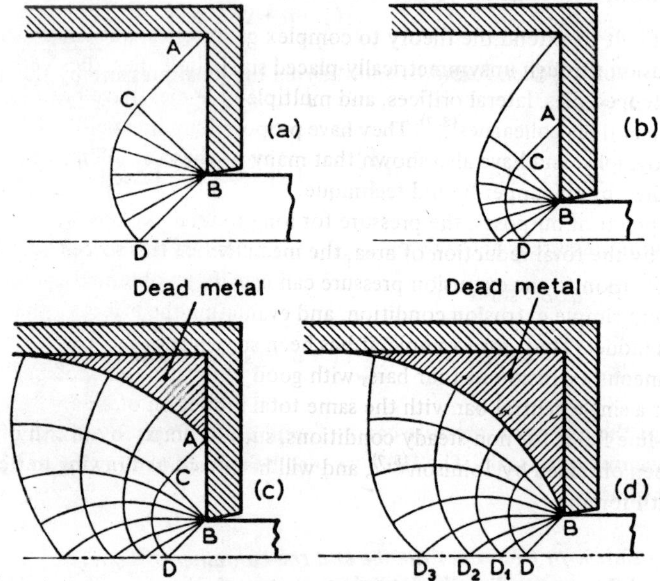

Figure 8.4 Slip-line fields for extrusion through a square die.

(*a*) Frictionless die and container ($r = \frac{2}{3}$).
(*b*) Frictionless die and container ($r > \frac{2}{3}$).
(*c*) Frictionless die with sticking friction at the container walls, showing a dead-metal zone.
(*d*) Sticking friction at die and container.

When there is a finite friction at the container wall, the slip-lines must meet the wall at the appropriate angle. For example, when there is sticking friction at the walls, as often happens if the container is not specially lubricated, the α and β slip-lines become normal and tangential. The field is then as shown in Figure 8.4*c*, with a dead-metal zone.

Friction at the die modifies the angles of the triangle *ABC*, as in drawing, until, with sticking friction, the slip-lines become tangential to the die face, or the dead zone fills the die as in Figure 8.4*d*. This diagram closely resembles the dead zones often found during extrusion with poor lubrication conditions.

If there is friction at the die but none at the container wall, the slip-lines meet the wall at 45°. A simple example of this has been considered in Chapter 5, §5.8 for 50% reduction in inverted extrusion, where the dead-zone boundary assumed was simply the radial exit boundary (see also Figure 8.5). Actually, the conditions for direct and inverted extrusion are formally the same when the container is smooth, but the deformation patterns observed in direct and inverted extrusions differ, and it is possible that a field of the type shown in Figure 8.4*a* would be a more suitable choice for inverted extrusion. Johnson(8.7) has extensively investigated these fields for a wide range of reductions of area and coefficients of friction.

8.5 Extrusion through unsymmetrical and multi-hole dies

It is difficult to extend the theory to complex geometries, but a number of examples of extrusion through unsymmetrically-placed single-hole dies, dies with oblique faces, stepped dies, lateral orifices, and multiple-hole dies, have been examined by Johnson and his colleagues(8.7). They have proposed slip-line fields and verified them by hodographs, and have also shown that many of these problems can conveniently be handled by the upper-bound technique.

For practical purposes, the pressure for longitudinal extrusion is determined mainly by the total reduction of area, the mean die angle, and the friction. A useful approximation to the extrusion pressure can usually be obtained by assuming an equivalent simple extrusion condition, and evaluating the effect of these parameters for this model. Thus, for example, it has been shown that the extrusion pressure for simultaneous extrusion of four bars, with good lubrication, is practically the same as that for a single central bar with the same total reduction of area.

Slip-line fields for non-steady conditions, such as occur at the end of extrusion, have been obtained by Johnson(8.7), and will be considered in Chapter 10 in connection with forging.

8.6 Metal-flow streamlines in extrusion

In the preceding sections, the slip-line fields and hodographs have been used entirely for calculation of pressures. They can also be used to determine the streamlines along which elements of the workpiece move in the deformation zone, and the velocities at any points on these lines.

As an example, we may consider the simple 50% extrusion of Chapter 5, §5.8.

The slip-line field and hodograph are reproduced in Figure 8.5, with 15° intervals in the fan region.

In the rigid billet, an element X moves parallel to the extrusion direction. As it crosses the boundary of the deformation zone at D, it undergoes a velocity change tangential to the slip-line, represented by the parallel vector bd in the hodograph. Its absolute velocity is then given by Od. Similarly, an element X^* crossing the boundary at E will have a velocity Oe in the deformation zone. Having entered the deformation zone, the element X will start to move to the right of D parallel to the hodograph vector Od, but when it reaches the radius OE its velocity will be parallel to Oe. As an approximation, the mean velocity vector parallel to Od' may be drawn from D to cut OE at E'. From the radius OE to the radius OF, the streamline may be drawn parallel to Oe', and so on. When the element reaches OB it will again experience a velocity discontinuity, parallel to BO, so that finally it moves parallel to the axis again. The discontinuous lines in the deformation zone may be smoothed out, to give a more realistic representation of the flow. Clearly, the accuracy is improved by taking smaller intervals.

It is usually of greater interest, from the point of view of internal strain or strain rate, to consider the distortion of planes perpendicular to the streamlines. The element X may be supposed to be at some position $x = x_0$ at the time $t = 0$. The velocity V along the streamline has components V_x and V_y, respectively parallel and perpendicular to the axis, which are presumed to be continuous single-valued functions of x. Then

$$V_x = \frac{dx}{dt}, \quad t = \int_0^x \frac{dx}{V_x} \tag{8.20}$$

From this relationship it is possible to determine either the time necessary for any element to reach the position x, or the positions of all the elements at any selected time after leaving the line $x = x_0$. The latter shows the deformation of a straight line scribed upon the side of the billet. The solution is conveniently obtained graphically.

Let us consider again the element X approaching the point D in Figure 8.5*a*. The velocity Od' as the element passes through the zone DOE will have a horizontal component $V_x = Od''$, as shown in Figure 8.5*b*. Similarly, in the zone EOF, $V_x = Oe''$, and so on. The numerical values in this example are marked on the diagrams. To obtain the integral in equation 8.20, the reciprocal of the horizontal velocity component is plotted against x, assuming constant velocity in each zone. The numerical values of $1/V_x$ for the streamline DE'–B' are given in Figure 8.5*b* and are plotted in projected positions in the lower part of Figure 8.5*a*. V_x is also plotted for information, but is not necessary for the calculation. The integral is then equal to the sum of all the areas $(1/V_x)_{d'} \times (x_{E'} - x_D)$ corresponding to each section traversed. Taking the dimension AB as 8 units, for example, and selecting the line $x = x_0$ at 1·0 unit from E, the area under the first section of the graph is $(x_D - x_0)/V_0 = 1{\cdot}2$, the second 1·88 x 1·095, and so on. The progressive increase in area with x is plotted in Figure 8.5*c*, for each of the elements crossing the inlet boundary at A, C, D, E, F, G, and B. A horizontal line representing a selected constant value of t is then drawn, and the respective x-intercepts are transposed to the original diagram for each element, measuring from $x = x_0$. Two such plots are shown in Figure 8.5*d*.

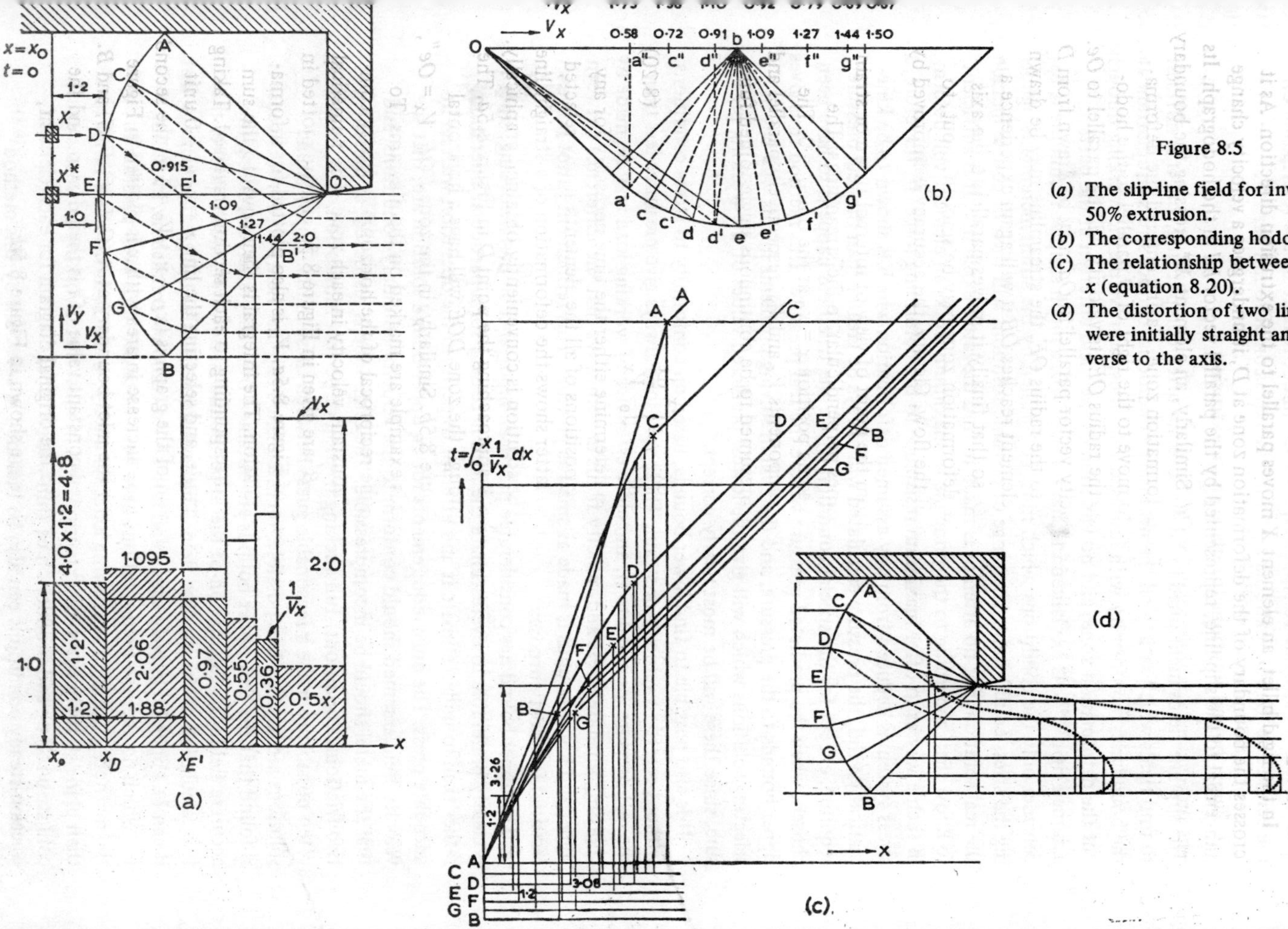

Figure 8.5

(*a*) The slip-line field for inverted 50% extrusion.
(*b*) The corresponding hodograph.
(*c*) The relationship between t and x (equation 8.20).
(*d*) The distortion of two lines which were initially straight and transverse to the axis.

The deformation predicted by this approach agrees quite well with experimental observations on extruded strip, except for the minor feature of the retardation on the centre-line. Similar patterns are found in extrusion of round bar[8.1].

8.7 Upper-bound solutions for plane-strain extrusion

The slip-line fields mentioned in §8.4 above are often complicated, and it is difficult to find fields which satisfy all the necessary conditions: correct stress equilibrium at the boundaries, conformity with imposed velocities, positive plastic working throughout the deformation zone, and no violation of the yield criterion in the parts assumed rigid[8.6]. As explained in Chapter 5, §5.10, a much simpler technique is to consider only the conditions which must be fulfilled by the strain increments, ignoring the stress equilibrium. The range of possible solutions will then of course include the real solution, but no attempt is made to find the latter explicitly. It is sufficient to know that, according to the principle of maximum work, an element will deform in such a way as to offer maximum resistance to the deforming force. If, therefore, a stress system is deduced from any assumed deformation which is shown to be kinematically valid, the pressure calculated will always be greater than the real pressure required, except for the special condition where the true field happens to have been chosen, when the predicted pressure will be the real pressure. This method thus gives upper bounds to the pressure, and the skill in solving problems in this way lies in selecting solutions which will give the lowest predicted values of the working pressure, since these will be most nearly correct.

It is also possible to find lower bounds, based on conformity to stress conditions, without regard to velocity or strain. These solutions always give pressures less than the real pressure, or equal to it, and close estimates of working load may be made by choosing separate stress and velocity fields in such ways that the difference between upper and lower bounds is made as small as possible. This refinement is not however essential for most purposes.

The upper bound has considerable practical importance in metalworking, since it is clearly preferable to overestimate the load necessary for a particular operation, rather than to underestimate it and attempt to perform the operation with an inadequate press. The only requirements for an upper-bound solution are, that the velocity field assumed should conform to the velocity boundary-conditions, and that the metal should be incompressible. The technique is restricted, at present, to two frictional conditions: full sticking friction or perfect lubrication. It is, however, often possible to make a reasonable guess between these limits. Upper-bound solutions are based, when possible, on established slip-line field solutions, because it is found that the best upper bounds are obtained from diagrams resembling the accurate field. The calculation of the upper bound is, however, much quicker than full calculation of the slip-line field. This technique has also the great advantage that it can be applied relatively easily and quickly to complex conditions, including those for which no slip-line field solutions have been obtained. It is even possible to apply the method to problems involving axial symmetry, which is much more important than plane strain in practical extrusion, and in most metalworking processes except cold strip-rolling and production of thin-walled tubes.

Non-steady conditions can also be included in solutions by this method[8.8].

8.7.1 *Upper-bound solution for strip-extrusion through tapered dies. Frictionless extrusion*

Consideration of the slip-line field for extrusion through frictionless tapered dies, (Figure 8.3), suggests a very simple approximation in terms of straight-line velocity discontinuities *AC* and *OC*, as shown in Figure 8.6.

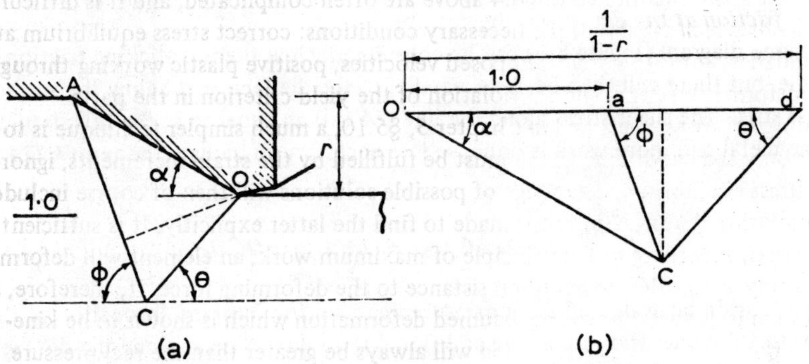

Figure 8.6

(*a*) An approximate straight-line analogy to the slip-line field for frictionless extrusion, giving an upper-bound solution.
(*b*) The hodograph for this solution.

An element of the billet reaches the discontinuity line *AC* with unit velocity along the axis, represented by *Oa* on the hodograph. On crossing *AC*, its velocity is changed suddenly, parallel to *AC*, so that it moves parallel to the die face *AO*. The corresponding lines in the hodograph are *ac* and *Oc*, drawn at angles ϕ and α respectively to *Oa*, completing the velocity triangle. The velocity of the element remains constant until it crosses *OC*, when it experiences a sudden change parallel to *OC*, represented by *cd*, at an angle θ to the horizontal, so that the final velocity has the appropriate magnitude *Od*.

$$Od = v_1 = h_b/h_a = 1/(1-r) \tag{8.21}$$

The force acting along *AC* is the product of the shear stress *k* and the area, that is kAC per unit width. The velocity along *AC* is u_{AC}, so the rate of working is $k\,AC\,u_{AC}$. Energy is also dissipated along the discontinuity *OC*, and the total must equal the rate of performance of work by the external force, as in equation 5.32.

$$\frac{dW}{dt} = \Sigma k \,.\, s \,.\, u$$

$$= k(AC\,u_{AC} + OC\,u_{OC})$$

If the extrusion pressure is *P*, and the billet moves with unit velocity, this equation becomes

$$P \,.\, \frac{h_b}{2} \,.\, 1 = k(AC\,u_{AC} + OC\,u_{OC}) \tag{8.22}$$

The lengths representing *AC, OC*, u_{AC}, u_{OC} can be measured directly on the diagrams in Figure 8.6, giving the value of *P* immediately. The best value is the lowest one which can be found, by choice of θ, and will usually be somewhat higher (5–20%) than the accurate solution given by a slip-line field.

8.7.2 *Upper-bound solution for strip extrusion through tapered dies. Sticking friction at the die*

The same diagram (Figure 8.6) can be considered when there is sticking friction at the die, but there will then be shear parallel to the die face, just below the surface of the strip. The shear stress along the line *OA* is then equal to *k*, the full strength of the metal and more work is done, so it is necessary to include the term $k\, OA\, u_{OA}$:

$$P.\frac{h_b}{2}.1 = k(AC\,u_{AC} + OA\,u_{OA} + OC\,u_{OC}) \tag{8.23}$$

This can again be evaluated from measurements on the diagram. Inspection of the lengths of the lines shows that the additional term due to shear along the die face will increase the extrusion pressure by about one half.

8.7.3 *Upper-bound solutions for strip extrusion through square dies*[8.9]

An upper-bound solution for a frictionless die and container is shown in Figure 8.7*a*. This closely resembles the slip-line field of Figure 8.4*a*. The upper-bound solution overestimates the pressure in this example by about 30%.

8.7.4 *Metal-flow streamlines deduced from upper-bound solutions*

The metal flow pattern is deduced from the lines of velocity discontinuity in upper-bound solutions, in a manner analogous to that used for the slip-line field in §8.6.

All the elements of metal move with one constant velocity while traversing each of the zones such as *BOC*. An element crossing the inlet boundary at *B* takes the velocities *Oc, Od*, and Oe_1 in sequence as it crosses the lines *CO, DO* and *EO*. The horizontal velocity components V_x are equal to Oc', Od', and Oe'_1. The reciprocals of these are plotted in Figure 8.7*c* over the appropriate ranges of horizontal distance *x*. The cumulative areas in this graph are evaluated and plotted in Figure 8.7*d* for the element crossing at *B*, and the procedure is then repeated for the elements

---

Figure 8.7

(*a*) An upper-bound solution for frictionless extrusion through 180° dies with $\frac{2}{3}$ reduction of area. The dotted lines represent distortion of the line $x = x_0$, derived from the intercepts in (*d*).

(*b*) The corresponding hodograph, with the values of V_x and $1/V_x$ appropriate to the zones *BOC* etc.

(*c*) $1/V_x$ plotted against *x*, assuming constant values through each section.

(*d*) A plot of $t = \int_0^x \frac{dx}{V_x}$, derived from (*c*), against *x*.

(*e*) Approximate temperature distribution in the extruded product.

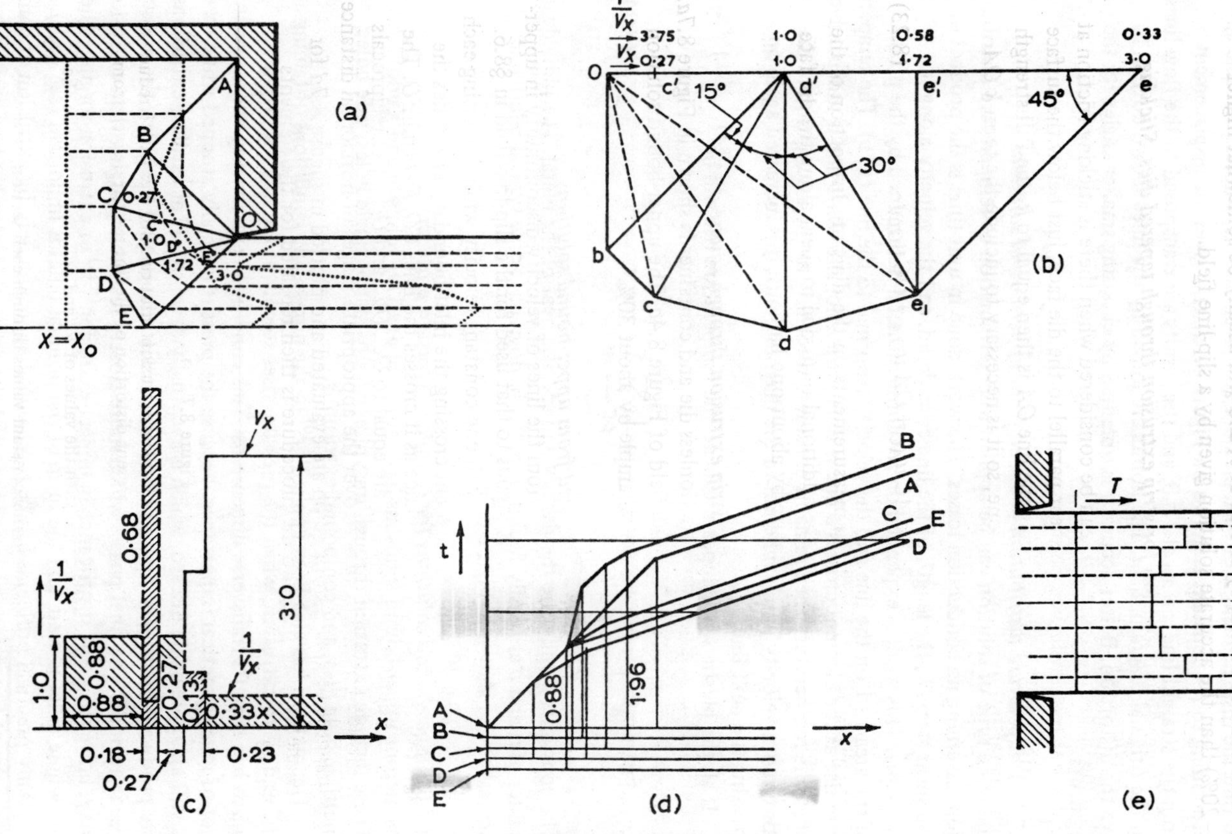

Figure 8.7 (*see foot of previous page for caption*).

crossing at *A, C, D*, and *E*. This gives $t = \int_0^x \frac{dx}{V_x}$ for each element, as in equation 8.20. To find the displacement x of each element at a given instant, a horizontal line corresponding to t = constant is drawn on Figure 8.7*d*. This gives the displacement from the straight line $x = x_0$ at $t = 0$, and the results are transposed to the flow lines on the original diagram, 8.7*a*. The points are joined by the lines shown dotted, to give the distortion. Better accuracy is obtained by choosing smaller angles for the segments.

8.7.5 *Temperature distribution deduced from upper-bound solutions*

The same type of solution can be applied to determination of temperature distribution, assuming no conduction losses. The work done in unit time is the product of the shear stress k, the length of the line considered, and the velocity along it ($k \,.\, s \,.\, u$). This work is expended on a volume of metal determined by the product of the area $s \,.\, 1$ on the line and the velocity v normal to the line ($s \,.\, v$). The temperature rise as the element crosses a line is thus proportional to ku/v and is independent of s. The temperature then remains constant, according to the assumption of discontinuous velocity jumps inherent in the diagram, until it is increased again on crossing the next line.

If all the plastic work appears as heat, the temperature rise will be

$$\Delta T = \frac{1}{\rho c} \sum \frac{ku}{v} \tag{8.24}$$

The constants ρ and c are the density and specific heat of the metal. The individual values of u and v are easily found from the hodograph. Some typical relative temperatures[8.9] are shown in Figure 8.7*e*.

8.7.6 *Upper-bound solutions for complex extrusion problems*

Numerous upper-bound solutions have been obtained by Johnson and Kudo[8.7]. These include extrusion through unsymmetrically-placed single holes, side holes, multiple-hole dies, and compound extrusions.

The method has also been applied[8.10] to the non-steady conditions obtaining at the end of an extrusion, when the pressure rises steeply as the billet shortens to much less than its transverse dimensions. This condition is interesting but is usually not of much practical importance, because the press is normally stopped while there is still an appreciable discard, which is eventually sawn off. If discardless extrusions are produced by using a dummy block of soft material behind the billet, the pressure rise is avoided. The final phase, which more closely resembles forging than extrusion, may, however, be of significance in impact extrusion[8.11] or extrusion-forging, where the whole billet is used, and the form of the residual billet is important (see §8.9).

Many practical problems can be represented approximately by extrusion of combinations of flat strips and round bars; and the estimation of redundant work in conditions of axial symmetry is of considerable importance, and this can be done by approximate slip-line field or upper-bound theory (§§10.12, 13.5).

8.8 Axially-symmetrical extrusion

As was seen in §8.2, load determination by the stress-evaluation method is straightforward for extrusion of round bar, but the redundant-work factor is too large for such solutions to be valid with practical die angles.

Attempts have been made to apply plasticity theory to problems of axial symmetry[8.12], but it is not yet known how to construct stress and velocity fields for this condition. Direct application of plane–strain slip-line fields has also been attempted, but the accuracy of the assumptions involved is unknown, and experimental checks suggest that considerable errors may be introduced.

There are, however, two useful methods available for load determination, one an adaptation of the upper-bound technique, the other a semi-empirical formulation based on two-dimensional slip-line fields.

8.8.1 *Application of upper-bound technique to axial symmetry*

The upper-bound method for axial symmetry has been developed by Kudo from his technique[8.8] for obtaining upper-bound solutions to plane–strain problems for which no slip-line fields exist. The plane–strain method may be illustrated for the same simple example of strip extrusion to 50% reduction through square dies. The slip-line field solution has already been discussed in Chapter 5, §5.8. For the upper-bound solution, the billet is divided into rectangular units, whose sides are assumed to move with constant speed in directions normal to their respective planes (Figure 8.8*a*).

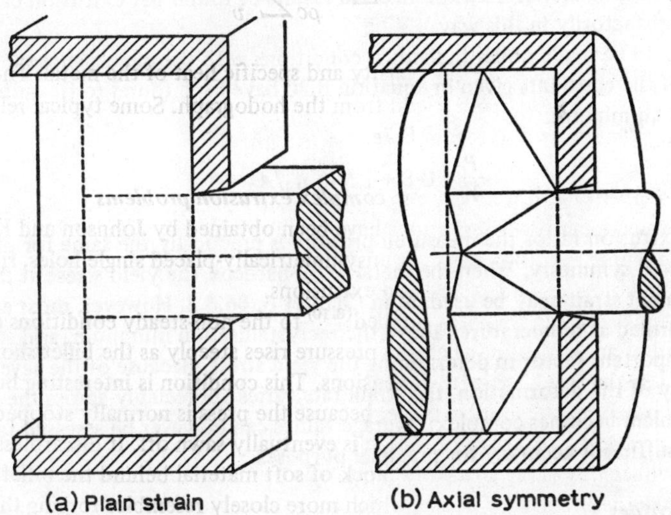

Figure 8.8(*a*) Plane–strain extrusion with square dies, considered by the upper-bound method for problems unsolved by slip-line fields. The billet is divided into rectangular units.

Figure (*b*) Division of a billet into cylindrical regions for estimation of an upper-bound load in extrusion of round bar.

The energy dissipated in the deformation of each unit is calculated, including the work done in shearing over sticking-friction interfaces where appropriate. The best combination of dimensions for the rectangles is the one which gives the least total energy dissipation. Velocity fields within these rectangles are then selected, again to give the least possible energy dissipation. The method requires some skill and practice, and the reader is referred to the original paper[8.8] for examples.

For problems involving axial symmetry, the approach is similar, but the billet is divided into cylindrical regions instead of rectangles, as in Figure 8.8*b*.

These units may again be divided by triangular velocity fields chosen for least energy dissipation. (See also Chapter 13).

8.8.2 *Semi-empirical method based on plane–strain slip-line field solutions*

The results of slip-line field theory for plane–strain extrusion with square dies can be summarised graphically[8.13]. They agree remarkably well with experiments designed to test them[8.14], and there is a similarity to results obtained from axially-symmetrical extrusions. The curve can be fitted quite well by an analytical expression with two empirical constants a and b.

$$\frac{P}{2k} = a + b \ln A_1/A_2 = a + b \ln \frac{1}{1-r} \tag{8.25}$$

The logarithmic term represents the contribution of homogeneous deformation.

Johnson[8.15] suggested applying an equation of this type to predict the extrusion pressure for axially-symmetrical conditions. His experiments with lead showed that, for a non-hardening metal, the experimental results of round bar extrusion can be described satisfactorily in this way.

Tests should be made under individual conditions with the metals chosen, but the values of the constants given in equation 8.26 have been found to fit quite well for lead and aluminium.

$$\frac{P}{\bar{Y}} = 0{\cdot}8 + 1{\cdot}5 \ln A_1/A_2 \tag{8.26}$$

With large extrusion ratios the measured pressure is practically the same for plane–strain and axial symmetry. When the metal work-hardens, the yield stress at the mean equivalent strain may be used, as in Chapter 6, §6.3.2. However, most extrusion is performed at temperatures above the recrystallisation minimum, and strain rate is an important factor in determining the yield stress. Because of the severe inhomogeneity of the deformation, the strain rate varies appreciably across the section and the problem becomes complex. Average values can however be assessed fairly easily with sufficient accuracy for practical purposes.

8.8.3 *Die profiles*

As well as the common flat (180°) dies and the conical profiles considered above, it is possible, for modest reductions, to design dies of improved efficiency[8.16]. Visioplasticity or upper-bound techniques are used to suggest profiles that give constant longitudinal strain or strain rate, or streamline flow. Such dies may also give low stress or minimum inhomogeneity of deformation, leading to reduction of defects.

An important characteristic of these dies is their low entry angle, which also encourages lubricant feed through. They are probably advantageous in hydrostatic extrusion, but tend to be too long for conventional extrusion with large reductions.

8.9 Special forms of extrusion

8.9.1 *Bridge dies*

Tubes of steel and other relatively hard metals are extruded over a mandrel which is inserted in the billet and moves forward with the ram. The tubing moves much faster than the mandrel and it is necessary to provide good lubrication between them. For the softer metals and alloys, particularly aluminium, it is possible to use a stationary mandrel held in the die by two or four lateral webs. The aluminium flows through the two or four apertures and reunites, where it has been divided by the webs, before emerging from the die as a tube. By this means, tubes can be formed from solid, unpierced billets. It is not practicable to lubricate the short mandrel in this process because it is inaccessible and because any lubricant is liable to interfere with proper welding of the metal surfaces when they reunite after passing the supporting webs. The mandrel acts somewhat like a sticking-friction piercing point. Other hollow sections can be produced by this technique.

8.9.2 *Impact extrusion*[8.11]

Very thin-walled tubes, such as are used for collapsible containers for toothpaste and many domestic commodities have been made by impact extrusion of aluminium since about 1920, though the principles were first demonstrated long before.

Two basic methods can be used. In the first, a slug of aluminium or other soft metal is coated with a lubricant and placed in a shallow die. It is then subjected to the high-speed impact of a cylindrical punch with a nose-piece giving the required clearance in the die. The metal flows up the sides of the punch in a thin film. The sequence may be very rapid and large quantities of these tubes can be produced in a short time. By suitable design, a shoulder or collar can be left in the die, so that the product is finished in one blow, except for threading the neck. The other method, usually known as the Hooker process, is a forward-extrusion technique basically similar to ordinary extrusion of tube with a mandrel attached to the pressure-pad, but it operates at very high speed. Cup-shaped blanks are preferable for this process. It requires less pressure for thick-walled tubing than the inverted extrusion, and is used for copper and brass tubes. Tubes with closed ends and tapering wall-thickness can be produced.

The mechanical properties of impact extruded products are good, and a considerable advantage of the process is that complex finished articles can be produced in one operation, without needing expensive machining operations. Aluminium extrusions up to about 110 mm ($4\frac{1}{2}$ in.) diameter and 350 mm (14 in.) long can be obtained in a wide variety of shapes, with integral fins or projections inside or outside. The residue of the billet may be formed into a flange or a heavy base, as shown in the selection of cold extrusions in Figure 8.9.

Low-carbon steel, and even some alloy steels, can be cold-extruded in useful sizes, and have good mechanical properties.

Recently, very high-speed processes of explosive violence have been introduced,

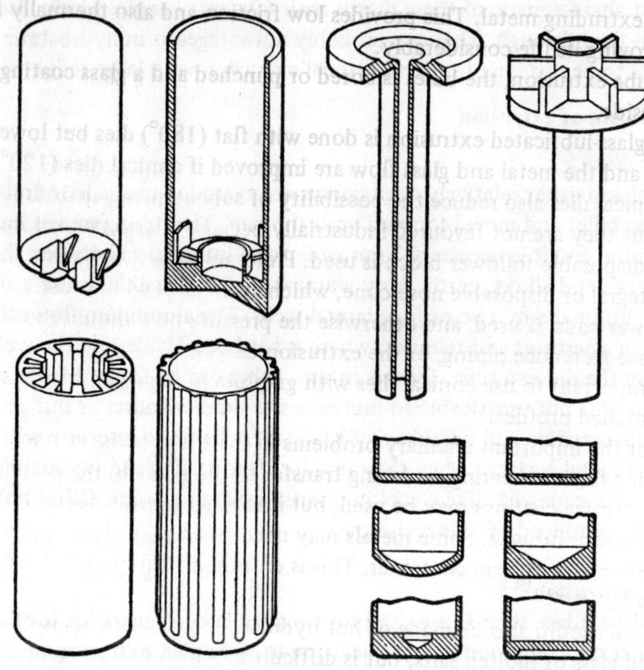

Figure 8.9 A selection of cold-extruded aluminium products, showing the versatility of the process.

(*Courtesy of Imperial Metal Industries (Kynoch) Ltd.*)

using compressed air or even actual explosives. These are very interesting from the point of view of the properties of the materials, and appear to be suitable for cold extrusion even of very brittle alloys.

8.10 Lubrication in practical hot extrusion

Hot extrusion makes heavy demands on lubrication. Graphite is effective, even at very high temperatures, because there is very little access for air, so the graphite does not oxidise appreciably. Long extrusions are possible if the lubricant is provided in the form of a pad that steadily softens or melts during the extrusion to produce a uniform thin coating (50 μm) as the metal passes through the die. Glass is the most widely-used pad material[8.2], though various inorganic salts with suitable melting points can be used[8.17]. Basalt is also suitable.

Typically the pad is compacted lightly from fairly fine glass powder, bonded with sodium silicate. It is made with the same diameter as the billet and may be about 10 mm thick. A central hole is provided with the same diameter as the extruded product. The billet is heated to the extrusion temperature, usually 1,200°C for stainless steel, and rolled over a tray of powdered glass to pick up a peripheral coating that lubricates its passage through the container. The face of the billet contacts the pad, which is placed in front of the die, so that the glass softens and is entrained with the extruding metal. This provides low friction and also thermally insulates the die, improving its life considerably.

For tube extrusion, the billet is bored or punched and a glass coating is provided on the inside.

Most glass-lubricated extrusion is done with flat (180°) dies but lower forces are required and the metal and glass flow are improved if conical dies (120°–150°) are used. Conical dies also reduce the possibility of subcutaneous defects being produced, but they are not favoured industrially because a large discard is involved unless a disposable follower block is used. Preferably the flow should also be initiated by an integral or disposable nose cone, which again adds to the cost.

Follower pads, if used, and otherwise the pressure pad, should show high friction on the face to reduce piping, or the extrusion defect.

It is important to use conical dies with graphite lubrication, which is less tolerant of mismatched profiles.

One of the important ancillary problems in extrusion is the avoidance of oxidation during furnace heating or during transfer of the billet to the press. Glass or other protective coatings may be used, but it is usual to heat the billets in a controlled-atmosphere furnace. Some metals may need to be canned, that is, sealed in a mild steel or other cheap container. This is extruded with the billet and subsequently removed.

Mention should also be made of hot hydrostatic extrusion, which is now possible with fluid glass or molten salts, but is difficult to control.

Bridge-die extrusion is a specialised technique involving conflicting lubrication requirements. It is used for tubes and other hollow products where close concentricity is required. A short mandrel is held as an integral part of the die by two or more radial webs, so that it is much more rigid than the conventional cantilever type of extrusion mandrel. The extruding billet divides over the webs but must reunite as it passes through the die. Good welds are produced in aluminium or lead because the surfaces are free from contamination, but the metal should not stick to the bridge webs. If these are lubricated, the subsequent homogeneous weld in the tube will be weakened. Good welds are also possible in copper and brass tubes, if care is taken to avoid oxide particles becoming entrained.

High energy-rate extrusion uses graphite in preference to glass, which requires time to soften.

8.10.1 *Hot extrusion of steels*

Temperature is an important parameter in hot extrusion. Carbon steels are extruded at 900–1,200°C. The higher temperature gives lower force but greater problems of oxidation. Die wear is also affected, in a complicated way involving temperature, force, lubricant film thickness, length and time of extrusion, as well as the possibility of oxide fragments being present. Speed and temperature interact in determining the film thickness in the melting-solid type of lubrication. Ram speeds of 50–100 mm/sec are often used.

Glasses are most commonly used for stainless steels at 1,100–1,200°C, being chosen roughly on the basis of softening point, though this is not directly relevant, to give a suitable coating thickness under the particular conditions. Graphite is undesirable because reaction to form chromium carbides is likely, leading to surface embrittlement. Either glass or graphite is used for carbon steels, glass giving the longer extrusions.

Removal of glass is troublesome. Hydrogen fluoride or molten sodium hydride baths are used. Both are hazardous and effluent disposal is becoming increasingly difficult. Methods have been suggested for removing glass without pickling, but these do not completely remove all glass. Graphite is removed by conventional acid pickles for carbon steels, but has to be filtered out of the bath.

8.10.2 *Copper alloys*

Copper can be said to be self-lubricating since at temperatures above about 600°C its oxide is softer than the metal and thus forms a good natural lubricant. Copper can therefore be satisfactorily hot extruded at 750–950°C with no further lubricant. Brasses have a high zinc contact in their oxides and show higher friction, so a lubricant is required. They are extruded at 750–850°C.

Graphite, usually applied as a dispersion, in grease, oil or resin, is widely used for extrusion of copper and copper alloys. Low-melting glasses can also be used, but care must be exercised since these usually contain lead. Barium chloride is a suitable melting solid[8.17].

8.10.3 *Aluminium alloys*

Aluminium is often extruded at 550–500°C with no lubricant, mainly to avoid entraining oxide and lubricant to form surface and subsurface defects because flat-faced dies are used. If tapered dies are used, oils or graphite can give good lubrication and low forces, but staining may result. Graphite is easily visible on shiny aluminium. Tapered dies also produce larger discards. Pickup can be a serious trouble. This is particularly important for complex extruded sections, such as glazing bars, that are subsequently anodised. Any pickup becomes very obvious.

The extrusion defect is important in aluminium, especially because square dies and high container friction produce high forward-extrusion force. As the billet becomes short, the full deformation zone for sticking friction is unable to develop (Figure 8.4*d*) and the force rises rapidly with further ram travel. It may then be easier for the billet metal to flow inwards over the pressure pad face and out along the axis, as in extrusion forging (§10.12). This produces a high outward velocity on the axis, leading to a cavity or pipe. It is strikingly enhanced if the pad face is lubricated, and may be somewhat reduced by improving mechanical restraint by circular grooves on the pad.

Die wear is another problem, necessitating good lubrication.

8.10.4 *Other alloys*

Titanium is commonly extruded at about 850°C with a suitable glass of fairly low softening point. Molten salts can be used or the billet can be canned in copper, which also completely protects it from oxidation. Heavily-graphited grease can be used directly on titanium or on the canned billets.

Beryllium is too brittle for most metalforming, because of its hexagonal structure, but it can be satisfactorily warm-extruded at 500–650°C using a graphite lubricant.

At higher temperatures it should be heated and extruded in sealed steel cans. The oxide, in air-borne form, is dangerous. The billets may sometimes be silver coated. Nickel and cobalt alloys are extruded at 1,100–1,200°C with glass lubrication, or sometimes with basalt.

Magnesium is self lubricating, like copper, and extrudes easily at 400–500°C. Molybdenum and tungsten require high temperatures, respectively about 1,900°C and 2,100°C. They are extruded with glass, which also protects against loss of molybdenum by formation of a volatile oxide. Graphite can also be used. Tungsten is sometimes canned in molybdenum.

8.11 Lubrication in practical cold extrusion

Cold extrusion and cold forging follow the same lubrication procedures as cold tube drawing, but the conditions are more severe because of the large increase in surface area produced during extrusion. Conversion coatings are essential, unless there is a major contribution from hydrostatic lubrication. The only alternative appears to be silver or tin coating. Pickup is a major limitation but it is important to reduce the forces as far as possible because of the inevitably high stress levels in cold extrusion. Nevertheless, as in deep drawing, the flow and fracture can be controlled by lubrication in selected areas. This is particularly important for correct die filling.

Hydrostatic extrusion inherently provides a lubricating film.

8.11.1 *Cold extrusion of steels*

Phosphate conversion coatings[8.18] are almost universally used for carbon steels and for such alloy steels as can be cold extruded, supplemented by dried soap or a graphited paste. A zinc phosphate is used, as in wire drawing, but at about 10 mg/m^2 (1,000 mg/ft^2) which is about three times as thick as for wire drawing. The soap may be a soluble sodium stearate, or an insoluble calcium stearate, but aluminium stearate is preferred by some operators. These all give satisfactorily low wear of the tools and a good surface finish on the product. Impact extrusion uses a phosphate and soap.

Cold heading is a less severe operation, for which phosphate and soap gives good results, but simple soap coating may be adequate. Graphited paste can also be used.

For hydrostatic extrusion, an oil of moderate viscosity with a high MoS_2 content (10%) is often used. Castor oil containing up to 20% MoS_2 in castor wax has also been recommended.

Stainless steels are oxalate coated and extruded with aluminium or zinc soap film.

8.11.2 *Copper alloys*

Copper is again relatively easy to extrude, using tallow, grease or lanolin. Sulphonated tallow or graphited grease can be used for brasses. Free sulphur must be avoided.

The commonest lubricants for copper-alloy cold-extrusion are zinc stearate films or soap-fat emulsions.

8.11.3 *Aluminium alloys*

The stresses are much lower for aluminium than for steels. At low speed lanolin or tallow give the best results, but sulphonated tallow is recommended for higher speed extrusion. Mineral oils with boundary additives can be used, but dry zinc stearate is widely used in industry for impact extrusion.

To obtain accurate die filling, the lubricant coating must be uniform. Thicknesses up to 0·02 mm may be applied, and the zinc stearate is not easy to control. Solution-deposited waxes have been favoured in some applications.

Hydrostatic extrusion can be used, with simple mineral oils containing boundary additives.

8.11.4 *Other alloys*

Titanium is the most important of the other alloys that are commercially cold-extruded. As in drawing, the stock may be lightly oxidised or anodised to prevent pickup, but fluoride phosphating is more reliable. The supplementary lubricant can be soap but a 10% graphite in resin dispersion gives more accurate die filling. Methacrylate and other polymer coatings are not suitable because they are too thick and uneven.

Lead is often cold extruded, and presents no lubrication problems. A fatty oil is suitable.

8.12 **Recent developments in extrusion**

8.12.1 *Theoretical contributions*

Extrusion, like wire drawing, is particularly suitable for theoretical study because the deformation is predominantly a steady-state process. It is however more interesting than drawing because the redundant strain, and consequently the distortion, is much greater and has more evident effects[8.16].

The process is also very convenient for visioplasticity experiments because the billet is short and the necessary halves are firmly held together by the inherent compressive stress. Extensive visioplasticity studies have been made, both of steady-state extrusion and of the changed deformation mode in the discard and the back end of the product. This technique is applicable to the related problems of extrusion-forging and coining. Both solid and hollow components can be studied[8.19].

The results generally agree well with slip-line field solutions, and show that the strain patterns in axi-symmetric flow correspond closely to those for plane strain.

Upper-bound solutions are very useful for predicting extrusion loads, especially for axisymmetric shapes[8.20], using the methods for subdivision into unit deforming zones. Upper bounds based on spherical deformation boundaries can be applied for square and other polygonal sections. These methods are less precise than slip-line fields for predicting local strains, but they can show major flow features, such as changes of deformation mode, the occurrence of an extrusion defect or the likelihood of central bursts or fishskin tearing.

There is a continuing interest in homogeneity of the structure and properties of extrusion products, especially for complex sections. These can often be related to die design, and die profiles can be calculated to reduce redundant work, improve uniformity of strain, or avoid defects[8.16, 8.20]. It seems likely that these theoretically-ideal dies will be of most use in hydrostatic extrusion. The finite-element method has also been applied to analysis of this process[8.22].

8.12.2 *Conventional and newer processes*

The use of glass and glass-like lubricants for hot extrusion of ferrous and other alloys has increased, but no new lubricants for cold extrusion have found general acceptance. Conversion coatings with a soap or similar supplementary layer still provide the best results at room temperature.

Hydrostatic extrusion is now well established[8.23], and shows wider commercial promise since techniques have been developed for higher speeds and particularly for continuous extrusion. In hydrostatic extrusion the conventional ram and pressure pad are replaced by a fluid that completely surrounds the billet. A high pressure is applied to the fluid so that the billet is forced through the die, effectively sealing the exit except for a fine leakage of fluid that acts as a lubricating film. The friction is very low, and it is possible to extrude into a pressurised chamber so that even brittle alloys can be worked. Warm and hot extrusion are possible[8.24].

Tubes and composite materials can be hydrostatically extruded, and the necessary forces can be further reduced by applying a front tension in an extrusion-drawing process.

Helical extrusion is a novel technique for the production of wire or tube, as described in §6.13.3. It is effectively a machining operation in which the incipient swarf is trapped and forced to escape through a small suitably-shaped orifice.

EXAMPLES

8.1 Draw the slip-line field for 2:1 extrusion with a 150° included-angle die, (*a*) with zero friction (*b*) with dead zone formation at the die and sticking friction on the container wall.

Solution. (*a*) The field resembles that for strip-drawing, shown in Figure 6.5. A 45° triangle is first drawn against the die face *AB*. Fan regions are constructed round the two singularities *A* and *B*, using 15° intervals. The field is then extended by drawing chords at $7\frac{1}{2}°$ to the radii. Chords inclined at 15° are then drawn from each intersection in turn, as explained in Chapter 6, §6.5.2. The field is built up until an intersection near the centre-line is reached. The angles ψ and ζ included in the fans *A* and *B* are selected so that the chords meeting near the centre-line are inclined at $52\frac{1}{2}°$ to it, since the real slip-lines should intersect it at 45°. The best values for ψ and ζ for a 15° mesh are thus found to be 15° and 90° respectively, but the nodal point then lies a little below the centre-line. To obtain a more accurate solution, the smooth slip-lines are drawn in, and further slip-lines are then constructed, using the same technique, with closer angular separation near the edges of the field. Thus $\psi = 10°$ gives an intersection above the centre-line. $\psi = 13°$ and $\zeta = 88°$ give a more accurate result.

It should be noticed that the angle of intersection of the exit β-line with the centre-line is $180 - (\theta + 45 + \psi - \zeta)$, θ being the inclination of the die face to the transverse direction (= 15°). For a valid field the slip-lines must meet the centre-line at 45°, so $\theta + \zeta - \psi = 90$ or, in this instance $\zeta - \psi = 75°$.

(*b*) The field with sticking friction resembles Figure 8.4*d*. The exit slip-line is straight from the die *B* to the centre-line, intersecting the latter at 45° in *D*. A fan centred on *B* with this radius is constructed with 15° sectors. The second line is extended to the centre-line at D_1 by a chord at $7\frac{1}{2}°$ and so on, adding a ring-shaped area to the field. This still fails to reach the container at normal incidence, and a further ring zone must be added from D_2. This almost satisfies the boundary conditions of 0 and 90° at the container and 45° at the centre-line. The dead zone is thus found to contain an angle 25° at the die and to extend about $3{\cdot}2a$ along the billet ($2a$ =

final strip thickness). To improve the solution, the intersections are joined by smooth curves and lines of closer spacing are drawn. A satisfactory solution is obtained with a field bounded by the α-line from D_3', the latter point being found from the β-line inclined at 26° to the exit β-line. The dead zone then starts with an angle 29°.

8.2 Draw the hodographs for 2:1 extrusion with a 150° included-angle die (*a*) with zero friction and (*b*) with dead zone formation, using the slip-line fields of Example 8.1.

Solution. (*a*) It is simplest to draw the hodograph for the 15° mesh first, though this implies a slightly larger extrusion ratio. The more accurate solution for the real boundary slip-lines can then easily be drawn in. (This procedure reduces the risk of numerical errors in the angles.) The hodograph starts with the line $Of = 1{\cdot}0$. In the vicinity of the centre-line there is a shear at 45°, represented by a line fF, as in Figure 6.5. Similarly on leaving this zone there is shear at 45° in the other direction and the final velocity is along the extrusion direction, Of produced to O'. The position of O' is determined by the extrusion ratio. The magnitude of the tangential velocity discontinuity is unchanged along the entry slip-line and is represented by radii of a fan centred on f in the hodograph. The radii are each parallel to the appropriate portion of the slip-line, so the fan includes an angle $\zeta = 90°$. The discontinuity on the exit line is also constant, so a second fan can be drawn about O', but this has only one sector $\psi = 15°$. The fans are then extended one intersection at a time as described in Chapter 6, §6.5.3. It will be found that the final absolute velocity along the die face is not accurately parallel to the die with the 15° mesh, but when the hodograph is smoothed out and the corrected fan angles 88° and 13° are used this is rectified.

(*b*) For sticking friction the hodograph may be started in the vicinity of the container wall. Again it is simplest to use first the 15° mesh, which approximately fills the actual plastic zone. Near the wall the chords meet at $7\frac{1}{2}°$ and $97\frac{1}{2}°$. The starting line $Oa = 1$ is followed by a line from a at $7\frac{1}{2}°$ to the normal, representing shear on the entry slip-line, which intersects a line at $7\frac{1}{2}°$ to Oa, representing absolute velocity parallel to the dead zone boundary. The magnitude of the discontinuity along the entry line is constant, so a fan of three radii is centred on a, giving the velocities in zones I, II, and III. An element leaving III, near the centre-line at D_3 will cross into the triangular zone where its velocity is parallel to Oa. This gives the intersection representing the velocity approaching D_2. The diagram is continued by considering the element from zone I crossing the chord into zone IV where its velocity must be parallel to the second chord of the dead zone boundary. This gives the first extension of the fan centred on a, and so the hodograph is built up, mesh by mesh. When D is reached and the diagram is completed it will be found to correspond to a reduction of area 2·2:1. The hodograph is then filled in by smooth lines joining the intersection and the corrected angles of the bounding slip-lines are used to draw further lines for the accurate solution.

8.3 What would be the maximum possible extrusion ratio for cold plane–strain extrusion of aluminium ($\bar{S} = 250$ N/mm²) through a frictionless 120° die using a 2,500 kN press taking 150 x 25 mm billets? (It is convenient to draw the slip-line field to a scale giving the final strip thickness $h_2 = 50$ mm, so the diagram occupies a space about 300 mm high and 250 mm wide).

Solution. Draw the approximate field by the chord method using $\Delta\phi = 15°$ intervals (see Figure 8.3). Since the die is frictionless and the semi-angle is 60°, the fan centred on the exit corner B contains an angle 60° or 1·05 radians. The die pressure between B and A (Figure 8.3) is thus $q = (1 + \alpha)\ 2k$ because p on BD is equal to k and on BC is $k + 2k\alpha$. Between A and the next intersection E_1 on the die face, p increases by $2k \,.\, 2\,\Delta\phi$, and so on.

The contribution of each zone to the extrusion force is

$$q_{AB}\,AB \sin\alpha + q_{AE_1}\,AE_1 \sin\alpha + q_{E_1E_2}\,E_1E_2 \sin\alpha, \text{ etc.}$$

$$= 2k(1 + 1{\cdot}05)\,(44{\cdot}5) + 2k(2{\cdot}05 + 0{\cdot}52)\,(43{\cdot}8) + 2k(2{\cdot}05 + 1{\cdot}04)\,(68{\cdot}8) + \ldots$$

the lengths being measured on the diagram ($\frac{1}{2}h_2 = 25$ mm).

Thus for a ratio 2·78:1, $\frac{P}{2k} = \frac{(2{\cdot}05)(43{\cdot}8)}{68{\cdot}8} = 1{\cdot}31$

4·5 :1, $\frac{P}{2k} = \frac{91+112}{112} = 1{\cdot}81$

7·75:1, $\frac{P}{2k} = \frac{204+212}{194} = 2{\cdot}14$

11·5 :1, $\frac{P}{2k} = \frac{416+382}{287} = 2{\cdot}78$

The maximum extrusion pressure available is 2,500/(150 × 25) = 0·67 kN/mm², corresponding to $P/2k = 0{\cdot}67/0{\cdot}25 = 2{\cdot}66$. The largest possible extrusion ratio is thus about 10:1.

(If the deformation were homogeneous the maximum ratio would be given by $\ln A_1/A_2 = 2{\cdot}66$, namely 14·2:1. With square dies, using equation 8.26,

$$2{\cdot}66 = 0{\cdot}8 + 1{\cdot}5 \ln A_1/A_2, \text{ giving } 3{\cdot}4{:}1)$$

8.4 If the billet of Example 8.3 were 300 mm long and the coefficient of friction against the container were $\mu = 0{\cdot}15$, what would be the initial contribution of container friction?

Solution. The contribution of container friction may be found approximately by the method of §8.2.2. For plane–strain conditions, axial equilibrium gives $h \, d\sigma_x = \mu \sigma_y \, dx$, leading to

$$(\sigma_{xc} - 2k) = (\sigma_{xb} - 2k) \, e^{\mu L/h}$$

The pressure is thus increased by a factor $\exp \frac{(0{\cdot}15)(300)}{25} = 6$.

8.5 Draw the slip-line field for 10:1 plane–strain direct extrusion (*a*) with a frictionless 180° die, (*b*) with a dead zone. (The second diagram occupies about 300 × 300 mm if the final strip thickness is chosen to be represented by h_2 = 50 mm.)

Solution. (*a*) The field resembles that shown in Figure 8.4*b*. A 15° mesh is drawn first starting with the exit slip-line at 45° to the centre-line. The fan centred on the exit edge *B* of the die is then drawn, extending to the 45° triangle on the frictionless die face. The field is then built up from this. This would fit accurately if the extrusion ratio were 9·4:1. A more accurate solution can be obtained by drawing the slip-lines smoothly between the intersections and then extending the field by about 1°, starting with the β line inclined at 31° to the exit slip-line, following this to its intersection with the centre line, and then drawing the appropriate α-line.

(*b*) The field resembles that shown in Figure 8.4*d*. Again the construction with a 15° mesh starts at the exit slip-line *BD*. The fan centred on *B* is first continued round to the die face, since if the slip-lines meet the die face they must do so at 90° and 0°. The radius at 15° to the exit line is extended by a chord at 7½° to meet the centre-line at D_1 and the field is increased by one annular zone of 15° meshes at a time. The fourth such zone, cutting the centre-line at D_4 and D_5 would meet the dead zone boundary at the correct angles (7½° to the normal and tangential) on the container wall if the extrusion ratio were 8·7:1. The dead zone would then include an angle 15° at the die edge.

To complete the solution the field is drawn with smooth lines between the intersections, and lines of closer spacing are drawn at the entry and dead zone boundaries to fulfil the conditions of sticking friction on the container for the required extrusion ratio. The dead zone is then found to include about 10° at the die edge.

8.6 Draw the slip-line field and hodograph for 4·3:1 extrusion through a frictionless square die, and from these draw the flow lines of metal passing through the plastic zone.

Solution. The slip-line field resembles that shown in Figure 8.4*b*. It is constructed in the same way as in Example 8.5*a*. A 15° mesh fits accurately. It is convenient to start the hodograph, following the procedure described in Example 8.2 using the approximate chord method. Finally the intersections are joined by smooth curves.

To draw the flow lines, the smoothed hodograph is used. The mean velocity in any mesh is then represented by the vector from the origin to the mid point of the appropriate hodograph section, as described in §8.6. Lines parallel to these are superimposed on the slip-line field traversing the meshes to which they refer, starting each one at one of the intersections on the entry slip-line, and following each through successive meshes, as in Figure 8.5. The intersections with the slip-lines are finally joined by smooth curves, which represent the flow of metal elements through the whole plastic zone.

8.7 Draw a slip-line field and hodograph for 3:1 extrusion with a frictionless 180° die, using 10° intervals in the slip-line field. Superimpose the metal flow lines on the slip-line field.

Solution. The slip-line field is of the same type as that shown in Figure 8.4*a*. The hodograph is started in the corner of the container where an element of approaching metal shears at 45° and then proceeds parallel to the die face. This determines the magnitude of the tangential velocity discontinuity along the inlet slip-line, which remains constant but changes direction with the slip-line until the centre-line is reached. The hodograph is completed by further shear at 45° on the exit line, giving a final longitudinal velocity 3·0.

Mean values of the absolute velocity in each 10° sector are found by joining the mid point of each in the hodograph to the origin. Lines parallel to these velocity vectors are transferred to the slip-line diagram systematically for each sector, starting at each intersection on the inlet slip-line. These lines are finally smoothed out to represent the actual flow of metal.

8.8 (*a*) Draw an upper-bound solution for 3:1 plane–strain extrusion with a frictionless 180° die. Compare the flow-lines predicted by this solution with those found in the preceding example from the slip-line field. Compare also the predicted values of extrusion pressure by the two methods.

(*b*) From the upper-bound solution, find the way in which a transverse grid scribed on the billet would become distorted in the extrusion process.

Solution. (*a*) A suitable upper-bound solution and hodograph are shown in Figures 8.7*a* and *b*. The velocity in any segment of the field is assumed constant. Thus the absolute velocity everywhere in *AOB* is represented in the hodograph by *Ob*, the velocity in *BOC* by *Oc*, and so on. An element crossing the entry boundary at *B*, for example, travels with velocity *Oc* during its passage from *B* to the next line *OC*. This direction is shown on the diagram by a line parallel to *Oc* drawn from *B* to cut *OC* in *C**. The next part of its passage is represented by a line parallel to *Od* drawn from *C** to cut *OD* in *D**. After traversing *BOE* the element emerges from the plastic zone across *OE* at *E**, and subsequently follows a path parallel to the centre-line. Similar paths are traced for selected elements crossing *AE* at suitable intervals.

(*b*) The horizontal components V_x of the absolute velocity in each segment of the field are first determined from the hodograph. These are 0 (*AOB*), 0·27 (*BOC*), 1·0 (*COD*), 1·72 (*DOE*) and finally 3·0 after crossing *OE*.

The reciprocals $(1/V_x)$ are plotted in the appropriate zones projected vertically below the field, as is described for the slip-line solution in §8.6. (Separate pieces of transparent paper may conveniently be used.) An arbitrary starting line is chosen to the left of *CD*, and the areas $\Delta x(1/V_x)$ beneath the reciprocal plot are determined for each segment. Elements crossing the entry boundary at *B, C, D*, and *E* each give a separate set of areas, to which may be added those for an element crossing midway between *A* and *B*.

The cumulative areas for each element are then plotted, on a single graph using the same projection, to show the variation of $\Sigma\Delta x/V_x$ with *x*. (If the initial diagram is drawn with smaller segments this approaches $\int_0^x \mathrm{d}x/V_x = t$.) To find the displacement of each element at a given time *t* after the assumed start, a horizontal line ($\Sigma\Delta x/V_x$ = constant) is drawn on this graph. The intersections of the line with the appropriate lines for the different elements crossing at *A*, *B*, etc. gives their *X*-displacements, which may be transferred to the upper-bound field directly as horizontal distances from the starting line. The points so obtained are joined to show the distortion of an initially straight grid. Several different values of t should be chosen.

The result should be compared with that of Example 8.7, which is a more accurate version of Figure 8.5*d*.

8.9 Estimate the pressure required to extrude aluminium curtain rail of I section 12 mm high with 6 mm wide flanges, all 1·6 mm thick, from 25 mm diameter bar stock.

Solution. The extrusion can be considered as simultaneous extrusion of three flat bars, two being 6 mm wide and 1·6 mm thick and one 8·8 × 1·6 mm. However, as stated in §8.8, if the extrusion ratio is large the extrusion pressure is sensibly that for extrusion of round bar of the same cross-sectional area. Thus, using equation 8.26,

$$\frac{P}{\bar{Y}} = 0{\cdot}8 + 1{\cdot}5 \ln A_1/A_2$$

in this example A_1 = 491 mm². The final area is 2 × 6 × 1·6 + 8·8 × 1·6 = 33·3 mm². A_1/A_2 = 14·7

$$\frac{P}{\bar{Y}} = 4{\cdot}83$$

The mean yield stress for aluminium (Example 2.3) is about 0·15 kN/mm² for heavy deformation at room temperature, so $P \approx 0{\cdot}7$ kN/mm² and the force is about 350 kN.

8.10 Estimate the largest possible extrusion ratio for cold extrusion of 25 mm diameter, 3 mm-wall mild-steel tube in a 10,000 kN press.

Solution. Assume $\bar{S} \approx 0{\cdot}65$ kN/mm², $\bar{Y} \approx 0{\cdot}56$ kN/mm². If the tube is 25 mm O.D., the final area is 207 mm². From the approximate equation 8.26,

$$\frac{P}{\bar{Y}} = 0{\cdot}8 + 1{\cdot}5 \ln\frac{A_1}{A_2};\quad \frac{10{,}000}{0{\cdot}56 A_1} = 0{\cdot}8 + 1{\cdot}5 \ln\frac{A_1}{207}$$

whence $\dfrac{A_1}{207} = \exp\left(\dfrac{11{,}905}{A_1} - 0{\cdot}53\right)$

A first attempt at solution can be made using the expansion $\exp(x) = 1 + x + \dfrac{x^2}{2!}$ + etc.

$$\frac{A_1}{207} \approx 1 + \frac{11{,}905}{A_1} - 0{\cdot}53;\quad A_1^2 - 37A_1 - 2{\cdot}46 \times 10^6 = 0$$

A_1 = 1,588, implying an extrusion ratio 7·6, but this is very inaccurate since x has a large value (6·9). The equation is however quite quickly solved by numerical approximation:

A_1	$\dfrac{A_1}{207}$	$\left(\dfrac{11{,}905}{A_1} - 0{\cdot}53\right)$	exp
A_1 = 1,588,	= 7·67;	= 6·97,	exp 6·97 = 1,060
= 3,000	= 14·5;	= 3·43	= 31
= 3,500	= 16·9;	= 2·87	= 17·6
= 3,600	= 17·4;	= 2·78	= 16·1
= 3,530	= 17·05;	= 2·84	= 17·15

This implies an extrusion ratio of about 17:1. A standard size may be selected and the pressure checked. Assuming the initial bore diameter of the billet to be 25 mm, to allow clearance for the mandrel, the O.D. would be given by

$$\frac{\pi}{4}D_1^2 - 491 = 17 \times 207;\quad D_1 = 71 \text{ mm}$$

If $D_1 = 70$ mm, $A_1 = 3{,}357$, $A_1/A_2 = 16{\cdot}2$

$$\frac{P}{\bar{Y}} = 0{\cdot}8 + 1{\cdot}5 \ln 16{\cdot}2 = 4{\cdot}98; \; P = 2{\cdot}78 \text{ kN/mm}^2, \; PA_1 = 9{,}357 \text{ kN}$$

(If $D_1 = 65$ mm, $A_1 = 2{,}827$, $A_1/A_2 = 13{\cdot}7$

$$\frac{P}{\bar{Y}} = 0{\cdot}8 + 1{\cdot}5 \ln 13{\cdot}7 = 4{\cdot}73; \; P = 2{\cdot}64 \text{ kN/mm}^2, \; PA_1 = 7{,}500 \text{ kN})$$

It should be noted that the calculation does not allow for the initial peak pressure. On the other hand, a somewhat lower pressure would be required with a conical die.

REFERENCES

8.1 Pearson, C. E. and Parkins, R. N., *The Extrusion of Metals*, 2nd Ed., Chapman & Hall, 1960.

8.2 Sejournet, J. and Delcroix, J., 'Glass lubrication in extrusion of steel', *Lub. Engng.*, 11, 389–396, 1955.

8.3 Hoffman, O. and Sachs, G., *Introduction to the Theory of Plasticity for Engineers*, McGraw-Hill, 1953.

8.4 Hill, R. and Tupper, S. J., 'A new theory of plastic deformation in wire drawing', *J. Iron and Steel Inst.*, 159, 353–359, 1948.

8.5 Johnson, W., 'Extrusion through wedge-shaped dies', *J. Mech. Phys. Solids*, 3, 218–223, 224–230, 1955.

8.6 Green, A. P., 'A theoretical investigation of the compression of a ductile material between smooth flat dies', *Phil. Mag.*, 42, 900–918, 1951.

8.7 Johnson, W. and Kudo, H., *The Mechanics of Metal Extrusion*, Manchester Univ. Press, 1962.

8.8 Kudo, H., 'Some analytical and experimental studies of axi-symmetric cold forging and extrusion', *Int. J. Mech. Sci.*, 2, 102–127, 1960.

8.9 Tanner, R. I. and Johson, W., 'Temperature distributions in some fast metalworking operations', *Int. J. Mech. Sci.*, 1, 28–44, 1960.

8.10 Johnson, W., 'The plane–strain extrusion of short slugs', *J. Mech. Phys. Solids*, 5, 202–214, 1957.

8.11 Various, 'Conference on the cold extrusion of steel', Sheet and Strip Metal Users' Tech. Assoc., *Sheet Metal Ind.*, 30, 445–524, 1953.

8.12 Hill, R., *The Mathematical Theory of Plasticity*, Oxford Univ. Press, 1956, p. 262.

8.13 Hill, R., 'The theory of the extrusion of metals', *Selected Govt. Res. Repts.*, 6, Part 1, 191–204, 1952 (H.M.S.O.).

8.14 Purchase, N. W. and Tupper, S. J., 'Experiments with a laboratory extrusion apparatus under conditions of plane–strain', *J. Mech. Phys. Solids*, 1, 277–283, 1952.

8.15 Johson, W., 'The pressure for cold extrusion of lubricated rod through square dies of moderate reduction at slow speeds', *J. Inst. Metals*, 85, 403–408, 1957.

8.16 Blazynski, T. Z., 'Optimisation of die design in the extrusion of rod', *Int. J. Mech. Sci.*, 13, 113–131, 1971.

8.17 Rogers, J. A. and Rowe, G. W., 'An investigation of factors influencing lubricant behaviour in hot extrusion', *J. Inst. Metals*, 95, 257–263, 1967.

8.18 James, D., 'Chemical conversion coatings', *Wire Ind.*, 36, 685–690, 1969.

8.19 Thomson, E. G., Yang, C. T. and Kobayashi, S., '*Mechanics of plstic deformation in metal processing,* McMillan, New York, 1965.

8.20 Avitzur, B., *Metal forming: Processes and analysis,* McGraw-Hill, New York, 1968.

8.21 Kudo, H., 'An upper-bound approach to plane–strain forging and extrusion – 1', *Int. J. Mech. Sci.*, 1, 57–83, 1960.

8.22 Iwata, K., Osakada, K. and Fujino, S., 'Analysis of hydrostatic extrusion by the finite element method', *J. Engg. Ind. Trans. ASME*, 94B, 697–703, 1972.

8.23 Pugh, H. Ll D. and Donaldson, C. J. H., 'Hydrostatic extrusion – a review', *Ann. C.I.R.P.*, 21, 166–186, 1972.

8.24 Fiorentino, R. J., 'Comparison of cold, warm and hot extrusion by conventional and hydrostatic methods', *Metall. Metal Form.*, 40, 15 et seq., 1973.

9

Rolling of Flat Slabs and Strip

9.1 Introduction

Rolling-mills are very widely used for both hot-working and cold-working of many metals and alloys, in a great range of sizes. Some ingots destined for steel-sheet production may weigh 20 tonnes (tons) or more, and be at least 0·3 m (12 in.) thick. Individual mills may produce 50,000 tonnes (tons) of steel slab each week[9.1], in widths up to 1·8 m (72 in.). Aluminium can be rolled in strips 4m (172 in.) wide[9.2] but on the other hand, aluminium foil may be only about 0·025 mm (0·001 in.) thick.

9.1.1 *Hot-rolling*

Heavy reduction of area by hot-rolling is one of the major methods of producing bars of rectangular, round, or more complex cross-section. It also improves the properties of cast metal, by breaking up the cast structure and refining the grain size, giving better homogeneity and greater strength and toughness. During rolling and most forging processes, in contrast to extrusion, the pressure and deformation are restricted to a small volume at any given time, so that it is possible to deal with very heavy ingots, using equipment of moderate load capacity. The ingot is, however, fed continuously through the rolls, and there is usually a good speed of production. This is vital, as in all hot-working, because the stock hardens rapidly as it cools. A limiting factor in speed could be the time required to transport the slab back to the entry side of the rolls, but this is obviated, for most sizes, by using reversing mills. Very short slabs are usually returned manually, making considerable demands on the operators.

A hot-rolling mill for producing slabs or blooms from ingots is known as a cogging mill, and usually has two large rolls, mounted vertically one above the other. It is consequently described as a 2-high mill, in contrast to the 4-high and more complex mills often used in cold-rolling. Rolls of special profile are used to produce hot-finished lengths of various sections: round, hexagon, channel, angle, I-beam. These are of considerable economic importance, because very large tonnages are produced, but the rolls are usually designed empirically, following a few general rules, since the problem, as in closed-die forging, is primarily one of form-filling, for which there is at present little theory. We shall not attempt to evaluate working loads for these complicated conditions, but rough approximations may be made from a knowledge of the theory for simple rolling of comparable cross-sections of flat stock.

9.1.2 *Cold-rolling*

Cold-rolling of wide strip is the most suitable of all metalworking processes for theoretical analysis, and the theory of cold-rolling has been extensively developed. It is usual to roll very large quantities of material of given size and properties, and long uninterrupted coils are processed. The loads thus remain steady, and are little influenced by starting and stopping behaviour, in contrast to the conditions found in such processes as deep-drawing and short-tube production. It is common for several rolling mills to work in tandem, and it is essential that the tension in the strip between mills is held within close limits to maintain an even gauge. As the thickness of the slab is reduced, its length increases and the speed of the outgoing strip may reach very high values, up to about 250 m/sec (5,000 ft/min) in a 'mile-a-minute' mill. It is possible to control the tensions manually, by adjusting the relative speeds of rolls in adjacent stands, but computers have recently been installed in at least one large rolling mill for this purpose. Because of these high speeds, and the large tonnage output from rolling mills, proper pass planning, determined by material properties, acceptance of the strip by the rolls, and mill load capacity, has considerable economic significance.

It will be shown later in the chapter (§9.4) that the acceptance of strip by the rolls, which fixes the maximum reduction of area per pass, is determined by the inter-stand tensions, the coefficient of friction and the radius of the rolls:

$$(\Delta h)_{\max} = \mu^2 R \tag{9.1}$$

For this reason, rolling is unusual in giving a better performance with a somewhat poorer lubricant. However, if the friction is too high the load becomes excessive. Again theory can help in reaching a compromise.

Although the maximum reduction of area per pass increases with roll radius, the load P for a given reduction also increases (§9.5)

$$\frac{P}{w} = R' \times \int p(R, \alpha)\, d\alpha \tag{9.2}$$

The roll radius is consequently often kept small, to avoid heavy loads and the need for massive equipment. This has the further advantage that the strip may be rolled to thinner final gauge with small roll diameters (§9.7.1)

$$(h)_{\min} = 0{\cdot}035\, \mu R S \text{ mm} \tag{9.3}$$

This limiting thickness can be very important for thin strip and foil production, especially in hard metals. Long, slender rolls tend, however, to bend under load, giving a variation in gauge across the strip, so small-diameter rolls are usually supported by large back-up rolls, giving a 4-high mill. For very thin strip and foil, this principle may be carried further, the back-up rolls themselves being supported, with actual work-rolls of only 25 mm (1 in.) or 50 mm (2 in.) diameter. Such cluster-mills are capable of producing strip 0·25 mm (0·010 in.) thick in such hard materials as zirconium and titanium alloys[(9.3)]. When the yield stress S is low, as in aluminium, larger roll diameters can be used. Aluminium foil is rolled in 2-high mills, even down to 0·025 mm (0·001 in.) thick. The thinner gauges of foil, for cigarette and chocolate wrapping, are usually folded over and rolled double. The rolls are screwed down so far that the gap would be completely closed, and the rolls even under pressure, in the absence of the foil. This is equivalent to a negative value of the initial roll gap.

The gauge is then insensitive to roll setting, and is controlled by the tension applied by a power-driven coiler. This also helps to flatten the foil.

The theory of strip-rolling makes it possible to understand the interactions of all these factors: strip thickness and hardness, reduction of area or draft, roll diameter, roll gap, front and back tension, speed, and coefficient of friction. This provides a qualitative appreciation of the rolling process, which is a good guide to the choice of plant and of operating conditions; as well as a quantitative prediction of rolling loads, from which detailed optimum rolling schedules may be prepared.

9.2 Elementary assessment of roll load

9.2.1 *Homogeneous deformation*

A simple estimate of the roll load in flat-rolling can be obtained by considering the process as homogeneous compression between well-lubricated platens. This has already been mentioned, as an example of an application of the assumption of homogeneous deformation for load determination (Chapter 4, §4.2). The platens are imagined to be of length L, measured in the rolling direction and equal to the projected length of the arc of contact. In the transverse direction they overlap the strip, which has a width w. The projected area of contact is thus Lw. Usually the yield stress of the product will be greater than that of the ingoing stock, because of strain-hardening, but in this approximate calculation it is permissible, at least for hard strip, to assume a mean value $\bar{Y}$. The load P necessary to compress the strip is then

$$P = \bar{Y}Lw \tag{9.4}$$

It is still usual to evaluate roll loads in tons per inch width, which gives a rapid comparison between different metals, so this equation may be written

$$\frac{P}{w} = \bar{Y}L \text{ per unit width} \tag{9.5}$$

From the geometry of Figure 9.1, it can be seen that the projected length of contact L can be expressed in terms of the roll radius R and the draft, or reduction in strip thickness, Δh.

$$L^2 = R^2 - (R - \tfrac{1}{2}\Delta h)^2 = R\Delta h - \tfrac{1}{4}(\Delta h)^2 \tag{9.6}$$

In practice, the roll radius is always much greater than the draft. Typical values for a small mill might be R = 200 mm (7 in.) and h = 1·5 mm (0·060 in.) so

$$L^2 \approx R\Delta h$$

$$\frac{P}{w} = \bar{Y}\sqrt{R\Delta h}$$

Actually, the majority of cold-rolling involves wide strip which is relatively thin, and there is negligible lateral spread, so it is more accurate to assume plane–strain conditions, for which the appropriate yield stress is $2k = S$, which exceeds Y by 15% because of the constraint (equation 3.27). The equation should therefore be written

$$\frac{P}{w} = \bar{S}\sqrt{R\Delta h} \tag{9.7}$$

This gives a lower limit for the roll load, because it ignores the contribution of friction. From an examination of a range of typical rolling passes, Orowan has suggested an allowance of about 20% for the friction:

$$\frac{P^*}{w} = 1{\cdot}2\bar{S}\sqrt{R\Delta h} \tag{9.8}$$

This formula is not, of course, strictly accurate for any given rolling schedule, but it is easy to memorise and is very useful for rapid estimation of approximate roll loads. It can also be used, for example, in estimating roll flattening by Hitchcock's equation (9.21).

9.2.2 *Work evaluation*

Consideration of the work performed does not lead so directly to a value of the working load in rolling as it does in drawing or extrusion, because the major load is a constraining force, and does not act in the direction in which the workpiece travels. It is, however, possible to evaluate the work done in one revolution by the applied torque, and to equate this to the work of homogeneous compression of the strip.

Assuming that the resultant force on the roll acts at the centre of the arc of contact (Figure 9.1), the torque will be $P \cdot \frac{L}{2}$, and the work done per revolution by each roll,

$$W_1 = 2\pi P \cdot \frac{L}{2}$$

The volume of strip rolled during one revolution will be $2\pi R \cdot h_a \cdot w$ (assuming no forward slip), after reduction from thickness h_b to h_a. The work of homogeneous deformation in plane-strain is thus, using equation 4.6,

$$W_2 = V\int_{\epsilon_b}^{\epsilon_a} S\, d\epsilon = V\bar{S}\ln\frac{h_b}{h_a} = 2\pi Rh_a w\bar{S}\ln\frac{h_b}{h_a}$$

Equating this to the work done by both rolls:

$$2\pi PL = 2\pi Rh_a w\bar{S}\ln\frac{h_b}{h_a} \tag{9.9}$$

Since

$$\ln\frac{h_b}{h_a} = \ln\frac{h_a + \Delta h}{h_a} = \ln\left(1 + \frac{\Delta h}{h_a}\right) \approx \frac{\Delta h}{h_a}$$

$$\frac{P}{w} = \frac{R}{L} h_a\bar{S}\frac{\Delta h}{h_a} = \frac{R\Delta h}{\sqrt{R\Delta h}}\bar{S} = \bar{S}\sqrt{R\Delta h}$$

This is the same result as is obtained more simply by considering the static deformation (equation 9.7). It is often known as the work formula for rolling.

To allow for friction more accurately, the distribution of pressure on the rolls must be considered, by evaluating the local stresses.

9.3 Roll-pressure determination from local stress-evaluation[9.4]

The procedure has a strong formal resemblance to that described in Chapter 6 for strip-drawing, but there are important differences. In drawing, the die usually remains stationary, and a tension is applied to the drawn metal. The important working load is the applied tension. Rolling mills, on the other hand, usually have power-driven rolls which feed the strip through without any tension being necessary. The important load is the vertical constraining force on the rolls, corresponding to the die load in drawing. Intermediate conditions are also quite common, since a moderate tension between stands of a rolling mill reduces the roll load, improves the flatness of the strip, and gives a useful control of gauge. Idle rolls may, on the other hand, sometimes be used on a draw-bench for cold-forming small sections, and on a push-bench for hot-forming tubes on a mandrel. In the characteristic strip-rolling process, however, the peripheral velocity of the driven rolls exceeds that of the strip, which is consequently dragged in if the friction is high enough. As the strip is reduced in thickness it elongates and increases its linear speed until at the exit it is travelling faster than the rolls, and the friction acts in the reverse direction. There is a *neutral point* within the roll gap, at which the surface velocity of the strip equals the peripheral velocity of the rolls, there is no slip, and the direction of the friction force reverses. This is an important characteristic feature of rolling.

The assumptions used in evaluation of roll pressure in cold strip-rolling are:

Plane–strain conditions
Homogeneous deformation
Constant magnitude of the friction coefficient
Constant circular arc of contact
Neutral point within the arc of contact
Negligible elastic deformation

The significance of these will be considered more fully (§9.4) after deriving the results. The analysis is reasonably straightforward but calculations of roll pressures can conveniently be performed directly from the basic differential equation using a small computer.

9.3.1 *Derivation and general solution of the differential equation*

Longitudinal resolution of the forces acting on an element A, of unit width, in the deformation zone, on the exit side of the neutral point (Figure 9.1), gives:

$(\sigma_x + d\sigma_x)(h + dh) - h\sigma_x$ due to longitudinal stress

$2 \cdot \left(p_r \dfrac{dx}{\cos\alpha}\right) \sin\alpha$ due to radial pressure on both rolls

$2 \cdot \mu \left(p_r \dfrac{dx}{\cos\alpha}\right) \cos\alpha$ due to friction against both rolls

For steady rolling, these must be in equilibrium:

$$h\, d\sigma_x + \sigma_x\, dh + 2p_r\, dx \tan\alpha + 2\mu p_r\, dx = 0 \qquad (9.10)$$

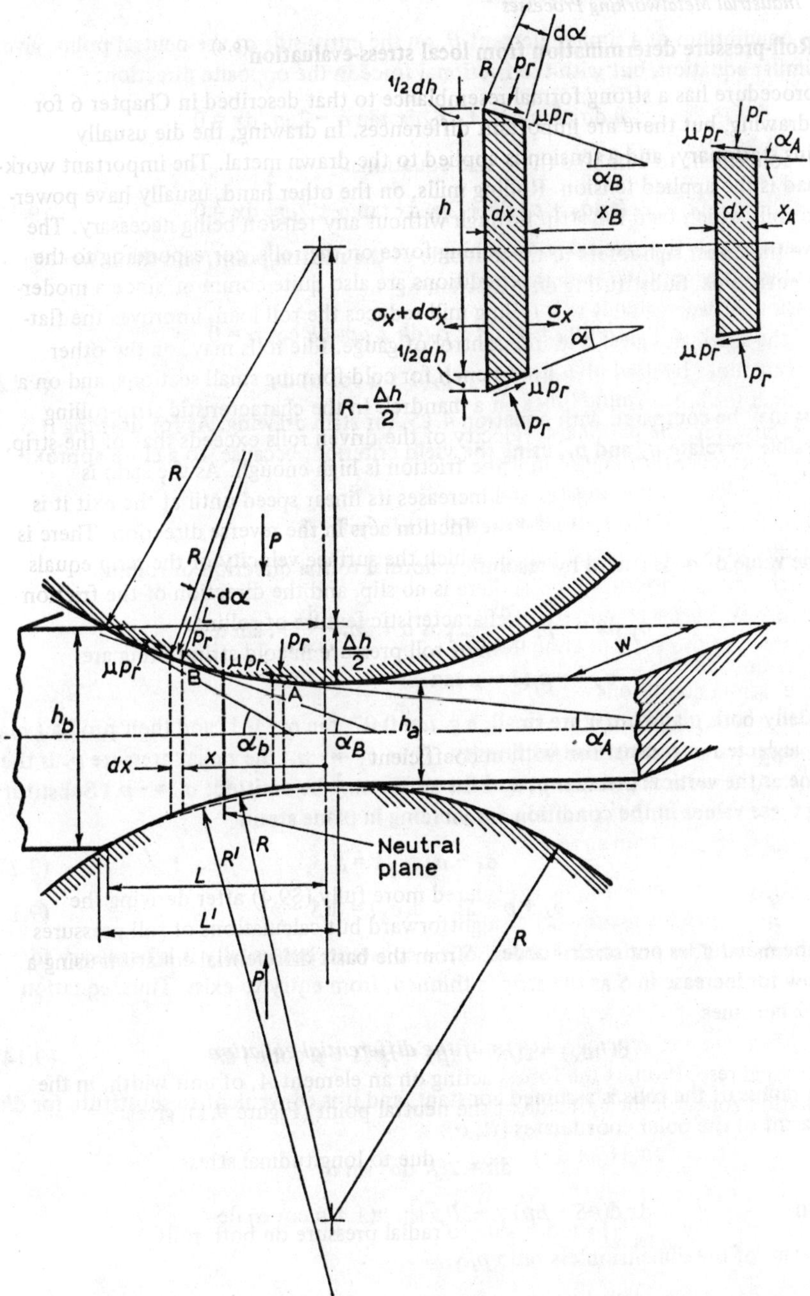

Figure 9.1 A section of the deformation zone in strip-rolling, showing the stresses acting on two elements of strip, one on each side of the neutral plane. The broken-line profile shows the deformation of the rolls to a radius R' under load.

The equilibrium of a similar element B, on the entry side of the neutral point, gives a similar equation, but with the frictional force in the opposite direction;

$$h\,d\sigma_x + \sigma_x\,dh + 2p_r\,dx\tan\alpha - 2\mu p_r\,dx = 0$$

It is convenient to combine these two equations:

$$h\,d\sigma_x + \sigma_x\,dh + 2p_r\,dx\tan\alpha \pm 2\mu p_r\,dx = 0 \qquad (9.11)$$

Here the upper sign refers to the exit side of the neutral point, and the lower sign to the entry side. Substituting $dh = 2dx\tan\alpha$,

$$h\,d\sigma_x + \sigma_x\,dh + p_r\,dh \pm \mu p_r\,dh\cot\alpha = 0$$

or

$$d(h\sigma_x) = -p_r(1 \pm \mu\cot\alpha)\,dh \qquad (9.12)$$

This may be compared with equation 4.15, for strip drawing. As for drawing it is possible to relate σ_x and p_r, using the yield criterion, because, to a close approximation,

$$\sigma_1 = \sigma_x, \text{ and } \sigma_3 = \sigma_y = -p$$

(The value of σ_y is found by resolution normal to the direction of rolling:

$$\sigma_y\,dx = -p_r\frac{dx}{\cos\alpha}.\cos\alpha + \mu p_r\frac{dx}{\cos\alpha}.\sin\alpha$$

$$\sigma_y = -p_r(1 - \mu\tan\alpha)$$

Usually both μ and $\tan\alpha$ are small, e.g. $\mu = 0{\cdot}07$, $\tan\alpha = 0{\cdot}1$, and their product may be neglected in comparison with unity. Thus $\sigma_y \approx -p_r$; the radial pressure p_r is the same as the vertical pressure p, and the suffix may be omitted: $\sigma_y = -p$.) Substituting these values in the condition for yielding in plane strain

$$\sigma_1 - \sigma_3 = 2k = S \qquad (3.27)$$

$$\sigma_x + p = S; \quad d(\sigma_x) = d(S - p) \qquad (9.13)$$

If the metal does not strain-harden, S is a constant; but usually it is necessary to allow for increase in S as the strip is thinned, from entry to exist. Thus, equation 9.12 becomes

$$d(h\sigma_x) = d(hS - hp) = -p(1 \pm \mu\cot\alpha)\,dh \qquad (9.14)$$

The radius of the rolls is assumed constant, and it is convenient to substitute for dh in terms of the polar coordinates (R, α):

$$dh = 2(R\,d\alpha)\sin\alpha$$

Then

$$d(hS - hp) = -2Rp\sin\alpha(1 \pm \mu\cot\alpha)\,d\alpha$$

In terms of the dimensionless ratio p/S,

$$\frac{d}{d\alpha}\left\{hS\left(1 - \frac{p}{S}\right)\right\} = -2Rp\sin\alpha(1 \pm \mu\cot\alpha)$$

$$hS\frac{d}{d\alpha}\left(1 - \frac{p}{S}\right) + \left(1 - \frac{p}{S}\right)\frac{d(hS)}{d\alpha} = -2Rp(\sin\alpha \pm \mu\cos\alpha) \qquad (9.15)$$

A simplification, suggested by Bland and Ford in 1948[9.5], allows this equation to be integrated directly. Under most circumstances, the variation in roll pressure with angular position in the roll gap is much greater than the variation in yield stress. Moreover, the change in the product hS will be smaller still, since S increases as h decreases. Thus the term $\left(1-\frac{p}{S}\right)\frac{\mathrm{d}}{\mathrm{d}\alpha}(hS)$ may usually be neglected in comparison with $hs\frac{\mathrm{d}}{\mathrm{d}\alpha}\left(1-\frac{p}{S}\right)$. This approximation is not valid when the rate of strain-hardening is high, as it often is for annealed metal, nor when high back-tension is applied, which, as we shall see in equation 9.20, reduces the variation of p/S over the contact. For most practical rolling in second and subsequent passes, the error involved will be only a few per cent. Making this approximation, equation 9.15 becomes

$$hS\frac{\mathrm{d}}{\mathrm{d}\alpha}\left(\frac{p}{S}\right)=2Rp(\sin\alpha\pm\mu\cos\alpha)$$

The angle of contact is also usually small, so further approximations may be made:

$$\sin\alpha\approx\alpha,\quad \cos\alpha\approx 1-\frac{\alpha^2}{2}\approx 1$$

$$h=h_a+2R(1-\cos\alpha)\approx h_a+2R\frac{\alpha^2}{2}$$

Then,

$$\frac{\mathrm{d}}{\mathrm{d}\alpha}\left(\frac{p}{S}\right)=2R\frac{p}{S}\frac{\alpha\pm\mu}{h_a+R\alpha^2} \tag{9.16}$$

$$\frac{\mathrm{d}\left(\frac{p}{S}\right)}{\frac{p}{S}}=\frac{2\,\alpha\,\mathrm{d}\alpha}{h_a/R+\alpha^2}\pm\frac{2\mu\,\mathrm{d}\alpha}{h_a/R+\alpha^2}$$

Both sides of this equation may be integrated, to give a general solution:

$$\ln\left(\frac{p}{S}\right)=\ln(h_a/R+\alpha^2)\pm 2\mu\,.\frac{1}{\sqrt{h_a/R}}\tan^{-1}\frac{\alpha}{\sqrt{h_a/R}}+\text{constant}$$

Introducing the symbol

$$H=2\sqrt{\frac{R}{h_a}}\,.\tan^{-1}\left(\sqrt{\frac{R}{h_a}}\,.\,\alpha\right) \tag{9.17}$$

$$\ln\left(\frac{p}{S}\right)=\ln\left(\frac{h}{R}\right)\pm\mu H+\text{constant}$$

Thus on the exit side of the neutral point,

$$\frac{p^+}{S}=c^+\,.\frac{h}{R}\mathrm{e}^{+\mu H} \tag{9.18a}$$

On the entry side,

$$\frac{p^-}{S}=c^-\,.\frac{h}{R}\mathrm{e}^{-\mu H} \tag{9.18b}$$

The values of the constants of integration may be found from the respective stress conditions at exit and entry.

9.3.2 *Rolling with no external tensions*

In the absence of front or back tensions, the longitudinal stresses at exit and entry must be zero; $\sigma_{xa} = \sigma_{xb} = 0$.

$$\frac{p_a}{S_a} = c^+ . \frac{h_a}{R} e^{\mu H_a}; \; H_a = 2\sqrt{\frac{R}{h_a}} . \tan^{-1}\sqrt{\frac{R}{h_a}} . \alpha_a$$

but
$$\alpha_a = 0, \; \sigma_{xa} = 0$$

Thus $H_a = 0$, and, from equation 9.13, $p_a = S_a - \sigma_{xa} = S_a$, so

$$1 = c^+ . \frac{h_a}{R}; \quad c^+ = \frac{R}{h_a}$$

At the entry:

$$\frac{p_b}{S_b} = c^- . \frac{h_b}{R} e^{-\mu H_b}; \; H_b = 2\sqrt{\frac{R}{h_a}} . \tan^{-1}\sqrt{\frac{R}{h_a}} . \alpha_b$$

Since $\sigma_{xb} = 0$, $p_b = S_b - \sigma_{xb} = S_b$, and

$$1 = c^- . \frac{h_b}{R} . e^{-\mu H_b}; \; c^- = \frac{R}{h_b} e^{+\mu H_b}$$

Equations 9.18 may thus be written:

On the exit side
$$\frac{p^+}{S} = \frac{h}{h_a} e^{\mu H} \tag{9.19a}$$

On the entry side
$$\frac{p^-}{S} = \frac{h}{h_b} e^{\mu H_b} . e^{-\mu H} = \frac{h}{h_b} e^{\mu(H_b - H)} \tag{9.19b}$$

The variation of roll pressure may be plotted as a function of angular position in the roll gap, from these equations.

The longitudinal stress at any position may readily be found from these equations, since

$$\sigma_x = S - p. \quad \frac{\sigma_x}{S} = \left(1 - \frac{p}{S}\right) \tag{9.13}$$

These graphs are often referred to in rolling as the *friction hill*, because of their shape (see Figure 9.2*a*). Either can be so described, but the roll pressure is usually implied, since roll load is the more important feature. This relationship between friction and roll load arises because the frictional contribution to the longitudinal stress increases with distance inwards from entry and exit, and this provides increasing resistance to the expansion of vertical sections under the vertical load. The roll load required to produce a given deformation is thus increased by the presence of a longitudinal friction-hill.

9.3.3 *Rolling with front and back tension*

In most production mills, the strip is fed through several rolling stands in tandem. To keep the strip flat, and also to control gauge, it is usual to maintain a tension in the strip between stands. Even single-stand mills usually employ power-driven coilers. Such tension has the further advantage of reducing the rolling loads, as may be seen by inserting the appropriate boundary conditions for front and back-tensions, t_a and t_b, in equations 9.18.

$$\text{At the exit:} \quad \sigma_{xa} = t_a, \ \alpha_a = 0, \ H_a = 0$$

The yield condition (equation 9.13) thus becomes

$$p_a = S_a - t_a$$

and the constant of integration in equation 9.18a is given by

$$\frac{p_a}{S_a} = 1 - \frac{t_a}{S_a} = c^+ \cdot \frac{h_a}{R} \, e^{\mu H_a} = c^+ \cdot \frac{h_a}{R}$$

$$c^+ = \left(1 - \frac{t_a}{S_a}\right) \frac{R}{h_a}$$

Similarly, at the entry,

$$c^- = \left(1 - \frac{t_b}{S_b}\right) \frac{R}{h_b} \, e^{+\mu H_b}$$

These constants correspond to those derived in the absence of tension, but are multiplied by a factor $\left(1 - \frac{t}{S}\right)$. The two expressions giving the roll pressure are consequently reduced by the applied tensions:

$$\left.\begin{aligned} \frac{p^+}{S} &= \left(1 - \frac{t_a}{S_a}\right) \frac{h}{h_a} \, e^{\mu H} && \text{(exit)} \\ \frac{p^-}{S} &= \left(1 - \frac{t_b}{S_b}\right) \frac{h}{h_b} \, e^{\mu(H_b - H)} && \text{(entry)} \end{aligned}\right\} \qquad (9.20)$$

These formulae may be expected to give reasonable accuracy for a large proportion of practical cold-rolling operations, but allowance should be made for roll flattening, as in equation 9.21.

9.4 Assumptions and applicability of the stress-evaluation method

9.4.1 *Discussion of the assumptions*

The above analysis makes a number of assumptions.

(i) *Plane–strain conditions.* In practically all cold-rolling of sheet and strip, the width w of the deformed metal is very much greater than its thickness h. It is therefore reasonable to assume plane strain, and the condition for yielding in plane strain can be applied to relate σ_x and p. This does, however, involve the subsidiary assumption that σ_x and p are principal stresses. Because the angle α is usually small, and μ is also low in normal cold-rolling, this is reasonable for most operations; but if the friction is high, the directions of the principal stresses may not lie exactly in those directions. The effects of high friction will be considered in §9.6.

(ii) *Homogeneous deformation.* In most cold-rolling, the metal will deform in a reasonably homogeneous manner, but if the friction is high, or a very light pass is taken, the surface layers may be more highly deformed than the central region, and it will not be possible to consider σ_x uniform over the section. Under these circumstances the derivation of the differential equation is not valid. High friction can be allowed for, as in §9.6, but the inhomogeneity involved in very light reductions, such as are used in temper rolling, cannot be included in this formulation. This special operation also involves other factors, which will be considered in §9.7.

(iii) *Constant coefficient of friction.* It is difficult to measure the coefficient of friction accurately in rolling, but a few attempts have been made to measure the local friction directly[9.6, 9.11]. However, the general agreement of theoretical load predictions with experiment (§9.6.1) suggests that the assumption of constant magnitude of μ throughout the roll gap is valid, provided again that the coefficient of friction is small. For high values of μ, and particularly for sticking friction, the modifications of §9.6 are necessary.

(iv) *Constant radius of curvature of the rolls.* In deriving equations 9.19 and 9.20, it was assumed that the rolls were rigid. In fact, the rolls will deform elastically to an appreciable extent, and allowance must be made for this. A sufficiently accurate correction is obtained by supposing that the rolls flatten to some greater radius R', but remain circular in profile, over the contact region. (See Figure 9.1.) The treatment is otherwise unchanged. Hitchcock[9.7] has given an equation for the changed radius under load, making this assumption.

$$R' = R\left(1 + \frac{c}{\Delta h} \cdot \frac{P}{w}\right) \tag{9.21}$$

where c is a constant depending on the elastic constants of the rolls. (This arises from Hertz' theory of elastic deformation of two cylinders in contact: $\frac{1}{R} - \frac{1}{R} = \frac{16(1-\nu^2)P}{\pi E x_b{}^2}$. It has the value 0·022 mm²/kN (3·34 x 10^{-4} in²/ton) for steel rolls. Thus for example, if $\frac{P}{w}$ = 4 kN/mm (10 ton/in.) and Δh = 0·88 mm (0·033 in.)

$$\frac{R'}{R} = 1 + \frac{0{\cdot}022}{0{\cdot}88} \times 4 = 1{\cdot}1$$

For harder strip, which would take a lower reduction in area under the same roll load, the roll flattening would be greater.

The deformed roll radius R' should be used in all accurate calculations of roll load. It may be estimated rapidly from equation 9.21, using the load derived from the approximate formula of equation 9.8. Successive approximations can of course be made, but the accuracy seldom warrants this additional computation.

(v) *Neutral point within the arc of contact.* This condition is satisfied in all practical rolling. If the neutral point does not lie within the arc of contact, the friction force acts entirely in one direction, and the rolls slip on the stock.

If the strip is required to enter the rolls unaided, the longitudinal component of the roll surface friction must exceed that of the radial pressure, at the entry:

$$\mu(p_r\,\mathrm{d}A)\cos\alpha_b^* > (p_r\,\mathrm{d}A)\sin\alpha_b^*$$

$$\mu > \tan\alpha_b^*$$

The maximum value $\alpha_b^* = \tan^{-1}\mu$ is often known as the *angle of bite* or gripping angle. Geometrically,

$$L \approx \sqrt{R\Delta h};\ \tan\alpha_b^* = \frac{L}{R - \dfrac{\Delta h}{2}} \approx \sqrt{\frac{\Delta h}{R}}$$

Thus the maximum draft is given by

$$\mu \geqslant \sqrt{\frac{\Delta h}{R}};\ (\Delta h)_{\max} = \mu^2 R \tag{9.22}$$

Large rolls and high friction allow heavy draft.

It is, however, possible to roll with greater draft if the strip is forced into the rolls. The limit with external pressure applied to the strip can be found, roughly, by considering the forces on the rolls. The rolls will start to slip when the average longitudinal component of the roll pressure equals that of the surface friction. For small angles

$$(p_r\,\mathrm{d}A)\sin\frac{\alpha_b^{**}}{2} \geqslant \mu(p_r\,\mathrm{d}A)\cos\frac{\alpha_b^{**}}{2}$$

$$\alpha_b^{**} \approx 2\tan^{-1}\mu$$

This *angle of nip* is thus twice the angle of bite. It is important when very heavy reductions of area are required. The rolling can then be started by tapering the front end of the strip.

The angle of bite can be used to find the coefficient of friction, by measuring the maximum draft at which the rolls will accept strip without it being pushed in. The method is not very accurate, and depends on the exact form of the front end of the strip, particularly if this carries a burr which increases the friction. It is, however, easy to make this test, and the results are fairly consistent if care is taken.

A more accurate but much more laborious method of estimating friction is to utilise the other limiting condition, with the neutral point at the exit. In these circumstances, the frictional force acts entirely in one direction. The total roll load P and torque T, per unit width, are then given by

$$P = \int_0^L p\,\mathrm{d}x;\quad T = \int_0^L (\mu p\,\mathrm{d}x)\,R = \mu R\int_0^L p\,\mathrm{d}x$$

$$\text{Thus } \mu = \frac{T}{RP}\text{, when the neutral point is at the exit} \tag{9.23}$$

For actual measurement, a back tension is applied, and increased until the emergent strip has exactly the same velocity V_2 as the roll periphery V_R, so that there is no forward slip, defined as $\frac{V_2 - V_R}{V_R}$. The magnitudes of T and P are then recorded. This is most conveniently accomplished by recording the forward slip, torque, and load simultaneously on a ciné-film as the back tension is increased, and analysing the data later[9.8].

(vi) *Negligible elastic deformation of the strip.* The strip is elastically compressed in the entry region before plastic deformation commences, and it recovers elastically after the point of minimum roll-spacing. Both of these zones increase the length of the arc of contact beyond the angle assumed for a rigid-plastic metal. The entry compression is usually negligible for all reductions above about 2%, but the greater yield stress and lower inclination of the roll surface at exit, make the effects of elastic recovery rather more important. These may be neglected only for passes exceeding about 10% reduction. Corrections can be applied for the influence of elastic deformation[9.9, 9.19], but for temper rolling, where the reduction in area may be 2% or less, they become too large and the theory is no longer useful (see §9.7.6).

(vii) *Low rate of strain-hardening.* A low value of $\frac{d}{d\alpha}(hS)$ is implied in the simplification of equation 9.15. This is usually a reasonable assumption for work-hardened metal, used in second and subsequent passes after annealing. For annealed strip, however, the errors may be appreciable, and good correlation with experimental measurements should not then be expected. If a computer is used to solve the differential equation, this error is not introduced. A second feature of annealed-strip rolling is that the surface finish will change appreciably during rolling, so that friction conditions may change. However, the majority of cold rolling is concerned with work-hardened metal, and accurate load prediction for annealed metal is often not very important, since the yield stress is lower, and maximum roll loads or roll-flattening are unlikely to be reached in a first pass.

(viii) *Low applied tensions.* The main purposes of applying front tension in a single-stand mill are to coil the strip satisfactorily, and to flatten the strip. It does however reduce the roll load for a given pass, and this may be of considerable importance in rolling very thin strip, because it reduces the roll flattening which sets a limit to the thinness which can be rolled. In these circumstances, the front- and back-tensions may be increased to the limits set by the yield stress in the ingoing- and outgoing-strip, or by the neutral point passing outside the arc of contact.

When very high tensions are applied, the variation of p over the arc of contact may not exceed the variation of S by as much as is implied by the approximation of equation 9.16. This is not very serious, since the correction for the friction hill, given by equation 9.19, will itself then be smaller, and its accuracy correspondingly less important.

Interstand tensions in multi-stand rolling mills will usually be well below these limits. They are varied slightly during rolling, for gauge control.

9.4.2 *Applicability of the stress-evaluation equations*

The above discussion suggests that equations 9.20 may be used with reasonable confidence for the majority of roll-load calculations. The main restriction is that the friction must be low, but this is realistic, since the coefficient of friction lies between $\mu = 0{\cdot}03$ and $\mu = 0{\cdot}09$ in practical cold-rolling. The equations will be less accurate for the first pass after annealing than for subsequent passes, and will not be accurate when high tensions are applied ($t \approx S$).

For temper rolling, when elastic deformation of the strip becomes important, a modified approach must be adopted, as described in §9.7.6.

When the friction is high, and particularly when there is sticking friction, the differential equation must be reconstructed, as in §9.6, without the assumption of homogeneous deformation.

9.5 Evaluation of load, torque and mill power for cold rolling

9.5.1 *Roll load*

The friction-hill equations (9.20), derived in §9.3, show the detailed variation of roll pressure with angular position in the roll gap. Allowance should always be made for the flattening of the rolls to a radius R' (equation 9.21), so these equations should be used in the form:

$$\left.\begin{aligned} \frac{p^+}{S} &= \left(1 - \frac{t_a}{S_a}\right)\frac{h}{h_a}\,\mathrm{e}^{\mu H}, \text{ on the exit side of the neutral point} \\ \frac{p^-}{S} &= \left(1 - \frac{t_b}{S_b}\right)\frac{h}{h_b}\,\mathrm{e}^{\mu(H_b - H)}, \text{ on the entry side} \end{aligned}\right\} \qquad (9.20)$$

where $H = 2\sqrt{\dfrac{R'}{h_a}}\,.\,\tan^{-1}\sqrt{\dfrac{R'}{h_a}}\,.\,\alpha, \quad R' = R\left(1 + \dfrac{c}{\Delta h}\,.\,\dfrac{P}{w}\right)$

The roll load is found by integrating, graphically, the appropriate values of the vertical pressure p. It has already been seen (equation 9.13) that, for practical values of μ and α, this pressure is equal to the radial pressure p_r.

From the geometry of the deformed rolls in Figure 9.1,

$$\mathrm{d}x = R'\,\mathrm{d}\alpha\cos\alpha \approx R'\,\mathrm{d}\alpha\left(1 - \frac{\alpha^2}{2}\right)$$

Since α is unlikely to exceed about 0·1 radians (6°), $\frac{1}{2}\alpha^2$ will be less than 0·005 and may be ignored. Thus

$$p\,\mathrm{d}x \approx pR'\,\mathrm{d}\alpha$$

The roll load per unit width of strip will then be

$$\frac{P}{w} = \int_0^{\alpha_b} p\,.\,R'\,\mathrm{d}\alpha$$

Assuming the deformed roll radius to be constant over the whole of the region of contact, the two arms of the pressure curve may be integrated separately:

$$\frac{P}{w} = R'\int_0^{\alpha_b} p\,d\alpha = R'\int_0^{\alpha_N} p^+\,d\alpha + R'\int_{\alpha_N}^{\alpha_b} p^-\,d\alpha \tag{9.24}$$

$$= R' \times \text{(area of curves plotted from equations 9.20)}$$

(The roll load is usually expressed in tons per inch width, so the units of the area $\int p\,d\alpha$ will be ton-radians per square inch).

9.5.2 *Roll torque*

The torque required for steady rolling will be the product of the net circumferential force and the distance from the roll axis to the circumference, in the deformation region. (See Figure 9.1.) Since the friction stress acts in opposite directions on opposite sides of the neutral point, the net circumferential force, per unit width, will be:

$$\frac{\mu P}{w} = \mu R'\int_{\alpha_N}^{\alpha_b} p^-\,d\alpha - \mu R'\int_0^{\alpha_N} p^+\,d\alpha$$

Although the rolls deform to a larger radius R', the distance from the surface to the centre will still be approximately equal to the original radius R. The torque, per unit width, is thus

$$\frac{T}{w} = \mu RR'\left\{\int_{\alpha_N}^{\alpha_b} p^-\,d\alpha - \int_0^{\alpha_N} p^+\,d\alpha\right\} \tag{9.25}$$

This value is not easily determined accurately, since it depends on the difference between two nearly equal areas.

An alternative method is to consider the moment of the resultant vertical force about the axis. If the perpendicular distance from the line of action of the resultant to the axis is some fraction, λ', of the length L' of the chord of contact with the elastically-deformed rolls,

$$\text{Torque} = \lambda' L' \times P \tag{9.26}$$

This neglects the contribution of horizontal forces, which deflect the resultant force slightly from the vertical. The fraction λ' is known as the *lever arm.*

$$\frac{T}{w} = \lambda' L' \frac{P}{w} = \lambda' L' R' \int_0^{\alpha_b} p\,d\alpha \tag{9.27}$$

$$(L' = R' \sin \alpha_b \approx \sqrt{R'\Delta h})$$

The lever arm has a value of about 0·5 in hot rolling, and about 0·45 for most cold rolling. Equation 9.26 may also be used for approximate calculation, using the value of $\frac{P}{w}$ determined by the modified work formula, equation 9.8.

9.5.3 *Mill power*

The power W_R supplied to each roll at an angular speed $\dot{\theta}$ is $T\dot{\theta}$ kN m/sec (kW) and in addition there will be friction at each of the roll-neck bearings. The load carried by each bearing is $\frac{1}{2}P$, giving a frictional drag $\frac{1}{2}\mu_n P$ and requiring a torque $\frac{1}{4}\mu_n Pd$, where μ_n is the coefficient of friction at the bearing, and d is the bearing diameter. Thus for two rolls, with four bearings, the total power will be

$$2W_R + 4W_N = 2T\dot{\theta} + \mu_n Pd\dot{\theta}$$

There will be losses in the motor and transmission, which may be allowed for by factors representing their efficiencies, η_m and η_t respectively.

The overall power requirement for the mill will thus be

$$W_M = \frac{1}{\eta_m \eta_t}(2T + \mu_n Pd)\dot{\theta} \text{ kW} \qquad (9.28)$$

In Imperial units, the speed is measured in revolutions per minute, N, and the torque is usually given in tons-inches, so the power is first converted to foot-pounds per second and then to horsepower, thus:

Horsepower for two rolls, with four neck bearings,

$$W_M = \frac{1}{\eta_m \eta_t}(2W_R + 4W_N) = \frac{1}{\eta_m \eta_t}(2T + \mu_n Pd)\frac{2\pi N}{60}\cdot\frac{2,240}{12}\cdot\frac{1}{550}$$

Usually the power required for the roll necks will be small, since the coefficient of friction of well-lubricated roll neck bearings[9.10] will be only 0·002 to 0·010. Typical relative values are given in Example 9.6.

9.6 Stress-evaluation for rolling with high friction

9.6.1 *Press-distribution measurements*

The derivation given in §9.3, originally proposed by Orowan[9.4] and simplified by Bland and Ford[9.5], assumes that the coefficient of friction is low and that the deformation is homogeneous. Comparison with experiments by Siebel and Lueg[9.11] shows that this leads to errors in the roll load, and particularly to large errors in the predicted pressure distribution, when the friction is high. The local pressure is difficult to measure and is not of direct concern in production, but this discrepancy emphasises a defect in the theory, and led to a more accurate theory, also proposed by Orowan[9.4], which is applicable to high-friction rolling.

Siebel and Lueg measured the local pressure by building a small pressure cell in the roll, and recording the variation as the roll rotated; a very refined technique, especially in 1933. Some results are shown in Figure 9.2.

These measurements establish the presence of a friction hill. It would naturally be expected that the peak would be rounded rather than sharp, in a real operation. For low values of coefficient of frction, there is good agreement with the stress-evaluation theory, assuming homogeneous deformation. There is, however, increasing departure from the predicted pressure values as the friction increases. For $\mu \approx 0{\cdot}4$, the peak pressure is much less than the predicted value.

A further anomaly appears from this result, since if the coefficient of friction

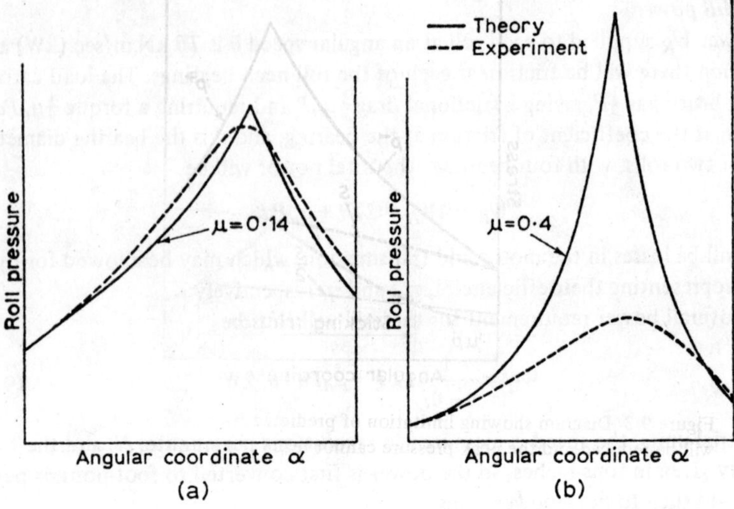

Figure 9.2 Local roll-pressure determinations by Siebel and Lueg[9.11] compared with the results of Orowan's theories[9.4]. (*a*) Low friction conditions (*b*) High friction. (*Replotted, by permission of the Institution of Mechanical Engineers.*)

remained at 0·4 over the whole arc, as assumed in the theory, the frictional stress in the region of maximum measured pressure (0·9 kN/mm^2) would be:

$$\tau = 0{\cdot}4 \times 0{\cdot}9 = 0{\cdot}36 \text{ kN/mm}^2$$

However, the yield stress in this region is about 0·2 kN/mm^2, and so the maximum shear stress which the metal can support (equation 3.27) is only

$$k = \frac{S}{2} = 0{\cdot}1 \text{ kN/mm}^2$$

As soon as this value is reached, the metal will flow by shearing beneath the surface. Thus a plot of roll pressure p_r and tangential stress would be expected to take the form of Figure 9.3. (See also Figure 10.1.)

There are two zones of slipping friction, near the entry and exit of the roll gap. These are separated by a region of sticking friction in which the strip surface moves with the roll surface, and the velocity change associated with the elongation is accommodated by shearing below the strip surface. This implies that the deformation can no longer be considered to be uniform across the strip. Because the lateral constraint is reduced, the pressure will not reach as high values as would be predicted on the assumption of constant coefficient of friction.

Thus two of the assumptions of the theory are invalidated by this experiment. The coefficient of friction is not constant throughout the roll gap, and the deformation is not homogeneous. Orowan suggested an evaluation, known as his 'Exact' method, in which these assumptions are not necessary.

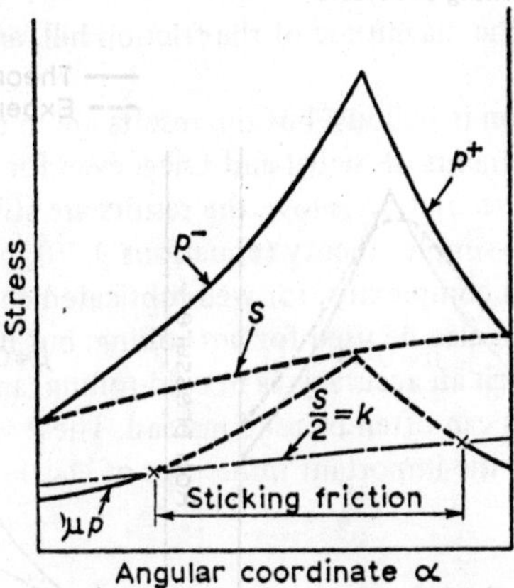

Figure 9.3 Diagram showing limitation of predicted frictional stress. The friction hill will be reduced and the peak pressure cannot reach the predicted value.

9.6.2 *Stress evaluation with partial sticking-friction*[9.4]

The remaining important assumptions are: plane strain, constant circular arc of contact, and negligible elastic deformation. The treatment starts, as in the low-friction theory of §9.3.1, by considering the equilibrium of an element in the deformation zone. The tangential stress is however represented by τ, which is understood to be equal to μp_r (with constant coefficient of friction μ, as before) in the two slipping-friction regions, but to be limited to $k = \frac{1}{2}S$ in the neutral zone where sticking friction occurs. Following the procedure leading to equation 9.12, equilibrium of the horizontal components of the forces acting on the strip requires that

$$\mathrm{d}(\sigma_x h) + p\,\mathrm{d}h \pm \tau \cot \alpha\,\mathrm{d}h = 0 \tag{9.29}$$

or

$$d(\sigma_x h) = -2R(p \sin \alpha \pm \tau \cos \alpha)\,\mathrm{d}\alpha$$

In the previous analysis, assuming homogeneous deformation, σ_x was considered uniform across the section, and it was then possible to relate σ_x and p as principal stresses by the condition of yielding in plane strain ($p + \sigma_x = S$).

Orowan, using an analogy between rolling and compression with inclined rough platens, for which a solution had been obtained by Nadai[9.12], suggested that a modified yield criterion could be adopted:

$$p + \sigma_x = mS \tag{9.30}$$

The value of m varies with the angle α and with the ratio $\mu p/k$. When $\mu = 0$ it has, of course, the value unity, but this falls to about 0·8 as the friction increases to the full sticking value. The physical reason for a reduction in mean yield stress is that the internal shear gives rise to a compressive component at the surface and a tensile component on the axis. Because of the latter, compression normal to the strip is facilitated, as by the application of external tension. This also limits the build-up of lateral constraint, and consequently the resistance to lateral expansion of elements under vertical load, which, as shown in §9.3.2, is the cause of the friction hill. Sticking-friction thus reduces the magnitude of the friction hill, and the maximum frictional stress developed.

Solution of the equation is tedious, but the results are in good accordance with the experimental measurements of Siebel and Lueg, even for coefficients of friction as high as $\mu = 0{\cdot}4$. When the friction is low, the results are still somewhat more accurate than those of the simpler theory (equations 9.20), but the accuracy does not warrant the additional complexity, for well-lubricated conditions.

The 'Exact' method can also be used for hot-rolling, but it is not usually necessary to know the loads to as high an accuracy as in cold-rolling, and one of the empirical or semi-empirical methods can often be used instead. These will be considered in §9.8. We shall first discuss the important influences of elastic deformation in cold strip-rolling.

9.7 The influences of elastic deformation in cold rolling

For purposes of theoretical examination, it is convenient to assume that the tools are perfectly rigid, and that the worked metal behaves as a rigid-plastic material; which means that any elastic deformation can be neglected in comparison with the large-scale plastic deformation imparted by the operation. This assumption gives adequate accuracy for most calculations of working loads. In many operations, such as wire-drawing and extrusion, the elastic stresses cause small changes in final dimensions, which can usually be ignored. In rolling, the tools deform elastically to a greater total extent because of their relatively large dimensions, and also because of the spring in the large housing. We have already seen (§9.4) that it is essential to allow for deformation of the rolls themselves in calculation of roll load, using equation 9.21. There are other consequences of elastic deformation which may often be of greater importance in rolling than this increase in load by, say, ten per cent.

9.7.1 *Minimum thickness in rolling*

When thin hard strip is rolled, it is found that it is not possible to reduce the thickness below a certain limit. Attempts to do so result in greater deformation of the rolls, without any plastic deformation of the strip. This is not obvious, but the cause may be described by starting with a simple analogy. Copper strip may easily be rolled between steel rolls, but if steel were fed into a pair of rolls made of copper it would be undeformed. This suggests immediately that relative yield stress (hardness) of rolls and strip is important, but does not show why the performance should be related to thickness. In fact if the interface were perfectly frictionless it would theoretically be possible to roll to infinitely thin strip (though it would of course then be necessary to feed the strip in because the rolls would have no bite). We have however seen (§9.3.2) that the friction causes an increase in longitudinal stress, which in turn increases the roll pressure – the friction hill. If the roll radius and the draft are kept constant, the area on which the frictional stress acts will also remain constant, but the cross-sectional area of the strip will decrease as the ingoing thickness is decreased. This results in an increase in the average longitudinal stress, and a corresponding increase in roll pressure, as the strip thickness is decreased. Thus a given draft requires an increasingly high load as the strip thickness is reduced, which is equivalent to attempting to deform metal of greater hardness. When a particular thickness is reached, the rolls will deform more easily than the strip.

If the draft is reduced, of course, the load will be reduced also and thinner strip can be rolled, but a limit is reached when no reduction at all is possible. Because the area of roll contact is less for a given draft if the roll radius is reduced, the limiting thickness will be smaller if a small roll is used. Also the harder the rolls, the thinner the strip which can be rolled. The higher the elastic modulus of the rolls the less the roll flattening, which is another factor increasing the contact area.

It is in fact found that the limiting thickness is very nearly proportional to these parameters:

$$h_{\lim} \propto c \,.\, \mu \,.\, R \,.\, \bar{S} \tag{9.31}$$

where c is the elastic deformation parameter of the rolls, as used in the flattening equation 9.21. For steel rolls, a useful formula[9.13] is,

$$\frac{\mu R S}{1860} \text{ in., or } 0{\cdot}035\ \mu R\bar{S} \text{ mm} \tag{9.32}$$

Consequently, to obtain the thinnest possible strip, the metal should be annealed; the coefficient of friction should be as low as possible, using polished rolls and a good lubricant; and small-diameter rolls should be used. To improve the resistance to deformation of the rolls themselves, it has recently been found possible to make rolls of tungsten carbide, which has a much higher Young's modulus than steel. Such rolls are very expensive, but have proved their worth in industrial production, especially in the Sendzimir mill (§9.9). Another way of reducing the load, and thus the roll deformation for a given pass, is to apply front and back tensions, but it may be dangerous to apply too much tension when the strip or foil is very thin. When all else has been done, it is still, of course, possible to roll in pairs or packs of greater numbers of foils. In production, it usually becomes uneconomic to roll strip long before the ultimate limit is reached.

9.7.2 *Bending of the rolls. Camber'*

As well as the flattening of the rolls in the contact region, the distributed load across the strip will produce a bending moment about the roll neck bearings, where the rolls are supported. The rolls are of course very short and stiff, in comparison with most structural beams, but they will nevertheless bend to a small extent, in the manner shown on an exaggerated scale in Figure 9.4*b*. This can be calculated with reasonable accuracy.

The result of such bending of the rolls is that the strip is slightly thicker in the centre than at the edges, by about 0·1 mm (0·004 in.). Even this is not acceptable for high-tolerance strip, and it may be corrected by grinding the rolls to a convex profile so that the combined effect of bending and longitudinal flattening of the rolls gives a flat interface and parallel strip (Figure 9.4*b*).

In practice, the correction made is slightly less than this[9.14], partly because perfectly-parallel strip tends to wander in the first stand of a mill, whereas a slightly crowned strip (not more than 0·025 mm is necessary) will remain central without applied back-tension. Another reason is that lateral variation in thickness causes cor-

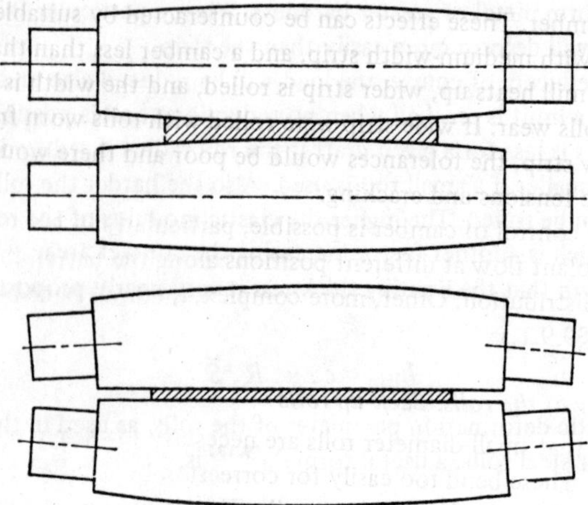

Figure 9.4 Roll bending and the allowance which can be made by forming an initial camber.

(*a*) Undeformed rolls under light load.
(*b*) Convex cambered rolls under calculated load.

responding but opposite variation in length, since the volume and width remain constant. Thus over-correction by roll cambering, which makes the strip thinner along the middle than at the edges, will make the central region too long. This results in a loose middle, which will tend to buckle, and tight edges. Application of tension to such strip between stands, or by a coiler, may then cause cracking at the edges, or even fracture right across. Loose edges are more easily levelled out without damage in subsequent passes, and are sometimes deliberately formed in the early passes, when very heavy overall reductions are required with no danger of edge-cracking.

Several formulae of varying complexity are available for calculation of camber. These are based on consideration of the bending of the roll as a short stocky beam, due to the bending moment about mid-points of the neck bearings, and to the shear forces. The central deflection thus contains two terms of the type

$$d = k_b \frac{Wl^3}{EI} + k_s \frac{Wl}{AG} \tag{9.33}$$

The values of the constants k_b and k_s depend on the relative dimensions of roll and strip[(9.15)], l is the effective length of the beam, W the roll load, E and G the elastic and shear moduli of the rolls and A the cross-sectional area of the barrel. For full-width strip, $k_b \sim 1{\cdot}0$, $k_s \sim 0{\cdot}2$, and for strip only half as wide as the rolls, $k_b \sim 0{\cdot}5$, $k_s \sim 0{\cdot}1$.

Even when the camber is correctly calculated and ground, it will be strictly accurate for one specified load and width only. In addition, the camber will usually increase at the start of a run because of the uneven temperature rise due to the central regions heating more rapidly than the ends of the rolls. This is most important in hot-rolling. On the other hand, wear is most severe at the centre, and this reduces the camber. These effects can be counteracted by suitable production planning, starting with medium-width strip, and a camber less than that required for full width. As the mill heats up, wider strip is rolled, and the width is then decreased again as the rolls wear. If wide strip were rolled with rolls worn from long production of narrow strip, the tolerances would be poor and there would be a serious danger of edge tensions and cracking.

Some local control of camber is possible, particularly in hot mills, by varying the amount of coolant flow at different positions along the barrel, thus altering the temperature distribution. Other, more complex, methods have been proposed. (See, for example, §9.9.1.)

9.7.3 *Bending of the rolls. Back-up rolls*

We have seen that small-diameter rolls are necessary for heavy reductions in thin strip (§9.7.1). These bend too easily for correction by camber, and they have to be supported by relatively large back-up rolls. This arrangement gives a 4-high mill, which is the most widely-used type of mill; though large slabbing-operations call for large-diameter rolls, and soft foils are also rolled in simple 2-high mills. When the working rolls are small in diameter compared with the back-up rolls, the stiffness is essentially that of the back-up rolls, which may be cambered in the same way as in the 2-high mill. The small work-rolls follow this camber during rolling.

For very thin strip it is necessary to use rolls which are so small in diameter that they can flex in the horizontal plane under the influence of the tangential forces, as well as bending under the major roll-load. For example, work-rolls 100 mm (4 in.) diameter and 1·5 m (60 in.) long might be used for 0·15 mm (0·005 in.) steel strip. To support such rolls adequately, two backing rolls are provided for each work roll, with their axes offset before and behind, forming a 6-high mill. These mills are not common, but the principle is used, and had been extended still further in the Sendzimir cluster type of mill (§9.9.1).

9.7.4 *Elastic distortion of the mill*

The mill housing itself, which is usually made of two box-type castings, can distort under heavy rolling loads. Housings are usually designed for rigidity, and the stress level is quite low, but owing to their length, the tensile extension of the side pillars and the bending of the top and bottom bridges combine to give an appreciable elastic extension. (See Example 9.5.) In addition to this, the application of the rolling load causes the chocks and bearings to bed down. Consequently the roll gap will increase during the rolling, as well as being increased slightly by the elastic roll-flattening already discussed. Apart from the small contribution of bedding-down, the deformation will be elastic, and the roll gap will increase linearly with load as shown in Figure 9.5*a*. The slope of this line is known as the mill modulus, and it is an important factor which must be taken into account when setting the roll gap for a given pass. (This is another reason for determining the roll load with reasonable accuracy before rolling.) If g is the roll gap before the strip enters, and h_a the outgoing thickness or gauge, the mill 'springback' s is simply

$$h_a = g + s$$

It is in fact possible to roll with a 'negative' roll gap, if the rolls are screwed down to a finite load before the strip enters and forces them apart.

The mill-springback graph approximates to a straight line, starting a little way along the gauge axis. It may be determined quite quickly for a given mill by measurement of load and springback, though it will vary slightly with the width of strip rolled.

Many of the operating characteristics of rolling mills depend on the elastic deformation, as can be appreciated by considering the two graphs representing the effect of load on the elastic distortion of the mill and the plastic deformation of the strip in the roll gap. The latter is obtained by using the friction-hill equation (9.24) to calculate the load required to roll strip of initial thickness h_b to various final thicknesses h_a. The resulting graph is usually similar in shape to that of Figure 9.5*b*.

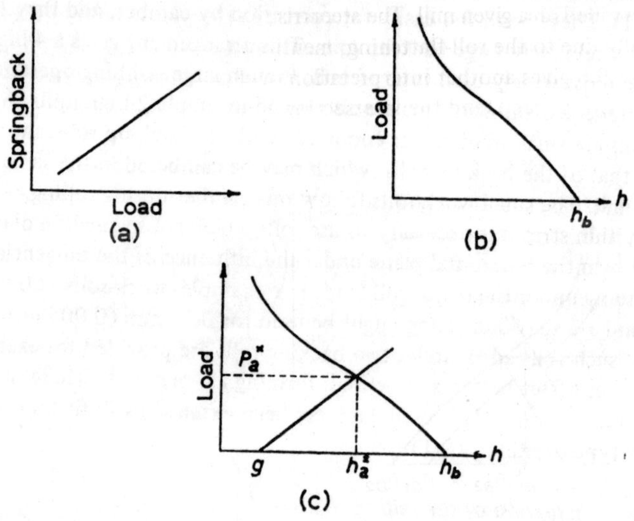

Figure 9.5*a* **Mill springback, which is the difference between rolled strip thickness and roll gap setting, plotted against load.**

Figure 9.5*b* **The relationship between the final thickness of strip h_a rolled from a thickness h_b, and the necessary roll load. This is often known as the plastic curve.**

Figure 9.5*c* **Combination of mill modulus graph and plastic curve shows the intersection determining the actual gauge produced for a given initial roll gap, and strip thickness h_b.**

Using strip of constant initial thickness h_b, the final thickness is decreased by reducing the initial setting of the roll gap g, but the gap will increase under load, as seen from Figure 9.5*a*, so that the final thickness exceeds g by an amount dependent on the mill modulus. This is easily evaluated from the graph by plotting the mill modulus line and the plastic curve together. As the load required by the reducing thickness increases, the real gap increases, and the strip thickness h_a* actually produced is found from the intersection of these lines, (Figure 9.5*c*). The actual load will be P_a^*. The influence of various parameters on the position of this intersection can be seen in Figure 9.6.

If the thickness of the ingoing strip is increased from h_{b1} to h_{b2} the plastic line will be moved along the axis to start at h_{b2} and the rolled strip thickness will be increased to h_{a2}, for a roll gap set initially at g_1. If the roll gap is opened to a value g_3, the mill-modulus line will be moved along the axis and the strip rolled from h_{b2} will be h_{a3}. The gauge of the final strip is also influenced by the mill modulus. A stiffer mill, with steeper line, will give less gap opening and consequently a thinner final strip h_{a4}.

Factors which decrease the load necessary for plastic deformation of the strip reduce the final strip thickness for a given setting on a given mill. Thus softer metal, better lubrication, and the application of tension, give thinner strip. We have already considered these influences in §9.7.1, in connection with the limiting thickness which can be rolled on a given mill. The steep rise in the plastic curve for small values of h is actually due to the roll-flattening, mentioned there as the main limiting factor. Figure 9.6 gives another interpretation of this. When h_b is very small, the plastic curve is very steep, and the intersection is less dependent on variation of g.

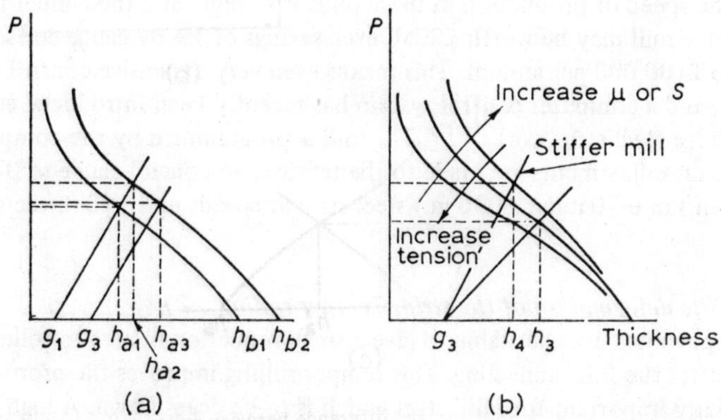

Figure 9.6 Mill characteristics shown by the elastic and plastic curves.

(*a*) The effects of increasing ingoing thickness, and roll gap, on final strip thickness.
(*b*) The effects of yield stress, friction, mill stiffness, and applied tension on final thickness.

Several parameters thus influence the outgoing gauge from a mill, and careful control is essential for the production of high-tolerance sheet or strip[(9.16)].

9.7.5 *Gauge control*

Longitudinal variation in gauge arises mainly from variations in thickness and hardness in the ingoing strip, due to lack of control in hot rolling or uneven annealing of cold-rolled strip. Changes in interstand tension and roll temperature also affect the gauge, and so do factors such as speed and lubricant which influence the elastic flattening of the rolls. A cyclic variation in thickness occurs if the rolls are not precisely concentric with their bearings. Normal cold-rolled sheet and strip is produced to close tolerances, demanded by the subsequent processing and by the economic wastage of strip exceeding the required thickness. Gauge control is therefore very important and the methods chosen must have very rapid response, for strip travelling at speeds up to 10 m/s or more. In principle, it is possible to use any of the parameters of Figure 9.6 for gauge control. The most obvious method is to change the roll gap. This is commonly applied in rolling heavy gauges, relying on manual control of the screw-down motors, and gives good results when used by a skilled operator. It is, however, less satisfactory for thin gauges and very high speeds. As the strip becomes thinner and harder, the final gauge becomes progressively less responsive to screwdown (Figure 9.6*a*). A more sensitive and more rapid technique, for thin strip, is to adjust the tension in the strip. In single-stand mills this is easily achieved by feeding a signal from a thickness-sensing device to the motor driving the coiling-drum, to alter its speed. Considerable savings have been made in the production of aluminium and other foils, for example, by using a radioactive source and a β-ray detector to control coiler speed and foil gauge[9.17]. In multistand steel strip mills the interstand tension is adjusted by altering the speed of one stand relative to another. This, of course, affects the whole mill, in which there may be six stands. Because the speed of production in these mills is so high, and the annual throughput of such a mill may be worth £20M, even savings of 1% by gauge control may amount to £100,000 per annum. This makes even very expensive control systems economic, and a computer control system has recently been introduced successfully in a very large steel-strip works[9.18]. The mill is programmed by the computer, and instantaneous adjustments are made to the tensions to control gauge to 0·015 mm (±0·0005 in.) in 0·70 mm (0·030 in.) steel strip at speeds up to 10 m/sec (2,000 ft/min.

9.7.6 *Elastic deformation of the strip. Temper rolling*

For some purposes, it is desirable to give a small reduction in area to rolled sheet and strip after the final annealing. This temper-rolling improves the proof stress, and is particularly important for mild steel which is to be deep-drawn. A high proportion of all rolled mild-steel sheet is, in fact, deep-drawn into car-body panels and other consumer goods. If annealed sheet is used, the sharp yield point (Figure 2.1) causes irregular yielding which is visible as Lüder's bands, and cannot easily be obscured by painting. The yield point itself can however be obliterated by rolling to a small reduction of 1–2% in area, immediately before subjecting the sheet to tension. If a period of more than about 24 hours elapses after temper-rolling, the yield point returns, but the sheet may remain usable for a week, or in cool conditions, longer.

Such light plastic deformation is comparable to the elastic deformation of the strip, and the latter may not be ignored. The assumption of rigid-plastic material is far from valid for these conditions, which may be dominated by the elastic deformation of the strip at entry and at exit. These regions in contact with the rolls augment the load directly and also increase the contact arc, adding to the friction hill. An additional longitudinal compression is set up, which is equivalent to a negative front-tension, again increasing the load. The normal theory leading to equation 9.24 cannot be used for temper-rolling, but an alternative method has been proposed[9.19].

9.8 Other methods of roll-load determination

A. *Cold-rolling*

The method described in §9.3 is simple in principle, and can easily be applied with the aid of a digital or analogue computer, but it is rather laborious for manual calculations. Some more rapid methods have been proposed, and these have also the merit of depending on empirical determinations on actual mills, rather than relying entirely on theory (see also §9.12.1).

9.8.1 *Cook & Parker method*[9.20]

This method requires the measurement of rolling loads for a number of annealed test-strips, on any single mill with a pair of well-finished rolls. The results are then plotted to show the variation of the ratio h_b/R with h_b^2/P_1 for various reductions of area. (h_b is the ingoing thickness, R the roll radius, P_1 the rolling load per unit width on annealed strip.) Then a first-pass curve can be plotted for any chosen initial thickness, showing P_1 as a function of reduction of area.

For multiple passes, which are usually required, an expression has been derived by Cook and Parker for the energy expenditure in any intermediate pass. The sum of these energy expressions for all the passes is then equated to the energy which would be required to perform the total reduction in one pass (if the mill were capable of this). The load for this hypothetical total reduction in one pass is derived immediately from the first-pass graph. The real first-pass load P is also taken directly from this graph, and the energy evaluated. A hypothetical combination of passes 1 and 2 then gives the energy for pass 2 and hence the load P_2, and so on through the sequence up to the full reduction.

This method can be applied quite rapidly to multiple-pass rolling schedules and involves only simple numerical calculations. Provided that the original measurements are made under appropriately comparable conditions, the predictions agree quite well with subsequent measurements.

9.8.2 *Bland and Ford graphical solution*[9.5]

In §9.2 a simple expression for the roll load was derived by considering an equivalent ideal compression

$$\frac{P}{w} = \bar{S}\sqrt{R\Delta h} \tag{9.7}$$

Allowance can be made for the friction by multiplying this by a function f. Bland and Ford suggested a function expressed in terms of the reduction of area r, and X.

$\left(X = \mu\sqrt{\frac{R'}{h_a}}\right.$, a combination which appears in their detailed theory given in §9.3.1$\left.\right)$.

Thus

$$\frac{P}{w} = \bar{S} \,.\, \sqrt{R'\Delta h} \,.\, f(X, r) \tag{9.34}$$

The values of f are given for various values of X and r in graphical form in their paper[9.5], together with the values of another function used for torque calculation. It is possible to extend the method to include applied tensions, as in §9.3.3.

9.8.3 *Ekelund's equation*[9.21]

An early analysis of the roll stresses, by Ekelund, led to his proposal of an equation for roll load, in 1927. Including the effects of roll flattening, this becomes:

$$\frac{P}{w} = \bar{S}\sqrt{R'\Delta h}\,.\left(1 + \frac{1{\cdot}6\mu\sqrt{R'\Delta h} - 1{\cdot}2\Delta h}{h_a + h_b}\right) \tag{9.35}$$

The expression in brackets is a function of friction, as in equation 9.34, but is given in simple analytical form. This equation gives remarkably good predictions of roll loads over a wide range of sizes and reductions. It can be recommended for general-purpose use in computing rolling schedules in a simple manner for single-stand mills without tensions. Like the other equations it can be corrected for tensions by the factor $(1 - t/\bar{S})$.

9.8.4 *C. E. Davies' method*

A simple method resembling Cook and Parker's, but suggested much earlier by Davies, is often used in practice[9.22]. The mean specific roll pressure, p_s, which is the roll load divided by the area of the contact between roll and strip, is determined empirically, assuming no deformation of the rolls, for a range of reductions. The values of p_s are then plotted against mean percentage total reduction for various values of h_b/R. The mean percentage total reduction is defined by Davies as the mean of the total reduction from the annealed state, at the beginning and at the end of the pass considered. This provides a chart for the particular metal rolled, but in this form it relates only to first passes. It is much more useful in practice to obtain the data initially from successive passes of say 30%, Figure 9.7.

The deviation from this in any desired rolling schedule for say 10% to 50% reductions per pass is then much smaller than the error introduced by considering multiple passes equivalent to a single heavy pass. To a first approximation, the chart may be scaled proportionally to allow for different yield stresses, but because friction may also be changed, it is preferable to establish an individual chart for each metal.

This method is seldom more accurate than the Ekelund equation 9.35, but it does not involve direct knowledge of the yield stress, nor of the coefficient of friction. It is therefore favoured where the only equipment available is the rolling mill itself. Ekelund's equation is more suitable for paper calculations when experimental time is not available on a mill.

B. *Hot-rolling*

9.8.5 *Ekelund's equation*

Ekelund actually developed his equation for hot-rolling first, taking account of the influence of strain rate on the mean yield stress. The remainder of the formula is the same as for cold-rolling, though of course the numerical value of μ will be higher.

$$\frac{P}{w} = \left\{\sigma + \frac{2V\eta\sqrt{\Delta h/R}}{h_a + h_b}\right\}.\sqrt{R\Delta h}\,.\left\{1 + \frac{1{\cdot}6\mu\sqrt{R\Delta h} - 1{\cdot}2\Delta h}{h_a + h_b}\right\} \tag{9.36}$$

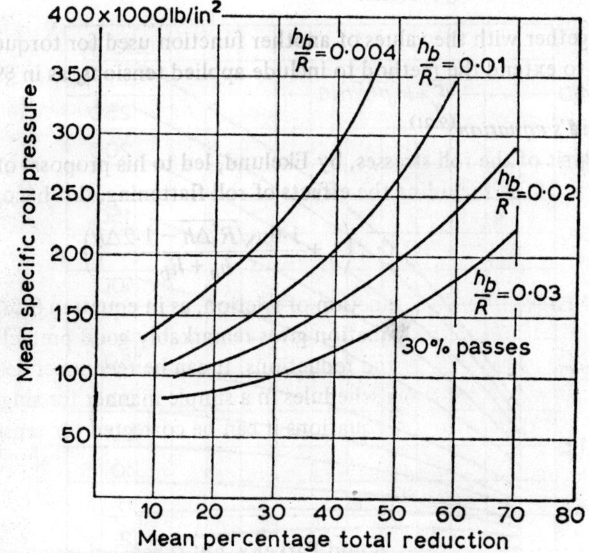

Figure 9.7 Roll pressure chart for typical mild-steel rolling. (*Replotted by permission of W. H. A. Robertson Ltd. and 'Metal Industry' Iliffe Production Publications Ltd.*)

He carried the empirical approach a stage further and gave formulae for the 'viscosity' η of the hot steel, the coefficient of friction μ, and even the 'static' yield stress σ, the latter in terms of percentage composition of constituent elements.

$$\eta = 1{,}373 - 0{\cdot}098T$$

$$\mu = 0{\cdot}84 - 0{\cdot}0004T \text{ (for billet temperatures } T \text{ in excess of } 700^\circ\text{C)}$$

$$\sigma = 100\eta(1{\cdot}4 + C + Mn + 0{\cdot}3Cr)$$

The temperature is measured in deg. C, the velocity V in mm/sec, and the yield stress in kN/mm². If however V is measured in inches per second, yield stress in lb/in² and dimensions in inches,

$$\eta = 0{\cdot}01\ (19{,}910 - 1{\cdot}42T)$$

the other equations remaining unchanged.

The general form of equation 9.36 is useful, though it is usually desirable to determine the mean yield stress directly for the temperature and strain-rate conditions required.

9.8.6 *Sims' method*

A solution of the same type has been given by Sims[(9.23)] in terms of the mean yield stress k (in our notation S) and a factor Q

$$\frac{P}{w} = k \,.\, \sqrt{R\Delta h} \,.\, Q \tag{9.37}$$

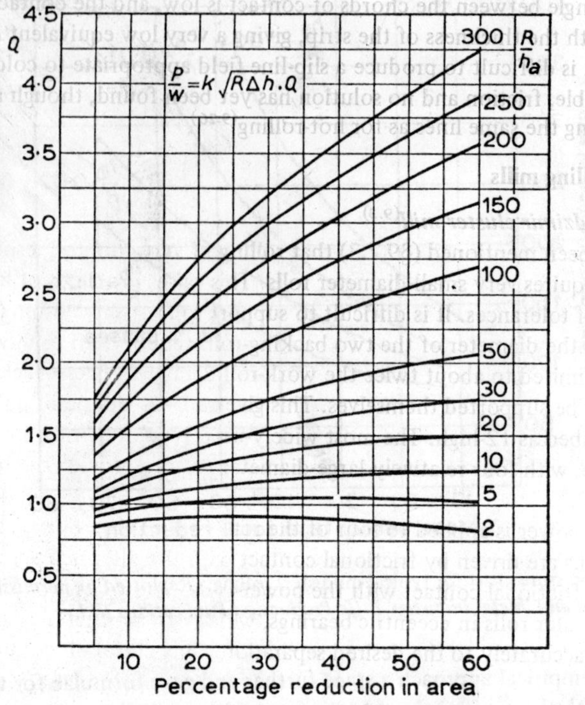

Figure 9.8 Larke's values of the Sims function Q. (*By courtesy of E. C. Larke and Chapman & Hall Ltd.*)

Figure 9.8 gives values of Q computed by Larke(9.24) on the basis of Sims' formulae for roll-pressure distribution. Larke concludes that the Sims function gives somewhat more accurate results than either the Ekelund equation or a similar correcting function proposed by Orowan and Pascoe(9.25).

Roll-flattening should be allowed for when appropriate, by the usual formula 9.21. Orowan and Pascoe point out that the ratio of width to thickness will influence the effective yield stress, and that allowance for a lateral friction hill must be made when very wide strip is hot-rolled.

9.8.7 *Alexander's slip-line field*(9.26)

Mention should be made of the slip-line field solution obtained by J. M. Alexander. From considerations of the metal flow and the boundary conditions, he constructed a slip-line field by the Prager method, assuming sticking friction over the whole of the arc of contact. The derivation is complex and the result is of considerable theoretical interest. It has been adapted by Johnson and Kudo(9.27), for use as an Upper Bound.

In passing, it may be remarked that the slip-line field type of solution is of more significance in hot-rolling than in cold-rolling, because the distortion, and so the redundant work, is greater. There is little redundant work in cold-rolling, as is indeed assumed in the Cook, Parker and Davies methods just described, because the effective included-angle between the chords of contact is low, and the contact is long in comparison with the thickness of the strip, giving a very low equivalent h/b ratio (Figure 5.6). It is difficult to produce a slip-line field appropriate to cold-rolling with low, variable, friction and no solution has yet been found, though attempts have been made along the same lines as for hot-rolling[9.26].

9.9 Special rolling mills

9.9.1 *The Sendzimir cluster mill*[9.3]

It has already been mentioned (§9.7.3) that rolling of very thin foil, especially of hard metals, requires very small-diameter rolls. This raises problems of lateral deflection and also of tolerances. It is difficult to support rolls of say 10 mm ($\frac{1}{2}$ in.) diameter, because the diameter of the two backing-rolls required for each work-roll is geometrically limited to about twice the work-roll diameter. These backing-rolls must therefore be supported themselves. This gives a 1–2–3 type of mill, which could be described as 12-high. The most widely-used type of Sendzimir mill involves one more stage, with four relatively large-diameter rolls to provide the final back-up, as in Figure 9.9.

The driving power is applied to four of the rolls in the second bank, and the actual work-rolls are driven by frictional contact with the intermediate rolls, which are, in turn, in frictional contact with the power-rolls. Screw-down is achieved by mounting the outer rolls in eccentric bearings, which can be rotated slightly to bring the work-rolls accurately to the desired separation. This motion is coupled to the

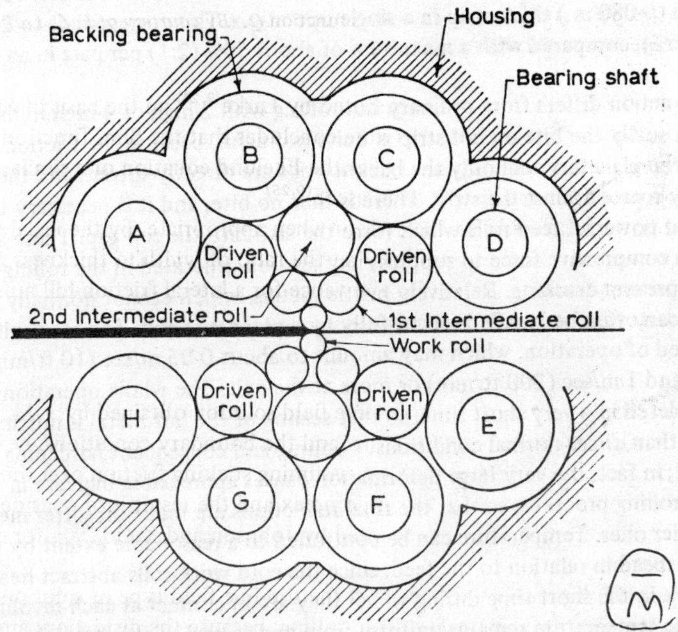

Figure 9.9 A cross-section through a cluster mill for cold-rolling thin strip. (*By courtesy of Sendzimir Ltd. and W. H. A. Robertson Ltd.*)

gauge-control device. Camber is adjusted at will during rolling by individual eccentrics supporting sections of the outer rolls.

Typical dimensions are (1 in. $\simeq$ 25 mm):

Work roll diameter	0·250 in.	1·125 in.	3·50 in.
Strip width	6 in.	48 in.	200 in.
Minimum strip thickness	0·0001 in.	0·001 in.	0·0035 in.

The strip may be rolled to 0·003 mm (0·0001 in.), with a tolerance held at 0·1 μm (±0·000005 in.) (five millionths) on the smallest mill. To reduce roll-flattening as far as possible, it is usual to make the work-rolls from tungsten carbide, which has a much higher Young's modulus than steel. Hard materials, including stainless steels, titanium-alloys, and nickel-alloys, may be rolled in these mills.

9.9.2 *The planetary mill*

The first planetary mill was patented by Sendzimir in 1948. These mills offer considerable advantages for hot-rolling, though the early versions encountered some installation and maintenance difficulties. The principle is quite simple, but the theory is complex in detail[9.28].

As shown in Figure 9.10, about twenty small-diameter work-rolls are spaced around one large backing-roll. As the whole system rotates, the small work-rolls come into contact with the hot slab surface, and roll down it. Each makes a small reduction along its arc of travel, but the combined effect is a very large reduction in cross-sectional area. For example a 50 mm (2 in.) thick slab can be hot-rolled to 2·0 mm (0·080 in.) thick strip in a single operation. This corresponds to 25:1 reduction (96%), compared with a maximum of about 50% (2:1) per pass in an ordinary mill.

The action differs from ordinary hot-rolling with 2-high or 4-high mills, in several ways. Usually the bite on hot strip is quite powerful, because the friction is high, but in the planetary mill only the backup-rolls are directly driven, and the work-rolls actually rotate against the strip. There is thus no bite, and it is necessary to provide a pair of powerful feed-rolls which force the slab into the roll gap. These supply enough compressive force to suppress any tension generated in the rolling operation, and so prevent cracking. Relatively brittle metals, which cannot normally be hot-rolled, can often be handled successfully by a planetary mill. The feed-rolls control the speed of operation, which may amount to about 0·05 m/sec (10 ft/min) ingoing speed, and 1 m/sec (200 ft/min) or more at the exit. The whole operation on a slab is completed in a very short time, so that scaling of the hot metal is much less serious than under normal conditions. Cooling, and consequent hardening, are also avoided; in fact, the very large deformation causes appreciable increase in temperature as rolling proceeds, so that the final reductions are made on softer metal than the earlier ones. Temperature can be controlled to a reasonable extent by altering the roll speed in relation to the feed, since the cold work-rolls abstract heat from the stock in the short time during which they are in contact at each revolution. Because the temperature remains uniform from end to end of the strip, the gauge tolerance can be held more accurately than on conventional hot-mills.

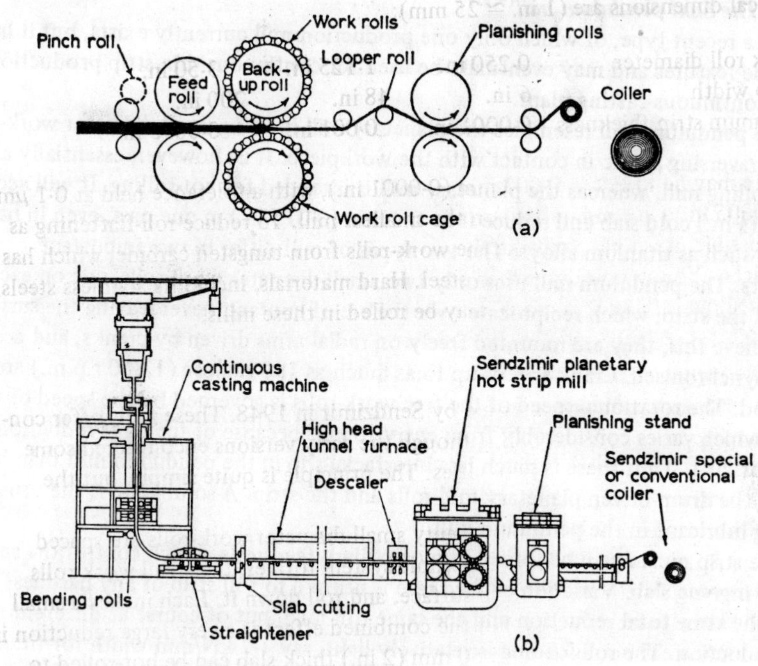

Figure 9.10

(*a*) A diagram of a planetary mill.
(*b*) A continuous casting plant combined with a planetary hot-rolling mill for continuous strip production. (*By courtesy of Sendzimir Ltd.*)

These considerable advantages are offset to some extent by disadvantages. One which received prominence at first was the formation of back-fin. This arises from a forging action as the small-diameter work-roll first contacts the hot slab. Some metal is displaced backwards, and may build up into a fin. This is found in practice to be small, at least with the common 75 mm (3 in.) diameter work-rolls. It can be further reduced by using feed-rolls which have slight corrugations parallel to their axis to improve the grip. The depressions formed in the slab surface provide spaces into which the metal can deform, instead of piling up in a fin. Planetary mills operating on square-ended slab may tend to produce separate strips from top and bottom edges for the first few inches. The strips then tend to open out on leaving the rolls, as is sometimes observed in conventional rolling. This defect has long been known as *crocodile cracking*. A good general account of a planetary hot-mill in actual operation has been given by Potter[9.29].

Recently Sendzimir have made a remarkable advance in steel strip production by coupling a planatary mill directly to a continuous casting plant[9.30]. The planetary mill has special advantages for this purpose; it accepts thick strip and imparts a very heavy reduction of area, and its speed of feed can easily be regulated to match the output of the continuous casting (see Figure 9.10*b*). This development may well have a very important influence in the industry.

9.9.3 *The Saxl pendulum mill*[9.31]

This is a recent type, of which only one production mill currently exists, but it has valuable features and may eventually be used for continuous thin-strip production from continuous-casting plant.

The pendulum mill resembles the planetary mill in using small-diameter work-rolls, traversing an arc in contact with the workpiece. It is, however, essentially a cold-rolling mill, whereas the planetary mill is intended for hot-rolling. It will accept 6 mm (¼ in.) cold slab and reduce it to 0·4 mm (0·015 in.) in one pass, even in hard metals such as titanium alloys. The two types of mill differ in two important respects. The pendulum mill uses only two small-diameter work-rolls, one on each side of the strip, which reciprocate over the arc of contact, never leaving the surface. To achieve this, they are mounted freely on radial arms driven by cranks, and accurately synchronised. Crank speeds up to as much as 180 rad/sec (1,800 r.p.m.) are planned. The rotational speed of the free work-rolls is governed by the speed of the strip, which varies considerably from entry to exit because of the large reduction, perhaps 20:1. Thus there is much less interfacial-slip in the pendulum mill than between the drum-driven planetary-mill rolls and the strip. A soluble oil is the usual rolling lubricant in the pendulum mill.

The strip must again be forced in by auxiliary feed-rolls, which control the speed of the ingoing slab. Variation of feed rate is utilised to roll strip of any hardness with the same total reduction and the same roll load, but of course at different rates of production. The roll loading is relatively light, say 1·2 kN/mm width for titanium alloy, and the mill itself is of light construction, considering the severity of the operation it performs. Because the process is rapid, as in the planetary mill, the temperature-rise due to deformation may be large, and the hardness of the finished strip may consequently be lower after a heavy reduction than after a more moderate pass.

The main disadvantage of the pendulum mill is its low output-rate, compared with a modern tandem-mill. Final speeds may be 0·3 m/sec (70 ft/min) and 10 m/sec (2,000 ft/min) respectively. On the other hand, interpass annealing time is avoided, and the low capital cost may permit two or more pendulum mills to be used in parallel. The mill is unsuitable for short slabs because of transient conditions at the ends, but bending and camber problems need not be severe, because the loading is light, and the rolls are supported in needle bearings; so very large widths can, in principle, be handled. There has been a tendency, as with the planetary mill, to form a 'washboard' surface on the final strip, but this can be overcome by an ingenious resilient pivot which provides a short length of parallel travel beyond the centre-line of the cams. This also provides very good gauge control.

9.9.4 *Continuous rotary-casting and rolling lines*[9.32]

An interesting continuous process for production of aluminium alloy strip has recently been developed by Stanat-Mann. The alloy is cast on to a slowly rotating wheel in the space between a peripheral groove, 200 mm (8 in.) wide x 16 mm (⅝ in.) deep, and a surrounding steel belt. The wheel is water-cooled so that the strip solidifies and after about 180° rotation may be fed off directly into a 2-high hot rolling mill. This is of conventional type, but is fitted with low-friction bearings to improve its efficiency at the low speeds (0·1 m/sec (15–18 ft/min)) required to match the casting wheel. At 9 mm (0·35 in.) thickness the strip is coiled and annealed and then cold rolled, in the same 2-high mill, to 3 mm (0·120 in.). It is finished on a 4-high mill at 3 m/sec (550 ft/min) as 16-gauge to 20-gauge (1–1·5 mm) strip, which is mainly used for making seam-welded tubing.

9.10 Lubrication in practical hot rolling

9.10.1 *Hot rolling of steels and other alloys*

Rolling is unusual in requiring a certain minimum friction to produce adequate bite to feed the stock through (equation 9.22). The demands on a hot-rolling lubricant are not severe, and cooling of the tools is usually all that is required for steels, nickel alloys or titanium. Graphite is sometimes dispersed in the coolant. Oxide scale plays a significant part in the wear of hot rolls and in the surface quality of the product. Slight scale is broken up in the rolling and can be tolerated, but excessive scale from soaking pits should be avoided or removed before rolling. The loose scale produced by a hot-rolling mill must be separated from the cooling water. As the coolant may be supplied at several thousands of gallons per minute (0·2 m^3/sec or more) this is a major problem. The water is led off in large flumes to settling pits, from which the scale is recovered as a high-grade ore. Usually the water is also contaminated by oil, from bearing leakage, which is removed by skimming the pits and further chemical treatment. Total-loss systems, in which all the coolant is discharged to a river or lake, have now disappeared, and rolling effluent is carefully controlled.

Copper alloys also are rolled with a coolant, usually a mineral-oil emulsion, though copper can satisfactorily be rolled dry on a laboratory mill.

Aluminium is exceptional. The oxide breaks up into islands exposing appreciable areas of metal that tend to adhere to the rolls. A coating of aluminium on the rolls is quickly established. With a suitable coolant, such as a fortified mineral-oil emulsion, this does no damage and even helps to increase the bite, but if no lubricant is applied the slab may finally wrap around the roll in a tightly adherent band. The stability of this coating is an important factor, since fragments may become embedded in the rolled product if they are detached. It is therefore important also that the emulsion should not deteriorate. Fortunately it is unlikely to become contaminated by bearing oil when the same emulsion can be used to lubricate fabric/resin bearings, but these can be used only at modest speeds. Most new high-speed mills use oil-lubricated bearings, and contamination is a continuing problem. On long aluminium-rolling lines the coolant may have to be changed every 6–8 weeks, involving very large quantities that cannot easily be disposed of. Great care must therefore be taken in the formulation and maintenance of hot-rolling emulsions for aluminium. Filtration is also important and presents serious problems because both centrifuges and activated filters are unacceptable for these emulsions.

Tungsten and molybdenum require very high temperatures for hot working (about 2,000°C) but can be warm rolled satisfactorily at 900–1,200°C. They may be canned, as for extrusion, or glass coated to prevent oxidation.

Rolling of rods and sections at high temperatures poses the same problems, but the lubrication conditions are often more severe because there is inevitably more slip between the roll and the stock.

9.11 Lubrication in practical cold rolling

As implied in §9.10, although superficially rolling does not involve severe frictional conditions, the stability of the lubricant can give serious problems. Very large quantities of coolant are used. 50 m^3 (10,000 gallon) tanks are by no means uncommon, and even on closed circuits up to 5 m^3 per hour may need to be added. Full-scale rolling mill trials are not lightly undertaken with experimental lubricants. Calcium ions in the water may combine with ionic emulsifiers, causing instability, so non-ionic emulsifiers are preferred, but even when deionised water is used, Al, Mg and Si ions may appear from the stock in light-alloy rolling.

The scale of operations in cold rolling is very large, and uniformity of the product is of paramount importance. The rolls still need to be cooled very reliably, primarily to avoid the development of uneven thermal camber that would produce thickness variations in strip (p. 228), and also to reduce wear, which in turn affects camber. The surface finish of the strip is directly dependent upon that of the rolls, and hence on wear. Another important surface feature is staining, which must be avoided both during the rolling process and during subsequent annealing.

The combination of properties needed in a rolling fluid, and the maintenance of stability in the very large quantities, make rolling lubricant development very difficult.

9.11.1 *Cold rolling of steels*

Palm oil is the most popular lubricant for mild steel strip rolling. Viscosity is important because there is usually a significant hydrodynamic contribution to the lubrication in rolling, but plain mineral oils are only moderately good. Some boundary action is essential, and even palm oil is known to improve during use, probably because free fatty acids are produced. Synthetic palm oils have been developed, based on tallow, but they have not won general acceptance.

It is important to remove palm oil, and indeed most rolling lubricants, before strip is annealed, or brown stains are likely to develop. Because the removal can never be complete, it is desirable to anneal in a free gas flow to minimise staining, which may of course arise from contamination by bearing oils as well as, or more than, from the coolant.

Because cooling is usually very important, many rolling mills use water-base fluids. The most elementary type of water cooling relies on a precoating of the strip with a fatty oil and application of plain water at the mill. This method raises serious disposal problems and is now obsolete. A variation is the mechanical dispersion of an oil phase in water but the same problems arise and these dispersions are used only in the highest speed mills. Almost all modern water-base rolling fluids are stable emulsions of oil in water, with a complex constitution of emulsifier and additives to prevent oxidation, corrosion, bacterial attack and foaming, together with boundary and sometimes E.P. agents. Some of the most important emulsions for steel strip are based on rapeseed oil and wax. These are stabilised with ionic emulsifiers and may be sensitive to the presence of multiple cations, especially Ca^{++} in hard water. Inversion in stagnant zones and consequent corrosion is always a danger with emulsions, but with care in design and supervision they can be used extensively.

For stainless steels better boundary lubrication is needed to prevent pickup. Mechanical dispersions, with more readily available fatty constituents to 'plate out' on the rolls, are still commonly used.

Special problems arise in cluster mills (p. 237), which require an oil for the bearings and the strip together. A mineral oil with boundary and E.P. additives, and an oxidation inhibitor, is used.

9.11.2 *Aluminium*

Aluminium, when exposed by fracture of its strong oxide, will adhere easily to a roll surface. This can seldom be completely avoided, but it can be controlled. The most common cold-rolling lubricants are mineral oils with boundary additives, such as 0·2% oleic acid. Emulsions are however being used increasingly for sheet rolling. A major problem with both is the production of stains. These may arise from residual organic constituents on the strip during the annealing process if insufficient air is circulating in the furnace, but may also be produced by polymerisation on the strip surface at the temperatures generated by the rolling process itself. Such brown or black stains are extremely difficult to remove. White stains from water residues or oxidation of the strip may also form during annealing. Recently a change to amine additives has been proposed, since these give much less staining but still good roll lubrication.

Considerable quantities of aluminium are rolled into foil. This requires highly polished rolls with a thin lubricant. Paraffin (kerosene) is used in a purified form, but staining during subsequent annealing can still be a serious problem. Additives, particularly sulphur, must be carefully avoided. Very thin foil is often rolled double. The outer surface is then bright and shiny but the inner surface is matt, due to differential deformation of the metal grains, enhanced by the presence of trapped lubricant. Care must be taken to avoid pinholes.

Attention should be given to the potential fire hazard from fine aluminium debris.

9.11.3 *Other alloys*

Copper and brasses are relatively undemanding. Mineral oils are adequate and boundary additives have little effect. Sulphur must be avoided because even small quantities cause staining. Emulsions, usually of mineral oil and a sulphonate, are widely used. For the higher surface quality on brasses, low-viscosity mineral oil with 20–30% fatty acid is used at relatively low speeds.

Nickel alloys are rolled, like stainless steel, using a mineral oil with boundary and E.P. additives.

Titanium adheres strongly to the roll surface, as it adheres to tools in all metalworking processes. It has a high flow stress and is preferably rolled hot, though it can be rolled on a cluster mill, using a mineral oil with castor oil or potassium palmitate additives.

9.12 **Recent developments in rolling**

9.12.1 *Theoretical contributions*

There appears to be much less interest in the theory of flat rolling than in other processes, perhaps because the existing formulae are adequate for practical calculations of roll force and power, while longitudinal strain is reasonably uniform. A very simple adaptation of equation 9.8 can be used in the form

$$P = \chi w\bar{S}\sqrt{R\Delta h} \tag{9.38}$$

If the load P, width w and height reduction Δh are measured for a given rolling pass, it is possible to calculate a value for χ from a knowledge of these and the hardness of the ingoing and outgoing strip, assuming constant roll radius R. The mean yield stress $\bar{S}$ is found from the hardness values, using equations 2.25 and 3.27.

This value of χ is then used to predict the roll load for the next pass, extrapolating the hardness values and allowing for measured mill springback. It has been found in laboratory exercises that this approximate method rapidly improves in accuracy as the number of passes increases and the reliability of the extrapolation improves. With no prior knowledge of the mill, for example, it is possible to roll a 10 mm slab to 1·0 mm ± 0·02 mm using only two or three more passes than the minimum theoretically possible without exceeding the safe load of the mill. The calculations are very quickly performed on a programmed calculator.

Upper-bound solutions can be applied to the rolling of sandwich and composite materials, but more remains to be explained about the incidence of delamination and also about crocodile and fish-tail cracking in single materials.

The rolling of sections presents important theoretical problems because, as in spread of flat slabs, the profiles depend on minor flow, in the transverse directions. Most rolling theory ignores this. Some attention is being given to these problems, mainly by semi-empirical methods based on visioplasticity. Similar analysis has also been applied to cross-rolling of tubes, discussed in §7.10.1.

9.12.2 *Conventional and newer processes*

The most important advance in conventional flat rolling is the introduction on a wide scale of automatic gauge control. This is associated in some instances with automatic non-destructive testing, to avoid processing unsound stock.

Rolling of composites introduces no new principles but it may require greater care in the selection of roll schedules and the preparation of the materials.

Direct rolling of powders is of more commercial interest than it used to be, because powders are the most convenient form in which metals can be reclaimed. As the costs of reclaiming are reduced by the introduction of more economic processes, powder rolling may be expected to become more attractive. A related technique of spray rolling has been developed, in which a thin strip is sprayed, for example, with steel powder, and then consolidated by rolling while still at a high temperature[9.33]. It may even be possible to spray oxide particles and to reduce them as part of the process. In principle it is possible in this way to make strip with chemical and mechanical properties graded across its thickness.

EXAMPLES

Use the stress-strain data of Chapter 2 examples where appropriate. Assume that rolling oil gives $\mu = 0{\cdot}05$ on copper strip, and $\mu = 0{\cdot}07$ on steel strip.

9.1 Calculate, by the Bland and Ford method, the roll load necessary to roll 250 mm wide

annealed copper strip from 2·50 mm to 2·00 mm thick with 350 mm diameter steel rolls. How much is the load increased if paraffin is substituted for rolling oil, raising the friction to $\mu = 0{\cdot}08$?

Solution. Equations 9.15–9.18 give the roll pressures:

$$\frac{p^+}{S} = \frac{h}{h_a} e^{\mu H}; \quad \frac{p^-}{S} = \frac{h}{h_b} e^{\mu(H_b - H)}; \quad H = 2\sqrt{\frac{R'}{h_a}} \tan^{-1} \sqrt{\frac{R'}{h_a}} \, . \, \alpha$$

This involves R', found from $R' = R\left(1 + \dfrac{0{\cdot}022}{\Delta h}\dfrac{P}{w}\right)$. The approximate value of $\dfrac{P}{w} = 1{\cdot}2\,\bar{S}\sqrt{R\,\Delta h}$ suffices for the roll flattening allowance. $\bar{S}$ is found from the σ/ϵ curve, using the area up to $\bar{\epsilon} = 0{\cdot}223$ (Example 2.3) giving $\bar{S} = 0{\cdot}18$ kN/mm².

$$\frac{P}{w} = 1{\cdot}2 \times 0{\cdot}18 \times \sqrt{175 \times 0{\cdot}5} = 2{\cdot}02 \text{ kN/mm}; \quad R' = 190 \text{ mm}$$

To find α_b: $h - h_a = R'\alpha_b^2$; $\alpha_b = \sqrt{\Delta h / R'} = 0{\cdot}0513$ rad.
The necessary values are then tabulated taking $\mu = 0{\cdot}05$, and p^+ and p^- are plotted against α. The area $\displaystyle\int_0^{\alpha_b} p\,d\alpha$ under this friction hill is found to be 10·9 N rad/mm². Then

$$\frac{P}{w} = R' \int_0^{\alpha_b} p\,d\alpha = 2{\cdot}07 \text{ kN/mm}; \quad P = 517 \text{ kN}$$

With $\mu = 0{\cdot}08$, a similar calculation gives $P \approx 550$ kN.

9.2 Compare the result of Example 1 with the value calculated by Ekelund's equation.

Solution. Equation 9.35 gives, for $\mu = 0{\cdot}05$.

$$\frac{P}{w} = 0{\cdot}18 \times \sqrt{190 \times 0{\cdot}5} \times \left[1 + \frac{1{\cdot}6 \times 0{\cdot}05 \times \sqrt{190 \times 0{\cdot}5} - 1{\cdot}2 \times 0{\cdot}5}{4{\cdot}50}\right]$$

$$\frac{P}{w} = 1{\cdot}82; \quad P = 455 \text{ kN}$$

For $\mu = 0{\cdot}08$,

$$\frac{P}{w} = 2{\cdot}01; \quad P = 502 \text{ kN}$$

9.3 What roll load is necessary to roll 150 mm × 6 mm copper strip, previously rolled 30%, with 20% reduction of area using 350 mm diameter steel rolls? What roll load would be necessary in a further 20% reduction-of-area pass? What would be the maximum deflection of the roll surface, assuming Hitchcock flattening?

Solution. The prior reduction by 30% corresponds to $\epsilon_{01} = \ln 1{\cdot}0/0{\cdot}7 = 0{\cdot}357$. In the first pass $\epsilon_{12} = \ln 6/4{\cdot}8 = 0{\cdot}223$; $\epsilon_{02} = 0{\cdot}580$. The mean yield stress for this pass $\bar{S}_{12} = (0{\cdot}35 + 0{\cdot}43)/2 = 0{\cdot}39$ kN/mm². Using the simple roll-load formula, equation 9.8,

$$\frac{P}{w} = 1{\cdot}2 \times 0{\cdot}39 \times \sqrt{175 \times 1{\cdot}2} = 6{\cdot}8 \text{ kN/mm}, \quad P = 1{,}020 \text{ kN}$$

In the next pass $\epsilon_{23} = 0{\cdot}233$, $\epsilon_{03} = 0{\cdot}813$, $\bar{S}_{23} = 0{\cdot}46$ kN/mm²

$$\frac{P}{w} = 1{\cdot}2 \times 0{\cdot}46 \times \sqrt{175 \times 0{\cdot}96} = 7{\cdot}1 \text{ kN/mm}, \quad P = 1{,}060 \text{ kN}$$

The roll flattening is given by equation 9.21

$$R' = R(1 + 0{\cdot}022 \times 7{\cdot}1 \div 0{\cdot}96) = R(1{\cdot}16)$$

If Δh is to be maintained, $R(1 - \cos\alpha_b) = R'(1 - \cos\alpha_b') = \frac{\Delta h}{2}$

$$1 - \cos\alpha_b' = \frac{0{\cdot}48}{175} = 0{\cdot}00274 \qquad \alpha_b' = 4{\cdot}3°$$

The maximum deflection is $R \sin\alpha \tan\alpha/2 = 0{\cdot}49$ mm.

9.4 What roll load is required to roll 500 mm x 2·50 mm mild steel strip, previously rolled, 30%, to 2·40 mm thick with 350 mm diameter steel rolls? What is the Hitchcock flattening (*a*) in the absence of tensions (*b*) with front tension equal to 150 N/mm² and back tension equal to 75 N/mm²? What is the thinnest gauge to which this strip can be rolled without tensions on this mill?

Solution. The prior reduction by 30% corresponds to $\epsilon_{01} = 0{\cdot}358$. In the reduction from $h_1 = 2{\cdot}50$ to $h_2 = 2{\cdot}40$, $\epsilon_{12} = \ln 2{\cdot}5/2{\cdot}4 = 0{\cdot}041$, so $\epsilon_{02} = 0{\cdot}399$. From the σ/ϵ curve $\bar{S} = (0{\cdot}69 + 0{\cdot}71)/2 = 0{\cdot}70$ kN/mm².

(*a*) Using the simple formula, equation 9.8,

$$\frac{P}{w} = 1{\cdot}2 \times 0{\cdot}70 \times \sqrt{175 \times 0{\cdot}10} = 3{\cdot}5 \text{ kN/mm}, \quad P = 1{,}760 \text{ kN}$$

The roll flattening is given by equation 9.21

$$R' = R(1 + 0{\cdot}022 \times 3{\cdot}5 \div 0{\cdot}1) = 1{\cdot}77R$$

(*b*) If tensions are applied, the equivalent yield stress is reduced at entry and exit, as in §9.3.3

$$S_b^* = 0{\cdot}69 - 0{\cdot}075 = 0{\cdot}615 \text{ kN/mm}^2$$

$$S_a^* = 0{\cdot}71 - 0{\cdot}150 = 0{\cdot}56 \text{ kN/mm}^2$$

The mean yield stress is thus reduced to 0·59 and the roll load becomes 1,480 kN and $R' \approx 1{\cdot}65R$.

For more accurate calculation, the method of §9.3 could be used. (See Example 9.1.)

(*c*) The limiting thickness that can be rolled without tensions on this mill is found from equation 9.32, assuming $\bar{S} \approx 0{\cdot}80$ kN/mm²

$$h_{\min} = 0{\cdot}035\mu R\bar{S} = 0{\cdot}035 \times 0{\cdot}07 \times 175 \times 0{\cdot}8 = 0{\cdot}34 \text{ mm}$$

9.5 A mill housing is 3 m high and is designed for a maximum tensile stress of 12 N/mm². Calculate the elastic extension of the housing under full load.

Solution. The elastic extension is found from the stress and the Young's modulus

$$e = \frac{\sigma}{E} = \frac{12}{20 \times 10^4} = 0{\cdot}6 \times 10^{-4} \text{ mm/mm}$$

The total stretch of the side members is consequently

$$3{,}000 \times 0{\cdot}6 \times 10^{-4} = 0{\cdot}18 \text{ mm}$$

This may be compared with the elastic deformation of the rolls calculated in Example 9.3 for a similar mill.

9.6 Determine the power required to drive the mill producing the strip in Example 9.1, if the emergent strip speed is to be 6 m/sec. Assume that the roll neck diameter is half the roll diameter.

Solution. The roll load, calculated in Example 9.1, is about 500 kN. The roll circumference is 1,100 mm, so the roll speed corresponding to a final strip speed of 6 /msec is 34·3 rad/sec.

The power consumed is found by equation 9.28:

$$2W_R = 2T\dot{\theta} = 2\lambda' L'P\dot{\theta}$$

$$= 2 \times 0{\cdot}45 \times \sqrt{190 \times 0{\cdot}5} \times 500 \times 34{\cdot}3 \text{ kN/mm/sec}$$

$$= 150 \text{ kNm/sec} = 150 \text{ kW}$$

$$4W_N = 4\mu_N \frac{Pd}{4}\dot{\theta}$$

Assume $\mu_N = 0{\cdot}005$; $4W_N = 0{\cdot}005 \times 500 \times 350 \times 34{\cdot}3$

$$= 30 \text{ kN m/sec}$$

If the overall efficiency of the power unit is about 80%,

$$\text{Total power} = \frac{1}{\eta_1 \eta_2}(150 + 30) = 225 \text{ kW}$$

9.7 What would be the maximum temperature reached by the strip as a result of the deformation in Example 9.1 if there were no heat losses? If the frictional heating is confined to zones 0·12 mm deep, what would be the temperature reached in these zones?

Solution. From the solution of Example 9.6, the power $2W_R$ is 150 kW. This is the power consumed by deformation and friction, 150 k joules/sec. The volume rolled per second is 6,000 x 250 x 2 mm³/sec, and the density of copper is $8{\cdot}9 \times 10^{-3}$ g/mm³, so the mass rolled per second is 26·7 kg/sec.

The specific heat of copper is 0·376 J/g/°C

$$150 = 0{\cdot}376 \times 26{\cdot}7 \times \Delta T; \quad \Delta T \approx 15°\text{C}$$

If the heat is confined to 0·12 mm thick zones at each surface, the effective volume is reduced in ratio 2 x 0·12 ÷ 2·0 = 0·12. Allowing 10% as frictional work

$$15 = 0{\cdot}376 \times 3{\cdot}20 \times \Delta T'; \quad \Delta T' = 12°$$

$$\Delta T + \Delta T' \approx 27°\text{C}$$

9.8 Suggest a suitable size and power of rolling mill for production of 750 mm wide 0·40 mm thick mild steel sheet in two passes from annealed sheet 0·80 mm thick.

Solution. The total strain $\epsilon_{02} = \ln 0{\cdot}8/0{\cdot}4 = 0{\cdot}693$. Two suitable passes might be $\epsilon_{01} = 0{\cdot}4$, $\epsilon_{12} = 0{\cdot}3$. Then, taking appropriate straight-line approximations to the stress/strain curve, $\overline{S}_{01} = (0{\cdot}74 + 0{\cdot}39)/2 = 0{\cdot}56$ kN/mm² $\overline{S}_{12} = (0.81 + 0{\cdot}71)/2 = 0{\cdot}76$ kN/mm². For the first pass $\ln h_0/h_1 = 0{\cdot}4$; $h_0 = 0{\cdot}80$, $h_1 = 0{\cdot}54$, $\Delta h_{01} = 0{\cdot}26$ mm. Thus $\Delta h_2 = 0{\cdot}14$ mm. Using equation 9.8 to find the approximate roll load,

$$P_1/w = 1{\cdot}2 \times 0{\cdot}56 \times \sqrt{R \times 0{\cdot}26}$$

$$P_2/w = 1{\cdot}2 \times 0{\cdot}76 \times \sqrt{R \times 0{\cdot}14}$$

Common L/D ratios for the rolls in such a 2-high mill lie between 1·5 and 2. Supposing D = 450 mm

$$P_1/w = 5{\cdot}14 \text{ kN/mm}$$

$$P_2/w = 5{\cdot}12 \text{ kN/mm}$$

These values are comparable with common practice. The maximum roll load is thus P_1 = 3,860 kN.

The limiting load for such short rolls is usually set by shear at the roll-neck shoulder. The roll neck diameter d is often equal to $\frac{1}{2}D$. The cross-sectional area of the roll neck is thus 4 x 10⁴ mm². A reasonable value for the shear strength would be 150 N/mm², so the load to fracture one neck would be 6,000 kN, giving a safety factor of about 3 to allow for stress concentration effects and fatigue.

The power input is derived from the torque (equations 9.27 and 9.28)

$$T_1 \approx 0{\cdot}45\sqrt{R\Delta h}\,P_1 = 0{\cdot}45 \times 7{\cdot}65 \times 3{,}860 = 13{\cdot}3 \times 10^3 \text{ kNmm}$$

(or recalculate using R').

$$2W_R = 2T\dot{\theta} = 26{\cdot}6\dot{\theta} \text{ kNm/sec} = 26{\cdot}6\dot{\theta} \text{ kW}$$

Add about 20% for power loss in the bearings (or calculate for an assumed μ) and allow 80% overall transmission efficiency. The motor input is thus $\approx 40\dot{\theta}$ kW. A reasonable size of motor for such a mill might be 450 kW. This would roll at $\dot{\theta} = 11{\cdot}2$, giving a strip exit speed of 2,520 mm/sec, $\approx 2{\cdot}5$ m/sec.

9.9 Use Ekelund's equation to determine the roll load and thence the mill driving power for hot rolling 150 x 100 mm billets to 125 x 100 mm with 300 mm diameter rolls at 6 rad/sec. Assume a mean yield stress of 70 N/mm² and neglect lateral spread.

Solution. Equation 9.36 gives

$$\frac{P}{w} = \bar{S}\sqrt{R\Delta h}\left\{1 + \frac{1{\cdot}6\mu\sqrt{R\Delta h} - 1{\cdot}2\Delta h}{h_a + h_b}\right\}$$

(*a*) Assume $\mu = 0{\cdot}4$ for hot rolling

$$\frac{P}{w} = 70\sqrt{150 \times 25}\left\{1 + \frac{1{\cdot}6 \times 0{\cdot}4 \times \sqrt{150 \times 25} - 1{\cdot}2 \times 25}{275}\right\}$$

$$= 4{,}287\ (1{\cdot}033)\ \text{N/mm}^2;\quad P = 443 \text{ kN}$$

Note that the frictional contribution is quite small, though μ is high, because of the large draft. In practice considerable lateral spread would be expected, which would increase the load.

(*b*) From equations 9.27 and 9.28

$$T = 0{\cdot}5 \times \sqrt{150 \times 25} \times 443 = 13{\cdot}56 \times 10^3 \text{ kN/mm}$$

$$2W_R = 2T\dot{\theta} = 162 \text{ kNm/sec} = 162 \text{ kW}$$

Allowing 10% for roll-neck friction, and 80% efficiency, the power input is 220 kW.

9.10 Estimate rolling schedules for cold rolling annealed copper and mild steel strips from 400 x 3·00 mm to 400 x 2·40 mm with 150 mm diameter rolls on a 4-high mill, if the roll load may not exceed 1,000 kN. Find the ratio of the power requirements for the two materials.

Solution. (*a*) Use the simple roll-load equation 9.8. If the operation is to be completed in one pass, the strain is $\epsilon_{01} = \ln 3{\cdot}0/2{\cdot}4 = 0{\cdot}223$. For copper $\bar{S}_{01} = 0{\cdot}18$ kN/mm² and for steel $\bar{S}_{01} = 0{\cdot}47$ kN/mm²

So for copper: $P = 400 \times 1{\cdot}2 \times 0{\cdot}18 \times \sqrt{75 \times 0{\cdot}60} = 580$ kN

for steel: $P = 1{,}510$ kN

Copper can thus be rolled easily in one pass, but steel requires more than one pass, since P may not exceed 1,000 kN.

In two equal passes for steel, $\epsilon_{01} = 0{\cdot}112$, $\epsilon_{12} = 0{\cdot}111$.

Then $\bar{S}_{01} = 0{\cdot}42$, $\bar{S}_{12} = 0{\cdot}56$ kN/mm²

$$h_0 = 3{\cdot}00,\ h_1 = 2{\cdot}68,\ h_2 = 2{\cdot}40,\ \Delta h_1 = 0{\cdot}32,\ \Delta h_2 = 0{\cdot}28 \text{ mm}$$

$$P_1 = 990 \text{ kN},\ P_2 = 1{,}300 \text{ kN}$$

Thus steel strip cannot be rolled to this size even in two passes. Three passes might be:

$$\epsilon_{01} = 0{\cdot}10 \quad h_1 = 2{\cdot}71 \quad \overline{S}_{01} = 0{\cdot}39 \quad P_1 = 873 \text{ kN}$$

$$\epsilon_{12} = 0{\cdot}07 \quad h_2 = 2{\cdot}53 \quad \overline{S}_{12} = 0{\cdot}54 \quad P_2 = 952$$

$$\epsilon_{23} = 0{\cdot}053 \quad h_3 = 2{\cdot}40 \quad \overline{S}_{23} = 0{\cdot}59 \quad P_3 = 884$$

The operation can just be completed in three passes with the mild steel strip.

(*b*) The power required is found from equation 9.28,

$$W_M = \frac{1}{\eta_1 \eta_2}(2W_R + 4W_N) = \frac{1}{\eta_1 \eta_2}(2T_R\dot{\theta} + 4T_N\dot{\theta})$$

which, for constant speed and friction conditions can be written

$$\eta_1 \eta_2 W_M = 2\lambda' L'P \,.\, \dot{\theta} + 4\mu_N \frac{P}{2} R\theta \,(\approx 0{\cdot}8\, W_M)$$

$$= (2 \times 0{\cdot}5\sqrt{R\Delta h} \times 1{\cdot}2 \times \overline{S}\sqrt{R\Delta h}\; w + 2\mu_N \times 1{\cdot}2 \times \overline{S}\sqrt{R\Delta h} \times wR)\,\dot{\theta}$$

$$= 1{\cdot}2\overline{S}(R\Delta h + 2\mu\, R\sqrt{R\Delta h})\,\dot{\theta} w$$

The following values can be calculated:

	$R\Delta h$	$2\mu\, R\sqrt{R\Delta h}$	$\overline{S}$	$\dfrac{\eta W}{w\dot{\theta}}$
Copper	45	5·0	0·18	9·0
Steel,				
pass 1	21·8	3·5	0·39	9·8
pass 2	13·5	2·8	0·54	8·8
pass 3	9·8	2·3	0·59	7·1

The power required is comparable for each pass, but the steel requires 3 passes.

REFERENCES

9.1 Ascough, H. H., 'Improvements in wide strip mills', *Sheet Metal Ind.*, **40**, No. 430, 117 et seq., 1963.

9.2 Anon, 'Widest rolling mill for aluminium strip – British Aluminium Co. Falkirk', *Metal Treatment and Drop Forging*, Dec. 1961, p. 514.

9.3 Sendzimir, M. G., 'The Sendzimir cold strip mill', *J. Metals*, **8**, 1,154–1,158, 1956.

9.4 Orowan, E., 'The calculation of roll pressure in hot and cold flat rolling', *Proc. Inst. Mech. Engrs.*, **150**, 140–167, 1943.

9.5 Bland, D. R. and Ford, H., 'The calculation of roll force and torque in cold strip rolling with tensions', *Proc. Inst. Mech. Engrs.*, **159**, 144–153, 1948.

9.6 Siebel, E. and Lueg, W., 'Measurement of stress in roll gap', *Stahl und Eisen*, **53**, (14), 346–352, 1933.

9.7 Hitchcock, J. H., 'Elastic deformation of rolls during cold rolling', ASME Research Publication *Roll Neck Bearings*, 1935.

9.8 Whitton, P. W. and Ford, H., 'Surface friction and lubrication in cold strip rolling', *Proc. Inst. Mech. Engrs.*, **169**, 123–140, 1955.

9.9 Ford, H., Ellis, F. and Bland, D. R., 'Cold rolling with strip tension, Part II: Comparison of calculated and experimental results', *J. Iron and Steel Inst.*, **171**, 239–245, 1952.

9.10 Underwood, L. R., *Roll Neck Bearings*, Iron and Steel Ind. Res. Council, 1943.

9.11 Siebel, E. and Lueg, W., 'Untersuchungen über die Spannungsverteilung im Walzspalt', *Mitt. K. W. Inst. Eisenforschung, Düsseldorf*, **15**, 1–14, 1933.

9.12 Nádai, A., *Plasticity*, McGraw-Hill, 1931.

9.13 Tong, K. and Sachs, G., 'Roll separating force and minimum thickness of cold-rolled strips', *J. Mech. Phys. Solids*, 6, 35–46, 1957.
9.14 Starling, C. W., *The Thoery and Practice of Flat Rolling*, London Univ. Press, 1962.
9.15 Cox, quoted by C. W. Starling, ref. 9.14, p. 57.
9.16 Sims, R. B., 'Automatic gauge control in rolling mills', *J. Inst. Metals*, 86, 289–302, 1957.
9.17 Anon., 'Strip mill with automatic gauge control', *Engineer,* Dec. 13, 1957.
9.18 Butterfield, M. H., *Computer Control of High Speed Cold Reduction Steel Strip Tandem Mills*, Internat. Systems Control Ltd., May 1962.
9.19 Bland, D. R. and Ford, H., 'Cold rolling with strip tension, Part III: An approximate treatment of the elastic compression of the strip', *J. Iron and Steel Inst*., 171, 245–249, 1952.
9.20 Cook, M. and Parker, R. J., 'The computation of loads in metal strip rolling by methods involving the use of dimensional analysis', *J. Inst. Metals*, 82, 129–140, 1953.
9.21 Ekelund, S., 'The analysis of factors influencing rolling pressure and power consumption in the hot rolling of steel', *Steel*, 93, Nos. 8–14, 1933 (translated from *Jernkontorets Ann*., Feb. 1927, by B. Blomquist).
9.22 Davies, C. E. quoted by L. R. Underwood, 'Rolling mills, methods of roll load and power calculation', *Metal Ind*., Feb. 27 et seq., 1948.
9.23 Sims, R. B., 'Calculation of roll force and torque in hot rolling mills', *Proc. Inst. Mech. Engrs*., 168, 191–200, 1954.
9.24 Larke, E. C., *The Rolling of Strip Sheet and Plate*, 2nd Ed., Chapman & Hall, 1963, p. 346.
9.25 Orowan, E. and Pascoe, K. J., *A Simple Method of Calculating Roll Pressure and Power Consumption in Hot Flat Rolling*, Iron and Steel Inst. Special Rept. No. 34, p. 124, 1946.
9.26 Alexander, J. M., 'A slip-line field for the hot-rolling process', *Proc. Inst. Mech. Engrs.,* 169, 1,021–1,028, 1955.
9.27 Johnson, W. and Kudo, H., 'The use of upper-bound solutions for the determination of temperature distribution in fast hot rolling and axi-symmetrical extrusion processes', *Int. J. Mech. Sci*., 1, 175–191, 1960.
9.28 Sparling, L. G. M., 'Calculation of rolling load and torque in a hot planetary rolling mill', *J. Mech. Eng. Sci*., 4 (3), 257–269, 1962.
9.29 Potter, D. McQ., 'Planetary hot mill: plant and practice at Habershon's', *Iron and Steel*, Oct.–Nov. 1957, 755–770.
Sendzimir, T., 'Planetary mill and its uses', *Iron and Steel Engng*., 35, 95–99, Jan. 1958.
9.30 Sendzimir, M. G., *16th Ann. Tech. Meeting of the Indian Institute of Metals, Jamshedpur*, Feb. 1963.
9.31 Saxl, K., 'The pendulum mill. Rolling mill design developed afresh', *Engineering*, 195, 494–495, 1963.
9.32 Anon., 'Continuous casting and rolling of aluminium alloy strip', *Metal Ind*., 5 Dec. 1963.
Brand, H. W., 'Progress with the rotary strip casting process', *Proc. Amer. Inst. Min., Met. Pet. Engrs., New York*, Feb. 1964.
9.33 Singer, A. R. E., 'Spray rolling of metals.' *Prod. Eng*., 51, 98–104, 1972.

10

Forging, Punching and Piercing

10.1 Introduction

Forging is a major method of shaping very large castings and may also be used to improve their metallurgical and mechanical properties. Large forces are involved, and the equipment is often very heavy. Several forging methods are available, but they may be classified broadly as either pressing or hammering. The former is preferred for homogenising large ingots, because the deformation zone extends throughout the cross-section. To reduce the overall load, the tools may compress only a small region at each stroke. In hammering the deformation is mainly restricted to the surface regions, except for small-section stock. Forces of several thousands of tonnes (tons) may be necessary, even though the yield stress is reduced as far as possible by working at high temperatures. Drop-forging, hammering and swaging can be performed with rather lighter equipment, though a drop-hammer itself may weigh several tons, and these techniques are widely used, particularly for producing a final shape. The problems of metal flow associated with the filling of a complex die shape still require a mainly empirical approach. These will not be discussed in this book, but they merit much more detailed study than they have hitherto received. It is also difficult to analyse the conditions of impact deformation in detail, though energy evaluations from similar slowly varying processes may often be helpful. For the purposes of this chapter, attention will be restricted to prediction of working loads in steady-state open-die forging with simple platens and punches. Constant yield stress, or an average value of the yield stress, will be assumed but it is important to remember that in hot-forging the yield stress will depend very much upon the strain-rate used. Conditions of high friction are usually found in hot-forging. On the other hand, lubrication is important in the cold-forging of steel and many non-ferrous alloys. Combinations of cold-forging and cold-extrusion are becoming increasingly important in industry for the smaller sizes of stock.

Punching uses tools which are small in comparison with the dimensions of the workpiece, and this operation is discussed separately as a special instance of forging. Piercing of a free block is analogous to deep punching, but piercing of a billet within a container is related in an inverse way to simple extrusion. Both of these will be discussed briefly.

Forging theory has been applied by B.I.S.R.A. to a fully-automatic process with an optimum schedule produced by a computer.

10.2 Determination of plane-strain compression load from local stress evaluation

10.2.1 *Low-friction conditions. Thin strip*

Figure 10.1*a* shows the stresses acting at any instant on a thin, wide plate compressed between parallel overhanging platens. It is assumed that the horizontal stress σ_x is uniform across the section and is a principal stress. Then horizontal equilibrium of the forces acting on unit width of an element on the right of the centre-line requires that

$$(\sigma_x + d\sigma_x)\,h - \sigma_x \,.\, h - 2\mu p \; dx = 0$$

On the left of the centre-line,

$$(\sigma_x + d\sigma_x)\,h - \sigma_x \,.\, h + 2\mu p \; dx = 0$$

These equations may be simplified and combined:

$$h \; d\sigma_x \mp 2\mu p \; dx = 0 \qquad (10.1)$$

The upper sign refers to the right-hand portion. Since σ_x is one principal stress, $-p$ is the other, and the two are related by the condition of yielding in plane strain (equation 3.27)

$$\sigma_1 - \sigma_3 = S$$

$$\sigma_x + p = S = 2k; \; d\sigma_x = -dp$$

Substitution in equation 10.1 gives a relationship between p and x:

$$hdp \pm 2\mu p \; dx = 0$$

$$\frac{dp}{p} = \mp \frac{2\mu}{h} dx \qquad (10.2)$$

This can be integrated directly.

$$\ln p = \mp \frac{2\mu}{h} x + \text{constant}$$

The constant of integration is found from the condition that the horizontal stress is zero at both edges, where $x = \pm\, b/2$. Thus at $x = b/2$,

$$(p)_{b/2} = c^- e^{-\frac{2\mu}{h} \cdot \frac{b}{2}} \text{ and } (p)_{b/2} = 2k - (\sigma_x)_{b/2} = 2k$$

$$c^- = 2ke^{+\frac{\mu b}{h}}$$

Thus on the right-hand side of the centre-line, $x = 0$,

$$\frac{p}{2k} = e^{\frac{2\mu}{h}\left(\frac{b}{2} - x\right)} \qquad (10.3)$$

The constant of integration has the same value on the left-hand side, since

$$(p)_{-b/2} = c^+ \, e^{+\frac{2\mu}{h}\left(-\frac{b}{2}\right)} = 2k$$

The pressure thus increases inwards, exponentially from $p = 2k$ at both edges, as shown in Figure 10.1*b*. The maximum value, at the centre, is

$$\left(\frac{p}{2k}\right)_{max} = e^{\mu \frac{b}{h}} \tag{10.4}$$

If μ is low, this may be written in approximate form

$$\left(\frac{p}{2k}\right)_{max} \approx 1 + \mu \frac{b}{h}$$

The mean pressure P on the platens is then

$$P \approx 2k\left(1 + \frac{1}{2}\mu \frac{b}{h}\right) \tag{10.5}$$

Thus in forging thin strip between parallel platens, as in rolling, the pressure distribution shows a 'friction hill'. Figure 10.1*b* may be compared with Figure 9.2. There is, however, an important difference. In rolling, the conditions are steady; but in forging, the geometry changes continuously as the operation proceeds, and the contact with the platens increases. The load at any instant can be found by integration of equation 10.3, but the progressive increase in load cannot easily be determined, since it depends on the spread of the workpiece, which in turn depends on the dimensions and on the friction.

When the workpiece is thicker, there is again a friction hill, but account must then be taken of the geometrical constraint, by the use of the slip-line field (§10.4).

10.2.2 *High-friction conditions. Thin strip*

If the friction is high, there is a limit to the shear stress at the interface, which cannot exceed the shear-yield stress of the metal, $k = S/2$. If this sticking-friction regime extends over the whole interface, the equilibrium equation 10.1 should be written

$$h\, d\sigma_x \mp 2k\, dx = 0$$

If σ_x and $-p$ are still principal stresses, the introduction of the condition of yielding gives

$$h\, dp \pm 2k\, dx = 0; \quad \frac{dp}{2k} = \mp \frac{dx}{h}$$

This may be integrated directly:

$$\frac{p}{2k} = \mp \frac{x}{h} + \text{constant } c$$

At the edges, $x = \pm b/2$, $\sigma_x = 0$, and $p = 2k$, so the constant of integration is the same for both sides of the centre-line:

$$c = 1 + \frac{b}{2h}$$

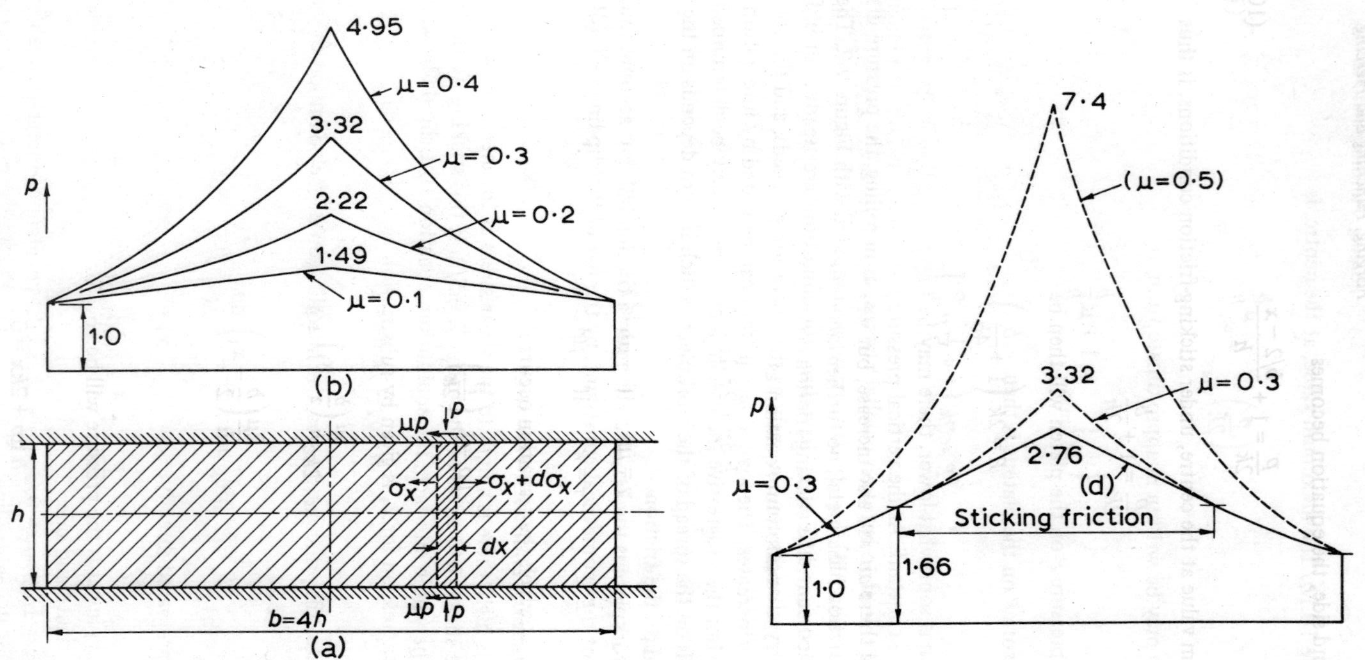

Figure 10.1

(*a*) The stresses acting on an element compressed under plane–strain conditions between parallel overhanging platens.
(*b*) The pressure distribution with low friction.
(*c*) The pressure distribution with sticking friction over the central region only. ($\mu = 0{\cdot}3$.)

On the right-hand side, the equation becomes

$$\frac{p}{2k} = 1 + \frac{b/2 - x}{h} \tag{10.6}$$

The maximum value at the centre, under sticking-friction conditions, is thus

$$\frac{p}{2k} = 1 + \frac{b}{2h} \tag{10.7}$$

The average pressure P on the platen will then be

$$P = 2k\left(1 + \frac{b}{4h}\right) \tag{10.8}$$

If the friction is somewhat lower, there may be slipping friction in the outer regions, with μ = constant. As the vertical pressure p increases inwards, the magnitude of the frictional stress μp will also increase, but when this reaches a value equal to k, the shear-yield stress of the metal, no further increase is possible. The central region will then be governed by the sticking-friction condition. This can be expressed formally by writing the equilibrium equation as

$$h\,\mathrm{d}p \pm 2\tau\,\mathrm{d}x = 0$$

where

$$\left.\begin{aligned} \tau &= \mu p \quad \text{if} \quad \mu p < k \\ \tau &= k \quad \text{if} \quad \mu p \geqslant k \end{aligned}\right\}$$

The point x_1 at which the transition occurs is found from the condition

$$\mu p = k; \left(\frac{p}{2k}\right)_{x_1} = \frac{1}{2\mu}$$

The analytical expression for x_1 is given by substitution in equation 10.3

$$\mu \,.\, 2ke^{\frac{2\mu}{h}\left(\frac{b}{2} - x_1\right)} = k$$

$$\frac{2\mu}{h}\left(\frac{b}{2} - x_1\right) = \ln\frac{1}{2\mu}$$

$$x_1 = \frac{b}{2} - \frac{h}{2\mu}\ln\frac{1}{2\mu}$$

From the point x_1 inwards the pressure will be governed by the equation for sticking friction. On the right-hand side

$$h\,\mathrm{d}p + 2kx = 0$$

$$\frac{p}{2k} = -\frac{x}{h} + \text{constant } c$$

At $x = x_1$, $\dfrac{p}{2k} = \dfrac{1}{2\mu}$, so the constant is

$$c = \frac{1}{2\mu} + \frac{x_1}{h}$$

Thus for values of x between 0 and x_1,

$$\frac{p}{2k} = \frac{1}{2\mu} + \frac{x_1 - x}{h} \tag{10.9}$$

An example for $\mu = 0{\cdot}3$ is given in Figure 10.1*d*.

10.2.3 *Inclined platens. Thin strip*

If the platens overlap the workpiece, as in the preceding sections, but are mutually inclined at a small angle, the equilibrium equation is derived in the same way. The solution resembles that for tube drawing with tapered dies (Chapter 7, §7.2) or for cold strip-rolling (Chapter 9, §9.3).

Instantaneously, if the platens are inclined at angles α_1 and α_2 to the central plane, horizontal force-equilibrium requires that, for the two sides of the centre-line,

$$hd\sigma_x + \sigma_x dh + p\,\frac{dx}{\cos\alpha_1}\sin\alpha_1 + p\,\frac{dx}{\cos\alpha_2}\sin\alpha_2 \pm \mu p\,\frac{dx}{\cos\alpha_1}\cos\alpha_1 \pm \mu p\,\frac{dx}{\cos\alpha_2}\cos\alpha_2 = 0$$

or

$$h\,\mathrm{d}\sigma_x + \sigma_x\,\mathrm{d}h + p\,\mathrm{d}x\,(\tan\alpha_1 + \tan\alpha_2 \pm 2\mu) = 0 \tag{10.10}$$

Substituting the relationship $\mathrm{d}h = \mathrm{d}x\,(\tan\alpha_1 + \tan\alpha_2)$, this becomes

$$h\,\mathrm{d}\sigma_x + \sigma_x\,\mathrm{d}h + p\,\mathrm{d}h \pm \frac{2\mu p\,\mathrm{d}h}{\tan\alpha_1 + \tan\alpha_2} = 0$$

It is convenient to make the substitution

$$B^* = \frac{2\mu}{\tan\alpha_1 + \tan\alpha_2}:$$

$$h\,\mathrm{d}\sigma_x + \sigma_x\,\mathrm{d}h + p(1 \pm B^*)\,\mathrm{d}h = 0 \tag{10.11}$$

If the inclination of the platens to the central plane is small, and the friction is low, σ_x and $-p$ may be assumed to be principal stresses (as in §9.3.1). Then, for plane-strain forging, the condition for yielding (equation 3.27) gives

$$\sigma_x + p = S = 2k;\ \mathrm{d}\sigma_x = -\mathrm{d}p,$$

so σ_x can be eliminated from equation 10.11:

$$-h\,\mathrm{d}p + (S \pm B^*p)\,\mathrm{d}h = 0$$

Taking the upper sign, this can be integrated directly if the metal does not strain-harden appreciably, so that S or $\overline{S}$ can be considered constant.

$$\frac{1}{B^*}\ln (S + B^*p) = \ln h + \text{constant}$$

or

$$S + B^*p = c^+ h^{B^*}$$

At the edge of the block, $h = h_a$, $\sigma_x = 0, p = S$, so

$$c^+ = S(1 + B^*)/h_a^{B^*}$$

Thus, in terms of the dimensionless ratio p/S

$$1 + B^* \frac{p}{S} = (1 + B^*)\left(\frac{h}{h_a}\right)^{B^*}$$

This may be written

$$\left(\frac{p^+}{S}\right)_x = \frac{1 + B^*}{B^*}\left(\frac{h}{h_a}\right)^{B^*} - \frac{1}{B^*} \tag{10.12}$$

It may be seen that this equation has the same form as the equation derived for die pressure in strip drawing (Chapter 4, §4.4.1):

$$\left(\frac{p}{S}\right)_x = 1 - \left(\frac{\sigma}{S}\right)_x = 1 - \frac{1 + B}{B}\left[1 - \left(\frac{h}{h_b}\right)^B\right] \tag{4.22}$$

The values of B^* and B are the same if $\alpha_1 = \alpha_2$, but the boundary conditions give $\sigma_x = 0$ at both sides in forging, and only at the entry in drawing. The relationship is true only for a small decrement during forging, since h_b and h_a change continuously

The other branch of the pressure-distribution curve in forging is found by considering the lower sign in equation 10.11.

$$S - B^*p = c^- . h_b^{B^*}, \quad c^- = S(1 - B^*)/h_b^{B^*}$$

$$\left(\frac{p^-}{S}\right)_x = \frac{1}{B^*} - \frac{1 - B^*}{B^*}\left(\frac{h}{h_b}\right)^{B^*} \tag{10.13}$$

There will be a net lateral thrust on the platens, due to the horizontal components of the normal and tangential forces. This may cause the workpiece to slip from the press if the platens are well lubricated, and is also significant in penetration by a wedge indenter or chisel, as will be seen in §10.7.

10.3 Determination by stress-evaluation of the load for forging a flat circular disc

The stresses acting on an element of a circular disc forged in the axial direction are shown in Figure 10.2.

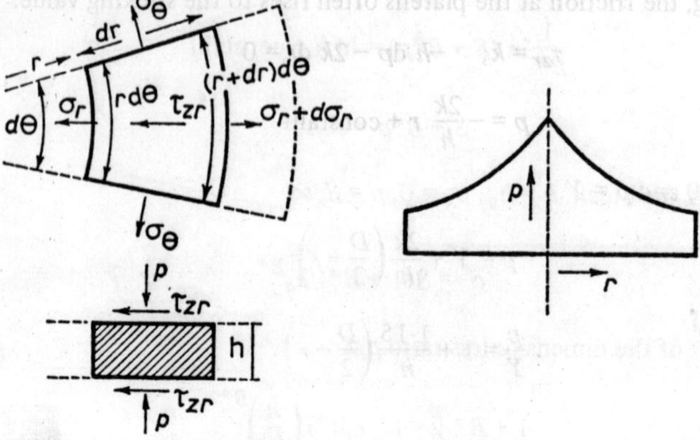

Figure 10.2 The stresses acting on an element of a forged disc.

Assuming that σ_r and σ_θ are constant throughout the disc thickness, radial equilibrium requires that

$$(\sigma_r + d\sigma_r)\, h\, (r + dr)\, d\theta - \sigma_r hr\, d\theta - 2\sigma_\theta h\, dr \sin\frac{d\theta}{2} - 2\tau_{zr} r\, d\theta\, dr = 0$$

$$\sigma_r hr\, d\theta + \sigma_r h\, dr\, d\theta + d\sigma_r hr\, d\theta - \sigma_r hr\, d\theta - \sigma_\theta h\, dr\, d\theta - 2\tau_{zr} r\, d\theta\, dr = 0$$

$$\sigma_r h\, dr + d\sigma_r hr - \sigma_\theta h\, dr - 2\tau_{zr} r\, dr = 0 \qquad (10.14)$$

It is reasonable to assume that σ_r, σ_θ and σ_z $(= -p)$ are the principal stresses, and it can be shown that the stress state is cylindrical, with $\sigma_r = \sigma_\theta$. Any yield criterion then gives (Chapter 3, §3.6.3)

$$\sigma_r - \sigma_z = Y; \quad \sigma_r + p = Y$$

Equation 10.14 thus becomes, with $\tau_{zr} = \mu p$ for sliding friction:

$$-h\, dp - 2\mu p\, dr = 0$$

$$\ln p = \frac{2\mu}{h} r + \text{constant}$$

The constant of integration may be found from the boundary condition that at $r = \dfrac{D}{2}$, $\sigma_r = 0$, $p = Y$, giving

$$\frac{p}{Y} = e^{\frac{2\mu}{h}\left(\frac{D}{2} - r\right)} \qquad (10.15)$$

Comparison with equation 10.3 shows that the pressure distribution for forging a disc is identical with that for plane-strain forging of a block of breadth b equal to the diameter D of the disc, provided that due allowance is made for the yield stress. The value Y should be used for axially-symmetrical conditions, and S for plane strain.

In hot-forging, the friction at the platens often rises to the sticking value. Then

$$\tau_{zr} = k; \quad -h\,dp - 2k\,dr = 0$$

$$p = -\frac{2k}{h}r + \text{constant}$$

At $r = D/2$, $\sigma_r = 0$ and $p = Y$.

$$p = Y + \frac{2k}{h}\left(\frac{D}{2} - r\right)$$

$$\frac{p}{Y} = 1 + \frac{1{\cdot}15}{h}\left(\frac{D}{2} - r\right) \tag{10.16}$$

10.4 Slip-line field solutions for plane–strain compression between parallel, frictionless platens

When the material to be forged is so thin that the platen breadth b greatly exceeds the strip thickness h, the method of local stress evaluation gives accurate results. For low values of coefficient of friction, the average pressure on the platen is given by equation 10.5:

$$P = 2k\left(1 + \frac{1}{2}\mu\frac{h}{b}\right)$$

If, for example, $b = 4h$ and $\mu = 0{\cdot}05$, $P \approx 1{\cdot}10 \times 2k$. In the limiting condition of perfect lubrication, $P = 2k$. This is in fact the basis of the plane-strain compression test described in Chapter 2, §2.6.2.

As the thickness of the material chosen is increased so that b/h is less than about 4, it becomes necessary to take account of the constraint imposed by the metal flow-pattern and the geometry. The application of slip-line field theory to determine the load under these conditions has been described in Chapter 5, §5.6, for various values of the ratio b/h. The pressure is equal to $2k$ for integral values of b/h greater than unity, but the necessity for the slip-lines to be curved causes a cyclic increase between each of those values (Figure 5.6). For practical purposes, these variations, which never exceed 4%, can usually be ignored, but they can be detected experimentally and furnish remarkable evidence of the validity of slip-line field theory.

If the thickness of the material is further increased however, so that h exceeds b, the slip-lines must always be curved, and the pressure rises steadily. The slip-line field (shown in Figure 10.4*a*) closely resembles that for frictionless strip-drawing shown in Figure 6.5, and the calculation proceeds in a manner similar to that of §6.5. In the limiting condition, the plastic zones at each platen are entirely separate, and deformation extends to the free surface alongside the platen. Bulge formation in strip-drawing arises in a similar way. Finally the problem is equivalent to indentation of a semi-infinite block by a flat punch, which requires a pressure $2{\cdot}57k$, as shown in §5.6.5. The full variation of compression-yield pressure with the ratio b/h is given in Figure 5.6. This may be compared with the dependence of the constraint factor f on the ratio c/d in strip-drawing, as in Table 6.1.

10.5 Slip-line fields for plane-strain compression between parallel platens with sticking friction

It is convenient to assume that the platens overlap the workpiece, to distinguish the operation from punching, which will be discussed in §10.7. An approximate slip-line field can be constructed by the method[10.1] described in Chapter 6, §6.5.2.

10.5.1 *Construction of the slip-line field.* $b/h = 3{\cdot}6$

Since sticking friction has been assumed, the slip-lines must meet the platen surface at 0° and 90° (see Chapter 5, §5.5.4). They must also meet the central plane at 45°, to avoid a resultant shear on the plane of symmetry. These conditions are fulfilled by two equal circular fans centred on the corners A and B, each having a radius $h/(2 \sin 45°)$, as in Figure 10.3*a*.

For ease of description, these fans may be divided into 15° segments, though 5° would give better accuracy. In Figure 10.3, the radii AC_1 and BC_1' are extended to the right by curves intersecting orthogonally and at 45° to the central plane, at D. The tangent to the slip-line AC_1D must therefore rotate clockwise by 15° between C_1 and D. To find the position of D it is easier to draw the chord C_1D rather than the true curve. This chord will be parallel to the mean of the tangents at C_1 and D, and is therefore drawn from C_1 at $7\frac{1}{2}°$ to the direction AC_1.

Similarly a chord is drawn from C_2 at $7\frac{1}{2}°$ to the direction AC_2, to cut another drawn from D at $52\frac{1}{2}°$ to the central plane. Their intersection determines the point D_1.

To find the position of D_2, a chord is drawn from D_1 at 15° to the direction of the chord DD_1, since both chords will be inclined at $7\frac{1}{2}°$ to the tangent at D_1.

The points E, E_1 and F are located in the same way. If the platen breadth has the value chosen, $b = 3{\cdot}6h$, F will be the centre-point. The field must always be symmetrical about the central plane, and also about the centre-line of the platens. To complete the field, the intersections are joined by smoothed curves, which will clearly be more accurate if 5° intervals are chosen.

The region bounded by the platen and the slip-line $C_3D_2E_1FE_1''D_2''C_3''$ remains rigid and moves with the platen. It can be regarded as a dead nose on the tool. If the platen breadth is increased to $b = 4{\cdot}56h$, the field can be extended so that D_3 becomes the limit of the plastic zone and the central rigid zone is bounded by $D_3E_2F_1G$, and its reflection. The solution may be extended in the same way, however wide the platens; at intermediate values of the ratio b/h the centre-line will fall on a slip-line at an intermediate angle, but there will always be a rigid zone in the centre of each platen (§10.5.4).

10.5.2 *Construction of the hodograph*

The compatibility of the chosen field with the velocity boundary conditions is verified by constructing the hodograph. This is most easily done by drawing chords to find the intersections, corresponding to the approximate slip-line field, and then joining these by smoothed curves.

The velocity solution is started at the central point F. Each platen is assumed to move towards the central plane with unit velocity, represented by a vertical line of unit length drawn from O in Figure 10.3*b*. As metal from the rigid zone adjacent to

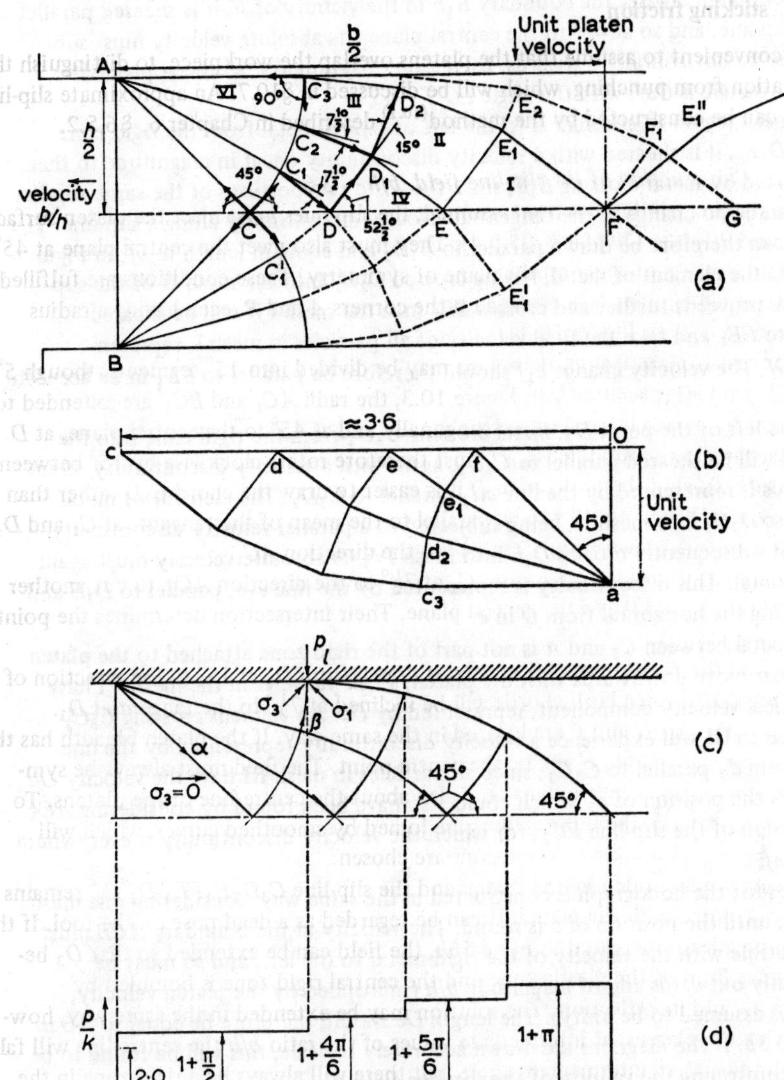

Figure 10.3 An approximate slip-line field solution, using a 15° network, for compression between overlapping platens with sticking friction.

(*a*) The initial stages in construction of the field, for $b = 3{\cdot}6h$.

(*b*) The hodograph construction for quarter of the field.

(*c*) The final smoothed slip-line field.

(*d*) The pressure distribution.

the upper platen crosses the boundary E_1F in the vicinity of F, it is sheared parallel to the slip-line, and so at 45° to the central plane. Its absolute velocity must subsequently be horizontally outwards, by reason of symmetry, so the appropriate velocity triangle is Oaf, with the angle $\angle Oaf = 45°$.

The next region to consider is to the left of E_1. As metal from the rigid zone crosses D_2E_1, it is sheared with a velocity discontinuity equal in magnitude to that experienced by metal crossing E_1F, since these are two portions of the same slip-line. There can be no change in the tangential velocity discontinuity along a slip-line. The line ae_1 can therefore be drawn parallel to E_1D_2 and equal in length to af, in Figure 10.3*b*. As the element of metal, having crossed D_2E_1 into the mesh II of the slip-line field, proceeds further and crosses EE_1 into the mesh I it will be sheared parallel to EE_1 and take the final velocity of all particles in mesh I, represented by Of. The velocity change e_1f should therefore be parallel to EE_1 in an accurate diagram.

To the left of the point D_2, metal crossing C_3D_2 from the rigid zone into the mesh III will be sheared parallel to C_3D_2, again with a velocity discontinuity equal to af. This is represented by the line ad_2 parallel to C_3D_2. The element of metal may then cross D_2D_1 into mesh II, being subjected to a parallel velocity discontinuity d_2e_1. If it subsequently crosses D_1E into mesh IV, its absolute velocity must again be horizontal. This discontinuity is represented by the line e_1e, parallel to D_1E and intersecting the horizontal from O in e.

The metal between C_3 and A is not part of the rigid zone attached to the platen but it must move downwards with the platen, so all particles in the mesh VI have unit vertical velocity-component, represented by Oa. An element crossing C_3C_2 from VI into III will experience a velocity discontinuity represented by the line drawn from d_2 parallel to C_3C_2, since all particles in mesh III have the velocity Od_2. This gives the position of c_3 which could also have been found by considering AC_3 an extension of the slip-line FC_3, for which the velocity discontinuity is everywhere equal to af.

The rest of the hodograph is constructed in the same way, considering one mesh at a time, until the position of c is found. The velocity at the boundary ACB must be compatible with the velocity of the rigid metal to the left, and so must be horizontally outwards and of magnitude b/h (multiplied by the platen velocity, which was assumed to be unity). The length Oc should therefore be equal to 3·6 in Figure 10.3*b*. If the diagrams are drawn accurately to scale, this will be found to be correct, confirming the validity of the slip-line field.

10.5.3 *Stress-determination from the slip-line field.* $b/h = 3{\cdot}6$

When the chosen field has been shown to be compatible with the stress and velocity boundary conditions, the stresses in the plastic zone may be evaluated using the Hencky equations (Chapter 5, §5.4).

The solution is started at the boundary ACB, where the horizontal principal stress σ_3 is equal to zero. Because the vertical stress σ_1 is compressive, σ_3 is algebraically greater than σ_1, and AC must be an α-line. The pressure p_C acting upon it is given by

$$\sigma_3 = -p_C + k = 0; \; p_C = k$$

which is constant along the length, because AC is straight. Between C and C_3 the β-line rotates anticlockwise through 45° ($\pi/4$), so the pressure at C_3 is found from the Hencky equation 5.7b

$$p_{C_3} - 2k\frac{\pi}{4} = p_C = k;\ p_{C_3} = k\left(1 + \frac{\pi}{2}\right)$$

This is constant over the portion AC_3 of the slip-line AF, which is straight and parallel to the platen surface. The vertical stress P acting over AC_3 is thus equal to the hydrostatic pressure

$$P_{AC_3} = p_{C_3} = k\left(1 + \frac{\pi}{2}\right) \qquad (10.17)$$

This may be seen from the Mohr circle, as the direct stress on the plane of maximum shear, or by considering the equilibrium of a cube element of side l parallel to a slip-line

$$-Pl = (\sigma_1 l \cos 45°) \cos 45° + (\sigma_3 l \sin 45°) \sin 45°$$

$$P = -\tfrac{1}{2}(\sigma_1 + \sigma_3) = \tfrac{1}{2}(p + k + p - k)$$

The pressure at D_2 is found from the 15° clockwise rotation ($-\pi/12$) of the α slip-line, using equation 5.7a

$$p_{D_2} + 2k\left(-\frac{\pi}{12}\right) = p_{C_3};\ p_{D_2} = k\left(1 + \frac{4\pi}{6}\right)$$

Similarly,

$$p_{E_1} + 2k\left(-\frac{\pi}{12}\right) = p_{D_2};\ p_{E_1} = k\left(1 + \frac{5\pi}{6}\right)$$

$$p_F + 2k\left(-\frac{\pi}{12}\right) = p_{E_1};\quad p_F = k(1 + \pi) \qquad (10.18)$$

The principal stresses and hence the vertical stress-component acting on the boundary C_3F can thus be found at each point along it. The mean vertical stress acting on the rigid zone boundary must be equal to the mean platen pressure over this region, so the latter can be evaluated, although it is not possible to calculate local pressures in the rigid zone. The pressure distribution is shown in Figure 10.3*d*.

10.5.4 ***General solution for sticking friction with parallel platens.*** ($b > h$)
If the platens are wider than the value $b = 3{\cdot}6h$ chosen in the above section, the plastic zone in contact with the platens will extend inwards to D_3, E_3 and so on. The slip-line field is continued by the same procedure as before and the pressures are

$$p_{D_3} = p_{D_2} + 2k\,.\,\frac{\pi}{12} = k\left(1 + \frac{5\pi}{6}\right) = 2{\cdot}62k$$

$$p_{E_3} = \qquad = k\left(1 + \frac{7\pi}{6}\right) = 4{\cdot}66k$$

etc., as in equations 10.18

The mean pressure over the central rigid zone is again calculated. If b is less than

3·6h but still greater than h, the rigid zone extends across the whole plate. Thus if $b = 1{\cdot}6h$, D lies on the centre-line of the platens. The vertical stress P is found by resolution of the forces acting along AC_1D. The following results are obtained:

b/h	1·0	1·6	3·6	6·6
$P/2k$	1·0	1·1	1·65	2·41

Hill[10.2] has shown that the average pressure over the plate can be represented with sufficient accuracy by the equation

$$\frac{P}{2k} = \frac{3}{4} + \frac{b}{4h} \text{ when } b > h \qquad (10.19)$$

This equation predicts somewhat lower average pressures than equation 10.8 which was derived from stress evaluation, but the difference becomes less significant as the ratio b/h becomes large and friction dominates the result.

If b is less than h, the operation can conveniently be considered as punching or indentation, which will be discussed in §10.7.

10.6 Compression with intermediate friction values

A slip-line field can be built up with the boundary condition that the slip-lines meet the platen surface at angles given by

$$\cos 2\theta = \frac{\mu q'}{k} \qquad (5.10)$$

assuming a starting value of q' first. The field resembles the solution found for sticking friction in Figure 10.3 but the hodograph is found not to conform to the velocity boundary-conditions except when b/h is integral. A solution has been obtained in a different manner by Prandtl[10.3].

10.7 Slip-line field solutions for plane-strain indentation or punching

If the platen breadth is small in comparison with the thickness of the workpiece, the forging pattern of the deformation is altered, and the process can be considered separately as one of indentation. Again the slip-line field is valid for infinitesimal displacements, but a full progressive solution has not been obtained.

10.7.1 *Indentation with a flat punch.* $h > b$

The slip-line field for a flat punch, shown in Figure 10.4, closely resembles the field of Figure 6.5 for strip-drawing, except that the tool angle α is zero.

It may be seen that the same solution holds if the central plane is replaced by a frictionless platen and only one indenter is used. If the strip thickness h exceeds 8·7b, the plastic zone does not reach the central plane, and the field approaches that for indentation of a semi-infinite block, which is discussed in detail in Chapter 5, §5.6.5. The indentation pressure for this limiting condition is given by equation 5.15:

$$P = 2k\left(1 + \frac{\pi}{2}\right) = 2k \times 2{\cdot}57 \qquad (10.20)$$

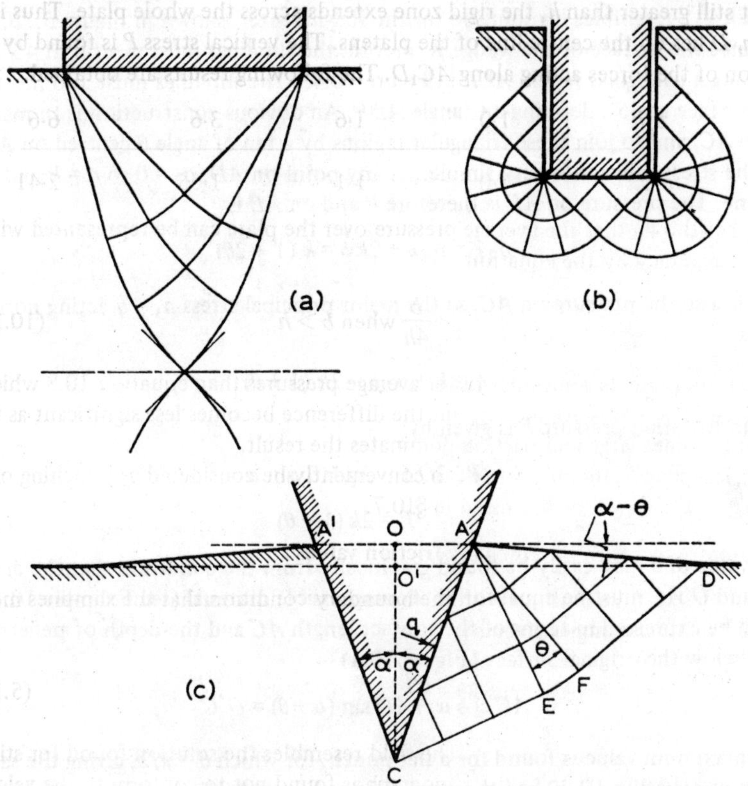

Figure 10.4 Slip-line field for indentation of a wide strip.

(*a*) Two flat punches, $b < h < 8{\cdot}7b$.
(*b*) Flat punch $h \sim \infty$. Deep groove.
(*c*) Smooth wedge. $h \sim \infty$.

10.7.2 *Deep penetration by a flat punch*

If the flat-faced punch, with suitable clearance at the sides, is envisaged at the bottom of a deep groove, as in Figure 10.4*b*, the stress boundary conditions will be the same as for initial penetration. A possible slip-line field is thus of the type shown, which closely resembles Figure 5.7*a*, except that the fan regions centred on the punch corners are developed through 180°. The pressure for deformation at the foot of a deep groove is therefore greater than that for initial deformation (equation 10.20) and is given by

$$P = 2k\,(1 + \pi) = 4{\cdot}14 \times 2k \qquad (10.21)$$

As mentioned in Chapter 5, this solution is appropriate to a rough indenter, and the area *ABC* can be considered as a dead zone or nose on the punch.

10.7.3 *Wedge-indentation of a semi-infinite block*

When a smooth wedge-shaped indenter enters a semi-infinite block, the metal is displaced upwards in the immediate vicinity. The slip-line field suggested and verified by Hill[10.2] resembles his solution for the flat indenter, shown in Figure 5.7*b*.

Since the surface of the wedge is smooth, the slip-lines must meet it at 45°, defining a triangular region *ACE* in Figure 10.4*c*. The slip-lines must also meet the free surface at 45°, defining a triangle *ADF*. An obvious construction is to make $AD = AC$, and to join these triangular regions by a fan of angle θ centred on A.

The stress solution is very simple. At any point on *AD*, $\sigma_3 = 0$ so $p = k$ and *AD* is a β-line. The pressure on *AF* is therefore *k* and on *AE* is

$$p_{AE} = p_{AF} + 2k\phi = k\,(1 + 2\theta)$$

This is also the pressure on *AC*, so the major principal stress $\sigma_1 = q$ acting normal to *AC* is

$$q = k + p_{AE} = 2k\,(1 + \theta)$$

The indentation pressure *P* is given by

$$P \,.\, 2OA = 2qAC \,.\, \sin\alpha$$

$$P = 2k\,(1 + \theta) \tag{10.22}$$

The value of θ may easily be found graphically from the condition that the areas $O'AD$ and $O'AC$ must be equal for an incompressible material (see Example 10.7), or it can be expressed in terms of the contact length *AC* and the depth of penetration $O'C$ below the original surface (Figure 10.4)

$$AC\cos\alpha - AD\sin(\alpha - \theta) = O'C \tag{10.23}$$

The maximum value is found for a flat punch, for which $\theta = \pi/2$, giving the same result as equation 10.20 and the same field as Figure 5.7*b*.

If there is friction at the wedge face, the slip-line field will have the same general shape, but the slip-lines will meet the wedge at the appropriate angle, given by equation 5.10.

Compression of a large wedge-shaped workpiece between flat platens[10.2] gives the same pressure

$$P = 2k\,(1 + \theta)$$

provided that the wedge angle α exceeds $\tan^{-1}\frac{1}{2}$, or about 27°. At this angle θ becomes equal to zero and $P = 2k$. The mode of deformation for steeper wedges is not known. On the other hand, as α approaches 180°, θ approaches 90°, and *P* approaches the limit $2k \times 2{\cdot}57$.

10.7.4 *Wedge-indentation of a finite strip*

If the strip thickness is only a small multiple of the breadth of the indentation, the slip-line field for indentation by two opposed wedge-shaped tools can be constructed in a manner similar to that described for strip-drawing (Figure 6.5). The result is the same for a single smooth-wedge indenter, the horizontal plane of symmetry being replaced by a flat frictionless anvil. The double-indentation system has been discussed by Johnson[10.4] in connection with wire-cutting by means of pliers. He has also considered compression by three symmetrically disposed platens, which represents the conditions of forging in a 120° V-anvil. There is then a central zone of undeformed metal, which is subjected to compression, in contrast to the outward movement of metal on the axis of a block subjected to two-platen compression (Figure 10.3). It has long been known that sounder forgings are produced by the V-anvil technique.

10.8 Piercing

Deep penetration by a flat punch has already been discussed in §10.7.2. It was shown that a wedge-shaped nose of dead metal tends to form, at least on a rough punch, and that the pressure can rise to about four times the normal yield stress.

When the billet is constrained by a container of width B, comparable with the breadth b of the punch, the conditions closely resemble those of direct extrusion in the presence of a dead-metal zone. The slip-line field for the latter process (Figure 8.4*d*) would be unchanged if the die were replaced by undeformed billet metal, and the rigid extruded bar were replaced by a flat-faced punch moving in the opposite direction, as in Figure 10.5.

If there is slipping friction at the container walls, the angles at which the slip-lines meet the walls must be adjusted appropriately.

Proceeding in the same way as in Chapter 8, the punch pressure will be found to decrease as the ratio of b/B decreases, as far as $b = \frac{1}{2}B$. With this ratio, the extrusion pressure is given by equation 5.23

$$\frac{P}{2k} = \frac{1}{2}\left(1 + \frac{\pi}{2}\right)$$

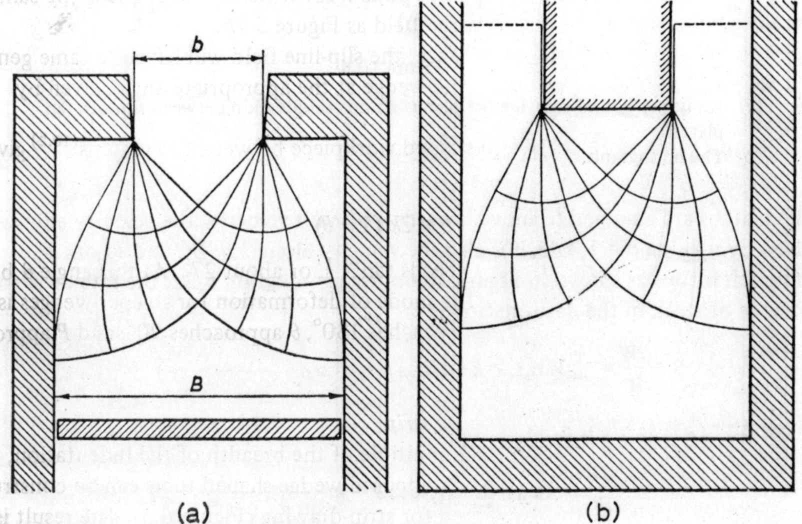

Figure 10.5 Slip-line fields for

(*a*) extrusion with a dead zone.

(*b*) piercing of a billet in a container, using a rough flat-faced punch.

The corresponding punch pressure will be twice this value

$$\frac{P}{2k} = \left(1 + \frac{\pi}{2}\right) \tag{10.24}$$

As the ratio b/B is decreased further the container will exert less influence and the pressure will begin to rise to the limiting value for deep indentation of a semi-infinite block given by equation 10.21.

$$\frac{P}{2k} = (1 + \pi)$$

10.9 Upper-bound solutions for compression with smooth platens

Bearing in mind the slip-line field for compression of thin wide strip between smooth flat platens (Figure 5.5), a simple upper-bound solution may be drawn as in Figure 10.6,

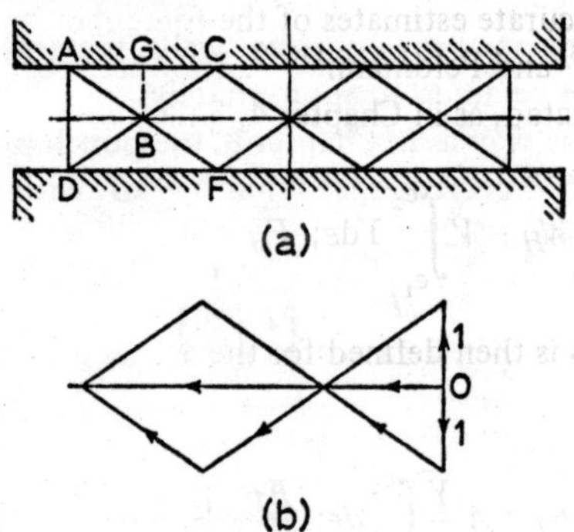

Figure 10.6

(*a*) An upper bound solution for plane–strain compression between smooth flat platens.
(*b*) The hodograph.

If the platens are assumed to move inwards with unit velocity, the velocity along AB is given by $u_{AB} \sin\theta = 1$, which is also the velocity along FB, BC and so on. The hodograph is thus as shown in Figure 10.6*b*. Using equation 5.32, the rate of performance of work in the deformation is

$$\frac{dW}{dt} = \sum \text{k.u.s.} = k\,(ABu_{AB} + BCu_{BC} - - - - -)$$

The length s_{AB} is $s_{AG}/\cos\theta$, so

$$\frac{dW}{dt} = ku_{AB} \sum s_{AG}/\cos\theta = k\,\frac{1}{\sin\theta} \cdot \frac{b}{\cos\theta}$$

The rate of performance of work by the applied pressure P is

$$\frac{dW}{dt} = Pb \,.\, 1$$

Equating these values,

$$Pb = \frac{kb}{\sin\theta\cos\theta} \text{ or } \frac{P}{2k} = \frac{1}{\sin 2\theta} \tag{10.25}$$

The pressure passes through minimum values, equal to $2k$, when $\sin 2\theta = 1$, or when $\theta = \pi/4$, which occurs when the ratio b/h is integral, as in the slip-line field solution (Figure 5.6).

The upper-bound solution for plane–strain indentation of thick metal with a single flat punch has been described in Chapter 5, §5.10.4. Indentation by a flat circular punch has been discussed by Levin(10.5).

10.10 A semi-empirical method for force calculations in extrusion-forging

Many complex components such as flanged tubes and shells can be made by a combination of forging with backward and forward extrusion. Some shapes of this type have already been described in Chapter 8, §8.9.2 under the heading of impact extrusion. Reasonably accurate estimates of the force required can be obtained by a method due to Siebel(10.6) and Feldmann(10.7). The force required for homogeneous deformation is first estimated, as in Chapter 4, §4.3,

$$W_H = V\int_{\epsilon_1}^{\epsilon_2} Y d\epsilon;\ F_H = A\overline{Y}\int_{\epsilon_1}^{\epsilon_2} d\epsilon \tag{4.7}$$

An equivalent yield stress is then defined for the actual process, with an efficiency factor η:

$$F_T = A\,\frac{\overline{Y}}{\eta}\int_{\epsilon_1}^{\epsilon_2} d\epsilon;\ \frac{F_T}{A} = \frac{\overline{Y}}{\eta}\,.\ln\frac{1}{1-r} \tag{10.26}$$

It will be seen that this equation is of the same form as the equations 6.38, 6.46 and 6.50 discussed in Chapter 6 in connection with drawing strip and rod. Nomograms have been prepared empirically for several basic geometrical types of forging, showing the values of η for different reductions of area in a variety of metals(10.8).

In some instances the efficiency factor has been calculated explicitly using a term involving $\mu \cot \alpha$ to allow for friction and a term dependent upon the geometry to allow for redundant work. Thus for forward extrusion of solid bodies, Feldmann suggests

$$\frac{p}{\overline{Y}} = \left[1 + \frac{\mu}{\alpha} + \frac{2}{3}\frac{\alpha}{\epsilon}\right].\ln\frac{1}{1-r} \tag{10.27}$$

For forward extrusion of hollow bodies the friction is doubled and the redundant work is reduced, as in tube drawing.

$$\frac{p}{\overline{Y}} = \left[1 + \frac{2\mu}{\alpha} + \frac{1}{2}\frac{\alpha}{\epsilon}\right].\ln\frac{1}{1-r} \tag{10.28}$$

Various combinations of forging and extrusion can be handled in this way(10.8), and the method is useful in practical cold-extrusion operations.

10.11 Application of theoretical analysis to automatic forging[10.9]

The British Iron and Steel Research Association have developed an automatic cogging forge based on computed optimum schedules. The prototype used a 200-ton (tonne) press to forge, for example a 20 mm (8 in.) square steel ingot into three adjacent sections, respectively 6 in., 4 in. and 3 in. square.

The cogging operation proceeds by steps along the workpiece, until the whole length has been compressed. Another pass is then taken, usually after rotating the workpiece by 90° about its axis. The compressive load for a single cogging-step was determined from two equations based on slip-line field analysis. For thin stock an equation similar to equation 10.19 was used; for thick stock a straight-line approximation to Figure 5.6 gave a constraint factor of the type

$$\frac{p}{2k} = 0{\cdot}8 + 0{\cdot}2\frac{h}{b} \tag{10.29}$$

Various combinations of squeeze ratio h_2/h_1, bite ratio b/w and shape factor h/w were programmed on a computer to find the minimum number of passes which could be used to produce the desired final shape, taking into account the lateral spread which was known as an empirical function of the bite. Various other factors, such as the increasing yield stress as the billet cooled, could also be included. From over 18,000 arbitrarily-chosen combinations of the variables, the 30 schedules requiring the least number of steps were printed out by the computer. Even within this selected group, the number of steps varied by a factor two, showing how difficult it would be to select an optimum schedule by trial and error.

The cogging steps and passes optimised by these calculations were set up for automatic action on an electrically-controlled hydraulic press. In the example quoted, the programme was completed to an accuracy better than average in 4½ minutes, which was about half the time that a skilled forging crew could achieve.

10.12 Extrusion – forging

Both hot and cold forging processes often involve flow of the metal in more than one major direction. Until recently, this type of problem was considered too complex for accurate analysis, though semi-empirical formulae were developed. (§10.10)

Chapter 13 describes some numerical methods that are now available for solution of complicated flow problems. As a basic example of compound flow, it is convenient to consider the well-defined geometrical configuration presented by a thick disc that is forged along its axis, between a flat platen and a die with a central orifice. This represents a common group of practical forging shapes, such as a wheel with a protruding solid hub. It has been studied by various methods, which can therefore be compared directly.

10.12.1 *Slip-line field solution for plane strain*[10.10]

It is convenient to consider first an accurate plane-strain solution and then to extend the results to axial symmetry, making approximate allowance for strain-hardening afterwards. Sticking friction is assumed.

Figure 10.7*a* shows a simple slip-line field for forging as in Figure 10.3. This

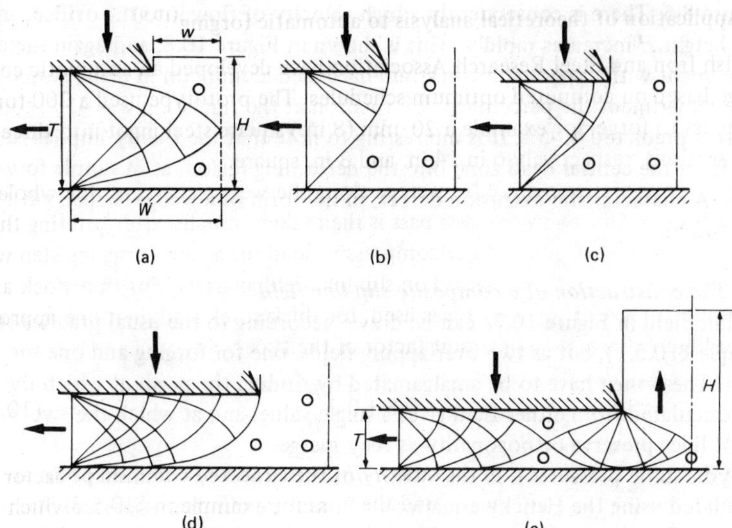

Figure 10.7 Slip-line fields for extrusion-forging with an orifice in one die.
(*a*) Stage I (*b*) Stage II (rigid core) (*c*), (*d*) Later Stage II
(*e*) Stage III forging and extrusion

applies to the present problem when the thickness T of the billet is large. As the die descends and deformation proceeds according to this mode, metal flows from the dead zones, ○, into the deformation zone. The total height H thus decreases.

As thickness decreases, the length of contact $(W-w)$ on the die increases, and a condition

$$(W-w) = T/\sqrt{2} \tag{10.30}$$

will be reached. It is then possible to draw a slip-line field as shown in Figure 10.7*b*. Further deformation in this mode, as in Figure 10.7*c*, requires less force than a continuation of deformation according to Figure 10.7*a*, as found by applying the Hencky equations. A transition from Stage I to Stage II is thus predicted, resulting in no further deformation of the central core. This is observed experimentally with strain-hardened aluminium, as shown in Figure 10.8, and also by metallographic examination. The percentage deformation at which this occurs in plane strain agrees very well with the theory, respectively 18% and 19·5% in a particular instance.

Further deformation continues in this way as shown in Figure 10.7*d*. The force required rises rapidly, as in simple forging with sticking friction, as the $(W-w)/T$ ratio increases. Eventually a condition will be reached at which a field of the type shown in Figure 10.7*e* requires less force than one of the rigid-core type. The construction of such a field is described in §10.12.2, but the load calculations follow the usual pattern, using the Hencky equations as in §10.5.3, starting from the two known free surfaces. A second transition is thus predicted, with some metal being forged and flowing outwards while the inner zone follows an extrusion type of field, as in Figure 8.5. There is consequently a high velocity of flow into the orifice, and the total height H increases rapidly. This is shown in Figure 10.8, and again there is close agreement with experiments on strain-hardened aluminium in plane strain. In a particular experiment the percentage deformation at the transition was 41%, compared with a predicted 43·5%. It is interesting to note that the theory implies flow of metal from the central dead zone into the deforming region, as in simple forward extrusion. A characteristic extrusion defect, in the form of a wedge-shaped cavity, is in fact found.

10.12.2 *The construction of a composite slip-line field*

The slip-line field in Figure 10.7*e* can be drawn according to the usual practice (see for example §10.5.1), but as two overlapping fields, one for forging and one for extrusion. These then have to be amalgamated by finding the point at which the pressure calculated from either field gives a single value, and at which the two families of lines preserve orthogonality as they merge.

The hydrostatic pressures p in the vicinity of the expected common point are first calculated using the Hencky equations 5.7, as for example in §10.5.3. Starting at the free surface on the left hand side, where $\sigma_3 = 0$ and $p = k$, a path is followed along the lines of the forging field, determining the value of p at each intersection of the network. Similarly p is calculated from the extrusion field, starting at the free surface in the centre. By interpolating slip-lines as necessary near the confluence, to improve the accuracy, a locus of points of equal pressure can thus be found, which must of course include the actual common point. This is located by plotting a second locus, of points at which the slopes of the lines from the two fields are equal. The two fields therefore meet where the isobaric and isoclinic loci intersect. Any slip lines outside the intersecting boundaries are no longer relevant, and are replaced by the two dead zones.

10.12.3 *Extension to axial symmetry*

The observed height changes in axially-symmetric forging of a disc follow exactly the same pattern as in plane–strain. Figure 10.8 is taken from axisymmetric specimens and clearly shows the three stages. In addition, the grain distortion seen in metallographic sections is very similar to that seen in plane-strain specimens, especially those of the Stage II.

Other evidence also suggests that the plane-strain slip-line field can be applied, with satisfactory accuracy for most purposes, to a diametral section of an axi-symmetric workpiece. For calculation of the hydrostatic pressures, the Hencky equations can be applied in their usual form, though this is not strictly a valid procedure.

As shown by Hill[10.2] (p. 279), the Geiringer equations for axial symmetry become

$$\begin{aligned} du - v\,d\phi + (u + v\cot\phi)\frac{dr}{2r} &= 0 \quad \text{along an } \alpha \text{ line} \\ dv + u\,d\phi + (v + u\tan\phi)\frac{dr}{2r} &= 0 \quad \text{along a } \beta \text{ line} \end{aligned} \tag{10.31}$$

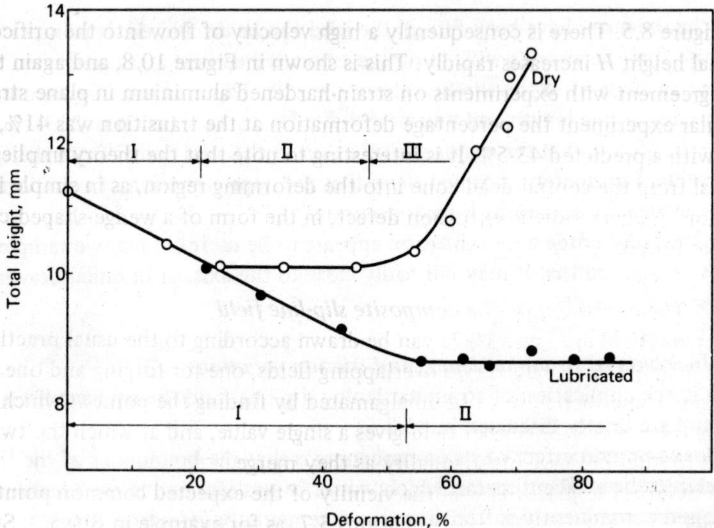

Figure 10.8 Experimental height measurements[10.10].

These are in accordance with the Levy-Mises equations (13.11) and it can further be shown, by introducing the von Mises yield criterion, that the stress equations are

$$\begin{aligned} dp + 2\tau\, d\phi + (\sigma + p - \tau \cot \phi)\, \frac{dr}{r} &= \frac{\partial \tau}{\partial s_\beta}\, ds_\alpha \text{ along an } \alpha \text{ line} \\ dp + 2\tau\, d\phi + (\sigma + p - \tau \tan \phi)\, \frac{dr}{r} &= \frac{\partial \tau}{\partial s_\alpha}\, ds_\beta \text{ along a } \beta \text{ line} \end{aligned} \quad (10.32)$$

If the shear stress τ is a constant, as in the Tresca hypothesis, these become

$$\begin{aligned} dp + 2\tau\, d\phi + (\sigma + p - \tau \cot \phi)\, \frac{dr}{r} &= 0 \text{ along an } \alpha \text{ line} \\ dp - 2\tau\, d\phi + (\sigma + p - \tau \tan \phi)\, \frac{dr}{r} &= 0 \text{ along a } \beta \text{ line} \end{aligned} \quad (10.33)$$

Further simplification can be obtained by introducing the Haar-Karman hypothesis $\left(\sigma + p = \pm \frac{Y}{2}\right)$ but this has been rejected by many workers on the grounds that it is arbitrary and does not allow a rigorous velocity field to be determined.

However, if r is large the contribution from the terms in $1/r$ is small, and to this approximation, equations 10.33 revert to the plane-strain Hencky equations (§5.4) and can be applied as such. The physical justification for this is that an element of a cylindrical body at a large radius can be regarded as a thin-walled tube. If this is imagined to be drawn in analogy to a close pass (§7.2.1) the circumferential strain will be negligibly small, and conditions approximate closely to plane-strain. If therefore the axisymmetric deformation can be compared to close-pass drawing rather than sinking, of a thin tube, the approximation is acceptable.

Once the pressures have been found, the load can be calculated. It is important to recognise that the stresses will act on the truncated conical areas generated by rotation of an element of the slip-line about the axis. This area must be used instead of the product of line length and unit perpendicular distance (as on p. 263). It will be seen, as in §5.8.4, that the major contribution to the load comes from the forces at large radius. Fortunately, it is just there that the approximation of equations 10.33 to the Hencky equations is most accurate.

This admittedly crude approximation appears to be useful in many examples, but it must be recognised that it may fail badly close to the axis, or in radial drawing conditions.

10.12.4 *Inclusion of strain-hardening and friction variations*

Advances in the application of strain-hardening slip-line field theory have been made recently and are briefly discussed in §5.8.5.

The most common effect of strain-hardening is that the boundaries of the field are spread into the adjacent metal, which would be undeformed in the idealised slip-line field, and consequently softer than the incrementally-deformed zones.

This has a special significance in the present example, because the whole central core is undeformed in Stage II, and the onset of stage III is consequently favoured. The transition therefore occurs at a lower deformation with annealed aluminium than with fully-hardened aluminium, as found experimentally.

Other slip-line fields can be drawn for various frictional conditions on the upper and lower surfaces. The most obvious effect is that high friction on these surfaces increases the force and encourages an earlier transition from Stage II to Stage III. Good lubrication delays the transition (Figure 10.8), and with zero friction the pressure on the die would remain close to $2k$, so Stage III would never occur.

10.12.5 *Upper-bound solutions using rigid-triangle velocity fields in plane strain*[10.11]

The workpiece is assumed to be divided into unit rectangular deforming regions, each being subdivided into rigid triangular zones (or sometimes double triangles). The size of each region is determined by the imposed geometry and by the condition that the calculated deforming force should be as low as possible, as in all upper-bound solutions (see §5.10). The subdivision into rigid triangles bounded by velocity discontinuities is again selected to minimise the force required. The method is described in §13.5, recognising that the field must be kinematically admissible. It is assumed that work is done against friction on the tool surfaces, by shearing on the boundaries between adjacent zones, and by shear along the velocity discontinuity lines. The calculations follow the principles given in §5.10, and equations such as 5.32 and 5.33 are built up.

It would of course be possible to obtain an upper-bound solution simply by dividing the workpiece into a few rigid triangles in an arbitrary way and applying equation 5.26 directly, but the optimising technique gives more detailed information and a lower, more accurate, load prediction. Figure 13.17 shows a suitable velocity field.

After minimising the rate of energy dissipation by suitable choice of zones and discontinuities, the total energy-dissipation rate is equated to the rate of performance of work by the postulated external load, which is thus explicitly determined. A series of loads can be predicted for a range of ratios of the thickness T to the width W.

It is found that the most suitable type of field, namely that giving the lowest rate of energy dissipation, depends on this ratio and on the friction conditions. The transitions from one type to another can be predicted using this criterion. For example, the fields shown in Figures 13.17*a* and *b* are the best for their respective geometric ratios. The similarity to the transitions predicted by slip-line fields is apparent.

Even the phenomenon of lift off, or, as it was called above, the extrusion defect, can be predicted because the lowest energy dissipation rate with large values of W/T calls for a vertical velocity away from the platen at the centre line.

The results described here refer to plane-strain extrusion-forging, but the same method can be applied to axi-symmetric conditions[10.12].

10.12.6 *Upper-bound solutions using deforming elements in axial symmetry*

In this type of solution, the workpiece is again imagined to be divided into suitable simple elements or zones, but these are not subdivided into rigid blocks[10.13]. Expressions are found for the major velocity components in each zone, and along the boundaries. The total rate of energy dissipation is then found from the internal energy dissipation rate for each zone, the frictional energy rate at the tool faces, and the dissipation along velocity discontinuities.

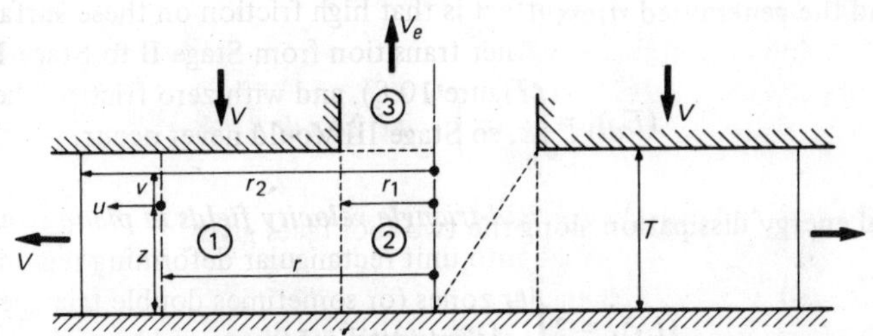

Figure 10.9 A field of deforming elements for forging of a boss on a disc[10.14].

Figure 10.9 shows a division of a disc into three suitable zones[10.14]. The total volume of the metal must remain constant and the velocity component across the boundaries must be continuous, though tangential velocity discontinuities will occur. Using these conditions, the radial velocity u at radius r and the axial velocity v at height z can be found, for example in zone 1.

The displacement of material in unit time due to the descent of the top die at velocity V is

$$\pi(r_2^2 - r_1^2)\, V$$

The loss from zone 1 into zone 2 is exactly equal to the loss from 2 into 3, because the metal is incompressible. This is given by the product of the exit velocity V_e and area:

$$\pi r_1^2 V_e$$

The flow across a boundary in zone 1 at radius r is

$$2\pi rTu$$

so

$$(u)_1 = \frac{\pi(r^2V - r_1^{\,2}V - r_1^{\,2}V_e)}{2\pi rT} = \frac{V}{2Tr}\left\{r^2 - r_1^{\,2}\left(1 + \frac{V_e}{V}\right)\right\} \tag{10.34}$$

The vertical velocity v decreases with height z, from $-V$ at $z = T$ to zero.

$$(v)_1 = -V\frac{z}{T} \tag{10.35}$$

These values allow the distribution of strain rate to be determined in this zone as a function of r and z, using the equations

$$\dot{\epsilon}_r = \frac{\partial u}{\partial r};\quad \dot{\epsilon}_\theta = \frac{u}{r};\quad \dot{\epsilon}_z = \frac{\partial v}{\partial z};\quad \gamma_{rz} = \frac{\partial u}{\partial z} + \frac{\partial v}{\partial r}$$

An effective or generalised strain-rate is next defined, as in equation 13.57 (cf. equations 13.30 and 13.17):

$$\dot{\overline{\epsilon}} = \left\{\frac{2}{3}\left(\dot{\epsilon}_r^{\,2} + \dot{\epsilon}_\theta^{\,2} + \dot{\epsilon}_z^{\,2} + \frac{\dot{\gamma}_{rz}^{\,2}}{2}\right)\right\}^{\frac{1}{2}}$$

The energy dissipation rate due to the deformation within zone 1 is then given in terms of $\dot{\overline{\epsilon}}$ and the generalised stress $\overline{\sigma}$

$$(\dot{E}_d)_1 = \int_0^T \mathrm{d}z \int_{r_1}^{r_2} \overline{\sigma}\,\dot{\overline{\epsilon}}\, 2\pi r\,\mathrm{d}r \tag{10.36}$$

The frictional energy dissipation along the two tool faces gives

$$(\dot{E}_f)_1 = 2\int_{r_1}^{r_2} (mk)\,|\Delta u|\,2\pi r\,\mathrm{d}r \tag{10.37}$$

where m is the friction factor ($\tau = mk$).

Along the discontinuity or boundary 1–2

$$(\dot{E}_b)_1 = \int_0^T k\,|\Delta v|\,2\pi r_1\,\mathrm{d}z \tag{10.38}$$

The total energy dissipation rate in this zone is

$$(\dot{E}_T) = (\dot{E}_d)_1 + (\dot{E}_f)_1 + (\dot{E}_b)_1 \tag{10.39}$$

The expressions for zone 2 are added to this, and the whole solution becomes algebraically complex, obviously requiring a computer, but no further principles are involved.

Reference to Figure 10.7*e*, or to other relevant background knowledge suggests that when the diameter $(2r_2)$ is large, there will be a flow divide in zone 1. The neutral radius is found from the condition that the radial velocity component is there zero.

This method too can give good load predictions, implicit in the calculations of the transitions, and detailed agreement with the experimental shape changes, shown in Figure 10.10.

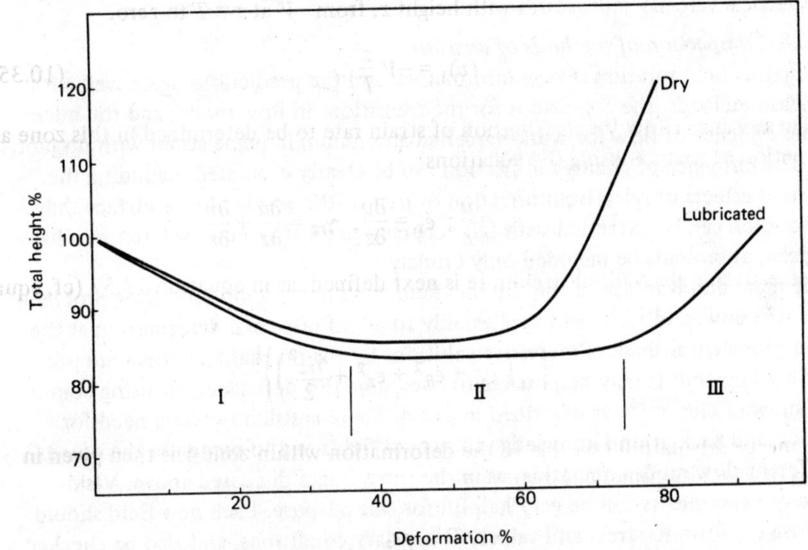

Figure 10.10 Total height as a function of flange reduction[10.14].

10.12.7 *Stress analysis in unit deformation zones*

Another method of calculation is based again on division of the workpiece into deforming zones, but the stresses are calculated by analysis[10.15] similar to that of Chapter 4.

It is assumed that the flow stress $\overline{\sigma}_i$ is constant within a given zone, which may be as narrow as desired, and that the appropriate frictional stress is added for each sequential zone in a form $\tau = f_i\,\overline{\sigma}_i$. Thus in a width Δb of parallel region bounded by two equally lubricated tool faces the stress σ_z normal to the platens increases according to

$$(\sigma_z)_{i+1} = (\sigma_z)_i + 2f\,\overline{\sigma}_i \cdot \frac{\Delta b}{h}$$

If the tool face is inclined, the expression includes the resolved components.

The method is essentially a summation of finite contributions replacing the integration used in Chapter 4 or §10.2. Curved surfaces can be included by approximation, and allowance can be made for local variations in flow stress $\overline{\sigma}$ due to local strain $\overline{\epsilon}$, strain rate $\dot{\overline{\epsilon}}$, and temperature. These can all be written into the computer program, so the technique is much more flexible than the solution of the differential equations previously used. It is particularly suited to complicated shapes, including the flash produced in closed-die forging.

The results are quite reliable for load determination if a suitable friction factor is selected, but at present they are less satisfactory for prediction of dimensions. The most appropriate flow model is again selected on the basis of minimum energy dissipation, assessed as minimum load to cause deformation. Constancy of volume and continuity of velocity across boundaries must be observed.

10.12.8 *Comparison of methods of analysis*

The slip-line field solution is very informative, and the predictions agree well with the measured loads, the conditions for the transitions in flow mode, and the micrographic evidence of flow for work-hardened aluminium in plane strain with no lubricant. The influence of changes in friction can be clearly evaluated, including the important effects of selective lubrication of part or the whole of one surface only. This solution can be extended with fair accuracy to axial symmetry but strain-hardening can, at present, be included only crudely.

The main disadvantage of the slip-line field analysis of compound-flow problems is that it is slow, and each field applies only to an infinitesimal deformation at the chosen geometrical shape. Progressive solutions by the manual technique are prohibitively long, but it may be possible to overcome this deficiency by using computer-drawn fields[10.16], as described in §13.4. There is still however a need for intuition and background knowledge to suggest fields and to recognise the possibility of different flow modes operating, as in the three stages discussed above. Visioplasticity experiments can be very helpful for this purpose. Each new field should of course conform to stress and velocity-boundary conditions, and also be checked to ensure that the yield stress is not exceeded in the postulated elastic regions, and that the rate of plastic working is everywhere positive.

The two upper-bound methods are essentially similar, though some familiarity with flow patterns is useful in making suitable subdivisions of the workpiece. The deforming-element method is the simpler of the two and correspondingly gives less detailed information. It is very useful for predicting loads and, to a lower accuracy, deformed shape. Some allowance for changes in friction and strain-hardening can be made, and progressive solutions present no problem.

It is possible to include curved boundaries, which considerably extends the scope of this method. The expressions become complex even for a single solution and experience in computing is needed.

The rigid-triangle method aims at greater precision by optimising the block shape and size and also optimising the subdivision. This requires more experience, to avoid very lengthy calculation, and is less flexible in its applications than the deforming-element method, but in skilled hands it can give very accurate predictions of load and shape, even including features such as the extrusion defect.

Neither of the upper-bound techniques can give accurate predictions of the detailed strain distribution or the shape of the dead zones, but both can be applied more rapidly than the slip-line fields. They are particularly useful for progressive solutions.

The stress analysis appears at first sight to be unnecessary, because the equations are at least as complicated to set up as for the upper-bound solutions, and the analysis does not pay attention to the details of flow. A visioplasticity study of a very similar problem[10.17] concluded however that the force predictions from stress analysis were more accurate than the upper bound or approximate slip-line field results. Shape change was not fully examined in this comparison. An advantage in principle of the stress analysis (or slab method) is that the elements considered in the equilibrium equations can be very narrow, so that it should be possible to include steep temperature gradients. These may be important in the late stages of forging when thin flash is formed.

For predicting homogeneity and shape changes, the slip-line fields appear to be most accurate.

10.13 Lubrication in practical hot forging

Open-die forging is often completed without any lubricant, especially for large ingots. The dies give longer service if they are cooled, and for more complicated shapes it is common to use a water spray containing graphite. In closed-die forging there is more need for a separating agent than for lubrication in the usual sense. Sawdust has traditionally been used, but graphite is the most common die lubricant now.

For warm working, which is increasing in popularity, there has been little specific development, and the lubricants used are adapted from hot-forging practice. Electrodeposition of lubricant films from molten salt baths has been suggested for warm working.

The deformation pattern of a billet and the filling of die cavities can be greatly influenced by the lubrication and the distribution of lubricant, but the methods used are almost entirely empirical. Graphited grease is applied locally by swabs where required.

10.13.1 *Hot forging of steels*

Colloidal graphite in water is the main lubricant, though sawdust is still widely used for coarse work and melting salts have some applications. Graphite may be applied in concentrations of about 20% but it is important to avoid high ash content, especially silica, which is abrasive. For stainless steels it is common to use graphite-in-oil suspensions, but all the carbon must then be removed to avoid the formation of chromium carbides on subsequent heating.

10.13.2 *Copper alloys*

Because copper oxide is soft at high temperatures, there is no serious lubrication problem. The die surface should be maintained in good condition to allow smooth flow, which is further facilitated by a graphite-in-oil lubricant.

Water-base graphite dispersions can also be used for brasses and bronzes.

10.13.3 *Aluminium alloys*

Aluminium, when exposed by fracture of the hard oxide, adheres readily to dies. Graphite in water or oil suspension can be used at billet temperatures 200–500°C, but advantage has been claimed for using a dry cadmium oxide and graphite mixture.

Extreme pressure additives are of little value, but aluminium stearate can be used at billet temperatures up to 400°C.

The lubrication can be improved by slightly roughening the surface to retain more lubricant, for example by immersion in hot 10% NaOH solution.

10.13.4 *Other alloys*

Titanium, as always, has a high tendency to adhere to tools. Glass gives good protection, but tends to freeze in die corners. Colloidal graphite in oil or water is commonly used, sometimes with resin bonding. Residual salt from salt-bath heating is effective but again tends to freeze in die corners. Silver plating is used in some industrial forging of titanium alloys.

Nickel can be lubricated with glass or tar, but graphite in oil is commonly used. Sulphur should be avoided. Tungsten and Molybdenum are self-lubricating if oxidised, but it is usually desirable to prevent oxidation. Glasses or graphite give good lubrication and protection.

10.14 Lubrication in practical cold forging

Cold forging is sometimes synonymous with cold extrusion, though it may be preferable to confine the latter term to processes with high extrusion ratios. The lubrication practices are the same as those described in §8.11 but the demands on the lubricant are less when the extension of the surface is smaller.

10.15 Recent developments in forging

10.15.1 *Theoretical contributions*

Various theoretical methods are now available for analysis of forces and deformation in forging. In open-die forging, *visioplasticity*[10.17] and *computer drawing*[10.16] of slip-line fields have increased the detailed knowledge of flow and distortion under various frictional conditions. It is not necessary to assume that either the coefficient of friction or the shear stress is constant over the whole of a tool face, and indeed it seems likely that the interfacial shear stress often increases towards the outer edges. This has little effect on the overall force required, but can influence the edge shape. In closed-die forging it can affect die filling.

By suitable approximations, slip-line field theory can be extended to axi-symmetric conditions, which are much more common in forging than plane–strain deformation[10.10]. Slip-line fields are also valuable in providing detailed analysis of flow and distortion in extrusion-forging, where quite different flow patterns may arise at different stages of the operation.

For approximate solutions to more complex problems, upper-bound methods are very suitable. Sufficiently accurate force calculations can be made without detailed knowledge of the flow and without model experiments. Two main methods are used, one assuming rigid zones separated by shear lines[10.11, 10.12], and the other dividing the workpiece into small suitably-shaped deforming blocks[10.13]. Both can be applied in axial symmetry, and the latter is suitable for problems involving curved tools.

An important contribution to the forging load with closed dies comes from the flash gutter, since the flow of metal there involves very high strain ratios and the force is superimposed on the major deformation force. Allowance can be made by the upper-bound method and also by a computer simulation based on stress analysis[10.15].

Finite-element analysis[10.18] applied to forging allows detailed calculations to be made of stress, strain and strain rate, and can also predict temperature distributions. An important aspect of the detailed numerical methods is that the material properties can be included, in the form of the stress/strain relationship. However, because the external shape changes considerably during forging, many separate calculations of incremental deformations are needed to represent a large deformation. This can soon outrun the capacity of even a large computer. A matrix method[10.19] reduces the computation and storage necessary. This also allows anistropic effects to be included.

These numerical methods (see Chapter 13) are very powerful and can be applied to complicated problems of cold heading or extrusion-forging, but they are at present mainly for academic use.

10.15.2 *Improvements in conventional processes*

Cold forging[10.7] is competitive with machining for large-quantity production, but it is unlikely to be favourable for small batches, or for any products that require final machining or grinding. There is consequently a strong interest in cold forming to high dimensional and surface-finish specifications. Much of this is associated with the design of rigid machines and tools, but the surface requirement is also related to the lubrication and wear of the tools. These are still not well understood for this process and much remains to be done. There is also a need for design and finishing techniques that will obviate the reliance on skilled hand-finishing methods.

Most forging is inherently noisy, especially all types of impact forging. Current and impending legislation on noise levels has initiated considerable activity in the reduction of noise by process design, machine design and, in the last resort, insulation. *Rotary forging*[10.20] is a recent development of known principles that has found commerical application for axi-symmetric heading, and is considerably quieter than hammering.

High energy-rate forging[10.21] has advantages for small components and for cropping of bar and tube sections. It can also be used for powder compaction and forging.

Warm forging[10.22] is increasing in popularity because it combines reduced yield stress with absence of oxide scale and the problems associated with excessive heat loss in hot working. There is however a lack of lubricants ideally suitable for warm working temperatures.

The organisation of forging plants for small components is important because of the rather low unit costs, and group technology offers economic advantages. Even fully automated factories are being considered in some countries[10.23].

10.15.3 *Newer processes*

Super-plastic alloys were originally considered primarily for sheet forming applications, but several super-plastic alloys are now available that are much harder and stronger at room temperature. These can be used as forging tools in some circumstances, though the tool life is very short[10.24]. They are particularly interesting for prototype and development dies, because the cost of production of each die is very much lower than for conventional tools.

Forging is also closely linked to metallurgy in thermomechanical treatment, where it is often the most convenient deformation process. Very high strength and good toughness can be produced in steels by this means (see §12.4.3).

Probably the most significant forging development is *hot isostatic compaction*[10.25]. Suitably graded and mixed powders are subjected to uniform hydrostatic pressure at high temperatures in an evacuated deformable container, which may be made of sheet metal or of a suitable glass. High density is achieved, with strong interface diffusion and bonding giving excellent uniformity of properties, especially in tool steels. The materials produced are tougher than conventionally pressed and sintered powder products. Complex shapes such as gas turbine blades can be made, close to their final dimensions which are finally trued by a coining operation. The powders can be mixed, to give gradations of mechanical or chemical properties, for example a hard wear-resistant surface with a tough backing. Composites, for example of tungsten-alloy fibres in a nickel-base superalloy, or ceramic, can also be produced by hot isostatic pressing. The same principle can be applied to diffusion welding of massive components, giving good fatigue properties.[10.26]

Another promising technique combines forging with casting. Metal flows into a die cavity as a liquid or paste and is then compressed at a temperature between the liquidus and the solidus, as it cools. The process resembles pressure die-casting, but involves much higher pressures. The castings made in this way have high density and are free from cavities, so the mechanical properties are improved. Close control of conditions is however necessary.

EXAMPLES

10.1 Evaluate the press capacity necessary for forging a 1 m long cylindrical bloom to hexagonal section with approximately 300 mm side, if the yield stress is initially 75 N/mm² but increases to 120 N/mm² at the end of the operation. Assume (*a*) that the bloom is partially lubricated so that $\mu = 0{\cdot}3$ (*b*) that there is no lubrication.

What maximum pressures would be expected?

Solution. A reasonable approximation is to assume that the load is equal to that necessary to cause yielding in a rectangular billet with $b = 300$ mm and $h = 2b \sin 60°$, under plane-strain conditions. Then, from equation 10.4, at the end of forging

$$\left(\frac{p}{2k}\right)_{max} = \exp(0{\cdot}3/1{\cdot}73) = 1{\cdot}19$$

The mean pressure is thus about $1{\cdot}1 \times 2k = 130$ N/mm². The contact area at the platen is 3×10^5 mm³ so the force is 39,000 kN.

With sticking friction, equation 10.8

$$\overline{p} = 2k\left(1 + \frac{0{\cdot}58}{4}\right) \approx 140 \text{ N/mm}^2;\ F = 41{,}000 \text{ kN}$$

$$p_{max} = 2k\left(1 + \frac{0{\cdot}58}{2}\right) \approx 155 \text{ N/mm}^2$$

10.2 What load is required to forge a 300 mm long 600 mm diameter cylindrical steel billet to 80% of its original length between flat platens (*a*) at room temperature with good lubrication ($\mu = 0{\cdot}05$) (*b*) at 900°C where $Y = 60$ N/mm² but there is sticking friction?

Solution. (*a*) Equation 10.15 shows that at the initial yielding

$$\left(\frac{p}{Y}\right)_{max} = \exp\left(\frac{\mu D}{h}\right) = \exp 0{\cdot}10 = 1{\cdot}105$$

When the cylinder has been reduced by 20% in height, to 240 mm, the area will have increased by 20% to maintain constant volume, so $D = 1{\cdot}1 \times 600 = 660$ mm

$$\left(\frac{p}{Y}\right)_{max} = \exp\left(\frac{0{\cdot}05 \times 660}{240}\right) = 1{\cdot}147$$

From Example 2.3, the yield stress for $\epsilon = 0{\cdot}223$ at room temperature is $S = 0{\cdot}61$, $Y \approx 0{\cdot}53$ kN/mm². So $p = 1{\cdot}074 \times 0{\cdot}53 = 0{\cdot}57$ kN/mm². The contact area for $D = 660$ is $34{\cdot}2 \times 10^4$ mm² so the steady forging load necessary would be $19{\cdot}5 \times 10^4$ kN.

(*b*) With sticking friction at the high temperature, from equation 10.16,

$$\left(\frac{p}{Y}\right)_{max} = 1 + \frac{1{\cdot}15}{240} \cdot \frac{660}{2} = 2{\cdot}58; \qquad \bar{p} = 60 \times 1{\cdot}79 = 107 \text{ N/mm}^2$$

$$F = 0{\cdot}107 \times 34{\cdot}2 \times 10^4 = 37 \times 10^3 \text{ kN}$$

10.3 Approximately what forging load would be required to transform a 1 m long, 1 m diameter cylindrical bloom into a square section of equal area in a hydraulic press? $Y = 60$ N/mm².

Solution. The cross-sectional area of the billet is $78{\cdot}5 \times 10^4$ mm² so the square would have sides equal to 886 mm. Assuming plane strain and sticking friction, equation 10.8 gives

$$\left(\frac{p}{2k}\right)_{max} = 1 + \frac{b}{4h} = 1{\cdot}25; \quad \bar{p} = 1{\cdot}12 \times 1{\cdot}15 \times 60 = 77 \text{ N/mm}^2$$

The area of platen contact is $88{\cdot}6 \times 10^4$ mm², so the forging load is 68×10^3 kN.

10.4 Draw the slip-line field and hodograph for plane–strain compression of a flat billet whose cross-section is 1 m x 165 mm. Assume (*a*) zero friction and (*b*) sticking friction.

Solution. The slip-line field for zero friction divides directly into six equal units, as in Figure 5.5*b*. The compression stress would therefore be exactly equal to $2k$. (*b*) With sticking friction the field resembles Figure 10.3. 15° intervals may be chosen at first. Using the same lettering as in Figure 10.3, the field can be drawn as far as $E_3F_2G_1H$, corresponding to a distance 475 mm from the platen edge. To complete the field a boundary α-line $E_3'F_2'G_1'H'$ is drawn at an angular spacing of about 4° further, across the β-lines constructed from F, G and H.

10.5 What pressure would be required for the compression of Example 10.4?

Solution. The stress solution is begun on AC, where $\sigma_3 = 0$ and $p = k$. Anticlockwise rotation along the β-line shows that the pressure along AC_3 is $p_{C_3} = k + 2k\frac{\pi}{4} = 2{\cdot}57k$. Between C_3 and D_2 the α-line rotates clockwise by 15° = $\pi/12$, so $p_{D_2} = k\left(1 + \frac{\pi}{2} + \frac{\pi}{6}\right) = 3{\cdot}1k$. Following the β-line D_2D_3 a further 15° gives $p_{D_3} = k\left(1 + \frac{\pi}{2} + \frac{2\pi}{6}\right) = 3{\cdot}62k$. Similarly $p_{E_3} = 4{\cdot}67k$, $p_{F_2} = 5{\cdot}2k$, $p_{G_1} = 5{\cdot}71k$, $p_H = 6{\cdot}24k$. The pressure at F_2' exceeds that at F_2 by $2k \times \left(4 \times \frac{\pi}{180}\right) = 0{\cdot}14k$, so $p_{F_2}' = 5{\cdot}34k$, $p_{G_1}' = 5{\cdot}85k$, $p_{H'} = 6{\cdot}38k$ and $p_{E_3}' = p_{E_3} + 0{\cdot}28k = 4{\cdot}95k$. To find the mean pressure over the dead zone, the vertical force components are summed: The vertical force F^* acting over the dead zone is

$$F^* = \int k \sin\psi \, ds + \int p \cos\psi \, ds$$

(since the stress acting normal to a plane of maximum shear is p, as seen in the Mohr circle).

Since k is constant, the first term is simply $k\frac{h}{2}$. The second is evaluated approximately as $\Sigma\bar{p}\ \Delta x$. Along the lines $E_3'F_2'$, $F_2'G_1'$, $G_1'H'$, the mean pressures are $5{\cdot}15k$, $5{\cdot}6k$ and $6{\cdot}12k$ respectively and the horizontal components of the lengths of these lines are 64·5, 67 and 55 mm, assuming a width of 250 mm for the drawing, representing the 1 m workpiece width. The sum of these products is $332k + 375k + 367k = 1074k$ to which must be added the contribution from the short length of line between H and the centre, $68k$. Thus

$$F^* = 68k + 1142k$$

The horizontal width of this half of the dead zone is 193 mm, so the mean vertical pressure over the dead zone is $6{\cdot}26k$. The total vertical force is

$$F = 2\left(p_{C_3}AC_3 + \frac{p_{C_3} + p_{E_3'}}{2}\ .\ C_3E_3' + F^*\right)$$

$$= 2(2{\cdot}57 \times 106 + 3{\cdot}76 \times 146 + 1210)k$$

$$= 2k \times 2031$$

The mean pressure $\dfrac{p}{2k} = \dfrac{F}{900} = 2{\cdot}25$.

10.6 A series of 8 straight-sided parallel grooves 6 mm wide with 6 mm separation is to be formed along a thick aluminium blank 100 mm wide and 300 mm long. If a suitable forging tool is used in a hydraulic press, what initial force would be necessary? How much would this increase by the time the punch had penetrated to a depth of 6 mm? Assume that the blank has previously been forged, so that the yield stress is sensibly constant at 150 N/mm².

Solution. The area of contact is $8 \times 6 \times 300 = 1{\cdot}44 \times 10^4$ mm². Assume that the plastic zones are initially separate, beneath each ridge, as implied by the Hill solution for indentation (Chapter 5, §5.6.5). Then the pressure is given by equation 5.15:

$$\frac{p}{2k} = 2{\cdot}57$$

The force W necessary to initiate yielding is

$$W = 2{\cdot}57 \times 0{\cdot}15 \times 1{\cdot}44 \times 10^4 = 5{\cdot}55 \times 10^3 \text{ kN}$$

After deep penetration, if the plastic zones remained separate, the pressure would increase to $4{\cdot}14 \times 2k$, according to equation 10.21, requiring a force $W = 8{\cdot}9 \times 10^3$ kN. It will be seen from Figure 10.4*b*, however, that the plastic zones would overlap slightly.

10.7 What force would be necessary to form 8 grooves of 60° included angle to the same depth in the blank of Example 10.6?

Solution. For separate grooves, equation 10.22 may be used: $p = 2k(1 + \theta)$. The value of θ is found geometrically. In Figure 10.4*c*, $AC = AD$, and for volume constancy the areas of triangles $O'AD$ and $O'AC$ must be equal. The perpendiculars from O' on to these sides must therefore be equal. A circle is drawn with centre O', touching AC, and tangents are drawn to it until one is found such that the intercept AD is equal to AC. This shows that $(\alpha - \theta)$ lies between 12° and 13°, so $\theta \approx 17{\cdot}5° = 0{\cdot}1\pi$ radians. If the grooves are 6 mm deep, $OA \approx 6 \tan 30 = 3{\cdot}4$ mm and the force required is $W = (1 + 0{\cdot}3) \times 0{\cdot}15 \times 2 \times 3{\cdot}46 \times 300 = 408$ kN for each groove, about $3{\cdot}3 \times 10^3$ kN in all, but the plastic zones will again interact.

10.8 Sketch an upper-bound solution for deep penetration of a 25 mm wide punch into a 150 mm wide billet. Find the force necessary to pierce such a billet under sticking friction conditions if $Y = 80$ N/mm. Assume plane strain.

Solution. The slip-line field of Figure 10.4 may be taken as a basis, suggesting the solution shown in Figure A10.1.

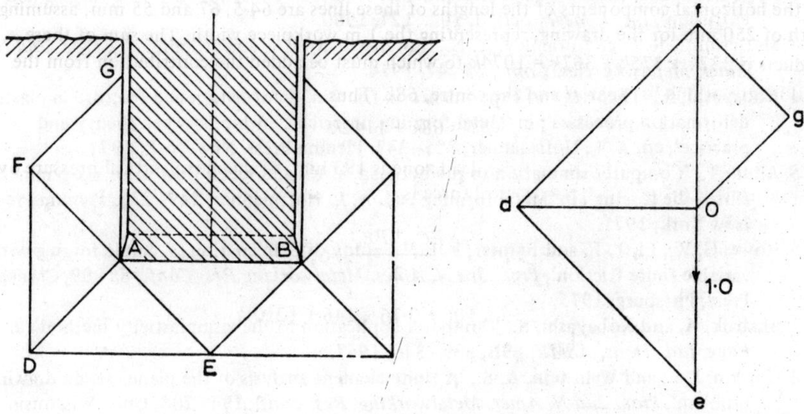

Figure A10.1 An upper-bound solution for deep piercing.

The hodograph is drawn assuming that the punch moves downwards with velocity 1·0. Then, for unit specimen thickness, using the left-hand side only,

$$\Sigma kus = k\,(AEu_{AE} + ADu_{AD} + DEu_{DE} + DFu_{DF} + AFu_{AF} + FGu_{FG})$$

$$= k\left(\sqrt{2}\times\sqrt{2} + \sqrt{2}\times\sqrt{2} + 2\times 1 + 2\times 1 + \sqrt{2}\times\frac{1}{\sqrt{2}} + \sqrt{2}\times\frac{1}{\sqrt{2}}\right)\frac{AB}{2} = 10k\,\frac{AB}{2}$$

This is equal to the rate of work performed by the applied force

$$(P \times 12{\cdot}5 \times t) \times 1{\cdot}0 = 12{\cdot}5 \times t \times 10k$$

If $Y = 80$ N/mm², $2k = 92$, $P = 460$ N/mm. The force required is thus not greater than 1,730 kN.

REFERENCES

10.1 Johnson, W. and Mellor, P. B., *Plasticity for Mechanical Engineers*, Van Nostrand, 1962.

10.2 Hill, R., *The Mathematical Theory of Plasticity*, Oxford Univ. Press, 1956, pp. 230, 215 and 222.

10.3 Prandtl, L., 'Anwendungsbeispiele zu einem Henckyschen Satz über das plastische Gleichgewicht', *Zeits. angew. Math. Mech*., 3, 401–406, 1926.

10.4 Johnson, W., 'The cutting of round wire with knife-edge and flat-edge tools', *Appl. Sci. Res. (A)*, 7, 65–88, 1957.

10.5 Levin, E., 'Indentation pressure of a smooth circular punch', *Q. J. Appl. Math*., 13, 133–137, July 1955.

10.6 Siebel, E. and Fangmeier, E., 'Untersuchungen über den Kraftbedarf beim Pressen und Lochen', *Mitt. K. W. Inst. Eisenforsch., Düsseldorf*, 13 (2), 1931.

10.7 Feldmann, H. D., *Cold Forging of Steel*, Hutchinson, 1961 (translated from *Fliesspressen von Stahl*, Springer, Berlin 1959).

10.8 Anon., *Kaltfliesspressen–Praktische Anwendung*, Verein Deutscher Ingenieure Arbeitsblatt. 5–3138.

10.9 Wistreich, J. G. and Shutt, A., 'Theoretical analysis of bloom and billet forging', *J. Iron and Steel Inst*., 193, 163–176, 1959; 'Automatic forging', *Iron and Steel*, July 1960.

10.10 Newnham, J. A. and Rowe, G. W., 'An analysis of compound flow of metal in a simple extrusion-forging process', *J. Inst. Metals*, 101, 1–9, 1973.

10.11 Kudo, H., 'An upper-bound approach to plane–strain forging and extrusion', *Int. J. Mech. Sci.*, **1**, 57–83, 1960.

10.12 Kudo, H., 'Some analytical and experimental studies of axi-symmetric cold forging and extrusion', *Int. J. Mech. Sci.*, **2**, 102–127, 1960.

10.13 McDermott, R. P. and Bramley, A. N., 'Forging analysis – a new approach', *Proc. 2nd N. Amer. Metalwkg. Res. Conf.*, 35–47, 1974.

10.14 Kobayashi, S., 'Theories and experiments on friction, deformation and fracture in plastic deformation processes', in 'Metal forming: interrelationship between theory and practice', ed. A. L. Hoffmanner, 325–347, Plenum Press, New York, 1971.

10.15 Altan, T., 'Computer simulation to predict load, stress and metal flow in an axi-symmetric closed-die forging', in 'Metal forming', ed. A. L. Hoffmanner, 249–273, Plenum Press, New York, 1971.

10.16 Rowe, G. W., Li, T. F. and Farmer, L. E., 'A study of deformation in simple forging with variable finite friction', *Proc. 3rd N. Amer. Metalworking Res. Conf.*, 85–99, Carnegie Press, Pittsburg, 1975.

10.17 Shabaik, A. and Kobayashi, S., 'Computer application to the visioplasticity method', *J. Engg. Ind. Trans. ASME*, **89B**, 339–346, 1967.

10.18 Gordon, J. L. and Weinstein, A. S., 'A finite element analysis of the plane–strain drawing problem', *Proc. 2nd N. Amer. Metalworking. Res. Conf.*, 194–208, Univ. Wisconsin–Maddison, 1974.

10.19 Shah, S. N. and Kobayashi, S., 'Rigid-plastic analysis of cold heading by a matrix method', *Proc. 15th Mach. Tool Des. Res. Conf.*, 603–610, 1974.

10.20 Slater, R. A., Barooah, N. K., Appleton, E. and Johnson, W., 'Rotary forging concept and initial work with an experimental machine', *Proc. Inst. Mech. Engrs.*, **184**, 577–586, 1970.

10.21 Various, *Proc. 3rd Int. Conf. Centre for High Energy Forming*, Vail, Colorado, 1971.

10.22 Berry, J. T. and Pope, M. H., 'Force requirements and friction in warm working operations', in 'Metal Forming' (Ed. Hoffmanner, A. L.), 307–324, Plenum Press, New York, 1971.

10.23 Takeyama, H., 'Automation developments in Japan', *Proc. 3rd N. Amer. Metalwkg. Res. Conf.*, 672–685, Carnegie Press, Pittsburg, 1975.

10.24 Saller, R. A. and Duncan, J. L., 'Stamping experience with superplastic alloys', *J. Inst. Metals*, 99, 173–177, 1971.

10.25 Brown, G. T., 'Powder forging: a review of the basic concepts and development prospects', *Powder Met.*, **14**, 121–143, 1971.

10.26 Paprocki, S. J. and Hodge, E. S., 'Hot isostatic compaction' Ch. XI of 'Mechanical behaviour of metals under pressure' (Ed. H. Ll. D. Pugh), Elsevier, Amsterdam, 1970.

11

Friction and Lubrication in Metalworking

11.1 Influences of friction in metalworking processes

The most obvious result of friction in general experience is that work has to be expended which would not otherwise be necessary. This is true in all metalworking processes. The higher the friction, the greater the load required to produce a particular deformation. Consequently, much attention has been given to the problems of producing low values of coefficient of friction. This is not, however, usually the major consideration in choosing a metalworking lubricant; elimination of the possibility of damage caused by metal transference from the workpiece to the tools is more important. Tool life can be prolonged both by reducing friction and by preventing metallic contact with the workpiece. However, if the lubricant film is too thick, a matt surface may result. To produce a bright surface, it may be necessary to sacrifice some lubricating efficiency. Some operations even require a certain minimum friction. In rolling of flat strip, for example (Chapter 9, equation 9.22), the rolls will skid if the friction is too low, and it is necessary to use a relatively poor lubricant to obtain the greatest possible reduction in area per pass. Apart from increasing external forces, frictional stress has an important influence on metal flow, and may cause serious inhomogeneity in the worked product, as well as surface cracks and other defects. Lubrication is therefore an important aspect of metalworking, and several factors need to be considered.

11.1.1 *Increases in forces attributable to friction*

The relative importance of the frictional contribution to load can be assessed by the analyses given in earlier chapters. Figure 11.1 shows some experimental results and calculations of drawing stress for various assumed coefficients of friction. The friction contributes about 10% to the load for an average heavy reduction of area (40%), with a good lubricant giving $\mu \approx 0{\cdot}05$. There is 20% contribution if the coefficient of friction is 0·1.

The frictional contribution is balanced against the increase in force due to redundant work, as explained in Chapter 6, §6.9, to find the optimum die angle, which is dependent upon the coefficient of friction.

The friction hill in rolling has been discussed in Chapter 9, and numerical examples are given on p. 245. The peak pressure may be considerably increased by friction, and increases in roll load by 10%–20% may be found even in cold-rolling, where the friction is usually low. In hot-rolling, the friction is much higher and the roll loads are further increased, but the effect on peak pressure is not so great as simple theory would predict, because of the limitation of sticking friction discussed in §9.6.

In cold-forging, load increases of 30% are found under common operating conditions.

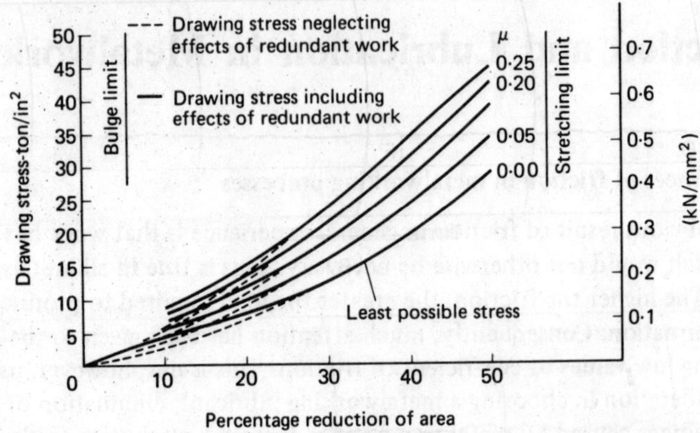

Figure 11.1 The variation of drawing stress with reduction of area for wide steel strip drawn through 10° dies. The lowest curve represents the least possible work of deformation. Friction increases the force required as shown for four friction coefficients. At low reductions the force is further increased by redundant work(11.12).

The effect of friction on the pressure trace obtained during extrusion has been discussed in Chapter 8, §8.2. The higher the friction, the greater the initial peak, and the steeper the fall in pressure as the billet length decreases. Under well-lubricated conditions, for example using melting-solid lubricants, the pressure may be almost constant once the initial peak has been passed.

A further influence of good lubrication is to provide smoother motion. Dry sliding may cause large fluctuations in force, particularly when the inertia of the system is low, as in machining. Figure 11.2 shows force-traces obtained during machining of mild steel. The lubricant reduces the average force and markedly smooths out the variations.

11.1.2 *Inhomogeneity of deformation produced by friction*

Inhomogeneity appears in two ways. The frictional stress causes a rotation of the directions of principal stress, which in turn determines which of the possible crystallographic planes are most favourably oriented for atomic slip to occur. The crystallographic orientation of the surface layers of the worked product will consequently be influenced by the friction, as can be shown by X-ray diffraction.

There is also a macroscopic inhomogeneity, observable in a grid scribed on a cross-section of the material before working, for example using a split wire, which is re-joined and then drawn through a die (Figure 11.3).

The surface layers are appreciably retarded by the frictional drag even in the presence of a lubricant. The additional surface strain produces additional hardening and sometimes cracking. Even very small cracks which are not easily visible may have serious consequences in fatigue or stress-corrosion life.

The general pattern of deformation may be completely altered by changes in

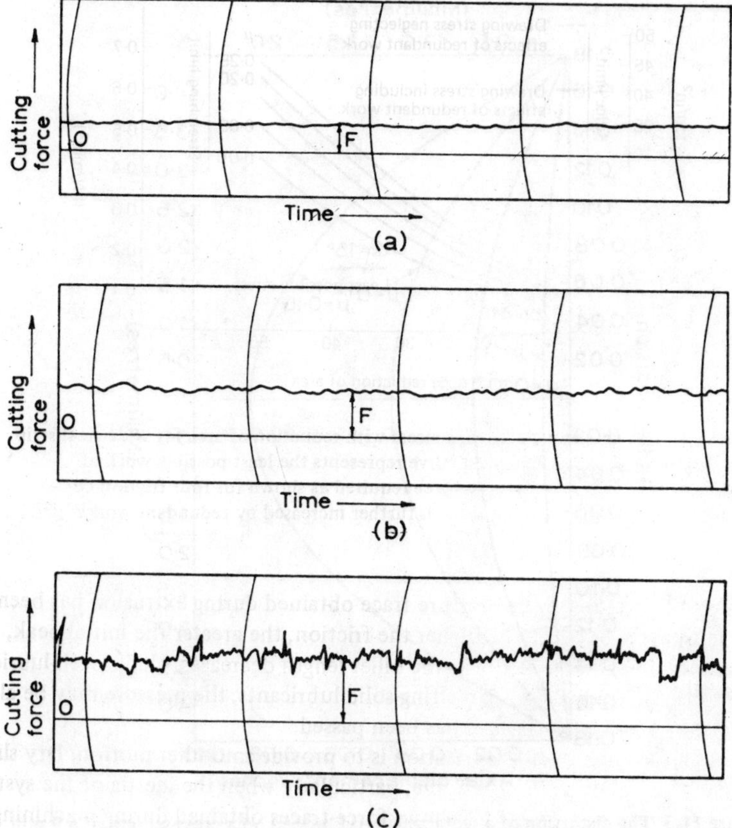

Figure 11.2 Cutting-force records obtained during machining of mild steel on a centre-lathe at 0·25 m/sec (50 ft/min).

(*a*) with a conventional cutting fluid.
(*b*) dry.
(*c*) in a vacuum.

(*E. F. Smart, Birmingham University*)

friction. For example, unlubricated extrusion often proceeds with the formation of a dead zone, which can be entirely eliminated by good lubrication. This has been discussed in Chapter 8, §8.4. Extrusion with a dead zone usually leaves a poor surface on the product, which may have traces of the original billet surface or its coating buried subcutaneously by transit over the dead-zone face. The built-up edge formed on a cutting tool under some conditions(11.2) is also associated with high friction. Again, the forces are increased and the surface finish of the product is poor. Good lubrication can reduce the size of the built-up edge, and considerably improve the performance and life of the tool.

In forging operations, the flow may be considerably influenced by friction(11.3), particularly in open-die forging. The barrelling produced by high friction in ordinary

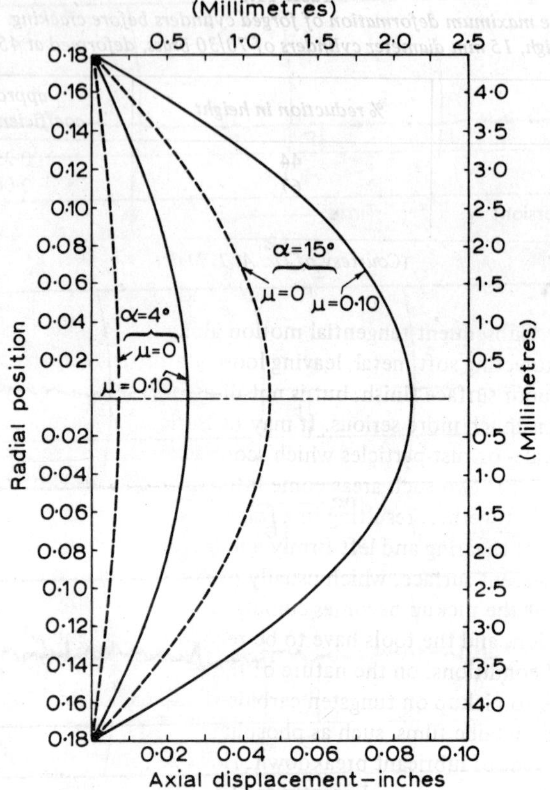

Figure 11.3 **The distortion of a rectangular grid scribed on a cross-section of a 9 mm (0·360 in.) diameter copper bar, resulting from a 36% pass through a die of 8° or 30° included angle. (*Calculated by a method due to Shield*[11.1])**

compression may introduce secondary tensile hoop stresses, which eventually limit the permissible reduction of area.

11.1.3 *Metal transfer*

By far the most serious result of inadequate lubrication is transference of workpiece metal to the tools. This 'pickup' occurs more readily with some materials than with others, but can seriously limit the range of possible reductions of area in a pass. A vital factor in most operations is the quantity of lubricant available in the working zone between tool and workpiece. If this is sufficient to fill all the depressions and to cover the major surface elevations, working can continue, but when the thickness of the lubricant film falls below this limit, there is danger of metallic contact followed by pickup. This can be demonstrated directly by measurement of lubricant film thickness using a radiotracer method[11.4].

Metal transfer may arise in two main ways. One is primarily associated with rough tool surfaces. If the lubricant film is depleted, workpiece metal may be forced into crevices in the tool, in the same way that a coining operation accurately replicates

Table 11.1
The maximum deformation of forged cylinders before cracking.
(20 mm high, 15 mm diameter cylinders of 70/30 brass, deformed at 450°C).

	% reduction in height	*approximate coefficient of friction*
Dry	44	0·25–0·3
Lubricated (graphite dispersion)	67	0·06–0·07

(Courtesy of Dr. A. T. Male)

the tool profile. Subsequent tangential motion along the tool face then tends to shear off the projecting soft metal, leaving loosely attached fragments behind. This usually gives a poor surface finish, but is not disastrous. The other type of transfer is adhesive and is much more serious. It may be initiated by a variety of causes, including small scale- or rust-particles which scour away the protective surface films, leaving bare metal. If two such areas come into contact under the working pressure they tend to weld together, resulting in a fragment of the workpiece being torn away by the subsequent shearing and left firmly adhering to the tool. The act of tearing exposes more nascent surface, which usually projects through the surrounding lubricant film, so that the pickup becomes cumulatively worse. This often terminates the working operation, and the tools have to be reconditioned. The adhesion depends, under practical conditions, on the nature of the materials; for example steels are much less liable to pickup on tungsten carbide than on tool-steel surfaces. The interposition of non-metallic films, such as phosphates, also helps to reduce or eliminate pickup in the event of lubricant breakdown. However, it should be remembered that many metalworking processes inherently require the formation of 40% to 50% of fresh metal surface during the deformation. In extrusion, the proportion is much greater. This nascent metal must be adequately protected immediately it is formed, or it will adhere to any tool material. The life of the tools, as well as the quality of the product, may be seriously affected if any adhesion occurs.

11.1.4 *Beneficial effects of friction*

It is not always desirable to reduce the coefficient of friction to a minimum, but even in these instances pickup should be avoided. The improvement in roll bite by increasing friction permits heavier drafts in cold-rolling of sheet and strip, as explained in Chapter 9.

$$(\Delta h)_{\max} = \mu^2 R \tag{9.22}$$

In push-bench manufacture of tube on a mandrel, it is advantageous to maintain a reasonably high friction on the mandrel, to carry part of the drawing force. In this way, the tensile stress on the leading end of the finished tube is reduced, and heavier reductions of area can be made without fracture (Chapter 7, §7.2.4). A somewhat analogous situation is found in deep drawing or pressing. To prevent the onset of necking in the walls while deep-drawing a cup, there should be high friction on the punch and low friction against the die. In extrusion, the back end of the billet tends to flow over the face of the pressure pad during the final stage when the extrusion pressure becomes too high. The rearward surface thus becomes entrained along the axis of an extruded bar, and even a hollow pipe may form. This feature is so common that it is usually referred to as *the* extrusion defect, without further qualification. The defective material has to be sawn off, or the extrusion must be terminated at an earlier stage, reducing the yield. The lower the friction at the face of the pressure pad, the easier it is for metal to flow inwards and the sooner the extrusion defect starts. This is particularly noticeable when a plain graphite dummy-block is used to obtain a complete, discardless extrusion[11.5]. Considerable lengths of rear-end pipe may then be produced, so the friction is deliberately increased by placing asbestos pads between the surfaces.

Obviously, high friction is desirable in the many varieties of tensile grips used in metalworking. Clutch mechanisms require a careful balance of high-torque transmission when fully engaged, and sufficient slip to provide smooth engagement without snatch, but most metalworking plant is directly driven.

11.2 Measurement of coefficient of friction

It is assumed in practically all metalworking theory that the tangential stress τ at the surface of the workpiece is directly proportional to the normal stress p, in the presence of a lubricant. A coefficient of friction $\mu = \tau/p$ can thus be defined, which is analogous to the familiar ratio of tangential and normal forces in elementary physics. The tangential stress τ is, however, limited to a value equal to the shear yield stress k of the metal itself (as discussed in Chapter 9, §9.6). Because the least value of the normal stress which can cause plastic flow is Y, the uniaxial yield stress, the maximum value of the 'coefficient of friction' for full sticking friction conditions is given by the ratio k/Y. These quantities are, of course, related in accordance with the yield criterion. Using von Mises' criterion, equation 3.22,

$$2k = 1{\cdot}155Y$$

Thus

$$\mu_{max} = \frac{k}{Y} = 0{\cdot}577 \tag{11.1}$$

The Tresca criterion gives

$$\mu_{max} = 0{\cdot}5 \tag{11.2}$$

which is convenient to remember, and is often quoted. The condition of constant coefficient is usually known as slipping friction or Coulomb friction, commemorating C. A. Coulomb's early investigations. The coefficient of friction may actually vary through a working pass, as the lubrication deteriorates due to thinning of the film and extension of the surface. Experimental studies suggest, however, that this is negligible for all well-lubricated operations. It may be assumed for practical calculations that the shear stress τ at a tool/workpiece interface, where the normal stress is p, is always given by

$$\tau = \mu p \tag{11.3a}$$

provided that $\tau < k$. Otherwise, there is sticking friction, and

$$\tau = k \tag{11.3b}$$

There is at present no generally accepted method of measuring the value of the coefficient of friction for given surfaces and lubricant. Various factors can influence the result, and it is essential that test conditions of surface geometry, chemical condition, lubricant film thickness, temperature, speed, environment and degree of deformation should match as closely as possible the actual conditions of the operation. This implies that the only strictly reliable friction data to use are those obtained from measurements during the operation considered. Fortunately, in most operations the frictional contribution to the applied force will not exceed 10–20%, so that approximate values of the coefficient, obtained from simulative tests, usually suffice. There is seldom a direct correlation with the well-known coefficients of friction obtained in laboratory tests in which the sliders are not deformed.

11.2.1 *Direct measurement of friction in metalworking*

The local value of the coefficient of friction cannot easily be determined. It is possible to insert two small probes in a large tool surface, such as a strip-forming roll. These are so designed that one is deflected by shear and the other by compression[11.6]. Even with this considerable complication, it is impossible to ensure that no local disturbance is introduced, and the method has not been applied to small tools such as wire-drawing dies.

All other methods involve measurement of an overall or average coefficient of friction μ. A value may be found from drawing wide strip with wedge-shaped dies of semi-angle α, by resolving the die load D and drawing force T, parallel and perpendicular to the die face, but small errors in measurement of D, T or α lead to large errors in the deduced value of μ. A more accurate method is to draw two parallel strips between curved dies, with a parallel-sided plug between the strips. Any desired degree of deformation may be imparted, and μ is found directly from the ratio T/D. (See reference 6.7.) Because the effective tool angle is so low, however, the lubricant film tends to be thicker on this plug than on a real die, and the friction is correspondingly lower.

11.2.2 *Coefficients obtained from correlation with theory*

The theoretical method described in the preceding chapters can be used to predict values of the working load for assumed coefficients of friction. Comparison with the measured load then shows immediately the value of μ during the operation. This approach is used, for example, in Figure 11.1.

It is easy to obtain an accurate measurement of the load for this purpose, but the theory assumes a knowledge of the yield stress, usually a mean yield stress. This may be a source of inaccuracy, particularly when annealed metal is used, because the initial rate of strain-hardening is high. In hot-working operations, strain-hardening is not important but the results are likely to be much less reliable, since the yield stress usually depends critically upon both temperature and strain rate.

Provided these variables can be adequately controlled, this method is valuable because it does accurately reproduce the flow and the geometrical conditions. The deduced coefficient cannot be as accurate as the load measurements because the frictional contribution is small, but satisfactory results have been obtained in strip-drawing, wire-drawing, rolling, and forging.

11.2.3 *Measurements depending upon shape change*

If the coefficient of friction can be deduced from a change in shape, the yield stress will not enter into the derivation, provided the material is homogeneous and there are no serious temperature gradients. Such methods are therefore in principle suitable for rapidly strain-hardening metals or for hot-working at high strain rates.

Hill[(11.7)] has proposed a method in which a flat rectangular lamina of length b, greater than ten times its width a, is compressed between flat overhanging platens. The friction influences the spread in the two major directions, and a simple relationship between the coefficient of friction and the shape change can be deduced. The test is, however, insensitive for coefficients exceeding about 0·05. Such values are usually found only when there is an appreciable hydrodynamic contribution to the lubrication, and are consequently dependent upon geometry and speed. A general correlation between the results of this test and most deformation processes is thus not to be expected.

Another test, suggested by Kudo and Kunogi[(11.8)] and developed by Cockcroft and Male[(11.9)] utilises axial compression of a ring between flat platens. When there is no friction, the ring deforms in exactly the same way as a disc, with both inside and outside diameters increasing in proportion to their distance from the centre. When there is a finite friction, the outer periphery is subject to greater restraint than the shorter inner one, and with sufficiently high friction it is energetically favourable for inward radial flow to occur, so that the internal diameter decreases. Thus, by measuring the ratio of internal and external diameters after axial compression of a ring of standard dimensions, it is possible to obtain a measure of the friction. When suitably calibrated, the test gives numerical values of μ. (See Figure 11.5.)

A further suggestion has been to use conical platens or indenters to compress a billet with conical depressions in its ends. The cone angle which produces no barrelling is equal to the friction angle, $\tan^{-1} \mu$. This method is rather tedious and not very accurate.

11.2.4 *Friction measurement in rolling*

The frictional conditions in rolling are unusual, since the direction of the frictional force reverses, and the interface is not subjected to pure slip. The coefficient of friction is usually low in cold-rolling, but it is desirable to be able to predict roll loads and interstand tensions more accurately in cold strip-rolling than in other processes. Very long lengths with closely controlled gauge tolerances are produced at high speeds in modern tandem mills, and too much empirical trial can be wasteful.

It is difficult to extract an accurate value of μ from comparison of measured forces with rolling theory, and the calculations involved are tedious unless a computer is used.

A method which has been used in research is to apply a steadily increasing back-tension until the neutral point coincides with the exit from the roll gap. Under these conditions the friction force acts in the same direction over the whole roll surface, and the coefficient of friction can be deduced from the ratio of the torque and the roll load, as shown in Chapter 9, §9.4.1.

The friction can also be calculated in terms of V_2, the exit speed of the strip, and V, the peripheral speed of the rolls. This method also is more suitable for research than for production calculations.

The simplest practical assessment of friction is to determine the maximum draft Δh which the rolls will accept. Then, as shown in §9.4.1,

$$\mu \approx \sqrt{\frac{\Delta h_{max}}{R}} \tag{9.22}$$

The maximum bite is increased if there is a burr on the leading edge of the strip, but with carefully prepared smooth edges the results are reasonably reproducible. This is a useful practical test.

11.3 The elements of friction theory

An understanding of lubrication requires some knowledge of the nature and cause of friction. The most comprehensive approach to the general problems of friction is that of Bowden and Tabor and their collaborators[(11.10)], based on the observation that practical metal surfaces are not smooth. When two such surfaces are placed in contact under a light load, they will touch at only a few relatively isolated surface asperities. The local pressure at these minute contacts will be very high, and in fact is found to be sufficient to cause plastic deformation. As the load is increased, the extent of these asperity contacts will increase, and fresh contacts will be made. If some average value of the yield stresses of all the contacts at a given instant is assumed to be $\bar{p}$, under a load W, then the sum of the projected areas of all the contacts will be given by

$$\Sigma A_x . \bar{p} = W \tag{11.4}$$

In the original formulation it was postulated that when dry metal specimens were used, cold welding would occur, and that the welded junctions so formed would have a mean shear strength $\bar{s}$ approximately equal to the shear strength of the metal. The frictional resistance then arises from the force F required to shear the junctions.

$$\Sigma A_x . \bar{s} = F \tag{11.5}$$

Since $\bar{s}$ and $\bar{p}$ are both related to the shear yield stress k of the metal, this immediately implies that F is proportional to W, so a coefficient of friction μ may be defined:

$$\mu = \frac{F}{W} = \frac{\bar{s}}{\bar{p}} = \text{constant} \tag{11.6}$$

This simple approach ignores the inter-relationship of the shear stress and the normal stress by the yield criterion. If von Mises' criterion (equation 3.19) is applied to a model of a frictional junction, the result is modified. Because the junction is deformed plastically by the initial application of load in a direction normal to the interface, further yielding will be produced by any additional tangential stress, however small. This flow causes the specimens to move closer together, increasing the contact area. The normal pressure is thereby reduced, if the load remains constant, but the tangential force which can be supported is increased. So the process continues; increasing tangential force produces increased contact and a stronger junction. The original load W is unchanged but the tangential force F may rise to very high values. The ratio F/W, which is no longer strictly a coefficient of friction because F and W are not directly related, may rise to very high values, between 10 and 100. Because the area of metallic contact increases, the strength of such a junction can also be very high when subjected to a pull normal to the surface. These effects can be followed in detail with very clean metal surfaces, demonstrating the reality of cold surface welding. Large forces, as much as ten times the original load, may be required to break such junctions apart[11.11].

Very small amounts of contaminant on the surface can, however, drastically reduce the shear strength of the interface, and the junction growth is terminated at an early stage. Detailed analysis shows that if the effective shear strength s_i of the interface is less than about 0·2k, junction growth is negligible. The coefficient of friction is then given by the ratio of the shear strength of the interface to the yield pressure of the metal.

$$\mu = s_i/\bar{p} \tag{11.7}$$

This resembles equation 11.6, but in the majority of practical conditions s_i is predominantly the shear strength of the surface film or of the lubricant. There may be local regions where the film is too thin to protect the metal, so that cold welding occurs. This is usually insufficient to influence the friction appreciably, but may be very important in initiating adhesive pickup. Once started, adherent transfer interferes with the lubrication of oncoming metal and becomes cumulatively worse.

This simple friction model is informative and gives results in accordance with much general experience, but when one specimen is deformed by more than a few per cent, as in all metalworking, the concept of local surface contacts must be modified. If the surfaces are dry the contacts start locally, but after a few per cent overall deformation there is sensibly complete geometrical conformity between the surfaces. The interfacial profile is then determined entirely by the contours of the hard tool. Any depressions in the tool will be filled by the relatively soft workpiece metal, and the projections so formed will be sheared off by subsequent tangential motion. Provided that the surfaces are sufficiently contaminated to prevent adhesion, this transferred metal will appear as loose transfer or debris, and will not be damaging. Indeed, the friction may be slightly reduced by the resultant smoothing of the tool surface. Lubricants can be trapped in such depressions and, depending on their compressibility, will tend to prevent the workpiece from coining into them, so that the transfer is considerably reduced. Lubricants can also be trapped in depressions in the soft metal. As these are compressed after entering the working zone, lubricant is steadily fed out exactly where it is required during the deformation. This can be very useful in providing a continuous supply of lubricant. Controlled surface roughness can thus be advantageous in metalworking[11.12], though it may be dangerous under light loads by causing local penetration of the thin lubricant film. Under both light and heavy deformation, the danger of adhesive transfer is increased if both surfaces are very smooth, because the scouring action involved in sliding makes it difficult to maintain unbroken lubricant films over large smooth areas in very close contact. Even when adhesive transfer does occur in metalworking, the area of contact cannot grow, so the apparent coefficient of friction is limited to the 'sticking friction' value (equation 11.3b) and cannot rise to the very large values observed under light loads. Sticking friction may arise from mechanical interaction or from the interaction of strong films such as oxides, and does not necessarily involve adhesive metal transfer.

Another feature which is important in lubrication is the surface temperature. The work necessary to overcome the friction is generated in a very thin layer, very rapidly, so the interfacial temperature will rise by an amount dependent on the friction and on the thermal conductivity of the materials. When the load-bearing contact is further restricted, as under light loads or where lubricant breaks down locally, the transient temperature rise due to deformation of an asperity junction may be very high indeed. Typical values given by Bowden and Tabor[11.10] are 300°C at 3 m/sec (600 ft/min) and 1,000°C at 10 m/sec (2,000 ft/min), recorded by an oscilloscope. In most metalworking the contact area is more uniform because of the large deformation of one member, so that lower temperatures would be expected at the same speeds. Many operations are performed at relatively low speeds, so the problem of extreme temperature rise is likely to be less serious. There will, however, always be a further temperature rise due to the heat generated by deformation within the metal, which may or may not have time to diffuse to the surface during the passage of the tool. In practice, there is not a strong dependence of temperature on speed in the lower speed range. However, surface temperatures of 200°C may be measured at moderate speeds such as 0·5 m/sec (100 ft/min). Many organic lubricants begin to fail if temperatures of this order are maintained for an appreciable time.

Metal cutting provides an extreme example of very high temperatures being generated over appreciable areas, because the strain in the workpiece is very large, the speed is very high, and the deformation zone and the chip have very small thermal capacity and poor thermal conductance to the nearest heat sink. Temperatures of 1,300–1,500°C may be reached at the rake face of a lathe tool cutting at 5 m/sec (1,000 ft/min).

11.4 Elementary principles of lubrication

Though the distinctions are not always clear, it is usual to consider lubrication under separate headings: hydrodynamic, boundary, extreme-pressure, and solid-film. The major function of a lubricant in metalworking is to prevent pickup. It is also desirable that a lubricant should reduce wear and reduce friction, but the latter is usually the least important of the three. Wherever possible, a film of lubricant should be maintained which is sufficiently thick to separate the surfaces completely. There is no advantage in exceeding this thickness. In the presence of a very thick film of lubricant, the individual crystals of the metal will deform freely, according to their crystallographic orientation, producing a matt appearance. This does not impair the mechanical or metallurgical properties of the finished product, but it is more difficult to inspect visually for flaws, and is usually regarded as undesirable. The highly burnished appearance often prized in metal-working is, on the other hand, obtained at risk of thinning the lubricant film close to the breakdown level.

11.4.1 *Hydrodynamic and thick-film lubrication*

Under certain conditions, thick films of lubricant may be established by the viscous forces acting on the fluid. This is encouraged by a small angle between the tool and the workpiece, and by high relative speed. For example, the viscous drag exerted on a lubricant by a copper wire travelling at high speed may force appreciable quantities into a low-angle die. If a carefully designed cylindrical nozzle surrounding the wire is sealed to the die it is possible to generate enough pressure in an oil by this means to deform a copper wire before it actually enters the die[11.13]. The same result can be obtained with steel wire if soap is used as the lubricant[11.14], provided that the starting problem can be overcome.

For more general applications, it is possible to increase the amount of lubricant entering the working space by trapping it in surface pockets formed either directly on the workpiece[11.12], or by suitable surface treatments such as phosphating. Table 11.2 shows one example of the effectiveness of such treatment, grit-blasting being used to form the surface traps. Another approach is to use a polymeric film with adequate strength and adhesion to the metal which will provide good protection from pickup though it may be desirable to reduce the friction in some instances by a supplementary film of soap or oil.

Table 11.2
Effect of surface preparation on cold drawing performance.

Lubricant	*Surface Preparation*	*Maximum reduction of area in one pass*		*Limitation*	*Surface roughness μ in C.L.A.*	
		Stainless Steel	*Mild Steel*		*undrawn*	*drawn*
Soap	Polished	7	13	Pickup	2	–
Soap	Abraded 600 emery	17	19	Pickup	10	–
Soap	Grit blasted (180 mesh grit)	58	56	Tensile fracture	50	10
Chlorinated oil	Grit blasted (180)	55	54	Tensile fracture	50	8
Castor oil	Grit blasted (180)	–	52	Tensile fracture	50	8
Mineral oil with 5% oleic acid	Grit blasted (180)	–	53	Tensile fracture	50	8

(1μ in = 0·025 μm)

11.4.2 *Boundary and extreme-pressure lubricants*

Some lubricants are remarkably effective as very thin films. Extensive experiments with small sliders under light loads[11.10] have shown that even monomolecular layers of some compounds will provide low friction, though such films are of course quickly worn away. These compounds are known as boundary lubricants. Liquid or solid fatty acids are particularly effective, but the reason for their effectiveness is that they react with a metal surface, to form a solid metallic soap. Indeed, on a non-reactive metal, the fatty acids are no better than paraffins of equal molecular weight. Long-chain organic compounds which are themselves solids have good lubricating properties, but are less strongly bound to metal surfaces than the reaction products.

In metalworking, fresh surface is created as the workpiece extends, and an important criterion of lubrication is that the new and highly-reactive surface should be protected by a lubricant film or reaction product, before it can contact the metal of the tool. Some materials, including stainless steels and titanium alloys, do not react easily with boundary lubricants, and these are found to be very prone to metallic pick-up.

One disadvantage of boundary lubricants is that they are usually organic compounds which decompose at temperatures of 250°C or less. For protection at higher temperatures, chlorinated organic compounds are used. These react to form solid chlorides such as $FeCl_2$ which decomposes at about 350°C. Such compounds were first developed for use in gears subjected to high pressures, and have become known as Extreme Pressure or E.P. additives, though their action is primarily dependent on temperature. They are particularly suitable for this application, because any local area on a gear tooth which becomes heated is exposed to the fluid for a long time between successive contacts. Chemical films can thus form quite easily. Many metal-working processes involve only one passage of short duration, for example about 20 milliseconds in rod-drawing at 0·5 m/sec (100 ft/min). Any reaction occurring after the workpiece leaves the tool is just corrosion, serving no useful purpose. The friction of a chloride film is usually higher than that of a true boundary film, so boundary and E.P. lubricants are often blended to provide low friction over as large a temperature range as possible. Sulphur compounds can also be used, forming sulphides such as FeS which decomposes at about 700°C. These are effective up to higher temperatures but the friction is somewhat higher still.

11.4.3 *Solid lubricants*

Any material of lower shear strength than the metal can in principle be used as a solid lubricant in metalworking. For example, a thin sheet of copper with $k \approx Y/2 =$ 150 N/mm^2 (10 ton/in^2) can be placed on a steel block in a cold-forging experiment. If the yield stress of the steel is 750 N/mm^2 (50 ton/in^2), the apparent coefficient of friction should be about 0·2, depending to some extent on the strain hardening. This can be confirmed experimentally (Figure 11.4).

Thin copper coatings have been used in some industrial processes, including steel-tube sinking. Lead is very soft and is one of the best lubricants known for cold tube-drawing. Thick films of lead are very tenacious and several heavy passes may be taken sequentially without relubrication. Solid polymers and waxes are also useful, and solid soaps such as sodium or calcium stearate are widely used in cold metal-working, with very good results.

Certain crystalline solids, particularly graphite and molybdenum disulphide, exhibit low friction up to high temperatures. Graphite dispersed in grease or tar is useful in hot-working of steels. Finely-divided lime or some other solids may be added to fluid lubricants to increase their viscosity and to provide mechanical protection for the surfaces.

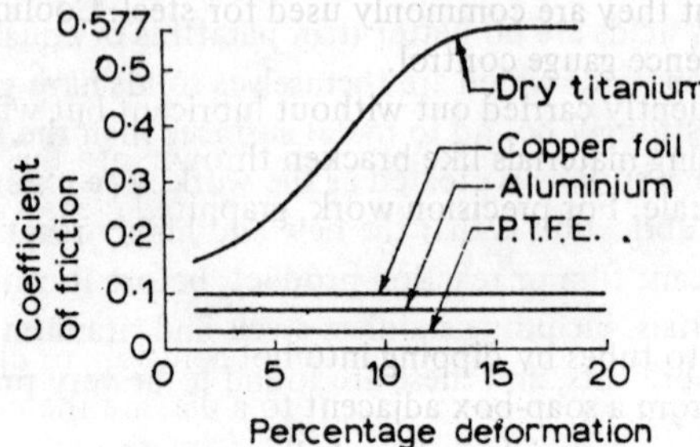

Figure 11.4 Coefficients of friction measured during forging of titanium using thin foils of low shear-strength solids between specimen and platen. The friction is constant until the foil is ruptured. Calculated coefficients of friction based on the mechanical properties of the titanium and the two metal foils are as follows:

Deformation %	*Yield Strength Y in Compression (N/mm²)* *Ti*	*Cu*	*Al*	$\mu = \frac{Y/2 \text{ copper}}{Y \text{ titanium}}$	$\mu = \frac{Y/2 \text{ aluminium}}{Y \text{ titanium}}$
10	700	155	85	0·11	0·06
20	930	250	110	0·13	0·06
30	115	300	115	0·12	0·05

(*A. T. Male*[11.15])

11.4.4 *Melting solids*

In recent years, use has been made of the principle of lubrication with a thin film of liquid generated by melting of a solid in contact with a hot workpiece. The effect is substantially the same as that of passing a hot knife through butter. One of the most successful applications has been in extrusion at about 1,000°C. A pad of compacted glass powder is placed between the die and a hot billet, and further glass powder is used to lubricate the billet surface in contact with the container. As extrusion proceeds, a uniform film of glass melts and passes through the die orifice, giving low friction, good surface finish and greatly improved die life[11.16].

Other melting solids, including ice[11.17] and many inorganic compounds, can be used at appropriate temperatures. At about 100–150°C a blend of polyethylene and wax provides good lubrication[11.18].

11.5 Examples of lubricants used in industrial metalworking

More detailed accounts of specific lubrication practice[11.21] are given at the ends of the respective Chapters 6–10.

11.5.1 *Lubricants for rolling*

Cold-rolling uses fluid lubricants of low viscosity. It is not desirable to have the lowest possible friction, and paraffin is often suitable for non-ferrous materials. Boundary lubricants may often cause staining on aluminium or copper alloys during subsequent annealing, but they are commonly used for steel. Cooling is an important factor, and may influence gauge control.

Hot-rolling is frequently carried out without lubricant but with a flood of cooling water, or with surprising materials like bracken thrown into the roll gap to generate steam and break up scale. For precision work, graphited greases or tar may be used.

11.5.2 *Cold-drawing*

Sodium soap applied to tubes by dipping into hot solution, or calcium stearate picked up by a wire from a soap-box adjacent to a die, are the most common drawing lubricants for ferrous materials and hard alloys. The coating is usually improved by prior deposition of a phosphate, oxalate or similar compound which helps to trap the lubricant and also provides additional protection from pickup. Various oils containing fatty or chlorinated additives and sometimes sulphur compounds are also successfully used, but in general require more careful control of die design and other features. Oils have the advantage of being easily applied at the tool face and not collecting grit or rust particles as easily as preformed soap-coatings. They are also easy to remove. For severe operations, polymeric coatings may give greater protection but are more expensive to apply and to remove. Lead is very effective but may involve toxicity problems. General health hazards, including dermatitis risks, must always be borne in mind in selecting metalworking lubricants.

11.5.3 *Forging*

Cold-forging has followed the pattern set by heavy drawing operations. Phosphating and sodium soap dipping is usually employed for ferrous materials, but soap alone often suffices for others. Hot-forging may be carried out without lubricant or with water sprays, or with some material such as sawdust intended to prevent adhesion, but graphite is widely used. Melting glass is sometimes used for forging, and especially for piercing operations.

11.5.4 *Extrusion*

Cold extrusion again follows the practice of cold-drawing lubrication for heavy passes. Lanolin is often used for the softer alloys. Various glasses are the most successful lubricants for very high temperature extrusions, including molybdenum at about 1,800°C, and stainless steels at about 1,200°C. Other melting solids can be used at similar and lower temperatures[11.19]. Even the natural oxides on certain metals provide some lubrication[11.20]. Graphite can be used for mild steel and stainless steel but is usually not capable of giving adequate lubrication over as great a length as glass.

11.5.5 *Cutting, drilling and other machining operations*

One of the major functions of a machining lubricant is to increase the life of a tool by reducing its temperature. Water has a much higher specific heat than oils, and so is favoured as a cutting fluid, but water emulsions are normally used in lathe operations to avoid corrosion. Under severe conditions at low speeds, such as drilling, tapping and broaching, it is advantageous to use neat oils containing boundary or extreme-pressure additives. Staining is again often important, particularly for non-ferrous materials. A feature of cutting fluids is that large quantities are circulated and may easily come into contact with the operator and with the expensive high-precision machine tool. Many commercial oils are available which avoid the problems of corrosion, staining, frothing, and toxicity or danger to health.

11.6 An assessment of simulative testing for lubricant evaluation

The simplest parameter to measure is probably the coefficient of friction, though this can seldom be measured directly. Simple pin-on-disc, four-ball and other machines using elastically-deforming specimens cannot give friction values that are relevant to metalworking. They fail to produce similar lubrication conditions and take no account of the important extension of the surface[11.21].

The plane-strain compression test relates to forging and, less directly, to rolling and wire drawing, but the thickness of the lubricant film is critically important. It does not usually show lubricant breakdown. The *ring test*, which does not rely on a force measurement or a knowledge of the yield stress of the material, is likely to be a better guide for forging lubricants, and gives a clear indication of failure of the lubricant. It is however again susceptible to changes in hydrodynamic factors in the lubrication. For more severe interfacial conditions where pickup is likely, the *twist-compression test* is suitable, see Figure on page 305.

Approximate simulations are sometimes possible. A parallel-sided strip may be drawn between two dies to simulate tube or wire drawing. Small strips can be rolled in a test mill. Circular cups can be drawn to simulate complex pressings. None of these is entirely satisfactory and it appears that there is no completely reliable substitute for tests in the real process. Nevertheless, an attempt can be made to summarise the most suitable tests for specific purposes.

11.6.1 *Rolling*

The coefficient of friction is important in rolling because it determines the angle of bite, as well as affecting the roll load. As shown in p. 295, this angle can be assessed easily by offering a short strip, with suitably bevelled leading edge, to the rolls under a light pressure. The roll gap is slowly increased until the strip is drawn in. Then

$$\mu = \sqrt{\Delta h/R} \tag{9.22}$$

A more laborious, but probably more informative test uses one strip that is rolled sequentially with gradually decreasing roll-gap setting. The exit thickness is plotted against the gap, and the test is continued until the curve becomes horizontal at the minimum attainable gauge. This does not give a direct value of μ, but the overall curve and the limiting thickness form a basis for comparison of lubricants. It has the advantage of being a real rolling process.

For some materials, especially aluminium, the formation of a roll coating is a dominant feature. This can be assessed only by actual rolling, though a twist compression test (§11.6.3) will give some guidance about the ability of a lubricant to prevent or reduce adhesion. There is no satisfactory test for roll wear, and very little information is available about the effects of lubricants on roll wear, though it is of considerable commercial importance.

Trials on full-scale rolling mills can be very expensive because of the large quantities of fluid used and the cost of lost time for production mills. If rolling lubricants must be assessed without access to any rolling equipment, it appears that the plane-strain compression test is the most suitable, but care must be exercised in relating the results to real conditions, especially to high-speed rolling.

11.6.2 *Extrusion*

There is no simulative test for lubrication in extrusion, because the geometric conditions are unique. Very large areas of new surface are created, and the lubricant can be applied only at the start, in the form of a coating or a pad, except in hydrostatic extrusion. Empirical correlations with rate of softening or melting have been proposed for the glass-type lubricants but no direct values can be deduced.

In trial extrusions on a press, it is important to distinguish the container friction from the die friction, which can be done by extruding billets of different lengths (§8.2.2). Temperatures in the billet and in the deformation zone may often be a critical factor, but full-scale trials are not expensive or difficult to conduct for extrusion lubricants. They may even be interspersed during production runs.

Changes of lubricant for hydrostatic extrusion are less easy, but still not expensive.

11.6.3 *Forging*

The assessment of an average coefficient of friction is easier for forging than for any other metalworking process. The ring test[11.9] provides friction data over the whole range from zero to sticking friction. It can give reliable results at room temperature and can be applied to hot forging without requiring a knowledge of the strain rate or temperature dependence of the yield stress.

A flat ring of outside and inside diameters D_o and D_i, and a height h is compressed axially. With low friction D_i increases, but when the friction is high a flow divide develops and the hole tends to close. The change in D_i is sensitive to the average friction. Any size of ring can be used, but it is convenient to keep $2D_i = D_o = 6h$, for which ratios calibration charts are available, as shown in Figure 11.5. The test is also valuable for determining the percentage deformation at which lubricant breakdown occurs. The two lubricants used for Figure 11.5 are indistinguishable in terms of coefficient of friction, but one (B) permits much heavier reductions to be taken.

It is possible to calculate effective values of μ after breakdown, but these are of less interest than the maximum deformation before breakdown.

For routine examination of forging lubricants, a simplified procedure can be used.

Five standard rings are used for each test, preferably with $D_o = 20$, $D_i = 10$, $h = 3{\cdot}3$ mm. These are compressed respectively to height reductions of 10, 20, 40, 55 and 70%, using suitable stops. The mean internal diameter of each is then assessed manually or with a photocell responding to the area of illumination, and the coefficients of friction are found by plotting onto the calibration chart. The lubricant is then classified according to two criteria, the value of the coefficient of friction at

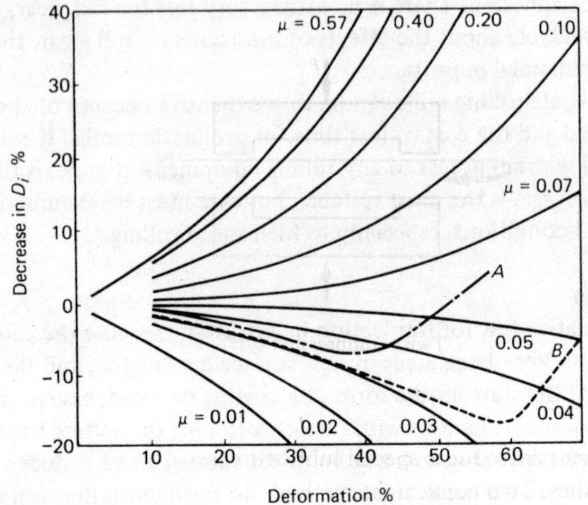

Figure 11.5 A calibration chart for the ring test, showing breakdown of two good lubricants.

20% reduction and the range of reductions within which the friction rises significantly, indicating breakdown of lubrication. Thus categories $5x - 1x$ correspond to $\mu_{20} < 0{\cdot}02$, $0{\cdot}021 - 0{\cdot}04$, $0{\cdot}041 - 0{\cdot}07$, $0{\cdot}071 - 0{\cdot}10$, $> 0{\cdot}10$ respectively, and categories $5y - 1y$ correspond to breakdown reductions $r_B > 70$, 55–69, 40–54, 20–39, 10–19%. The final rating of the lubricant is presented as $(mx + ny)$, where x and y are weighting factors chosen by experience to give the most realistic representation of the forming process to be used. In stretch forming, for example, low friction is the dominant factor, so x is chosen to be large, say 5, while $y = 1$. In contrast, for a coining operation pickup is likely to be more important so $x = 1, y = 3$ may be chosen.

The numerical values m and n can easily be stored in a computerised data bank and retrieved with any desired weighting factors x and y. It is of course possible to refine the integral steps if desired, and in principle other characteristics such as staining, corrosion, bacterial degradation, disposability, etc. can be quantified and stored in a similar way. It is suggested that this method makes interpretation and transfer of lubricant data much simpler and more useful to industrial operators.

The same approach can, of course, also be used for other processes, for example wire drawing, where a set of dies of progressively smaller diameter takes the place of the set of forging reductions.

For very severe conditions, where it is known that pickup is likely to occur, the forging ring test is not sufficiently sensitive. A twist compression test[11.21] is then preferred. This is performed in a compression machine with a flat plate that is rotated between two annular-faced specimens. An average friction value T/Pr can be recorded, but the qualitative inspection of the platen surface is often more informative about the amount and type of pickup.

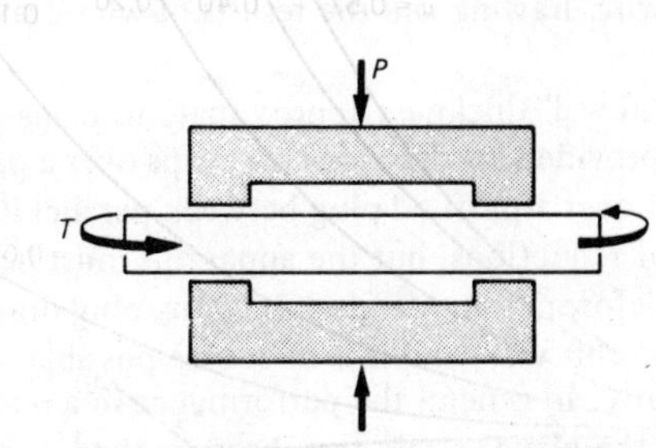

Twist-compression test

Closed-die forging introduces special lubrication problems that are not simulated by the flat-faced dies. Two convenient methods are available. One produces a central boss on a flat disc by forging with a hole in one die. The total height from the top of the boss to the back of the disc after a defined disc compression is a measure of the lubricant performance. If a slot is used across a diameter, instead of the central hole, the height and profile of the web produced are both informative. The appearance of the disc surface in both tests is indicative of the extent of hydrostatic or hydrodynamic lubrication, relative to boundary lubrication. Any scratches on the surface may be due to lubricant breakdown.

11.6.4 *Wire drawing*

There is no satisfactory simulative test for wire drawing. The plane-strain compression test can be related to wire drawing if care is taken to interpret the contributions of hydrodynamic lubrication, which differ greatly between the two systems. It is however easy to construct an elementary wire-drawing machine for test purposes. At low speeds, as used for rod drawing, even a short length of wire, say 1 m, will suffice to establish steady conditions, but for higher speeds a coil of wire can be drawn by a simple bull block. Care must always be taken to avoid breakdown at the shoulder of the reduced front section or tag. This should have a gradual curvature, and may be grit blasted to improve the lubricant throughput.

As suggested for the ring test in forging, it is convenient to categorise lubricants by sets of coefficients of friction and breakdown reduction values. For initial sorting of lubricants, a set of four dies suffices, respectively with high and low die angles, say 20° and 4°, and high and low reductions of area, say 20% and 40%. The categories can then be stored, perhaps in a computer data bank, and recalled with suitable weighting factors, including other lubricant properties of stability, staining, etc. It is necessary to use one of the theories of wire drawing to deduce a coefficient of friction from a measured drawing force, but any assumptions made, such as constancy of μ, are the same for all lubricants and do not affect comparative results. The mean yield stress of the wire can be determined by tensile tests on undrawn and drawn wires, or from an equation of the type $\sigma = B\epsilon^m$. For more detailed comparison, a larger set of dies can be used.

11.6.5 *Tube drawing and wall ironing*

Tube sinking resembles wire drawing and the results of wire-drawing tests can be applied.

Close-pass reductions in wall thickness approximate to plane-strain conditions. A suitable simulation is provided by drawing two strips over a parallel-sided plug and between two dies, or over a profiled plug between parallel dies. Either method can be used for a range of reductions, but the apparatus must be robust enough to avoid significant elastic deformation. Mandrel drawing, plug drawing and wall ironing can all be simulated in this way, and it is of course possible to lubricate the two sides of the strip differently. In general the performance in a real die is likely to be somewhat better than in the plane–strain test, because there is no side leakage in a cylindrical die.

11.6.6 *Sheet pressing and deep drawing*

In these operations the friction conditions vary from point to point. The *cupping test* is frequently used, but this was developed for testing sheet materials and is not very sensitive to changes in good lubrication conditions. It appears in several forms.

To find the limiting drawing ratio (L.D.R.), a set of blanks of different diameters is drawn with a given punch and the maximum blank diameter that can be drawn without failure is recorded as a ratio to the diameter of the punch. A flat-faced punch is used.

A similar test with a hemispherically-ended punch is indicative of lubrication during stretching. The better the lubricant, the more nearly the position of fracture approaches the pole.

The force required to draw a blank of given diameter may be recorded continuously to show the variation during different parts of the operation. More simply, the maximum force at a given depth of drawn can be recorded, but there is no direct way of separating the frictional contribution, and the methods are insensitive.

More elaborate equipment has been used to measure draw force, blankholder force and blankholder friction separately, for research purposes.

REFERENCES

11.1 Shield, R. T., 'Plastic flow in a converging conical channel', *J. Mech. Phys. Solids*, 3, 246–258, 1955.

11.2 Heginbotham, W. B. and Gogia, S. L., 'Metal cutting and the built-up nose', *Proc. Inst. Mech. Engrs.*, 175, 892–905, 1961.

11.3 Feldmann, H. D., *Cold Forging of Steel*, Hutchinson, 1961. (Translated from *Fliesspressen von Stahl*, Springer, Berlin, 1959.)

11.4 Golden, J., Lancaster, P. R. and Rowe, G. W., 'An examination of lubrication with soap, using radioactive sodium stearate', *Int. J. Appl. Rad. Isotopes*, 4, 30–35, 1958.

11.5 Sukolski, P., Brit. Pat. No. 853,210.

11.6 Siebel, E. and Lueg, W., 'Untersuchungen über die Spannungsverteilung im Walzspalt', *Mitt. K. W. Inst. Eisenforschung, Düsseldorf*, 15, 1–14, 1933.

11.7 Hill, R., 'On the inhomogeneous deformation of a plastic lamina in a compression test', *Phil. Mag.*, 41, 733–744, 1950.

11.8 Kudo, H., 'An analysis of plastic compression deformation of a lamella between rough plates by the energy method', *Proc. 5th Jap. Nat. Cong. Appl. Mech.*, 5, 75–78, 1955.
Kunogi, M., 'On the plastic deformation of the hollow cylinder under axial load', *J. Sci. Res. Inst. Japan*, 30 (2), 63–92, 1954.

11.9 Male, A. T. and Cockcroft, M. G., 'A method for the determination of the coefficient of friction of metals under conditions of bulk plastic deformation', *J. Inst. Metals*, 93, 38–46, 1964.

11.10 Bowden, F. P. and Tabor, D., *The Friction and Lubrication of Metals*, Part I, 2nd Ed., 1954; Part II, 1964, Oxford Univ. Press.

11.11 Bowden, F. P. and Rowe, G. W., 'The adhesion of clean metals', *Proc. Roy. Soc. A*, 233, 429–442, 1956.

11.12 Lancaster, P. R. and Rowe, G. W., 'A comparison of boundary lubricants under light and heavy loads', *Wear*, 2, 428–437, 1958/9.

11.13 Christopherson, D. G. and Naylor, H., 'Promotion of fluid lubrication in wire drawing', *Proc. Inst. Mech. Engrs.*, 169, 643–653, 1955.

11.14 Tattersall, G., *Hydrodynamic Lubrication in Wire Drawing*, BISRA Reports MW/D/46, 47, 48, 1959.

11.15 Male, A. T., *Ph. D. Thesis*, Birmingham, 1962.

11.16 Sejournet, J. and Delcroix, J., 'Glass lubrication in extrusion of steel', *Lubric. Engng.*, 11, 389–396, 1955.

11.17 Wallace, J. F., 'Stretch-forming control by phase change lubrication', *Metal Ind., Lond.*, 97, 415–418, 1960.

11.18 Rogers, J. A. and Rowe, G. W., 'Reinforced wax as a phase-change lubricant in metal-working', *J. Inst. Metals*, 92 (3), 95, 1963.

11.19 Rogers, J. A. and Rowe, G. W., Brit. Pat. No. 1,106,361, 1964.

11.20 Blazey, G., Broad, L., Gummer, W. S. and Thompson, D. B., 'The flow of metal in tube extrusion', *J. Inst. Metals*, 75, 163–184, 1948.

11.21 Schey, J. A. (Ed.), Metal deformation processes: frcition and lubrication', Marcel Dekker, New York, 1970.

12

Metallurgical Factors in Metalworking

12.1 Introduction

The major purpose of metalworking is to produce material of specified size and shape, but this would be of little value if the product did not have the required properties, particularly strength and ductility. Usually these are determined by the use for which the material is destined, but they may also be dictated by the process itself.

For example, equation 9.32 shows that strip cannot be rolled to less than a certain thickness, on a given mill, if its yield stress is too high. A limit is set in sheet forming or wire drawing by tensile instability (equations 6.25 and 2.13). Rolling and extrusion may be impossible in certain temperature ranges because of hot shortness. Forging may require high ductility to avoid cracks forming under secondary tension.

It is important to give due attention to the metallurgical nature and properties of a workpiece, although some pure metals are so ductile at room temperature that they can be worked almost indefinitely, even to translucent thinness with beaten gold.

The metallurgical condition, strength and ductility of a product are primarily determined by the composition and the amount and distribution of impurities or macroscopic defects in the stock material, but they can be modified to a large extent by thermal treatments. Failure occurring at a late stage of multiple working processes can be very expensive, so the incoming material should be adequately checked and its treatment controlled throughout the processing.

Large billets should be inspected at an early stage, preferably after casting, to reveal cracks, blowholes or other sources of weakness that might be difficult or impossible to remove later.

Some internal cavities will, however, weld together during forging, especially if they are free from oxygen. One example of overall inspection is provided by aluminium slabs 6 x 1 x 0·15 m that are regularly inspected by an automatic ultrasonic scanning method after the first rolling pass, before being rolled into sheet. Cold-forging billets are also carefully inspected, because the high local tensile strains occurring in many cold-forging operations impose severe ductility conditions. The first 'dumping' compression of a slug is itself a good test of foregeability. Much time and expense can be saved by total inspection of tubes at an early stage, conveniently by high-speed eddy current testing. Recent developments in non-destructive testing by automatic means have proved economic in many areas, by reducing the reject rate and the processing of faulty material, especially in titanium alloys, heat-resistant and high-strength alloys that are difficult to work even under favourable conditions.

It is also wise to check the metallurgical state of the workpiece. Segregation of constituents within the composition limits, as well as segregation of inclusions, can have serious effects. If the carbides in tool steels, for example, are left in bands instead of being uniformly distributed, serious weakness can develop in the tools, causing expensive or even dangerous fractures. The presence of certain elements, especially sulphur and phosphorus, can greatly reduce the ductility of steels.

Assuming that the conditions of the tools and the workpiece are satisfactory, metallurgical control is exercised, through thermal treatment, for two principal purposes: to maintain satisfactory working properties and to impart the required final properties to the product.

12.2 Hot, cold and warm working

The theoretical study of metalworking places great emphasis on forces, or working loads, which suggests that, in general terms, the workpiece should always be as soft as possible. Much smaller equipment can then be used, with evident saving of capital and tooling costs. The implication is that all workpieces should be heated – a philosophy upheld by village blacksmiths. Indeed, it is true that some operations would be totally impossible with cold materials. A rotor forging requiring a 10,000 tonne press, even when heated to 1,100°C, could hardly be produced otherwise. The further major advantage of *hot working* is that there is no strain-hardening and the ductility of the metal is maintained over an extensive range of deformation.

There are, however, serious disadvantages. It is expensive to heat the stock, and frequent reheating may be necessary to maintain the working temperature. It is also difficult to heat large volumes of metal uniformly, so the temperature, and hence the mechanical properties, may vary through a billet. This can cause trouble, for example, in extrusion where even a uniformly-heated billet will cool more rapidly on its underside when placed in a horizontal extrusion-press container. The furnace itself may also have appreciable temperature gradients from top to bottom or end to end. Scaling, or the formation of thick oxides, is an additional problem, but inert or reducing-atmosphere furnaces are expensive and completely scale-free billets may be harder to lubricate. (See Chapter 11.) Oxidation may lead to surface segregation and embrittlement, for example by residual copper in some steels.

It is more difficult to hold dimensional tolerances in hot working, because of temperature changes in the workpiece, the tools and the whole equipment. The combination of high stress and high temperature contributes to increased wear and deformation of the tools, often seriously supplemented by the abrasive action of hard oxide fragments. Thermal fatigue may damage the surface at low stress levels.

For these reasons, hot working is usually confined to the first stages of processing, where large sizes of stock are encountered but the dimensions are not critical and the surface quality does not impose a limitation. Dimensions in hot working are often specified at ± 12%. There is also considerable advantage in hot working as the first stage after casting. Continuous recrystallisation occurs during the deformation at hot working temperatures, so the original cast structure eventually disappears, leaving more uniform crystals with a greatly improved ductility, more suitable for subsequent cold working. Impurities, especially the damaging non-metallics, are broken up and more uniformly distributed. In addition, minor cracks and blowholes or porosity can be welded up, if free from oxygen, by the predominantly compressive stresses involved in most hot working. Even the secondary tensile stresses may be useful in revealing cracks or surface defects that might otherwise appear only much later.

12.2.1 *Recrystallisation*

The transition from hot to cold working is not sharply defined, but the basic difference is that in hot working there is no strain hardening, because recrystallisation occurs simultaneously with the deformation[12.1]. This is governed by the rate of diffusion, mainly the diffusion of *vacancies* within the lattice, which is strongly temperature dependent. A coefficient of diffusion can be defined as

$$D = D_0 \exp(-E/RT) \tag{12.1}$$

showing that there is an activation energy E for the process and an exponential dependence on the absolute temperature T. For most metals D_0 lies between 10 and 1,000 mm^2/s and $E \approx 20\,RT_m$. R is of course the gas constant. Experimental measurements of the activation energy (about 50 kcal mol^{-1}) suggest that an atom can diffuse about 1 mm in four months at a temperature just below the melting point in simple metals, compared with about 10^{-4} mm in the same time at room temperature[12.2]. As the rate of diffusion increases, the locking of *dislocations* decreases (§12.2.3) and the consequent strain-hardening diminishes. There is, however, an additional effect in that the activation energy for the highly-disordered regions near grain boundaries is much lower. Grain boundaries can therefore migrate relatively rapidly (100 or 1,000 times as fast as pure self-diffusion), producing recrystallisation.

In commercial metals and alloys the grain boundaries are pinned by inclusions and adsorbed solute atoms. As a rough guide, it can usually be assumed that recrystallisation will become rapid enough to produce softening within the time scale of a metalworking operation at about one third to one half of the melting point,

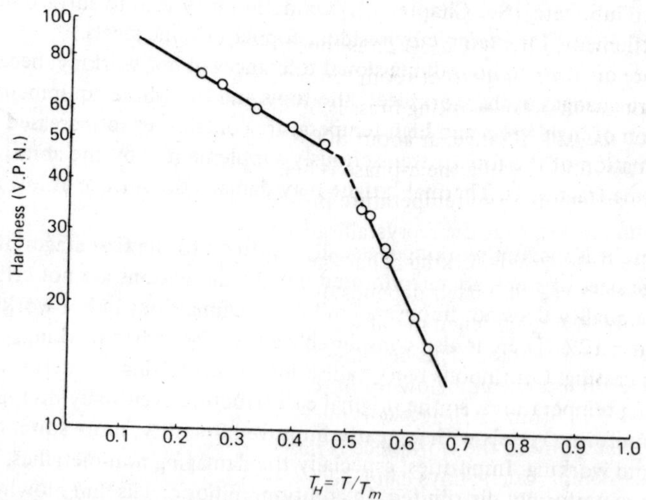

Figure 12.1. The variation of hardness with temperature for copper.

T_m, expressed in kelvin. Figure 12.1 shows the variation of hardness of copper with temperature. There are two approximately linear dependencies, intersecting at about $0{\cdot}5\,T_m$.

The recrystallisation temperature, and consequently the transition from cold to hot working, is thus related to the melting point of the metal in question, not to room temperature(12.3,12.4). Lead, for example, melts at 600 K, so $0{\cdot}5\,T_m = 300$ K, about room temperature, but it strain-hardens slightly and then recrystallises slowly on standing. It also exhibits room-temperature creep.

12.2.2 *Hot-working characteristics*

Some lead alloys, such as 49·5 Bi, 27·3 Pb, 3·1 Sn, 10·1 Cd, melt below 100°C and so should be well into their hot-working range at room temperature, but they happen to be brittle.

Brittleness can also appear in some important engineering alloys within certain elevated temperature ranges. Mild steel and low-carbon steels work-harden and strain-age simultaneously in the range 150–300°C, especially when they contain high nitrogen, and their ductility falls. This blue brittleness will be better understood if the section on dislocations and strain-ageing is read first (§12.2.3). Mild steel should therefore not be mechanically worked at these temperatures.

The phase structure of an alloy can also seriously affect ductility(12.5, 12.6). Figure 12.2 shows the equilibrium diagram for copper-zinc alloys. Compositions below about 30% Zn solidify as α solid solution but compositions with about 45% Zn solidify as β solid solutions.

The strength and ductility of α-brass both increase with zinc content, but the β-phase reduces ductility while further increasing strength. Beyond about 50% Zn a brittle γ-phase appears, which is of no mechanical use.

β-phase is hard and brittle at temperatures below about 470°C, but then undergoes a transition in which the Zn atoms in the lattice become disordered. At high temperatures, for example 800°C, the disordered β-alloy is much easier to work than α-brass.

The α-brass, typified by 70 Cu/30 Zn is thus preferred for cold working. It shows good ductility but strain-hardens at room temperature and can be annealed at about 600°C. The commonest hot-working brass is 60 Cu/40 Zn. It can be grain-refined by heating into the fully β range, at about 800°C, and shows its best properties if deformed at 700–600°C while the α-phase is being deposited. This gives a fine-grain structure and good room-temperature properties.

These examples show that the recrystallisation temperature is a useful but only general guide to hot or cold-working behaviour. A further complication is that the recrystallisation temperature itself depends upon the degree of prior working. A highly-disordered, strain-hardened metal will recrystallise at a lower temperature than when it is lightly worked. This becomes particularly evident if the recrystallisation temperature is lowered below room temperature, as happens for example in certain aluminium bronzes. Heavy working keeps such alloys soft, partly also because the temperature of the workpiece is raised, while light working allows strong strain-hardening. Such alloys can be cut only when the action is severe and continuous. Any rubbing serves only to increase the hardness.

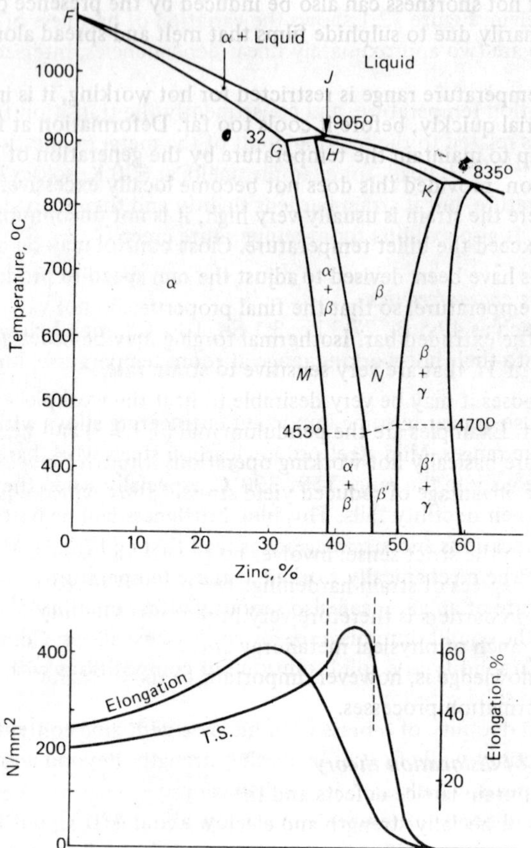

Figure 12.2 The equilibrium diagram for brasses [12.10].

The speed of deformation is also involved. High speeds that do not allow sufficient diffusion to occur will effectively strain-harden and thus cold-work the stock, even at elevated temperature. This is, however, offset by the additional local heating generated by the plastic deformation. Finally, it should be recognised that the deformation is seldom uniform, as seen in the slip-line fields, so that local strain, strain-hardening, strain-rate and temperature will all vary through the workpiece. In extreme cases there may be local melting due to adiabatic shear, especially at very high strain rates, such as occur in machining.

It is evident that hot working depends upon many factors, but in general terms the range of hot working extends from about the recrystallisation temperature to a little below the melting point for a pure metal. Usually the minimum hot-working temperature is higher for an alloy then for a pure metal, so the range is reduced, as the melting point is also lower. If the alloy contains a low-melting constituent or eutectic, as for example in leaded brass, the high-temperature cohesion may be seriously reduced, especially if that constituent is present as a grain-boundary film.

This condition of hot shortness can also be induced by the presence of impurities. In steels it is primarily due to sulphide films that melt and spread along the boundaries.

Because the temperature range is restricted for hot working, it is important to deform the material quickly, before it cools too far. Deformation at fairly high rates will also help to maintain the temperature by the generation of heat due to plastic deformation, provided this does not become locally excessive. In extrusion, for example, where the strain is usually very high, it is not uncommon for the exit temperature to exceed the billet temperature. Close control may be important and feed-back systems have been devised to adjust the ram-speed to produce a uniform desired product temperature, so that the final properties do not vary unduly from front to back of the extruded bar. Isothermal forging may be necessary for some alloys, especially of Ti, that are very sensitive to strain rate.

For some purposes it may be very desirable to heat the workpiece by the plastic deformation itself. Examples are the pendulum mill (§9.9.3) and helical extrusion (§6.13.3). Both are basically hot-working operations, though the workpieces are initially cold. The advantage of reduced yield stress is thus gained without the cost of furnace heating.

Cold working, in the strict sense, involves no recrystallisation and is characterised by greater or less degrees of strain hardening. The dislocation pattern and the presence of dislocation barriers is therefore very important. The full study of dislocations is a major branch of physical metallurgy and well beyond our present scope. An elementary knowledge is, however, important to the understanding of metal behaviour in deformation processes.

12.2.3 *Elements of dislocation theory*

All real crystals contain lattice defects and these can have a profound influence on certain properties, especially strength and electrical characteristics(12.7). We have already seen the importance of vacant lattice sites, or vacancies, in diffusion (§12.2.1). At cold-working temperatures dislocations dominate the plastic yielding, though other defects such as stacking faults and annealing twins are important in special instances and can influence dislocation movement.

For an elementary introduction to microplasticity a brief account of dislocations will suffice.

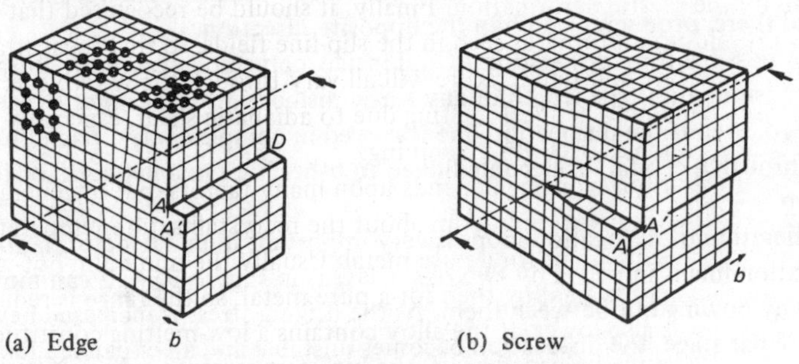

Figure 12.3. Diagram of dislocations

Dislocations are two-dimensional, or line, defects running through a crystal. Two important types can be distinguished. An *edge dislocation,* sketched in Figure 12.3*a*, arises when there is an additional row of atoms in the plane on one side compared with that on the other side of the dislocation. Conventionally, if the upper plane has the greater number, the dislocation is designated positive in a diagram and shown as ⊥.

Part of the crystal has in effect slipped over the rest in producing such a step, while the remainder has not yet slipped. The boundary between the two regions is the line *AD*, which is the edge dislocation. The direction of slip and the magnitude are given by the Burgers vector *b* of the dislocation, in this case one atomic spacing to the left.

A *screw dislocation* is formed if the Burgers vector is parallel to the dislocation line. The helical nature of the crystal surface left by the dislocation is apparent if a path is traced from the atom *A* over the surface to *A'*, one atomic spacing further along.

The usual mode of plastic deformation in a metal is by slip of one block over another along definite crystallographic planes, known as slip planes, usually those containing a high atomic density. It can be shown that such slip in a perfect crystal would require a shear stress approximately equal to one tenth of the shear modulus, hence about 1–5 kN/mm^2. Strength of this order can in fact be found in metal whiskers, which contain no *mobile dislocations,* but in all ordinary metals the shear strength is much lower. The reason is that the slip can occur much more readily by the gliding of dislocations. To a first approximation the forces opposing motion of the dislocation will balance the crystallographic forces assisting its motion, so very little external force is required to move it by glide in one direction or the other.

Movement of a dislocation out of its plane is more difficult, but can occur by migration of atoms from lattice sites near the dislocation into positions on the half plane of the dislocation, leaving vacancies in their original positions. This is clearly a diffusion process but is known as *dislocation climb*. It occurs most readily at high temperatures and is important in annealing. It can also allow dislocations to pass impurity atoms or precipitates that would otherwise impede their motion on the slip plane.

Edge dislocations can similarly act as *sinks* of vacancies. If, in contrast to the climb process considered above, a vacancy diffuses to the half plane, it may be annihilated there, producing climb in the opposite direction.

It is not possible for a dislocation to end inside a crystal, because the Burgers vector is constant along the length of any single dislocation and cannot suddenly become zero, even if the dislocation line is curved in complex ways. Each line must thus run through a crystal, or become linked to other lines in a network, or form a closed loop.

The generation of dislocation loops is very important in metal deformation[12.8]. If a dislocation line is impeded by two obstacles, as in Figure 12.4, it can move forward only by bowing out between them. As the driving stress is increased beyond the semicircular stage, the dislocation becomes unstable and an expanded loop forms, doubling back on itself. Eventually a complete loop separates, leaving a new line to repeat the process indefinitely, unless a back stress is produced by

Figure 12.4 The operation of a Frank-Read source of dislocations

obstructions elsewhere. The initial line may extend between impurity atoms, for example, or may be part of a network, so such sources are very common. If in fact these or similar *sources* did not operate, cold working would drive the dislocations out of the metal and result in a decrease rather than an increase in hardness. Such sources have been directly photographed in various materials[12.9].

The reason for strain hardening is basically that the dislocations interact and impede further glide. If one dislocation is obstructed by an obstacle, such as a grain boundary, an impurity atom, a precipitate or a *sessile dislocation* (one that can move only by climb), it will exert a back stress on a following dislocation of like sign, impeding its glide. An array of *piled-up dislocations* (Figure 12.5) will thus be formed, until the stress on the leading dislocation is sufficient to initiate slip on a new plane, or to produce climb around the barrier, or in extreme cases to initiate a crack.

When the obstacle is a grain boundary, the high stress induced by the pile-up will usually activate a new source so that slip proceeds by dislocation movement in the adjacent crystal. Grain boundaries therefore play an important part in yield, and it is found from more detailed analysis of dislocation stresses and movement that the yield stress σ_y is inversely related to the square root of the grain size, in the Petch equation

$$\sigma_y = \sigma_i + kd^{-\frac{1}{2}} \tag{12.2}$$

where σ_i is the glide resistance within the crystal and k is a factor depending on strain ageing.

One of the striking successes of dislocation theory, apart from some direct electron-microscopic evidence of dislocations and their movement, and other photographic confirmation, is the explanation of *upper yield-point* behaviour, principally in mild steel.

When carbon or nitrogen atoms are present in an iron lattice they cause a distortion which affects the hydrostatic (dilation) and shear components of the stress field[12.7]. At high temperatures solute atoms can migrate to the dislocation line

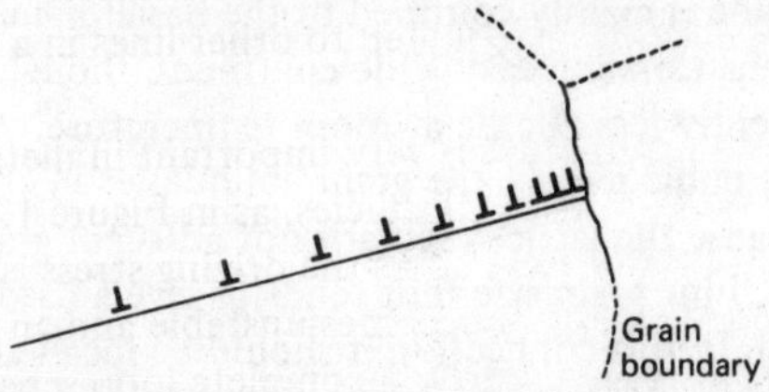

Figure 12.5. Dislocation pile-up at an obstacle such as a grain boundary.

and once there are unable to leave it without further activation. When the concentration of solute atoms becomes high enough in the vicinity of the dislocation the *atmosphere* of solute atoms will condense to a single line parallel to the dislocation line at the position of maximum binding, about two atomic spacings from the core of the edge dislocation. The stress required to free the dislocation line is thus high, but once the restraint is overcome the dislocation can glide freely. The upper yield point of annealed mild steel is consequently followed by a regime of *easy glide* at a much lower yield stress. If the specimen is allowed to age, say for a few hours at room temperature, the diffusing atoms will again pin the dislocations and an upper yield point will again occur, but if the specimen is reloaded immediately, yield proceeds at the lower stress. Once a sufficiently large strain has been imposed the strain-hardening mechanisms dominate, but at critical temperatures and rates of loading a succession of forming atmospheres and breaking away from them will produce a serrated stress-strain curve with sequential yield points. This happens at room temperature with some quenched aluminium alloys and at about 300°C with mild steel, depending on the strain rate. It is responsible for the phenomenon of blue brittleness in low-carbon steels.

12.2.4 *Cold working*

The plastic flow of metal during cold working is now seen as primarily dependent upon the movement of dislocations, vacancies and other defects.

In some instances the effect is dramatic. Lightly cold-worked mild steel will exhibit stretcher strains or *Lüders bands* that are directly related to the upper yield point and subsequent easy glide described above, because the flow is confined to localised regions. These stretcher strains appear as surface markings that are particularly undesirable in stretch-formed or pressed mild steel sheets[12.10].

For most operations, apart from stretching, the deformation is much greater. The main effect of dislocation movement is then to produce strain hardening. As we have seen above, the pinning of a short section of line may give rise to a dislocation source that can be operated by application of stress. The moving dislocations encounter many obstacles and also interact among themselves, so many sources are operated during deformation. As this proceeds, the interactions will become more numerous and more complex, so the hindrance to dislocation glide rapidly increases. In cubic crystals, where there are many possible slip planes, the three-dimensional intersections of dislocations are very important. Sessile dislocations may be formed and intersecting screw dislocations may cause *jogs*, which are steps in the dislocation line capable of moving only along the screw dislocation. Hexagonal metals, where glide is mainly confined to the basal plane, do not show this three-dimensional interaction and easy glide continues, though these metals (Zn, Cd, Mg) are also inherently less ductile at room temperature.

In general, with the cubic metals, the grain boundaries in crystal aggregates severly hinder slip because the lattice regularity of adjacent grains is disturbed in the transitional region. Fine grain size thus tends to give a harder, tougher and more uniform material. In this connection it should be recalled that cold working is frequently followed by or interspersed with annealing processes and that re-crystallised grain size is related to prior cold working, which may not be homogeneous. Many cold-worked products are, however, used without further treatment. The improved yield stress, maximum tensile strength and fatigue resistance may then be valuable[12.10].

The surface topography of cold-worked products is usually very much superior to that of hot-worked stock. Judicious choice of the final operation can leave a compressive surface stress, which also contributes to improved fatigue life. Tolerances are usually much stricter for cold-worked products, but the dimensions are much easier to control in the absence of temperature variations and surface scale. Local temperature gradients generated by the deformation can however lead to problems. The normal high finish also imposes more severe restrictions on the lubrication.

12.2.5 *Warm working*

It is possible in some instances to compromise between hot and cold working by operating at an intermediate temperature[12.11]. Such *warm working* may offer advantages in sheet forming and wire drawing, in utilising the reduction in yield strength and increase in ductility of some alloys at moderate temperatures, without the expense and associated problems of high-temperature heating.

The temperature should be low enough to allow sufficient strain-hardening to improve the properties, but high enough to avoid excessive forces and the possibility of tensile fracture. Usually the surface produced is good and there is less abrasive and deformation wear of the tools.

Warm working is particularly useful for hard heat-resistant alloys that cannot be worked cold but are very complex and may contain phases that would melt or soften unduly at high temperatures. It is not, however, usually suitable for age-hardening alloys, though there is some interest in warm working steels at temperatures between the blue-brittle range and the recrystallation temperature. Medium carbon and nickel-chrome steels can be worked between 400°C and 700°C.

12.2.6 *Superplastic alloys*

Certain two-phase alloys have been found to be capable of very large extension without fracture, in specified temperature ranges, above about $0{\cdot}3\,T_m$ K. The essential metallurgical feature is that the grain size is very small, below 5 μm, which must not be allowed to increase by recrystallisation, although the temperatures are sufficient for recrystallisation. It is probable that the two-phase structure is required mainly for this reason, and eutectics or eutectoids are suitably fine, though some single-phase metals with very fine grain size also show superplasticity[12.13]. The high elongation becomes possible because of an extremely high sensitivity of flow stress to strain rate.

The power-law relationship between flow stress and strain is given in equation 2.14 for annealed metals at room temperature

$$\tau = B\epsilon^m$$

At high temperature, creep processes are important in which the flow stress depends upon strain rate, often represented as

$$\tau = C\dot{\epsilon}^n \tag{12.3}$$

Here n is the strain-rate exponent, which often has a value of about 0·3. In superplastic alloys n takes a high value, 0·5–0·8, which approaches the value 1·0 that would apply to a Newtonian fluid (equation 12.31). They thus approximate to highly viscous materials like molten glass or heated polymers.

In a tensile test the limit to elongation is set by the tensile instability or onset of necking (equation 2.13). This involves high local strain and consequently a high local strain-rate. The flow stress of a superplastic alloy is therefore raised when necking starts, and the instability is arrested, transferring strain to another section of the specimen. Elongations up to 20x or more can be obtained; moreover the flow stress required remains very low, though of course it is dependent on the mean strain rate.

Examples of superplastic alloys are: Zn, 22% Al; Al, 33% Cu; Fe, 25% Cr, 6·5% Ni.

These alloys can be formed by processes analogous to glass blowing and to polymer forming, including shaping of sheet by atmospheric pressure. They are very cheap to work and may offer considerable savings, for example in the production of prototype dies.

12.2.7 *Workability*

Some metalworking processes such as extrusion, and especially hydrostatic extrusion, are essentially compressive, while others such as stretch forming are almost entirely tensile. Workability therefore has to be considered for each process individually, but some general concepts can be formulated because the limit is usually set by some form of cracking, attributable to direct or secondary tensile stress. Open-die forgings, for example, may be critically dependent upon peripheral cracking.

It is difficult to measure workability except in terms of the actual process with all its variables of material, dimensions, temperature, friction and shape. Nevertheless, laboratory tests on small specimens often have to be used because of the costs of full-scale trials. Use is then made of empirical correlations between the selected test and the known performance in practice. Hot-torsion testing, for example, can give a good indication of the suitability of steels for rotary piercing and tube manufacture.

Certain factors are now well established in connection with shape. For example, a rounded edge is much more prone to cracking in hot or cold rolling than a flat edge. Similarly, cracks tend to start in the middle of the barrelling that occurs during forging. The most obvious explanation is that these are due to the circumferential tension in a cylindrical billet that is forged with high friction, but the problem is more complex than that.

Cracks often form at 45° to the direction of maximum secondary tension. In rolling the cracks are usually transverse to the direction of rolling but at 45° to the plane, suggesting a shearing. This type of failure is also illustrated in Figure 5.1, which was used as a basis for considering slip-line fields. Such fracture along slip lines is characteristic of metals that have been heavily deformed or are otherwise in a relatively brittle condition. The line of fracture may sometimes be curved, following a slip-line closely.

A different form of localised deformation and fracture occurs at very high strain rates when local melting is produced by *adiabatic shear*, the heat produced cannot diffuse away rapidly enough, so the temperature in the shearing zone rises, making further shear there progressively easier.

In addition, fracture may occur under conditions of approximately *hydrostatic tension*. The slip-line field for wire drawing (Figure 6.7), for example, predicts that the hydrostatic stress p on the centre line is tensile. This arises from the condition that with zero back-pull the average stress on the entry boundary to the slip-line field must be zero, but the stress in the vicinity of the die is compressive. With an applied back-pull, of course, the central tension is increased. Under some circumstances arrowhead cracks are found, in so-called cuppy wire, at intervals along the axis.

It is clear that no simple property test will satisfactorily predict the performance under these widely differing conditions, but it is found that the percentage reduction in area measured in a tensile test gives some guide to the cold workability of a material and that this can be related to its suitability for a given operation. There is evidence[12.14] that the *workability* in a process can be defined as a product of two factors, f_1 representing ductility of the material and f_2 inherent in the process:

$$W = f_1\,(\text{material}) \times f_2\,(\text{process})$$

though a better equation can be set up if it is recognised that the basic ductility also affects the flow of metal in the process and hence f_2, and friction is also involved:

$$W = f_1\,(\text{material}) \times f_2\,(\text{process, material, friction}) \tag{12.4}$$

The value of f_2 has to be determined experimentally, but can then be used for various materials. If f_1 is taken as the reduction of area in a tensile test, f_2 may be typically 1.4 for edge cracking in rolling.

A similar approach can be applied to hot working, but it becomes less reliable as the local temperatures vary, especially when, as in the brasses considered above, the ductility of one phase differs markedly from that of another in an alloy. In general the most important precaution is to avoid hot-shortness conditions, but this may be difficult if impurities are allowed to segregate before or during the working.

12.3 Defects in metalworking

Many of the important macroscopic defects have been mentioned before but can be briefly summarised here in relation to the workability factor.

12.3.1 *Defects characteristic of individual processes*

In open-die forging a common cause of inhomogeneity is the incomplete penetration of deformation through the billet. This is clearly explained by the slip-line field, which for a light pass on thick stock takes the form of the indentation of a semi-infinite block (Figure 5.7). In hammer forging the penetration of deformation is even less because of the inertial effect of the deforming and surrounding metal.

Surface cracks may appear if there is hot shortness or the working temperature is too low for adequate diffusion. Even when the original composition of the billet is correct, sulphur from furnace gases may often cause hot shortness in iron or nickel alloys. The tensile cracks may sometimes be avoided by redesign of dies to produce compressive stress. This can also be applied to central cavities, which may otherwise be subjected to a hydrostatic tensile stress because of the slip-line field configuration (Figure 10.4), as in drawing. Axial blow-holes in castings may thus expand in forging, but they can be closed by using a three-platen compression, which gives a central compressive stress. Heading operations may also produce central tensile stresses. Closed-die forging is, of course, almost entirely compressive, but cracks can be produced by inhomogeneous flow near the flash gutter and in other areas. Residual stresses (see §12.3.2) are not usually serious in forgings as a result of the process itself, but may arise from uneven cooling, especially if the billet is quenched.

Rolling defects may remain from the original billet in the form of internal blow-holes and surface irregularities. The latter may be rolled over, possibly trapping scale, and persist as laps. The non-metallic inclusions from the casting tend to appear as long stringers, which may produce weakness. This is particularly damaging when the stress system produces a tension in the thickness direction, usually in hot rolling a thick slab, when a large crocodile crack may be produced, separating the slab into two sections. Surface damage may occur if the rolls are not kept smooth and clean, but the main problem in rolling is usually to establish uniform thickness and flatness. Elaborate feed-back control systems may be used for gauge control. Temper rolling is a useful process, already mentioned, for removing upper yield-point behaviour. Because the deformation does not penetrate the strip. as in light wire drawing, the surface is elongated and this produces a compressive residual stress. Normal rolling, however, extends the surface less than the centre, and a residual tensile stress is generated.

In extrusion the deformation is usually far from uniform and the inhomogeneity is strongly influenced by the die profile and state the of lubrication, as shown by the slip-line fields (Figure 8.4). In extreme instances the retardation of the surface layers may cause intermittent sticking and pressure build-up resulting in fir-tree cracking, especially in hot-short billets. The most common defect is known simply as the extrusion defect, arising from reverse flow, across the ram face rather than along the container. This may form a central pipe, often containing oxide from the back face of the billet. Such a change in flow pattern for a short residual billet can be seen from the appropriate slip line fields (Figure 10.7). An inclusion of scale from areas on the cylindrical surface may occur when a dead zone forms in the die corner. Again the slip line field shows the flow lines across the dead zone, which may become completely separated from the billet as skin material flows inwards along the streamline. The surface quality of the product can be improved as well as removing this entraining effect, by using a conical die, which does not allow a dead zone to form (Figure 8.3). Sometimes the surface oxide is deliberately left in the container by cutting a 'skull' with an undersized pressure pad. The formation of a skull may also occur inadvertently if the surface of the billet is too greatly chilled, especially if, as in some brasses, there is a consequent phase change to a harder constituent such as β brass. Temperature variations, due to cooling in the container and heating in the plastic zone, tend to cause wide variations in structure and in properties from back to front of the product in some sensitive alloys, but much less in pure metals. A common feature of the microstructure of extrusions in aluminium, titanium or nickel alloys is a ring of large grains due to recrystallisation being rapid after a critical deformation. Again this problem is alleviated by the more uniform deformation pattern produced by tapered dies.

The defects in rod and wire drawing can also be associated with the deformation and stress system predicted by slip-line fields, but depend upon the ductility of the stock. Central cracks may be formed under the hydrostatic tension on the axis and are liable to be dangerous if not detected. Upper-bound solutions have also proved useful in analysing cuppy wire. Surface defects are usually due to scoring by hard particles of grit or oxide or to local breakdown of lubrication. The latter is particularly important for stainless steel, nickel alloys and titanium, all of which have a tendency to adhere to the tools in the form of pickup. With very light reductions and high die angles, it is possible to induce bulge formation ahead of the die, as predicted by the slip line fields (page 137) and, in extreme cases, to shear or machine off surface fragments. The detailed form of these follows the slip-line field boundary closely, giving another example of fracture along a shear line in a relatively non-hardening alloy. The residual stress pattern depends upon the plastic deformation distribution. Light reductions cause deformation in the surface only, and result in residual surface compression along the wire with a radial tensile stress on the axis, falling to zero at the surface. Heavier passes, as normally taken, produce a tensile longitudinal surface stress and a compressive radial stress on the axis. Both produce circumferential compressive stress. The residual stress pattern in tubes is more complex. In sinking, where the diameter is reduced with no internal support, the longitudinal surface stress is tensile on the outside and compressive on the inside. The ratio of sinking to drawing, the pass geometry and the lubricating conditions will all affect the residual stress in ordinary plug drawing.

The major defect in sheet drawing is local thinning, which may lead to tensile cracks, In compressive areas, such as the flange or some corners, the sheet may wrinkle due to two-dimensional buckling. Lubrication plays an important part in sheet forming, and may significantly alter the uniformity of deformation. The residual stresses, which can cause appreciable springback and shape change, are also modified by lubrication, as well as by the geometric form of the tools. Surface defects such as scratches and local pickup can occur when lubrication is inadequate, but if the surface is free or is too well lubricated it may exhibit an orange-peel appearance, due to unrestrained deformation of individual grains according to their orientation and available slip-planes. The detailed slip behaviour associated with the upper yield-point tends to produce localised easy-glide bands that appear as stretcher strains or Lüders bands. They can be avoided by eliminating the upper yield point, usually by deforming slightly (2%) in temper rolling. The sheets must then be used within a short period or, as explained above (p. 316), the yield point will return after room-temperature diffusion of the carbon or nitrogen in the steel. Another defect directly attributable to slip processes in earing. Four or sometimes six ears may be formed on a drawn cup, due to orientation of the grains in rolled sheet, allowing easier yielding in some directions than others[12.15]. The theory of anisotropic yielding is beyond the scope of this book. Briefly, the yield criterion can be written in the form

$$F(\sigma_y-\sigma_z)^2+G(\sigma_z-\sigma_x)^2+H(\sigma_x-\sigma_y)^2+2L\tau_{yz}^2+2M\tau_{zx}^2+2N\tau_{xy}^2=\text{constant} \quad (12.5)$$

where F, G, H are determined experimentally in uniaxial tension and L, M, N are determined experimentally in shear. This equation of course reduces to von Mises equation when there is no anistropy. In terms of principal stresses, the anistropic yield criterion is

$$F(\sigma_2-\sigma_3)^2+G(\sigma_3-\sigma_1)^2+H(\sigma_1-\sigma_2)^2=1 \quad (12.6)$$

If the yield stresses measured experimentally in simple tension are Y_1, Y_2 and Y_3,

$$2F=\left(\frac{1}{Y_2}\right)^2+\left(\frac{1}{Y_3}\right)^2-\left(\frac{1}{Y_1}\right)^2$$

$$2G=\left(\frac{1}{Y_3}\right)^2+\left(\frac{1}{Y_1}\right)^2-\left(\frac{1}{Y_2}\right)^2$$

$$2H=\left(\frac{1}{Y_1}\right)^2+\left(\frac{1}{Y_2}\right)^2-\left(\frac{1}{Y_3}\right)^2 \quad (12.7)$$

The corresponding Levy-Mises equations for anisotropic material[12.16] are

$$\mathrm{d}\epsilon_x=\mathrm{d}\lambda\{H(\sigma_x-\sigma_y)+G(\sigma_x-\sigma_z)\},\ \text{etc.} \quad (12.8)$$

In some circumstances, preferred orientation may be deliberately introduced to give improved drawability in sheet forming.

12.3.2 *Residual stresses*

Residual stresses are important in many products of metalworking[12.4], as seen in the preceding section. They can cause distortion, either directly or after subsequent heat treatment, and can seriously affect fatigue life and resistance to stress corrosion[12.17].

Residual stresses arise wherever there has been inhomogeneous plastic deformation, but it is important to recognise that they are elastic stresses due directly to differences in elastic strain and cannot exceed the yield stress of the material, or the yield stress of the softer component of a pair.

The general problem of three-dimensional residual stresses becomes very complex, but three simple examples can be chosen to illustrate their origins.

(*a*) *Unequal elongation of two adjacent bars*

If two bars of equal length L are elongated to strains ϵ_1 and ϵ_2, the final length difference will be

$$(\epsilon_1-\epsilon_2)L$$

and if this is small, the two bars can be brought to a common intermediate length by elastic compression and elastic tension respectively (Figure 12.6*a*).

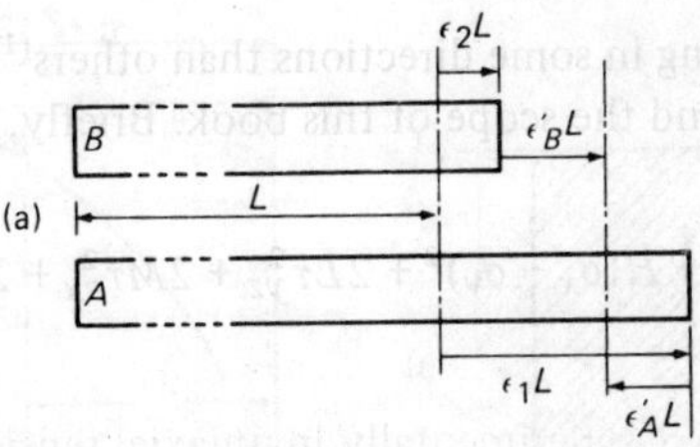

Figure 12.6(a) Diagram showing elastic strains that produce residual stresses

Thus

$$(\epsilon_1 - \epsilon_2)L = -\epsilon_A' L + \epsilon_B' L = \left(-\frac{\sigma_A'}{E} + \frac{\sigma_B'}{E}\right)L \tag{12.9}$$

The corresponding elastic stresses σ_A' and σ_B' must be related by the condition that there is no external force

$$\sigma_A' A_A + \sigma_B' A_B = 0 \tag{12.10}$$

So

$$-\frac{\sigma_A'}{E}\left(1 + \frac{A_A}{A_B}\right) = \epsilon_1 - \epsilon_2, \tag{12.11}$$

assuming that there is no bending.

If the difference in length is large it can be accommodated by plastic flow, which will not raise the stress level except by a small increase in the yield stress, which is the limiting stress level for residual stress.

After a long time at room temperature, or a shorter time if heated, the residual strain and hence the residual stress will be relaxed by processes involving the diffusion of crystallographic defects (§12.5.1).

The unequal deformation situation often arises in metalworking. In temper rolling, for example, the deformation is mainly confined to the surface regions, though these are of course not sharply distinguished from the bulk. The surface is thus elongated more than the core, so the residual surface stress is compressive.

In general, the sign of the residual stress is opposite to the sign of the plastic strain causing it.

Similar arguments can be applied to composite materials, in which the local strains must be compatible and the local forces in balance.

(*b*) *Compression with unequal prestrain* (Figure 12.6*b–d*)

In general in a real situation, this could apply to further compression of a disc machined from the specimen shown in Figure 10.7*d* where the yield stress in the central region is less than that of the remainder. If the area A_A is prestrained to ϵ_1 and A_B to some larger value ϵ_2, the total strains after a further compression ϵ_3 will be $(\epsilon_1 + \epsilon_3)$ and $(\epsilon_2 + \epsilon_3)$, with corresponding yield stresses Y_A and Y_B, Figure 12.6 (c) The area A_A of the block would recover by Y_A/E is unrestrained, and A_B by the larger amount Y_B/E. The compatibility equation for equal length is, Figure 12.6 (d).

$$-\epsilon_A' h_A + \epsilon_B' h_B = h_3\left(1 + \frac{Y_A}{E}\right) - h_3\left(1 + \frac{Y_B}{E}\right) \tag{12.12}$$

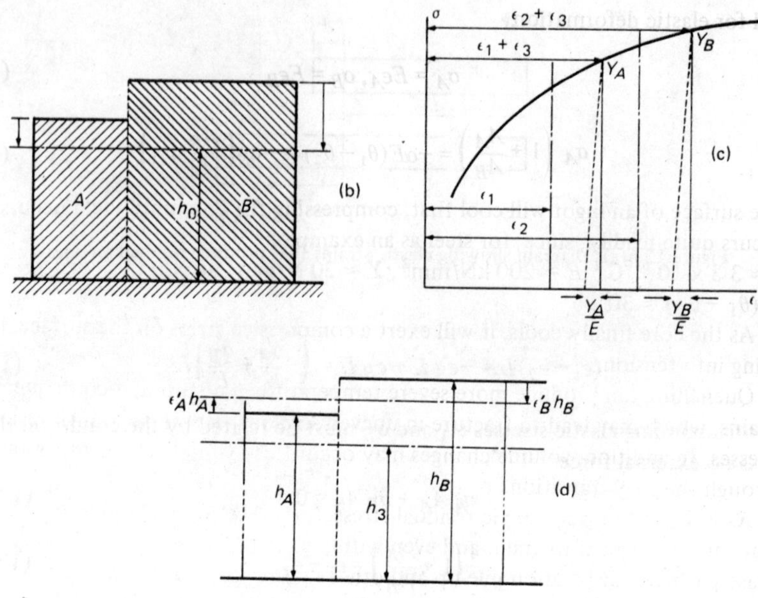

Figure 12.6 (b–d) Residual stresses arising from unequal prestrain

Neglecting second order small quantities, $h_A = h_B = h_3$

$$-\epsilon_A' + \epsilon_B' = (Y_A - Y_B)/E \tag{12.13}$$

Using the condition of zero external force and the modulus relationship again

$$-\sigma_A' + \sigma_B' = Y_A - Y_B \tag{12.14}$$

$$\sigma_A\left(1 + \frac{A_A}{A_B}\right) = -(Y_A - Y_B) \tag{12.15}$$

(c) *Thermal residual stresses*

Even slow cooling of large ingots can introduce high local stresses, leading to plastic flow and reversed residual stresses.

Suppose that a block of length L cools through θ_1°C, while an adjacent block cools through θ_2°C. The respective lengths, if free, will then be

$$L(1 - \alpha\theta_1) \text{ and } L(1 - \alpha\theta_2)$$

If $\theta_1 > \theta_2$ and they are constrained to be of equal length, A must be extended and B contracted by strains ϵ_A and ϵ_B, respectively. For compatibility

$$+\epsilon_A L - \epsilon_B L = -(\alpha\theta_1 L - \alpha\theta_2 L) \tag{12.16}$$

Because there is no external force

$$\sigma_A A_A + \sigma_B A_B = 0 \tag{12.17}$$

and for elastic deformation

$$\sigma_A = E\epsilon_A, \sigma_B = E\epsilon_B \tag{12.18}$$

$$\sigma_A\left(1 + \frac{A_A}{A_B}\right) = -\alpha E(\theta_1 - \theta_2) = -\sigma_B\left(\frac{A_B}{A_A} + 1\right) \tag{12.19}$$

The surface of an ingot will cool first, compressing the core until yield occurs. This occurs quite readily, since, for steel as an example,
$\alpha \approx 3{\cdot}3 \times 10^{-6}$ °C^{-1} $E \approx 200$ kN/mm^2, $Y \approx 20$ N/mm^2 at 900°C,
so $(\theta_1 - \theta_2) \approx 30°$.

As the core finally cools, it will exert a compressive stress on the surface, itself going into tension.

Quenching can produce more severe temperature gradients and consequent strains, which may lead to fracture in uneven sections, as well as to high residual stresses. In addition, volume changes may occur, for example in steels passing through the α–γ transition.

As well as the macroscopic residual stresses, there will usually be microstresses that vary from grain to grain, and even within grains, due to precipitated second-phase particles, dislocation pile-up and other factors. Such microstresses can be detected by the broadening of the X-ray diffraction lines.

Residual strains are not always harmful and may even be useful, for example in straightening of tubes by plastic bending, or in roller levelling of sheet.

Springback is a common phenomenon that must be allowed for in sheet metal working and some other operations.

Grinding and cutting operations usually induce large plastic compressive strains in the surface and hence residual tensile stresses, which can be very damaging in fatigue. Heavy reductions in wall thickness during tube drawing, in contrast, tend to produce compressive circumferential residual stresses.

12.3.3 *Measurement of residual stresses*

The most common method for assessing residual stresses, which cannot be measured directly, is to release part of the stress and to measure the resulting shape or dimension change.

Consider a cylinder of length L, with a compressive residual stress in the outer layers. A light cut is made over the surface, reducing the cross-sectional area from A_1 to $A_1 - \Delta A_1$. The bar must still be under zero external force, so the force initially supported by the removed skin must be redistributed over the core,

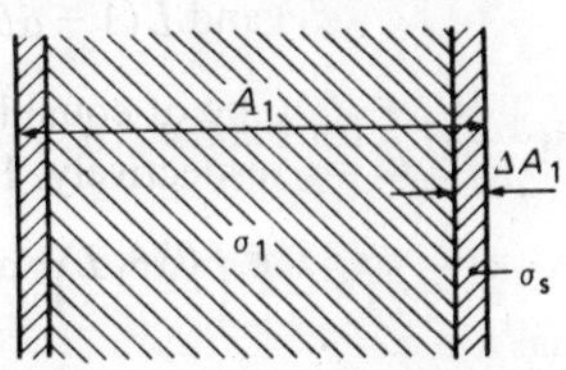

so that

$$\sigma_s \, \Delta A_1 = (A_1 - \Delta A_1)\sigma_1 \tag{12.20}$$

The core therefore changes in length elastically by an amount ΔL_1 given by

$$\frac{\Delta L_1}{L_1} = \frac{\sigma_1}{E}$$

Thus

$$\sigma_S = \frac{(A_1 - \Delta A_1)}{\Delta A_1} \frac{E \Delta L_1}{L_1} \approx A_1 E \frac{(\Delta \epsilon_1)}{(\Delta A_1)} \tag{12.21}$$

If the change in length is measured accurately after each cut, the surface stress can be calculated. As the cylinder is progressively machined away, of course, the resulting stress is increased, so the effect of each outer layer must be subtracted from the stress measured for its successor.

$$(\sigma_S)_n = A_n E \frac{(\Delta \epsilon)}{(\Delta A)_n} - \sum_{1}^{n-1} A_m E \frac{(\Delta \epsilon)}{(\Delta A)_m} \tag{12.22}$$

This is not very accurate, but shows the principle. Usually empirical methods are used, giving comparative values.

If a cylinder is progressively bored out, instead of machining the outside, both radial and longitudinal strains can be measured. For tubes, a longitudinal cut is made and the contraction or expansion of the tube diameter is noted. Flat material can conveniently be examined by grinding one side lightly, or preferably etching it away uniformly, and recording the subsequent curvature of the specimen due to the unequal longitudinal stresses across the section.

More detailed information can be obtained, at greater expense, by utilising X-ray diffraction. This is limited to surface stresses because X-rays do not penetrate more than about 0·01 mm in metals, but for consideration of fatigue life this is a critical region. The principle is to compare the atomic spacing d′ in strained and unstrained (d) crystals, using *Bragg's law*

$$n\lambda = 2d \sin \theta \tag{12.23}$$

where λ is the wavelength of the X-rays, n the diffraction order and θ the measured angle. The strain normal to the surface, if a normal incidence beam is used, is

$$\epsilon_3 = \frac{d' - d}{d}$$

But

$$-\epsilon_3 = (\sigma_1 + \sigma_2) \frac{\nu}{E} \text{ for biaxial stress } (\sigma_3 = 0)$$

so

$$\sigma_1 + \sigma_2 = -\frac{E}{\nu} \left(\frac{d' - d}{d} \right) \tag{12.24}$$

It is usual to make two determinations with two different angles of incidence, using a more elaborate analysis of the same type but generalised with direction cosines. Sophisticated automatic dual X-ray goniometers with built-in computation are now available for detailed residual stress analysis. They are used, for example, in examination of ground surfaces of aerospace components.

12.4 In-process heat treatment

For hot working, as seen in § 12.2, the necessary temperature is high enough to cause recrystallisation and the only heat treatment necessary is likely to be for homogenising and stress-relieving. Homogenising is required for large castings, to redistribute the impurities. It consists simply in holding the casting at a sufficiently high temperature, usually above annealing temperatures, for a long time, usually several hours or even days. Stress-relieving is also a diffusion process and will be discussed in § 12.5.

In cold working, however, it is very important to reheat the stock at intervals during sequential passes, to remove the effects of strain. There are two important reasons; one to restore the ductility of the metal, and the other to recover its low yield stress. A stainless steel tube, for example, may be annealed four times as it is drawn from 40 mm to 5 mm diameter, and again many times if it is further drawn to hypodermic needle sizes. Each annealing is expensive, especially if controlled atmospheres are used, and may be accompanied by removal and replacement of lubricant-carrier coatings. It is therefore important to assess correctly the amount of strain that can be imparted between annealing processes, including of course any redundant strain.

Still considering tube drawing as an example, the limiting reduction in area per pass is in principle calculable from equation 7.12 – in the example given, about 51%. In practice, to allow for variations of material and ingoing dimensions, and especially lubrication, passes not exceeding 35% are usually taken on a fixed plug. Three such passes will produce a total strain $\epsilon_{03} = 3 \times 0{\cdot}43 = 1{\cdot}29$, or about 73%. Four would produce 82%.

It is unfortunately not possible to specify on good theoretical grounds the strain or percentage reduction at which a given material should be reannealed, but practical experience suggests that fractures will be more frequent if annealing is not carried out after reductions of about 80% in ferrous materials, though greater total deformation is possible with the copper alloys. Following this guideline, and recognising that stainless steel is an expensive material, three 35% passes might be chosen. More commonly, the first reduction after annealing would be larger than the others, for three reasons. In the first pass, the final yield stress S_1 (equation 7.11) would be higher than the mean yield stress $\bar{S}$(p. 117) because of the high initial slope of the stress–strain curve, allowing a heavier reduction in area. Also, the ductility would be greater, so any additional strain, due perhaps to misalignment, would be less damaging. Finally, the lubrication would probably be better in a first pass, partly because of the lower pressure, but more importantly because of the greater difficulty of lubricating smooth drawn surfaces.

12.4.1 *Annealing*

In the simplest instance, the objective of in-process hear-treatment is to increase ductility and to reduce yield stress[12.10]. This is easily achieved with pure metals by heating them to a temperature sufficiently above about half the melting point T_m, so that diffusion occurs readily. The annealed metal can then be cooled at any rate, even by quenching, if no phase changes occur. Since prior working itself reduces the recrystallisation temperature, the minimum temperature for annealing will be lower for heavily-worked materials. The heating time is also important. It must be sufficient, but any additional time may produce undesirable grain growth, as well as wasting money.

With multi-phase alloys, attention has to be paid to the equilibrium diagram and to the times necessary for the attainment of equilibrium, or acceptable approaches to it. The alloy must be brought into the desired metallurgical condition with an appropriately fine grain size. The most common metalworking alloys are the plain carbon steels, so it is convenient to choose these to illustrate in-process heat-treatments in detail.

The iron-carbon equilibrium diagram is well known and the relevant portion of it is reproduced in Figure 12.7, from 0–1%C content. The melting point of pure iron is 1,537°C (1,810 K), so $\frac{1}{2} T_m$ = 632°C, and recrystallisation begins at about 500°C.

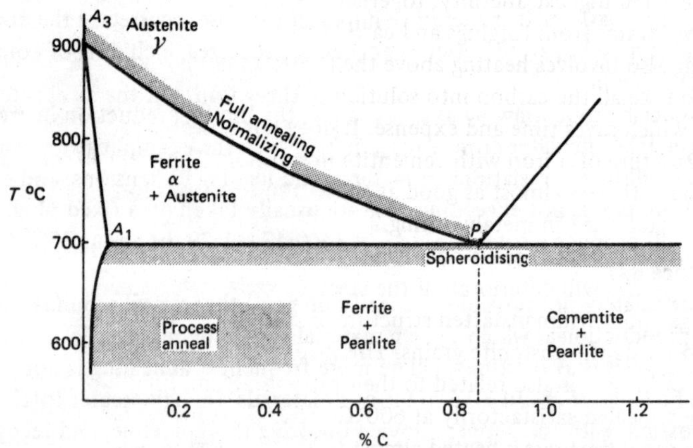

Figure 12.7 Part of the iron-carbon equilibrium diagram(12.10)

The softening process most commonly used for worked mild steel, which contains perhaps 0·15% is *process annealing,* also called *sub-critical annealing.* This involves heating to 500–650°C, followed by soaking at that temperature for an appropriate time, say 20 minutes for a 10 mm section, and cooling at a specified rate, probably over some hours. This gives considerable diffusion and recrystallisation of the ferrite matrix, which will have been severely strained in the preceding working. The pearlite will remain in elongated islands.

Prolonged heating, for a day or more, usually at slightly higher temperature, causes the pearlite to contract and *spheroidise.* This imparts greater ductility, which is often useful in cold extrusion and forging but must be undertaken with special care, otherwise cementite may form at the grain boundaries leading to intergranular fractures during cold working. It is also, of course, expensive.

It may be convenient to anneal large batches of mild steel, using perhaps 12 hours soaking at 670° followed by cooling for a few days to obtain full penetration of the heat, but it is difficult to ensure uniformity in the specimens from different parts of the furnace in *batch annealing.*

For wire and sheet steel there is advantage in *continuous annealing.* The stock is passed slowly through a furnace tunnel of sufficient length to allow heating and cooling; with total times of about 20–30 minutes. The product is uniform and sufficiently soft for further drawing, but the time does not allow significant grain growth. *Flash annealing,* at much higher temperatures for short times, can also be used, even with a throughput at common wire drawing speeds.

No transformations occur in sub-critical annealing, but in *full annealing* the steel is taken above the upper critical point A_3. It is applied to steels with less than 0·8%C; that is the hypoeutectoid steels which, on cooling from the austenitic state, first form ferrite. This is shown by the line A_3P on Figure 12.7. The steel is heated into the austenitic range, for example to 910°C for a 0·1%C steel, and very slowly cooled in the furnace after an appropriate soaking time. This gives the lowest possible yield stress and highest ductility, together with a refined structure. It is also used to remove strains from forgings and castings.

Normalising also involves heating above the ferrite solubility line A_3P for a sufficient time to take all the carbon into solution in the γ iron, but the steel is then cooled in air, which saves time and expense. It does not however produce the full equilibrium structure of α iron with cementite in solution. The yield stress and ductility are nevertheless almost as good as those resulting from full annealing, so normalising is often used in metalworking.

It should be noted that there is no advantage in heating to higher temperatures, indeed the properties will deteriorate if the steel is overheated. The austenite crystals then grow large and a Widmanstätten structure arises on slow cooling as the ferrite is precipitated inside the austenite grains. This gives a low ductility.

Annealing of brasses is also related to their phase diagram, Figure 12.2. The α brasses can be annealed satisfactorily at 600°C–750°C. At higher temperatures the grains coursen and if the brass is heated almost to the solidus temperature a 'burnt' structure is produced, with very poor properties. In α-β brass the relative proportions of α and β can be altered by changing the cooling rate and it is even possible to suppress the precipitation of α crystals by quenching from about 800°C. This is undesirable during metalworking, as it increases the strength and reduces the ductility.

12.4.2 *Heat-treatable precipitation-hardening alloys*

These alloys, of which duralumin is a good example, form an important class of materials that should be considered in connection with process heat-treatment[12.4].

The metallurgical changes are complex, but can be briefly outlined with respect to an Al 4% Cu alloy[12.10]. The initial stage of heat treatment is to raise the temperature sufficiently for the copper to go into solution, and to hold it long enough for all the copper to be dissolved. Figure 12.8 shows that there is a falling solubility curve and this is characteristic of all precipitation-hardening alloys. Typically the

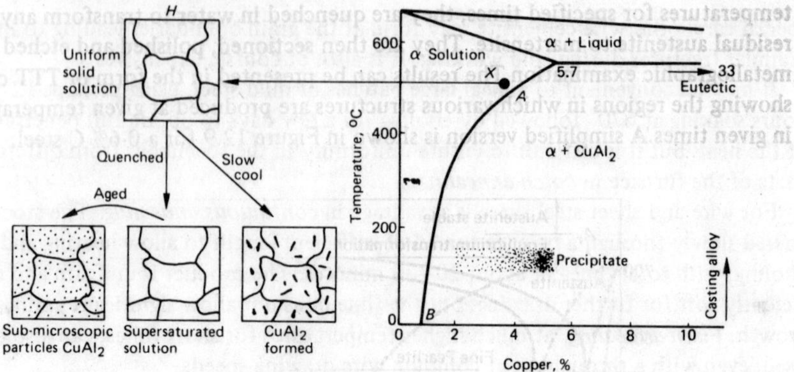

Figure 12.8. Part of the equilibrium diagram for Al-Cu alloys(12.10).

Al 4% Cu alloy would be solution treated at 500°C. If it was then cooled very slowly a compound Cu Al_2 would be precipitated as relatively coarse particles, only 0·5% Cu being retained in solution at room temperature, giving a ductile product.

If, however, the alloy is quenched from 500°C, a supersaturated solution results. This is of course unstable and Cu Al_2 will precipitate slowly, in very fine dispersion, over a period of many hours at room temperature, gradually hardening the alloy. This age hardening will be discussed in more detail later (p. 333). It can be greatly accelerated by heating at temperatures up to 200°C. Because the increased hardness is also accompanied by reduced ductility, the alloy should not be worked in this condition. It is much better to anneal the alloy, at say 450°C, when all the precipitate will be in the form of Cu Al_2 particles which are too large to interfere significantly with the dislocation movement (see §12.5.2).

A comparison can be demonstrated in rolling, where a 7·5 mm thick slab could be rolled to 1·5 mm in five passes in the annealed condition but required twelve passes with the same load limitation if solution treated and quenched. Moreover, the latter cracked badly and was quite useless after rolling.

When the high strength of duralumin is required, the alloy is first worked in the soft condition and subsequently heat-treated. This will be further discussed in §12.5.

12.4.3 *Thermo-mechanical treatments*

Recently there has been considerable interest in producing very high strength steel by judicious choice of working temperature[12.18]. The best known of these thermo-mechanical treatments (TMT) are ausforming and marforming. It is important to recognise that these processes depend upon metastable conditions, which of course do not appear on the equilibrium diagrams, so it is necessary to consider the *time-temperature-transformation* (TTT) curves[12.2].

The structure formed during continuous cooling of steel from the fully austenitic range (above the upper critical point A_3) can be studied through isothermal transformations as functions of time. Small specimens are quenched to selected temperatures below A_1, for example in molten tin. After holding at these intermediate temperatures for specified times, they are quenched in water to transform any residual austenite to martensite. They are then sectioned, polished and etched for metallographic examination.The results can be presented in the form of TTT curves showing the regions in which various structures are produced at given temperatures in given times. A simplified version is shown in Figure 12.9 for a 0·6% C steel.

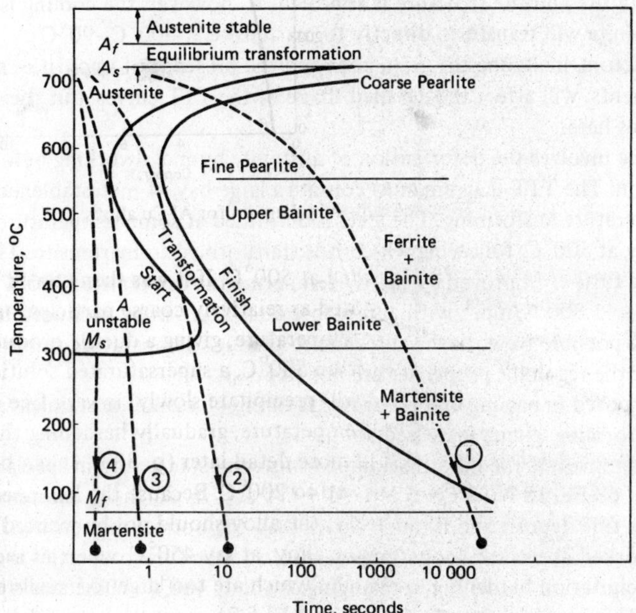

Figure 12.9 TTT curves for a 0·6%C steel

The equilibrium transformation is represented by the band A_s–A_f. The austenite starting temperature A_s is that at which austenite can just form due to heating, but does not increase perceptibly with time. The austenite finishing temperature A_f is the maximum at which any ferrite can continue to exist.

At temperatures close to 700° C the austenite forms some ferrite and also produces carbides (Fe_3C) by a process of nucleation and growth. The incubation period for formation of nuclei is relatively long and the areas of ferrite and carbide, deposited together as pearlite, grow to large sizes.

At about 500° C the nucleation rate is greater but the grain growth is slower so a finer pearlite forms. Between 500° C and 350° C the initial nuclei are ferrite, but cementite then precipitates, forming finer needles of bainite, in fact upper bainite which is distinguished from lower bainite. The latter forms at temperatures between 350° C and 250° C.

At temperatures below 250° C the austenite cannot transform to pearlite and bainite by diffusion processes, but changes by internal shearing to martensite, a distorted body-centred cubic structure.The shearing transformation occurs at very high speed and is therefore not time-dependent. The amount transformed depends on the elastic energy to produce shearing and is determined primarily by the temperature. Martensite begins to form at the temperature M_s (about 300°C for 0·6°C steel) and is complete at M_f, about 80°C.

With continuous cooling, the rate of fall of temperature affects the structures produced. The cooling curve can be superimposed on the TTT curve, as shown dotted. During slow cooling the transformation to ferrite and pearlite is completed at a high temperature and the structure is annealed. If, however, the cooling is very fast, the austenite will transform directly to martensite at 300°C–90°C.

Various factors, including the grain sizes and the presence of impurities and alloying elements, will affect the detailed shape of the TTT curves, but these need not concern us here.

Ausforming involves the deformation of austenite by metalworking before its transformation. The TTT diagram must contain a large bay of metastable austenite for low-temperature ausforming. The steel is deformed at some temperature between A_1 and M_S, say at 500°C, following which it is transformed to martensite. The properties can be further improved by lightly tempering the martensite (300°C). Strengths up to 1,800 N/mm^2, with acceptable ductility, can be produced in this way. It is also possible to work the austenite in the high-temperature stable range, above A_S, but the resultant properties are not so good.

It may be noted in passing that cryogenic forming of austenitic stainless steel is a type of ausforming, giving increased strength.

Marforming involves the deformation of transformed austenite in special maraging steels, e.g. 68 Fe, 18 Ni, 8 Co, 5 Mo, Al + Ti trace, in which the martensite is produced in a fine deposit and is much softer than the normal martensite. This material is worked at or near room temperature, and the final properties are developed by precipitation hardening in maraging at about 480°C. These steels have a very good fracture toughness. Ordinary steels can be strengthened usefully by imparting a small deformation, not exceeding a few per cent, while the steel is in a tempered martensitic condition, and then retempering.

12.5 Post heat-treatment

Worked materials usually have some degree of inhomogeneity and some residual stress. Even hot-worked alloys will often require reheating to homogenise the structure and to relieve stresses of mechanical and thermal origin. Cold-worked alloys may require softening, to restore sufficient ductility, for example in rolled sheet that needs to be further deformed, or in bars or tubes that need to be bent. The strain hardening in cold-worked stock may, however, be useful in giving higher strength, so only a stress-relieving process or no further treatment is needed. Products of either hot or cold working may also be heat-treated to produce optimum properties, for example in dispersion-hardening alloys.

12.5.1 *Stress relieving*

As shown in §12.3.2, residual stresses can easily result from metalworking and may have values up to the yield stress. They can be substantially reduced by moderate heating, which slightly reduces the yield stress, but more significantly enhances the diffusion of point defects. This allows edge dislocations to climb out of their slip planes and to be annihilated. Dislocations of like sign may also become aligned in walls forming small-angle sub-grain boundaries. This *polygonisation* evens out the dislocations and produces a slight softening. It is accompanied by a large change in electrical resistivity and in stored energy, because of vacancy diffusion.

Such *recovery* should be distinguished from *recrystallisation*, which involves nucleation and growth, totally replacing the strained lattice with effectively stress-free grains. This occurs at the higher temperatures associated with annealing. Stress relieving temperatures may be typically 500°C for carbon steels and 250°C for brass. Times of about one hour are commonly used, but compromises have to be reached with other requirements. The grain structure is unaffected, but there is evidence that polygonisation provides the preliminary stage for recrystallisation.

12.5.2 *Improvement of properties*

Post heat-treatment of steels is usually directed towards the attainment of maximum strength compatible with adequate ductility. The greater hardness is obtained by rapidly quenching from the fully austenitic range to produce martensite, as seen in § 12.4.3. This is then tempered by moderate heating to improve the ductility of the highly-distorted structure without serious reduction in yield stress. The temperatures usually lie in the range 450°C to 250°C, the higher tempering temperatures giving tougher material at the expense of hardness. Temperatures of 150–250°C may be used where extreme hardness is needed, but these only provide some stress relief and encourage decomposition of any retained austenite. At higher temperatures, diffusion becomes significant and tempering begins to merge with annealing. In some steels, tempering may decompose the martensite into a fine dispersion that provides age-hardening effects, but this is too complex to discuss here[12.2].

A better-known *age-hardening* process occurs in duralumin type alloys, as already briefly discussed (§ 12.4.2) in connection with in-process heat-treatment. The post-treatment for obtaining optimal properties is somewhat more subtle. Considering the Al 4% Cu alloy again, with reference to Figure 12.8, we can consider the sequence of events after quenching a solution-treated sample, if it is held at say 150°C over a period of days.

The copper atoms rejected from the supersaturated solution first aggregate into very small and very thin platelets, perhaps only 10^{-5} mm across, forming what are known as Guinier-Preston zones (GP1). These are then increased in size by diffusion of further copper atoms into them, forming GP 2 zones, also known as θ'' precipitates. These do not quite fit into the aluminium lattice, but do fit in certain directions so are not entirely separate from it. Significant distortion of the lattice is thus caused, setting up large strain fields which impede the movement of dislocations so that the yield stress is significantly raised, even though the precipitate is too small to see in an optical microscope. Further continuation of heating at the same temperature produces another intermediate precipitate θ', again very small in size, and the hardness and yield stress reach a maximum value. Beyond this stage over-ageing occurs, in which the precipitate becomes a much coarser one, θ, which has the composition Cu Al_2 and loses coherences with the Al lattice. The strain fields are thus reduced and the precipitation-hardening effect diminishes. The fine precipitates also have an influence on the strain-hardening behaviour and, as explained earlier, should be avoided during the working processes.

The properties of such alloys can be controlled over a fairly wide range by suitable heat-treatment after working. It may even be possible to obtain significant improvement by room-temperature ageing. For example, aluminium alloy rivets are sometimes closed by cold forging in the quenched condition, which is sufficiently soft and ductile for the purpose, and are then allowed to age-harden by precipitation of the θ' phase at room temperature during a period of several days. The strength can be appreciably increased and even doubled. Obviously care has to be taken to avoid ageing before the rivets are closed, so they are quenched shortly before use.

12.5.3 *Recrystallisation after metalworking*

The oriented grain structure of worked material can be restored to a more uniform type by diffusion at annealing temperatures, but it should be remembered that the recrystallisation is itself affected by prior working. In general terms, cold-worked material, having a higher energy content, will recrystallise more rapidly than less-worked material[12.19]. These are however certain critical ranges, very obvious in some alloys, in which prior strain produces very large grain growth on subsequent recrystallisation. This occurs at about 11% deformation in aluminium and is also found in titanium and nickel alloys. It is readily seen in cross-sections of extruded products, where the inhomogeneity of deformation gives a wide range of strains, appearing as a ring of coarse grains. In forged sections the large grains are likely to appear in the lightly-deformed dead zones. These areas will be weaker than the main body, but can be eliminated by suitable overall working of the stock or by making special preforms. Slip-line field theory can contribute to the prediction of local strain and the avoidance of the critical regions (Figure 10.7).

12.6 Tool materials

12.6.1 *Solid tools*

The basic requirement of a metalworking tool is that it should retain its dimensions. Clearly, it should not deform plastically and it will therefore have a higher yield stress than the workpiece, or in other words it will be harder. Tools will of course deform elastically and §9.7 shows that this can be important, for example in rolling, where flattening and bending of the rolls and stretch in the mill housing affect the flatness and thickness of the strip. Equation 9.31 shows that elastic deformation of the tool sets a limit to rolling. In other operations, such as tube and bar drawing, the die will expand slightly under load, but the effect is constant and is easily compensated.

For very hard materials the modulus of elasticity is significant. In rolling, the high E value of tungsten carbide reduces flattening and allows thinner strips to be rolled. It is also critical in the special situation of a cold forging punch,which acts like a strut and may be limited by failure due to buckling at a critical load P_1 specified by the Euler formula

$$P_1 = \frac{\pi^2 EI}{L^2} = \frac{\pi^2 EA}{(L/k)^2} \tag{12.25}$$

It may also be important in the bolstering of dies, which involves shrinking a ring on the outside to improve the tensile strength of the die, because *E* differs considerably between tungsten carbide and steel, so the stresses will differ at a given strain. Otherwise the modulus is not important in metalworking tools; moreover it is approximately the same for all steels.

Increasing hardness in tool materials also, in general, gives improved wear resistance. Diamond is the hardest known material and it is in fact used very effectively for making drawing dies with very small holes for fine wire drawing. Such dies show very little wear, but are eventually re-used several times by being bored out to successively greater diameters. The compound next in hardness to diamond is the artificial cubic modification of boron nitride, but the most commonly used hard tool materials are *cemented carbides,* usually tungsten carbide bonded with cobalt. Large dies, many centimetres in diameter, can be readily produced, though they usually require bolstering to provide an initial high compressive stress (p. 337), because the material is more brittle than steels. WC – Co is an extremely useful tool material and is widely used for cutting tools as well as forming tools. It is made in many shapes and sizes by powder metallurgical techniques and can readily be pierced and contoured by electric discharge machining. It has very good wear resistance and provides the longest life of any common tool material. A major feature is that it combines high hardness with an acceptable toughness, which is lacking in ceramic materials. Various compositions containing WC, TiC, TaC, etc. are available. A general rule is that increasing Co content reduces hardness but increases toughness, and finer grain size increases wear resistance. 11% Co with 1 μm grain size is widely used.

Ceramic tools can be used with advantage in cutting, and especially of course in grinding, but have so far found little application in metalforming because of their brittleness. Some sintered alumina tools have been used and they show very good abrasive wear resistance combined with resistance to corrosion and to pickup. It is of course very important to avoid shock loading with these materials[12.19].

The next major group of tool materials is the *tool steels*[12.20, 12.21]. These are obviously very hard and are difficult or impossible to forge. They contain many constituents and even hot working is very restricted because of the possibility of some phases being taken too close to their melting temperatures. They can be pierced and shaped by electric discharge machining, grinding and to some extent by electrochemical machining.

Hot-working tool steels should in general be as hard as possible, with enough room-temperature toughness to avoid cracking. They usually contain tungsten, or molybdenum. A typical tool steel for hot forging is H 21:

0·3 C 0·3 Mn 10 W 3 Cr 0·3 V 0·3 Mo

It can be oil quenched from 1,150°C and tempered at 570°C. A harder one is H 12:

0·55 C 0·8 Mn 1·1 Si 5 Cr 1·5 Ni 1·5 Mo 1 V

A common hot-extrusion die steel is

0·35 C 5 Cr 1·5 Mo 1·5 W 0·4 V

This is reasonably resistant to thermal crazing.

In addition nickel cast-irons are used, for example in tube drawing at high temperatures and for mill rolls. Ni – Hard contains

2·8 C 0·5 Si 1·2 Mn 4·5 Ni 0·75 Cr 0·25 Mo

It is used in a martensitic condition.

Abrasion by oxide scale is very damaging to hot-working tools and the scale should be removed, as far as possible, immediately before the workpiece is presented to the tool. Mechanical or water-jet descaling is usually employed.

Cold-working tools do not need to retain their strength up to high temperatures, but do require a high room-temperature hardness because of the relatively higher yield stress of the workpiece. The hardest useful alloys are the cobalt alloys of the Stellite type but these are too brittle for metalworking. There are various tool steels available that provide high hardness, usually attributable to their constituent carbides of Cr, V and Mo or W. They also have good toughness and wear properties. A typical composition is

1 C 1 Mn 1·2 W 0·5 Cr

This can be oil quenched from 800°C and lightly tempered at 200°C.

High-speed steels are used for some applications where abrasion-resistance, high compressive strength and toughness are vitally important, for example in cold-extrusion punches. It has recently been shown that these steels strain-harden very rapidly over a small plastic strain range. In cold extrustion the punch always deforms slightly in the first few operations, because of the extreme pressure developed in cold extrusion or forging of steel, so this property is very valuable. The high-speed steels are complex but usually contain high tungsten or tungsten together with molybdenum. Two representative compositions are

T 1 type: 0·75 C 4·1 Cr 18 W 1·1 V
M 2 type: 0·85 C 4·1 Cr 6·4 W 5·1 Mo 1·9 V

They are mainly used for cutting but can themselves be formed in the annealed condition. They are softened at 850°C and slowly cooled. After forging they should be reheated to 680° and air cooled to relieve the stresses and homogenise the structure. The complex phase composition requires great care in heat treating to avoid cracking. Quenching may be done in two stages, cooling slowly to 600°C in a salt bath and then quenching in oil. Multiple tempering is also desirable, for example heating to 400°C and cooling, followed by a further heating to 350°C and cooling.

Size plays an important part in tools. Not only is the cost greatly increased with large sizes, for example in tungsten carbide, but the alloys may simply not be available because of production difficulties, including heat treatment. Bolstering has already been mentioned for carbide dies. It allows much smaller pellets of carbide to be used by transferring some of the stress to a softer and cheaper surrounding material. The principle is well known, and is the same as that used in reinforcing gun barrels[12.17]. The theory is developed from the Lamé equations for elastically loaded cylinders:

$$\sigma_{rr} = \frac{pr_o^2}{r_1^2 - r_o^2}\left(1 - \frac{r_1^2}{r^2}\right)$$

$$\sigma_{\theta\theta} = \frac{pr_o^2}{r_1^2 - r_o^2}\left(1 + \frac{r_1^2}{r^2}\right) \qquad (12.26)$$

where σ_{rr} and $\sigma_{\theta\theta}$ are respectively the radial and circumferential (or hoop) stresses induced at radius r by a pressure p acting in a cylinder of internal radius r_o and external radius r_1.

It can be seen that in a single cylinder there is little advantage in making r_1 much greater than about $3r_o$, since the maximum value of $\sigma_{\theta\theta}$, at $r = r_o$, is given by

$$\sigma_{\theta\theta} = p \,.\, \frac{1 + (r_1/r_o)^2}{1 - (r_1/r_o)^2} \qquad (12.27)$$

The strength is greatly increased if a second tube is shrunk onto a relatively thin tube so that the stress at r is artificially raised. It is possible in this way to design a bolstered die so that not only is the stress across the interface always compressive, but also the circumferential stress at the outer radius of the carbide insert never becomes tensile, even when the bar or tube is drawn through it. In fact with tungsten carbide a small tensile strain is permissible without fracture.

The shrink fitting needed to produce the initial large compressive hoop stress is achieved by making the inner diameter of the bolster slightly smaller, by a precisely known amount, than the outer diameter of the die. The bolster is then expanded by heating through several hundred °C, until it can just be fitted over the die in a rapid pressing operation. When it cools it induces compressive hoop stress in the die, as explained in § 12.3.2, and tensile hoop stress in the bolster. Alternatively, the die pellet may be forced into the bolster at room-temperature, using a slight taper to generate the necessary interference, and the die may first be cooled in liquid nitrogen.

12.6.2 *Coated tools*

The bulk hardness of metalforming tools is, as has been seen above, very important in preventing them from deforming plastically. Extreme hardness is, however, associated with brittleness so the second advantage of hardness, in reducing wear, may not be fully utilised. Various attempts have been made to increase the surface hardness of dies to improve their life.

One method, already mentioned in connection with high speed steels, is to allow a small amount of deformation which produces a large strain hardening. However, this is of limited application. Another possibility is to utilise a martensitic type of transformation induced by strain, as in Hadfield's Mn steel. This has been used effectively in certain aluminium bronze alloys for the forming of titanium. The wear rate is appreciable for such tools, but there is no pickup of titanium on the bronze.

A simple method, used in cold tube drawing, is to plate a steel die with hard chromium. This considerably increases the wear resistance, as well as reducing pickup, and has the further advantage that the worn surface layer can be stripped off and replaced quite easily.

It is also possible to coat steel dies with ceramic or hard-metal facings by flame-spraying or plasma-spraying. Such treatments have been used but require great care to produce smooth, coherent and adherent coatings that will not chip or spall off in use. It is important that the substrate should be hard enough to deform only elastically, but even then the elastic strains produced by thermal or mechanical means may not be compatible, leading to detachment of the film. Air-hardening steels, which precipitation harden at room temperature, are preferred as substrate material, to avoid softening due to the inevitable high temperature of deposition by spraying.

A gentler heating is involved in vapour plating, where the coating is formed by double decomposition reactions in surrounding vapour. Titanium carbide, for example, can be deposited on moderately heated steel through decomposition of $TiCl_4$ and a hydrocarbon vapour. Such coatings are brittle but very wear resistant and tend to reduce risk of pickup. Ion implantation involves no heating.

A further possibility, exploited recently in cutting tools, is to produce a composite sintered material. For example, a wear-resistant but brittle TiC surface can be produced with a tougher but somewhat softer WC–Co supporting material.

12.7 The properties and forming of polymers

The general family of polymeric materials offers a very wide range of chemical, physical and mechanical properties. In recent decades these materials have found widespread industrial and domestic applications, in many instances replacing metal items. It is important to consider the forming processes used for polymers, together with the properties influencing and being influenced by their manufacture.

The major forming techniques superficially resemble those used for metals, but differ in important respects because of the material properties. The theoretical treatment of polymer shaping is more complex and much less developed than for metals. Only a very brief account will be attempted here and the reader is referred to some of the specialist textbooks for further information[12.22–12.24].

12.7.1 *The nature of simple polymers*

The simplest organic polymers are hydrocarbons, though others may contain hydroxyl, fluorine or other atoms or groups of atoms. They are formed of basic units, or *mers,* which can share electrons and thus form repeated groups within a single large molecule. The plastics in common use are high polymers with very large molecular weight. The reason for this is that the primary C–C bond strength is very much larger than the cohesive bond strength between molecules (hundreds of k joule/g mol, compared with a few k joule/g mol). Neighbouring molecular chains of short length thus slide over one another relatively easily and the strength of the solid is low, compared with that of high molecular-weight polymers in which the long chains have stronger interaction, especially when they are mechanically interlinked or densely packed. The limiting strength is, of course, that of the primary homopolar valance bonds.

In order to form such long molecules with repeated covalent bonding, it is necessary for the *monomer* to contain a double bond. For example, the plastic known as polythene is derived from ethlene C_2H_4.

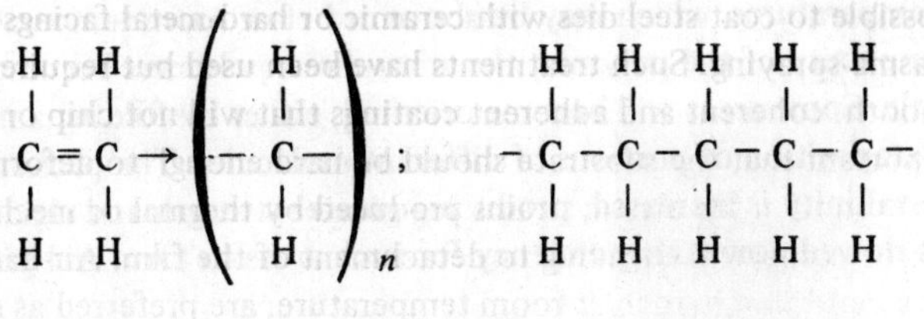

$$\begin{array}{c} H \quad H \\ | \quad\; | \\ C = C \\ | \quad\; | \\ H \quad H \end{array} \rightarrow \left(\begin{array}{c} H \\ | \\ -C- \\ | \\ H \end{array} \right)_n ; \quad \begin{array}{c} H \quad H \quad H \quad H \quad H \\ | \quad\;\; | \quad\;\; | \quad\;\; | \quad\;\; | \\ -C-C-C-C-C- \\ | \quad\;\; | \quad\;\; | \quad\;\; | \quad\;\; | \\ H \quad H \quad H \quad H \quad H \end{array}$$

monomer mer polymer

This material may exist in many solid forms. The molecular weight, depending of course on the number of repeating units (C = 12, H = 1), may be anywhere between 100 and many thousands. Thd individual molecules may be linear or branched or cross-linked to others or themselves.

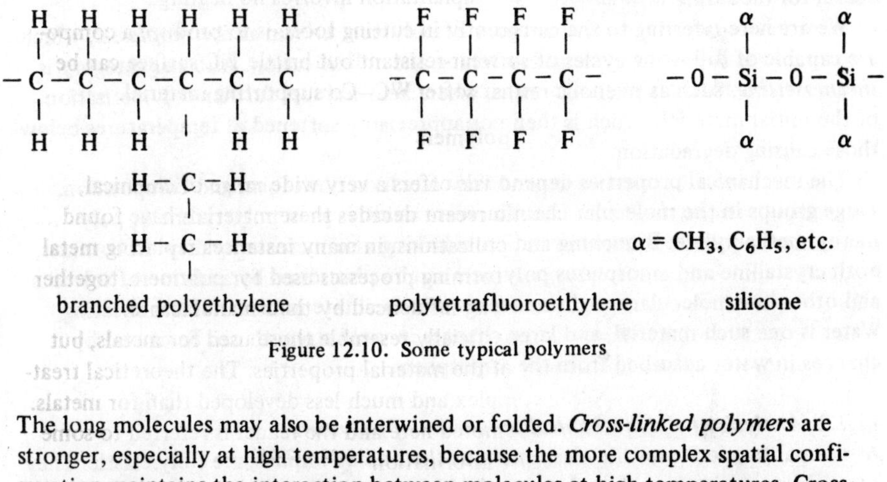

$$\begin{array}{c} H \quad H \quad H \quad H \quad H \quad H \\ | \quad\;\; | \quad\;\; | \quad\;\; | \quad\;\; | \quad\;\; | \\ -C-C-C-C-C-C \\ | \quad\;\; | \quad\;\; | \quad\;\; | \quad\;\; | \quad\;\; | \\ H \quad H \quad H \quad\;\; | \quad\;\; H \quad H \\ H-C-H \\ | \\ H-C-H \\ | \end{array} \qquad \begin{array}{c} F \quad F \quad F \quad F \\ | \quad\;\; | \quad\;\; | \quad\;\; | \\ -C-C-C-C- \\ | \quad\;\; | \quad\;\; | \quad\;\; | \\ F \quad F \quad F \quad F \end{array} \qquad \begin{array}{c} \alpha \qquad\quad \alpha \\ | \qquad\quad | \\ -O-Si-O-Si- \\ | \qquad\quad | \\ \alpha \qquad\quad \alpha \end{array}$$

$\alpha = CH_3, C_6H_5$, etc.

branched polyethylene polytetrafluoroethylene silicone

Figure 12.10. Some typical polymers

The long molecules may also be interwined or folded. *Cross-linked polymers* are stronger, especially at high temperatures, because the more complex spatial configuration maintains the interaction between molecules at high temperatures. Cross-linked polyethylene will, for example, withstand boiling water, whereas straight-chain polyethylene becomes soft.

If the H-atoms are replaced by F-atoms, the basic strength of the carbon chain is unaffected, but the interaction between chains is reduced. Polytetrafluoroethylene has a bulk strength comparable to that of polyethylene, but a remarkably low friction. It is used in dry lubrication (p. 300).

The backbone chain can be formed from other atoms, including the silicon-oxygen family (Figure 12.10). These silicones contain 0-atoms linking Si-atoms to which are attached organic groups such as methyl or phenyl radicals. They are stable at appreciably higher temperatures than the organic polymers.

Polymers can exist in three main physical states: melt, amorphous and crystalline. At high temperatures the lower straight-chain polymers behave like high-viscosity liquids, or melts. Those of higher molecular weight exist in a tangled or amorphous conditions. They are very flexible at high temperatures, behaving like rubbers. Some retain this property at room temperature and are known as *elastomers,* but then have little relevance for forming.

At lower temperatures, which may be above or below room temperatures depending on the complexity of the polymer, the amorphous polymers become glassy and relatively brittle. Some areas will become oriented, either by folding a molecule upon itself or by juxtaposition of portions of different molecules. The packing in these regions of crystallinity is increased, producing greater strength and rigidity. As the cooling rate is slowed down, the tendency for such areas to form and grow will be enhanced.

Some polymers, particularly nylon, have all the crystallite regions oriented in approximately the same direction, giving a further increase in strength.

For good forming properties, the polymer should be heated into the amorphous or melt condition. Temperatures up to about the glassy transition temperature can be used for normal applications of polymer products; this may be 70°C. Temperatures from 120° up to melting point are suitable for forming. The viscous state is useful for moulding.

We are here referring to the category of polymers known as *thermoplastic,* which are capable of following cycles of softening and solidification. Another category is *thermosetting,* such as phenolic resins, where heat induces further polymerisation of the initial material, which is then not appreciably softened at temperatures below those causing degradation.

The mechanical properties depend very much on both form and composition. Large groups in the molecular chain increase rigidity, as in polystyrene or polymethylmethacrylate. Branching and cross-linking also increase rigidity. In contrast, both crystalline and amorphous polymers can be plasticised by addition of solvents and other low molecular-weight materials that facilitate inter-molecular shearing. Water is one such material, and large changes in yield strength can be caused by changes in water adsorbed from the atmosphere.

12.7.2 *Analysis of the mechanical properties of polymers*

At normal temperatures the elastomers, as their name implies, are fully elastic. They have a large strain range, but show marked hysteresis. The size of the hysteresis loop, and consequent loss of energy in cyclic deformation, increases with temperature.

Most polymers behave viscoelastically. At high rates of loading they respond elastically, but they then relax in a creep-like behaviour as the long molecules glide over one another. A simple model is given by a spring and dashpot system (Figure 12.11).

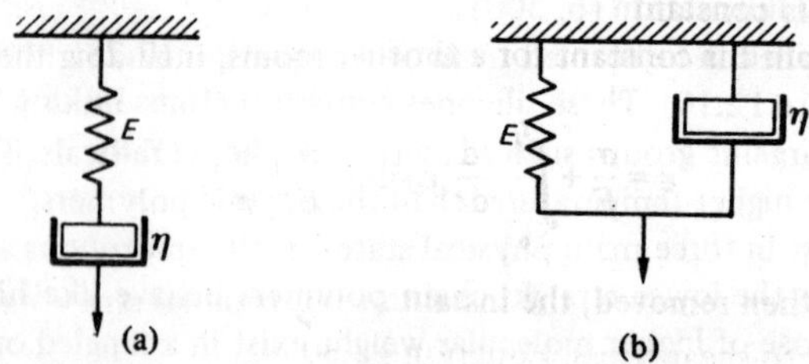

Figure 12.11. Simple mechanical analogues of polymer deformation

The total strain ϵ is the sum of the elastic strain ϵ_s and the unrecoverable strain ϵ_d, in the model shown in Figure 12.11 *a*.

$$\epsilon = \epsilon_s + \epsilon_d \tag{12.28a}$$

and the same is true of the strain rates

$$\dot{\epsilon} = \dot{\epsilon}_s + \dot{\epsilon}_d \tag{12.28b}$$

For the elastic component it is not possible to define a single constant of proportionality between stress and strain, but a secant modulus can be calculated from the stress-strain curve over a given range

$$E_s = \sigma_s/\epsilon_s$$

If it is assumed that the dashpot strain follows a Newtonian viscosity law, the coefficient of viscosity is defined as

$$\eta = \tau / \frac{\partial V_x}{\partial x} \tag{12.29}$$

and the velocity gradient $(\partial V_x/\partial x)$ can be regarded as a shear strain that increases continuously with time for as long as the stress is applied. Thus

$$\frac{d\gamma}{dt} = \frac{\tau}{\eta} \tag{12.30}$$

In terms of the direct stress and strain we can write

$$\dot{\epsilon}_d = \sigma_d/\eta \tag{12.31}$$

so equation 12.28b becomes, with a single stress σ applied,

$$\dot{\epsilon} = \frac{1}{E}\frac{d\sigma}{dt} + \frac{\sigma}{\eta} \tag{12.32}$$

Two conditions may be considered:

(i) If the applied strain rate is constant at $\dot{\epsilon}_c$

$$\frac{d\sigma}{dt} + \frac{E}{\eta}\sigma - E\dot{\epsilon}_c = 0$$

giving σ in the form $\sigma = A + Be^{xt}$

$$\sigma = \dot{\epsilon}_c \eta \{1 - \exp(-E\epsilon/\dot{\epsilon}_c \eta)\} \tag{12.33}$$

since $t = \epsilon/\dot{\epsilon}_c$ if $\dot{\epsilon}_c$ is constant

(ii) If the stress applied is constant for a time t_1, equation 12.28a gives

$$\epsilon = \frac{\sigma}{E} + \int_0^{t_1} \frac{d}{dt}\epsilon_d \, dt = \frac{\sigma}{E} + \frac{\sigma}{\eta} t_1 \tag{12.34}$$

and if the stress is then removed, the instantaneous residual strain will be

$$\epsilon_r = \frac{\sigma}{\eta} t_1 \tag{12.35}$$

If then the length of the strained polymer is held constant, the stress will relax in accordance with the equation

$$\frac{1}{E}\frac{d\sigma}{dt} + \frac{\sigma}{\eta} = 0 \qquad (12.36)$$

$$\sigma = \sigma_o \exp(-Et/\eta)$$

This can be written in the form

$$\sigma = \sigma_o \exp(-t/T) \qquad (12.37)$$

suggesting that $T = \eta/E$ is a relaxation time, the time for the stress to decay to $1/e$ of its original value. Thus for times much shorter than η/E the polymer behaves like an elastic solid, and for times much greater than this, like a viscous fluid. Metals also behave in a similar way during high temperature creep.

Such a model is useful in understanding the response of polymers to stress, but it is oversimplified. In particular it assumes an instantaneous elastic response, but in fact the uncurling of the long polymer chains gives a delayed elastic response. This can be considered equivalent to the parallel spring and dashpot model shown in Figure 12.11b. The basic equation is now written in terms of the force equilibrium.

$$\sigma = \eta \frac{d\epsilon}{dt} + E \qquad (12.38)$$

which has the solution

$$\epsilon = \frac{\sigma}{E}(1 - \exp(-Et/\eta)) \qquad (12.39)$$

The response is thus an exponential extension until the stress is removed, followed by an exponential recovery to the original length.

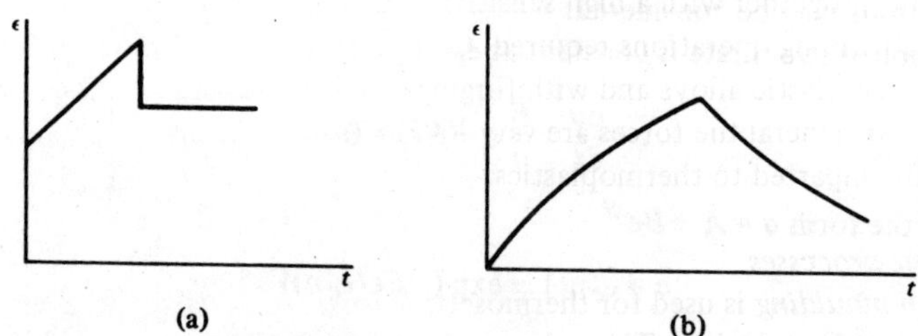

Figure 12.12. Strain response of simple polymer models, when load is applied, held and then removed.

The responses for the series model (Maxwell) and the parallel model (Voigt) are shown respectively in Figures 12.12*a* and *b*. The Maxwell element relaxes stress and the Voigt element relaxes strain.

Both these models relate to Newtonian viscosity, but for most polymers the velocity gradient $\partial V_x/\partial x$ is not directly proportional to shear stress. Equation 12.30

should then be written

$$\frac{d\gamma}{dt} = k\tau^n \tag{12.40}$$

for a non-Newtonian liquid. It is often convenient to approximate to the curve represented by this equation, using an expression

$$\frac{d\gamma}{dt} = \frac{\tau - \tau_o}{\eta} = \frac{\tau'}{\eta} \tag{12.41}$$

which effectively replaces it by two straight lines,

$$\frac{d\gamma}{dt} = 0 \text{ at } \tau < \tau_o$$

and

$$\frac{1}{\tau'}\frac{d\gamma}{dt} = \text{constant for } \tau > \tau_o$$

Thus τ_o is a critical stress to initiate flow, which then follows a viscous law of the simple type. This approximation was proposed by Bingham, and polymers are often regarded as *Bingham solids,* but the 'yield stress' τ_o has little physical meaning. It should not be confused with the low-temperature, low-speed yield point that marks the onset of measurable permanent deformation in some polymers; the Bingham solid approximation then applies to the non-Newtonian flow beyond that stress.

At high temperatures, the viscous flow dominates.

12.7.3 *The forming of polymers*

The major limitations of polymeric materials are their relatively low strength and lack of rigidity, together with a high sensitivity to temperature, but these factors facilitate the forming operations required. Comparisons can be drawn with the forming of superplastic alloys and with forging of alloys in a state between solidus and liquidus. In general the forces are very low in polymer forming and very large strains can be imparted to thermoplastics.

(*a*) *Moulding processes*

Compression moulding is used for thermosetting resins, especially the phenolic resins for which it was first devised. The polymer, in the form of briquettes or fairly large particles, is heated in a mould until it is effectively in a molten state. Pressure is then applied and the temperature is maintained long enough to cause full polymerisation, after which the product is cooled and ejected. This process is inherently slow but can be considerably speeded up by heating the material in a separate container, from which it flows into the mould cavity and is subjected to pressure only when the mould is full. In this *transfer moulding* several moulds may be fed from one container. Some modern resins set much more rapidly than the phenolics.

Injection moulding is commonly used for thermoplastic materials, though it can also be used for thermosetting resins. It somewhat resembles transfer compression moulding but operates semi-continuously or continuously. Small pellets are fed into a heating container, using either a ram or in more modern machines usually a screw feed. The temperature of this container, or injection cylinder may be 150°C–350°C and a pressure is developed that forces the softened polymer into a water-cooled mould. The mechanical and physical conditions are complex, since all the polymers have high viscosity and poor thermal conductivity. It is difficult to homogenise the material and the temperature, and also to transmit the pressure. The helical screw can help in these respects. Formulae have been suggested for the filling and cooling times required, but these are mainly empirical. A typical cycle time is about 10 seconds.

The moulding processes produce very little scrap material and can be used to make objects with large wall thicknesses, without introducing significant internal stresses. There is little wear of the mould, so very large numbers of objects can be made from one mould.

In *blow-moulding* the softened polymer, usually in the form of tubing, is introduced to the mould and then blown against the walls by internal air pressure, as in glass blowing.

(*b*) *Extrusion*

Extrusion is used for the production of intermediate-stage material suitable for subsequent further forming, and also for making long fibres, rods, tubes and special sections. In *melt extruders* the materials is pre-melted and flows into the press. *Plasticising extruders* accept polymer granules, which are then moderately heated and extensively worked. This plasticity is often accelerated by addition of suitable solvents. The early ram-type presses have now been replaced by screw-fed machines, operating continuously.

Extrusion is also suitable for coating wires or sheet continuously. The wire is simply fed through the die at a suitable speed with the extruding polymer. It is possible to replace the wire in the die by a stationary nozzle through which air is blown. When correctly adjusted, this produces a uniform tube of large diameter and very thin wall. The polymer chills rapidly and may be coiled as lay-flat tube, or slit for sale as sheet.

(*c*) *Calendering*

Calendering is an interesting variant on the common process of rolling. The ingoing and outgoing thicknesses are not however related, as they are for metals (Chapter 9). Instead the flow is governed by viscous behaviour and the stock material can be of any shape. The exit thickness is greater than the roll gap and is determined by the velocity continuity equations for the very viscous material, which must be compatible with uniform velocity in the rigid final product. In fact a stable solution cannot be produced for a strictly Newtonian fluid, but non-Newtonian fluids and Bingham-type plastics behave in this way. Heated rollers are used, and thick plastic sheets are normally produced by calendering.

(*d*) *Casting, Sintering, etc.*

Casting of clear thermosetting resins, with suitable catalysts, is well known for decorative preservation of flowers and insects. It is also useful in electrical components.

The procedure for making reinforced or laminated shells, widely used for boats and many other purposes, is known as *contact moulding.* This may be manual or automated, but is outside the scope of this book.

Thermoplastic materials can also be cast and heavy PVC (polyvinylchloride) objects are often made in this way.

Sintering is an unnecessarily expensive technique for most polymers, but is a satisfactory method of producing PTFE, which decomposes at high temperatures and cannot be satisfactorily moulded.

(*e*) *Sheet forming*

Many plastics articles are formed from sheet. The processes resemble those for metals but require very low forces; even atmospheric pressure may suffice.

Drape forming involves heating the sheet to a moderate temperature. It is then clamped at the edges and stretch-formed over a die. One of the problems encountered is that the areas first touching the die will be chilled and remain thicker than the rest. This trouble can be minimised by blowing hot air between the sheet and the die.

Vacuum forming also uses a heated sheet, but the space between the sheet and the die is evacuated so that the air pressure on the other side forces the soft plastic into close contact with the die. Again the problem of local cooling is important and more uniform results can be obtained if the sheet is pre-formed by an intermediate pressing operation.

It is interesting to note that these sheet-forming techniques can also be used for superplastic metallic alloys (§12.2.6), whose high-temperature properties are very similar to those of heated polymers, though they are much harder at room temperature.

REFERENCES

12.1 Bailey, A. R., *A text book of metallurgy,* 2nd ed., Macmillan, 1960.

12.2 Cottrell, A. H., *An introduction to metallurgy*, Edward Arnold, 1967.

12.3 van Vlack, L. H., *Elements of materials science,* 2nd ed., Addison-Wesley, 1966.

12.4 Dieter, G. E., *Mechanical metallurgy,* McGraw-Hill, 1961.

12.5 Rhines, F. N., *Phase diagrams in metallurgy,* McGraw-Hill, 1956.

12.6 Brick, R. M. and Phillips, A., *Structure and properties of alloys,* McGraw-Hill, A.M., 1949.

12.7 Cottrell, A. H., *Dislocations and plastic flow in crystals*, Oxford University Press, 1953.

12.8 Read, W. T., *Dislocations in crystals,* McGraw-Hill, 1953.

12.9 Smallman, R.E., *Modern physical metallurgy,* Butterworth, 1962.

12.10 Rollason, E. C., *Metallurgy for engineers,* 4th ed., Edward Arnold, 1973.

12.11 Fiorentino, R. J., 'Comparison of cold, warm and hot extrusion by conventional and hydrostatic methods', *Metall. Metal Form* **40,** 15 *et seq.*, 1973.

12.12 Berry, J. T., and Pope, M. H., 'Force requirements and friction in warm working operations' in *Metal Forming,* ed. Hoffmanner, A. L., Plenum Press, 307–324, 1971.

12.13 Johnson, R. H., 'Superplasticity', *Met. Rev.* **15,** 115–134, 1970.

12.14 Cockcroft, M. H., and Latham, D. J., 'Ductility and workability of metals', *J. Inst. Metals,* **96,** 33–39, 1968.

12.15 Hill, R., *Mathematical theory of plasticity*, Oxford University Press, 1950.
12.16 Backofen, W. A., *Deformation Processing*, Addison-Wesley, 1972.
12.17 Cottrell, A. H., *The mechanical properties of matter*, Wiley, 1964.
12.18 May, M. J. and Latham, D. J., 'Thermomechanical treatment of steels', *Metal Treat*, 23, 3 *et seq.*, 1972.
12.19 Alexander, J. M. and Brewer, R. C., *Manufacturing properties of materials*, van Nostrand Rheinhold, 1963.
12.20 Smithells, C. J., *Metals reference book*, Butterworth, 1962.
12.21 Various, *Metals Handbook, properties and selection of materials*, Am. Soc. Metals, 1961.
12.22 Kinney, G. F., *Engineering properties and applications of plastics*, John Wiley, 1957.
12.23 Winding, C. C., and Hiatt, G. D., *Polymeric materials*, McGraw-Hill, 1961.
12.24 Ogorkiewicz, R. M. (ed.), *Thermoplastics: Properties and Design*, John Wiley, 1974.
12.25 Ward, I, M., *Mechanical properties of solid polymers*, Wiley Interscience, 1971.

13

Numerical Methods in Metalworking Theory

13.1 Introduction

In many practical situations the flow patterns, together with the associated strain-rate and temperature distributions, are much more important than the actual forces required. Only a small proportion of operations usually need the full power of the machine, though local tool stresses may in some instances impose limits to an operation. Slip-line field theory provides a powerful technique for solving metalworking problems in terms of deformation and local stress distribution as well as overall forces, but it has two important deficiencies. Individual solutions are time-consuming and detailed material properties cannot be included directly in the theoretical solutions.

It has been shown in Chapter 8 that the calculation of distortion in extrusion is quite straightforward using the slip-line field and hodograph, but it is lengthy. Partly for this reason and partly because academic solutions can most easily be checked by measuring forces, there has been a tendency for slip-line fields to be used mainly for calculation of working loads, which fails to exploit their full potential. It should, in principle, be possible to predict ductility, grain size distribution and other features of the product from a knowledge of the local strains. The latter can be obtained from the distortion patterns and the time required can be greatly reduced by programming a computer to draw the slip-line fields and the hodographs.

This is particularly valuable when applied to non-steady processes, such as forging, in which the geometric form changes continuously throughout the operation. A set of sequential slip-line fields is then required, each differing from its predecessor in a predictable quantitative way. Once the general equations and the boundary conditions have been established, multiple solutions of a progressive nature can be generated rapidly and of course it is not necessary for the intermediate fields actually to be drawn.

Similarly, a range of fields for increasingly severe friction conditions, or for a set of tool angles, can be provided. The results can be presented numerically, but the advantage of ready comprehension is retained if selected slip-line fields are plotted out by the computer. This is particularly useful for instructional purposes. It is interesting to recall that the current methods of manual drawing of slip-line fields were introduced as a simplification of the more tedious point-to-point techniques used 25 years ago[13.1]. The latter can be directly adapted for computer solutions.

In addition to the use of a computer simply to produce slip-line fields and their hodographs, it is possible to operate entirely on algebraic and trigonometric equations representing stress and velocity continuity and boundary conditions. Upper-bound procedures can be used in which the stress boundary conditions need not be met, or more precise solutions can be obtained using the virtual work principle or variational techniques. These are capable of giving very detailed information, in matrix form, about the strain-rate, strain and stress distributions. They are not limited by the restrictive assumptions of slip-line field theory, and with suitable modifications and approximations can include the dependence of the flow stress of the material on strain-rate, strain and temperature. They can also be extended to axisymmetric and other geometric shapes.

At first sight these methods appear formidable, but the complication is largely in the sheer number of equations to be handled. If it is firmly recognised that this is a question of the size of the computer installation and the patience of the programmer, rather than of mathematical complexity, the principles can be seen to be quite logical and can be clearly understood. Some further development of basic equations is, however, necessary as a background. These relate three-dimensional stress to strain in elastic deformation and to strain-increment in plastic deformation.

13.2 Stress–strain relationships for elastic–plastic solids

The Levy-Mises and Prandtl-Reuss equations are important and can be derived quite simply.

13.2.1 *Elastic deformation*

We define the Young's modulus $E = \sigma/\epsilon$ and the shear modulus $G = \tau/\gamma$ in the usual way.

In three dimensions the stress and strain are related by three equations of the form

$$\epsilon_x E = \sigma_x - \nu\sigma_y - \nu\sigma_z \tag{13.1}$$

because the strain in the x-direction due to a stress in the y-direction is opposite in sign to the direct strain and is related to it by Poisson's ratio ν.

For the shear stress and strain

$$\tau_{xy} = G\gamma_{xy} \quad \text{or} \quad \tau_{xy} = 2G\left(\frac{\gamma_{xy}}{2}\right) \tag{13.2}$$

It is convenient to work in terms of the half strain $\gamma_{xy}/2$, because this avoids rotation of the axes as strain progresses.

Change of size, due to uniform or hydrostatic pressure, has to be distinguished from change of shape, due to shearing.

The hydrostatic stress, causing dilatation, is the mean stress and because of its symmetry it is also known as the *spherical stress component* σ_m.

$$\sigma_m = \frac{\sigma_x + \sigma_y + \sigma_z}{3}, \text{ sometimes written } \sigma'' \tag{13.3}$$

The shearing stress, causing distortion, is the difference of the given stress from the mean and is known as the *deviatoric stress component,* σ'.

$$\begin{aligned} \sigma'_x &= \sigma_x - \sigma_m \\ \sigma'_y &= \sigma_y - \sigma_m \\ \sigma'_z &= \sigma_z - \sigma_m \end{aligned} \tag{13.4}$$

The strain component associated with change of size is, for example in the x-direction,

$$(\epsilon_x)_{\text{size}} = \frac{1}{E}\{\sigma_m - \nu(\sigma_m + \sigma_m)\} = \frac{(1-2\nu)}{(E)}\sigma_m \tag{13.5}$$

and the strain component associated with change of shape is

$$(\epsilon_x)_{\text{shape}} = \frac{\gamma_{xy}}{2} = \frac{\tau_{xy}}{2G} = \frac{\sigma_x - \sigma_m}{2G} \tag{13.6}$$

The total elastic strain is thus

$$(\epsilon_x)_E = \frac{\sigma'_x}{2G} + \frac{1-2\nu}{E}\sigma_m \tag{13.7}$$

For convenience, especially in computerised solutions, these relationships are often written in suffix notation, for example

$$\begin{aligned} &\sigma_{ii} = \sigma_{xx}\,(=\sigma_x) \qquad \epsilon_{ii} = \epsilon_{xx} = \epsilon_x \\ &\sigma_{ij} = \sigma_{xy}(=\tau_{xy}) \qquad \epsilon_{ij} = \epsilon_{xy} = \gamma_{xy} \\ &\sigma'_{ii} = \sigma_{xx} - \sigma_m \end{aligned} \tag{13.8}$$

In this way, equation 13.7 can be written

$$(\epsilon_{ij})_E = \frac{\sigma'_{ij}}{2G} + \frac{1-2\nu}{2G}\delta_{ij}\,\sigma_m \tag{13.9}$$

where $\sigma_m = \frac{1}{3}\sigma_{ii}$, implying $\frac{1}{3}(\sigma_{xx} + \sigma_{yy} + \sigma_{zz})$, and δ_{ij} is referred to as the *Kronecker delta,* such that $\delta_{ij} = 1$ when $i = j$, and $\delta_{ij} = 0$ when $i \neq j$.

13.2.2 *Plastic deformation*

In plastic deformation it is the increment in strain, $d\epsilon$, that is related to the stress σ' causing distortion, not the total strain as in elastic deformation.

Reuss assumed a direct proportionality between stress and plastic strain-increment.

$$d\epsilon_p \propto \sigma'; \quad d\epsilon_x = \sigma'_x d\lambda$$

(Other possibilities exist, for example St. Venant assumed a proportionality between strain-rate and deviatoric stress

$$\dot{\epsilon}_p \propto \sigma'; \quad \sigma'_x = \lambda \frac{d\epsilon_x}{dt} \tag{13.10}$$

which is more characteristic of polymeric materials or metals at high temperatures.)

Using the Reuss assumption, we can write, for plastic deformation

$$\frac{d\epsilon_x}{\sigma_x - \sigma_m} = d\lambda = \frac{d\epsilon_y}{\sigma_y - \sigma_m} = \frac{d\epsilon_z}{\sigma_z - \sigma_m} = \frac{d\gamma_{xy}}{\tau_{xy}} \tag{13.11a}$$

In suffix notation this becomes

$$(d\epsilon_{ij})_p = \sigma'_{ij}\,d\lambda \tag{13.11b}$$

This group of equations, 13.11a or 13.11b, is known as the *Levy-Mises equations* or sometimes as the *flow rule.* It should be noted that $d\lambda$ is instantaneously a positive constant of proportionality, but may vary, for example due to hardening, as deformation proceeds. Since there is no volume change in plastic deformation, we have another important relationship that is often used:

$$(d\epsilon_{ii})_p = (d\epsilon_x + d\epsilon_y + d\epsilon_z)_p = 0 \tag{13.12}$$

13.2.3 *Combined elastic and plastic deformation*

If the elastic and plastic components of deformation are combined, retaining the important separation of dilatation and distortion, equations 13.9, 13.11 and 13.12 give

$$\begin{aligned} d\epsilon'_{ij} &= (d\epsilon_{ij})_p + (d\epsilon_{ij})_E = \sigma'_{ij}\, d\lambda + \frac{d\sigma'_{ij}}{2G} \\ d\epsilon_{ii} &= (d\epsilon_{ii})_p + (d\epsilon_{ii})_E = \;\; 0 \;\; + \frac{1-2\nu}{E}\, d\sigma_{ii} \end{aligned} \tag{13.13}$$

These are the *Prandtl-Reuss equations.* It is necessary also to define the transition between fully elastic and elastic/plastic deformation, which can be done using the von Mises yield criterion (equation 3.19 gives a simplified version):

$$\sigma'_{ij}\,\sigma'_{ij} = 2k^2 \tag{13.14}$$

For most metalworking, apart from detailed consideration of residual stresses, it is possible to ignore the small elastic strains in comparison with the much larger plastic strains, so the Levy-Mises equations (13.11) suffice.

It is of interest to compare the Lévy-Mises equations for plastic deformation directly with the equations for elastic deformation. Equation 13.11 can be written

$$\begin{aligned} (d\epsilon_x)_p &= d\lambda(\sigma_x - \sigma_m) = d\lambda\left\{\sigma_x - \frac{\sigma_x + \sigma_y + \sigma_z}{3}\right\} \\ &= \tfrac{2}{3} d\lambda\{\sigma_x - \tfrac{1}{2}(\sigma_y + \sigma_z)\} \end{aligned} \tag{13.15}$$

which may be compared with equation 13.1 for elastic deformation, giving

$$(d\epsilon_x)_E = \frac{1}{E}\{d\sigma_x - \nu(d\sigma_y + d\sigma_z)\}$$

There is a formal relationship between $\frac{2}{3}\,d\lambda$ and $\frac{1}{E}$, while ν is associated with the value $\frac{1}{2}$ in plastic deformation. This corresponds to the condition for no volume change in plastic deformation:

$$\begin{aligned} dV = d\epsilon_x + d\epsilon_y + d\epsilon_z &= \frac{1}{E} d\sigma_x - \nu(d\sigma_y + d\sigma_z) \\ &+ \frac{1}{E} d\sigma_y - \nu(d\sigma_z + d\sigma_x) \\ &+ \frac{1}{E} d\sigma_z - \nu(d\sigma_x + d\sigma_y) \end{aligned}$$

$$dV = \frac{(1-2\nu)}{E}(d\sigma_x + d\sigma_y + d\sigma_z) = \frac{(1-2\nu)}{E} \cdot 3\,d\sigma_m$$
$$= 0 \text{ if } \nu = \tfrac{1}{2}$$

13.2.4 *Generalised stress and strain*

It is often convenient to express the stress system that will just cause yielding, in terms of a generalised stress. This can be defined with reference to the von Mises equation. In terms of principal stresses, equation 3.19 determines the criterion for yielding, which can be written in the form

$$(\sigma_1 - \sigma_2)^2 + (\sigma_2 - \sigma_3)^2 + (\sigma_3 - \sigma_1)^2 = \text{constant} = C\bar{\sigma}^2$$

with $\bar{\sigma}$ as the generalised stress. The constant can then be set to give $\bar{\sigma} = \sigma_1$ in simple tension:

$$\sigma_1{}^2 + 0 + \sigma_1{}^2 = C\bar{\sigma}^2; C = 2$$

Thus

$$\bar{\sigma} = \sqrt{\tfrac{1}{2}\{(\sigma_1 - \sigma_2)^2 + (\sigma_2 - \sigma_3)^2 + (\sigma_3 - \sigma_1)^2\}} \tag{13.16}$$

In a similar way, a generalised strain increment can be defined, for plastic strain

$$(de_1 - de_2)^2 + (de_2 - de_3)^2 + (de_3 - de_1)^2 = C(d\bar{e})^2$$

and the constant chosen so that $d\bar{e} = de_1$ for simple tension, where because $dV = 0$ and $de_2 = de_3$, $de_1 = -2de_2 = -2de_3$.

$$C(d\bar{e})^2 = (de_1 + \tfrac{1}{2}de_1)^2 + (-\tfrac{1}{2}de_1 - de_1)^2 = (\tfrac{9}{4} + \tfrac{9}{4})\,de_1{}^2$$

$$d\bar{e} = \sqrt{\tfrac{2}{9}\{(de_1 - de_2)^2 + (de_2 - de_3)^2 + (de_3 - de_1)^2\}} \tag{13.17}$$

These generalised values can be inserted in equation 13.11, by considering uniaxial tension again, for which

$$\sigma_m = \tfrac{1}{3}\sigma_x,\ \sigma_x - \sigma_m = \tfrac{2}{3}\sigma_x,\ \bar{\sigma} = \sigma_x$$
$$de_x = d\bar{e}$$

so

$$d\lambda = \frac{de_x}{\sigma_x - \sigma_m} = \frac{d\bar{e}}{\tfrac{2}{3}\sigma_x} = \tfrac{3}{2}\frac{d\bar{e}}{\bar{\sigma}}$$

This must be generally valid as a value for the constant, so

$$de_x = \tfrac{3}{2}\frac{d\bar{e}}{\bar{\sigma}}(\sigma_x - \sigma_m) = \frac{d\bar{e}}{\bar{\sigma}}\{\sigma_x - \tfrac{1}{2}(\sigma_y + \sigma_z)\} \tag{13.18a}$$

which is sometimes a convenient form in which to use the flow rule.

It is also useful to recognise that

$$d\dot{e}_x = \frac{d\dot{\bar{e}}}{\bar{\sigma}}\{\sigma_x - \tfrac{1}{2}(\sigma_y + \sigma_z)\} \tag{13.18b}$$

13.3 Visioplasticity

The technique known as visioplasticity combines experimental measurement of metal flow with subsequent calculation of stresses. It can conveniently be applied to solve problems in plane strain or axial symmetry[13.2].

13.3.1 *Technique*

The metal flow is observed by sectioning a billet on a plane containing the direction of major deformation. The surfaces are then ground flat and polished on both sides of the cut. They must be in close contact when the billet is reassembled, so it may be necessary to start with oversized halves from two separate billets to ensure a good fit. One face is etched with a regular grid of circles or squares and the distortion of the grid is observed with a microscope or on an enlarged photograph after a small increment of deformation has been given to the billet. A detailed etching procedure is described in §13.4.5.

The displacement and hence the velocity of each element of the grid can thus be assessed, giving results such as those shown in Figure 13.1.

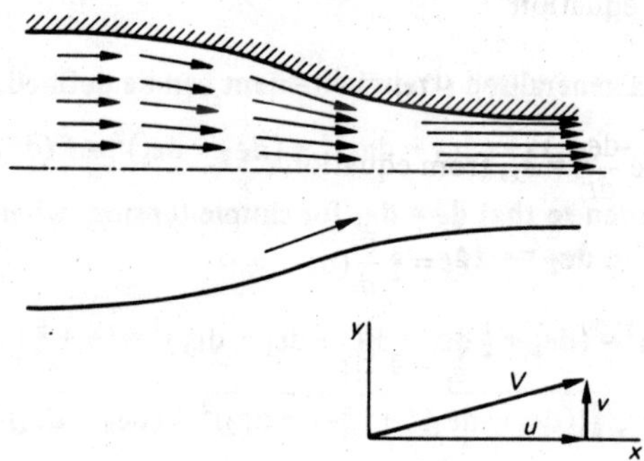

Figure 13.1. Instantaneous velocity vectors deduced from the distortion of an etched grid.

The velocity vector at each point can be analysed into its components u and v in the x and y-directions. If the intervals are sufficiently small, smooth curves can be drawn for a chosen element, showing the flow lines. It is also possible to plot graphs showing the variations of u and v with the position co-ordinates x and y. The slope of the appropriate velocity/position curve gives the strain rate at a chosen point:

$$\dot{\epsilon}_x = \frac{\partial u}{\partial x}; \ \dot{\epsilon}_y = \frac{\partial v}{\partial y} \qquad (13.19)$$

The shear-strain rate is given by

$$\dot{\gamma}_{xy} = \frac{\partial v}{\partial x} + \frac{\partial u}{\partial y} \qquad (13.20)$$

These experimental strain rates can then be converted into stresses, using the Levy-Mises equations, for plane-strain or axisymmetric conditions.

13.3.2 *Determination of stresses*

(*a*) *Plane strain*
As was shown in Example 5.1, the force equilibrium of an element can be described in the form of stress equations. For plane-strain conditions

$$\frac{\partial \sigma_x}{\partial x} + \frac{\partial \tau_{xy}}{\partial y} = 0 \tag{13.21a}$$

$$\frac{\partial \sigma_y}{\partial y} + \frac{\partial \tau_{yx}}{\partial x} = 0 \tag{13.21b}$$

$$\frac{\partial \sigma_z}{\partial z} = 0 \tag{13.21c}$$

The stress can be related to the strain increment and hence to the strain rate using the Levy-Mises equations derived above, in the form

$$d\epsilon_x = \tfrac{3}{2} \frac{d\bar{\epsilon}}{\bar{\sigma}} (\sigma_x - \sigma_m) \tag{13.18a}$$

and the related shear equation

$$d\gamma_{xy} = 3 \frac{d\bar{\epsilon}}{\bar{\sigma}} \tau_{xy}$$

In terms of strain rate $\dfrac{d\epsilon_x}{dt} = \dot{\epsilon}_x$, from equation 13.18b,

$$\dot{\epsilon}_x = \tfrac{3}{2} \frac{\dot{\bar{\epsilon}}}{\bar{\sigma}} (\sigma_x - \sigma_m) \tag{13.22a}$$

$$\dot{\epsilon}_y = \tfrac{3}{2} \frac{\dot{\bar{\epsilon}}}{\bar{\sigma}} (\sigma_y - \sigma_m) \tag{13.22b}$$

$$\dot{\epsilon}_{xy} = \dot{\gamma}_{xy} = 3 \frac{\dot{\bar{\epsilon}}}{\bar{\sigma}} \tau_{xy} \tag{13.22c}$$

Thus

$$\dot{\epsilon}_x - \dot{\epsilon}_y = \tfrac{3}{2} \frac{\dot{\bar{\epsilon}}}{\bar{\sigma}} (\sigma_x - \sigma_y)$$

$$\sigma_x = \sigma_y + \tfrac{2}{3} \frac{\bar{\sigma}}{\dot{\bar{\epsilon}}} (\dot{\epsilon}_x - \dot{\epsilon}_y) \tag{13.23}$$

from which σ_y can be determined by differentiating with respect to y and combining the result with equation 13.21b and the differentiated form of 13.22c.

$$\frac{\partial \sigma_x}{\partial y} = \frac{\partial \sigma_y}{\partial y} + \tfrac{2}{3} \frac{\partial}{\partial y} \left\{ \frac{\dot{\epsilon}_x - \dot{\epsilon}_y}{\dot{\bar{\epsilon}}/\bar{\sigma}} \right\}$$

$$\frac{\partial \sigma_y}{\partial y} = - \frac{\partial \tau_{xy}}{\partial x}$$

$$\frac{\partial \tau_{xy}}{\partial x} = \tfrac{1}{3} \frac{\partial}{\partial x} \left\{ \frac{\dot{\gamma}_{xy}}{\dot{\bar{\epsilon}}/\bar{\sigma}} \right\}$$

Combining these,

$$\frac{\partial \sigma_x}{\partial x} = \tfrac{2}{3}\left[\frac{\partial}{\partial y}\left\{\frac{\dot{\epsilon}_x - \dot{\epsilon}_y}{\dot{\bar{\epsilon}}/\bar{\sigma}}\right\} - \tfrac{1}{2}\frac{\partial}{\partial x}\left\{\frac{\dot{\gamma}_{xy}}{\dot{\bar{\epsilon}}/\bar{\sigma}}\right\}\right] \qquad (13.24)$$

which can be integrated to give a value fot the stress

$$\sigma_x = \tfrac{2}{3}\int_{y_1}^{y}\left[\frac{\partial}{\partial y}\left\{\frac{\dot{\epsilon}_x - \dot{\epsilon}_y}{\dot{\bar{\epsilon}}/\bar{\sigma}}\right\} - \tfrac{1}{2}\frac{\partial}{\partial x}\left\{\frac{\dot{\gamma}_{xy}}{\dot{\bar{\epsilon}}/\bar{\sigma}}\right\}\right] dy + [\sigma_x]_{y=y_1} \qquad (13.25)$$

This requires a knowledge of $\bar{\sigma}$, the generalised stress, which can be found for a given material from the stress/strain curve, usually fitted to a form similar to that of equation 2.14,

$$\bar{\sigma} = A(\bar{\epsilon})^n \quad \text{or} \quad \bar{\sigma} = \sigma_0\left(1 + \frac{\bar{\epsilon}}{\bar{\epsilon}_0}\right)^n$$

in which $\bar{\epsilon}$ is determined using equation 13.17.

The constant of integration $[\sigma_x]_{y=y_1}$ is however a function of x and further manipulation is required to reduce this to a single value. Differentiating equation 13.25 with respect to x:

$$\frac{\partial \sigma_x}{\partial x} = \tfrac{2}{3}\int_{y_1}^{y}\left[\frac{\partial^2}{\partial x \partial y}\left\{\frac{\dot{\epsilon}_x - \dot{\epsilon}_y}{\dot{\bar{\epsilon}}/\bar{\sigma}}\right\} - \tfrac{1}{2}\frac{\partial^2}{\partial x^2}\left\{\frac{\dot{\gamma}_{xy}}{\dot{\bar{\epsilon}}/\bar{\sigma}}\right\}\right] dy + \frac{d}{d_x}[\sigma_x]_{y=y_1}$$

But from equation 13.21a

$$\frac{\partial \sigma_x}{\partial x} = -\frac{\partial \tau_{xy}}{\partial y}$$

and from equation 13.22c

$$\frac{\partial \tau_{xy}}{\partial y} = \tfrac{1}{3}\frac{\partial}{\partial y}\left\{\frac{\dot{\gamma}_{xy}}{\dot{\bar{\epsilon}}/\bar{\sigma}}\right\}$$

Thus at $y = y_1$, where the integral is zero,

$$\frac{\partial [\sigma_x]_{y=y_1}}{\partial x} = -\tfrac{1}{3}\frac{\partial}{\partial y}\left\{\frac{\dot{\gamma}_{xy}}{\dot{\bar{\epsilon}}/\bar{\sigma}}\right\}$$

$$[\sigma_x]_{y=y_1} = -\tfrac{1}{3}\int_{x_0}^{x}\left[\frac{\partial}{\partial y}\left\{\frac{\dot{\gamma}_{xy}}{\dot{\bar{\epsilon}}/\bar{\sigma}}\right\}\right]_{y=y_1} dx + [\sigma_x]_{\substack{y=y_1\\x=x_0}} \qquad (13.26)$$

Substituting this into equation 13.25,

$$\sigma_x = \tfrac{2}{3}\int_{y_1}^{y}\left[\frac{\partial}{\partial y}\left\{\frac{\dot{\epsilon}_x - \dot{\epsilon}_y}{\dot{\bar{\epsilon}}/\bar{\sigma}}\right\} - \tfrac{1}{2}\frac{\partial}{\partial x}\left\{\frac{\dot{\gamma}_{xy}}{\dot{\bar{\epsilon}}/\bar{\sigma}}\right\}\right] dy - \tfrac{1}{3}\int_{x_0}^{x}\left[\frac{\partial}{\partial y}\left\{\frac{\dot{\gamma}_{xy}}{\dot{\bar{\epsilon}}/\bar{\sigma}}\right\}\right]_{y=y_1} dx + [\sigma_x]_{\substack{y=y_1\\x=x_0}} \qquad (13.27)$$

If therefore the single value of stress at $x = x_0, y = y_1$ is known, the value of σ_x at any point can be found. The integration can be performed graphically, or by fitting a polymonial expression to the flow function, or by a fully numerical method using the equation in finite-difference form.

The first two methods will be illustrated with reference to an axisymmetrical problem[13.3], which is more frequently encountered in practice than the plane-strain problems.

(*b*) *Axial symmetry*

A similar analysis of the stresses can be made for axial symmetry if $\dot{\epsilon}_x$ is replaced by $\dot{\epsilon}_z$, $\dot{\epsilon}_y$ by $\dot{\epsilon}_r$ and $\dot{\gamma}_{xy}$ by $\dot{\gamma}_{zr}$. It is necessary also to introduce the hoop-strain rate $\dot{\epsilon}_\theta$, but this can be found if $\dot{\epsilon}_r$ and $\dot{\epsilon}_z$ are known, because there is no volume change (equation 3.12).

The equation corresponding to 13.27 is

$$\sigma_z = \tfrac{2}{3}\int_{r_1}^{r} \left[\frac{\partial}{\partial r}\left\{\frac{\dot{\epsilon}_z - \dot{\epsilon}_r}{\dot{\bar{\epsilon}}/\bar{\sigma}}\right\} - \frac{\dot{\epsilon}_r - \dot{\epsilon}_\theta}{\dot{\bar{\epsilon}}/\bar{\sigma}} - \tfrac{1}{2}\frac{\partial}{\partial z}\left\{\frac{\dot{\gamma}_{zr}}{r\dot{\bar{\epsilon}}/\bar{\sigma}}\right\}\right] dr - \tfrac{1}{3}\int_{z_0}^{z}\left[\frac{\partial}{\partial r}\left\{\frac{\dot{\gamma}_{zr}}{\dot{\bar{\epsilon}}/\bar{\sigma}}\right\} + \frac{\dot{\gamma}_{zr}}{r\dot{\bar{\epsilon}}/\bar{\sigma}}\right]_{r=r_1} dz + (\sigma_z)_0 \tag{13.28}$$

13.3.3 *Solution of the stress equation*

(*a*) *Graphical method*

As stated in §6.3.1, four graphs are drawn, of the axial velocity u against axial distance z, radial velocity v against radial distance r, and u against r and v against z. The slopes of these curves gives the strain rates

$$\dot{\epsilon}_z = \frac{\partial u}{\partial z};\ \dot{\epsilon}_r = \frac{\partial v}{\partial r};\ \dot{\gamma}_{zr} = \frac{\partial v}{\partial z} + \frac{\partial u}{\partial r} \tag{13.29}$$

Figure 13.2 shows some typical values for an axisymmetric extrusion of lead with a reduction ratio 3.

From these it is possible to plot the strain rates, using equations 13.29, as in Figure 13.3. The next step is to evaluate the generalised strain rates using the equation, similar to 13.17,

$$\dot{\bar{\epsilon}} = \sqrt{\tfrac{2}{9}\{(\dot{\epsilon}_z - \dot{\epsilon}_r)^2 + (\dot{\epsilon}_r - \dot{\epsilon}_\theta)^2 + (\dot{\epsilon}_\theta - \dot{\epsilon}_z)^2\} + \tfrac{1}{3}\dot{\gamma}_{zr}^2} \tag{13.30}$$

and then to integrate this along a flow line to find the generalised strain $\bar{\epsilon}$, from which the generalised stress $\bar{\sigma}$ can be deduced, using a flow stress curve of the type

$$\bar{\sigma} = A\bar{\epsilon}^n \tag{13.31}$$

All the quantities in equation 13.28 are thus known. The local slopes and the integrals can be found from suitable graphs, so that σ_z is expressible at all points in terms of $(\sigma_z)_0$, which is for example zero on the exit line.

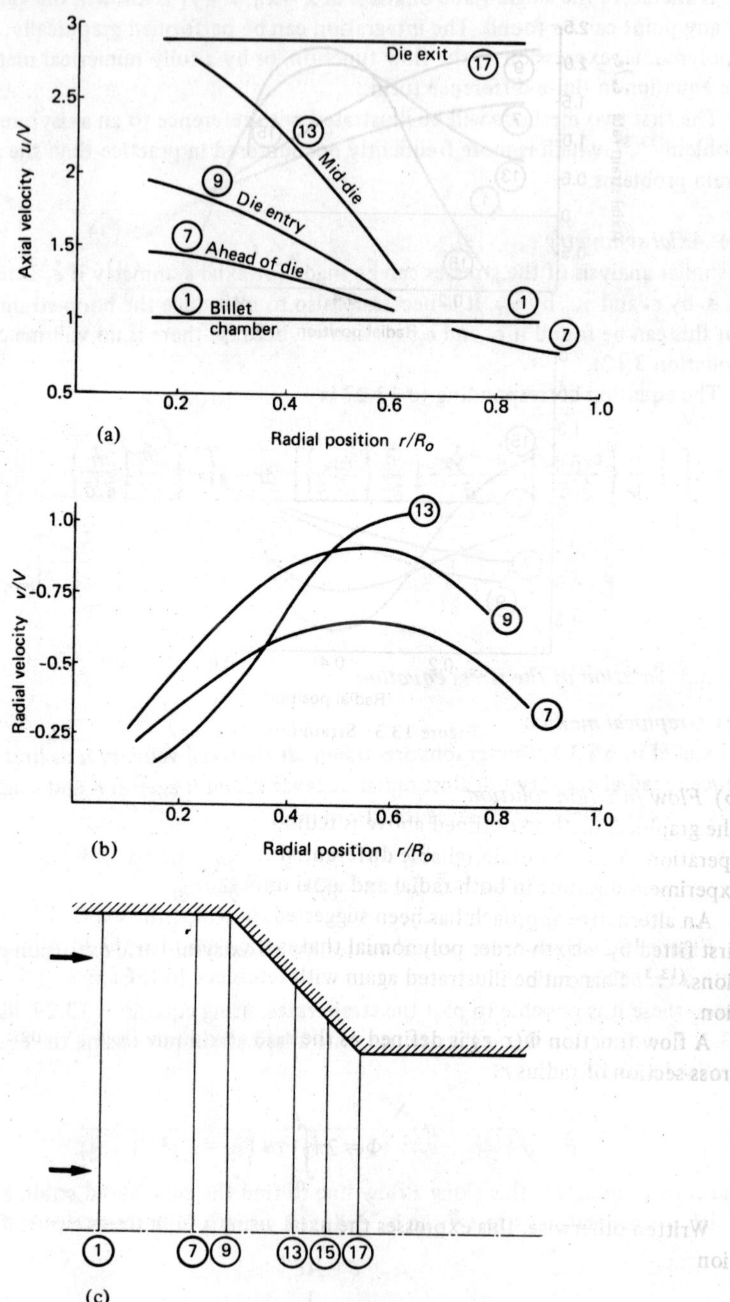

Figure 13.2 Velocity components in axial and radial directions, as functions of radial position, for various axial-section positions[13.3].

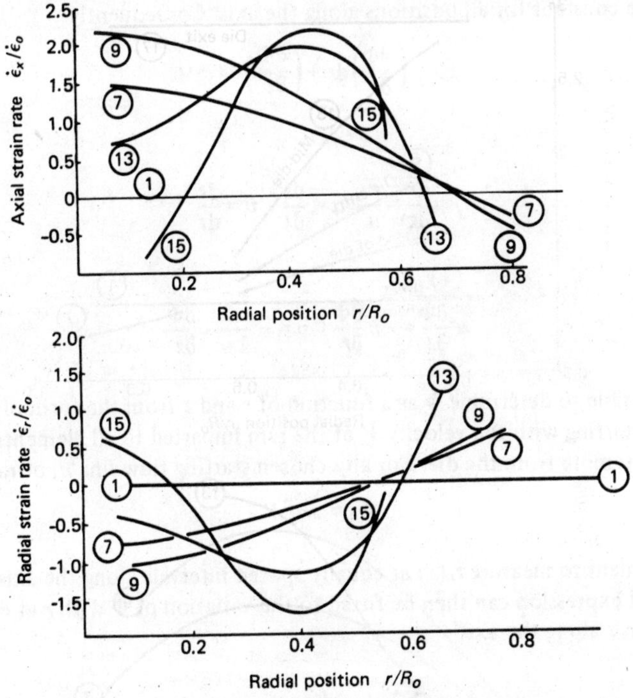

Figure 13.3. Strain-rate variations.

(b) Flow function solution

The graphical method outlined above is tedious, involving considerable manual operation. Also, to obtain reliable differentials it is necessary to smooth the experimental results in both radial and axial directions.

An alternative approach has been suggested, in which the experimental data are first fitted by a sixth-order polynomial that is then used for the remaining computations.[13.3] This can be illustrated again with reference to the axi-symmetric extrusion.

A flow function $\Phi(r, z)$ is defined as the rate of volume flow through a circular cross-section of radius r

$$\Phi = 2\pi \int_0^r ru\,\mathrm{d}r \tag{13.32}$$

Written otherwise, this expresses the axial velocity u in terms of the flow function

$$u = \frac{1}{2\pi r} \cdot \frac{\partial \Phi}{\partial r} \tag{13.33}$$

Now there is no volume change in plastic deformation, so the flow function must be constant along the constraining container and die surface, and it can be assumed that in steady flow the total volume passing within the bounds of given stream lines will also be constant for all positions along the axis. Consequently

$$d\Phi = \left(\frac{\partial \Phi}{\partial r}\right)dr + \left(\frac{\partial \Phi}{\partial z}\right)dz = 0 \tag{13.34}$$

and

$$\frac{dr}{dz} = \frac{v}{u};\ u = \frac{dz}{dt},\ v = \frac{dr}{dt}$$

so

$$u\frac{\partial \Phi}{\partial z} = -v\frac{\partial \Phi}{\partial r};\ v = -\frac{1}{2\pi r}\cdot\frac{\partial \Phi}{\partial z} \tag{13.35}$$

It is possible to determine Φ as a function of r and z from the form of a given flow line, starting with the velocity V of the ram imparted to all elements of the workpiece remote from the die. For any chosen starting flow line n, of radius r_n

$$\Phi_n = \pi r_n^2 V \tag{13.36}$$

It is convenient to measure $r_n(z)$ at equally spaced intervals along the axis, z. A polynomial expression can then be fitted to the variation of Φ with r at each individual section z along the axis

$$\Phi = \sum_{1}^{6} A_m r^m \text{ at } z = \text{constant} \tag{13.37}$$

The axial and radial velocities u and z are then found by numerical differentiation of the function according to equations 13.33 and 13.35 and these can be processed as strain rates following the method of § 13.3.3(a) using the basic polynomial.

However, the calculation of stress is usually less significant than the evaluation of strain distribution. The flow function can also be used to compute the axial strains and thus the distortion.

13.3.4 *Evaluation of distortion in extrusion*

Once the axial velocity variation along a flow line has been established, it is a simple matter to evaluate the deformation pattern of an initially rectangular grid in much the same way as is described in §8.6.

As in equation 8.20, for axial velocity u,

$$u = \frac{dz}{dt};\ t - t_0 = \int_0^z \frac{dz}{u}$$

It is convenient to use finite differences

$$\Delta t = \frac{\Delta z}{u} \tag{13.38}$$

At some arbitrary axial position z_1, from equations 13.33 and 13.37,

$$u = \frac{1}{2\pi r}\frac{\partial \Phi}{\partial r} \quad \text{and} \quad \Phi = \sum_{1}^{6} A_m r^m$$

so

$$u = \frac{1}{2\pi r}\sum_{2}^{6} A_m m r^{m-1}$$

which can be substituted into equation 13.38 to find the time intervals between successive z positions.

The results can then be added to find the distance travelled along different flow lines in a given time t_1, as done graphically in Figure 8.7.

13.3.5 *Assessment of visioplasticity*

The visioplasticity method provides information in great detail and is valuable for determination of the distributions of strain rates directly from experiment. The derived values of strain are accurate and a detailed knowledge of stress distributions can also be obtained. The results usually agree very well with the observed distortions of grids marked on the samples and the predicted loads are very close to the measured values, at least when the temperatures and strain rates are low enough for reliable stress–strain curves to be used.

The disadvantage is that an experiment must be performed first. The method is thus accurately deductive but not predictive. It can however be used for detailed investigations and for checking simpler predictive techniques, such as upper bounds and approximate slip-line fields.

13.4 Use of a digital computer to draw slip-line fields

Slip-line fields have been applied to many plane-strain problems(1.11), as described elsewhere in this book and in many publications, but they have been restricted in general to steady-state conditions or to instantaneous incremental deformations. In principle it has long been recognised that progressive slip-line fields can be drawn, using the hodographs to determine the shape after each incremental deformation. It is, however, very laborious to undertake such drawing manually and little attention has been given to this aspect of plasticity. Digital computers offer the possibility of producing progressive solutions or solutions for a range of geometric or frictional conditions(13·4, 13·21). The application of numerical methods to slip-line fields is not new(13·1); indeed it preceded the graphical method but fell into disuse.

The well-known slip-line field for compression of a wide strip between flat parallel platens with no lubricant, described in detail on p. 261 offers a simple example for the use of a computer to draw a slip-line field.

13.4.1 *Computer drawing of the slip-line field for plane-strain compression*

Figure 13.4*a* reproduces part of the field. As in the manual solution, this is constructed by first locating all the nodal points, or intersections. The arcs between the nodal

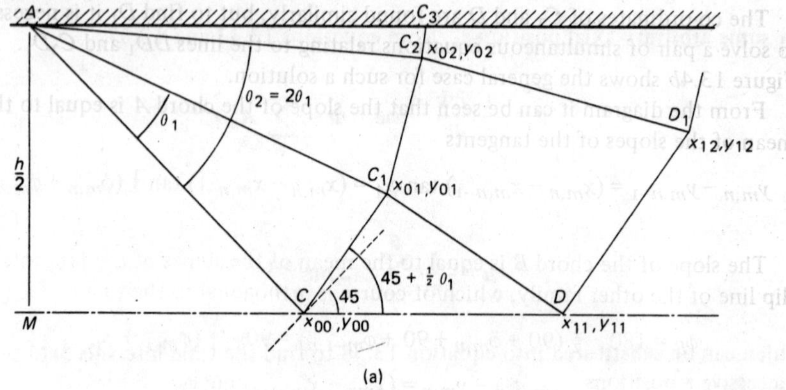

Figure 13.4(a). Part of the slip-line field, showing the chords and nodal points.

points are regarded as being replaced by chords (see § 13.4.2). The accuracy obviously improves as the segmental angle is decreased and it is much easier to take small increments in angle when a computer is used.

The points are labelled systematically, following a sequence from $m = 0$ onwards for the β lines and corresponding n values for the α lines.

The point M on the horizontal centre line may be selected as the origin of the (x, y) co-ordinates. Then the corner A has the co-ordinates $(0, +h/2)$ and the first intersection C of the slip-line field on the centre line, (x_{00}, y_{00}) is at $(+h/2, 0)$. The intersection C_1 (x_{01}, y_{01}) is defined by the incremental angle θ_1. The tangent to the slip line on the centre line at C must be at $45°$, so the angle of the chord CC_1 to the centre line will be $(45 + \theta_1/2)°$. The co-ordinates of C_1 are thus

$$x_{01} = \frac{h}{2} + AC\theta_1 \cos\left(45 + \frac{\theta_1}{2}\right) \tag{13.39}$$

$$y_{01} = AC\,\theta_1 \sin\left(45 + \frac{\theta_1}{2}\right)$$

and $AC = \sqrt{2}\,h/2$.

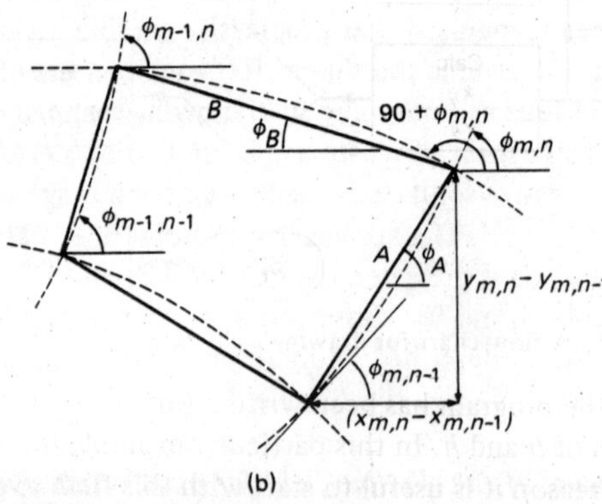

Figure 13.4(b). The co-ordinates of the nodal points of a general slip-line network.

The co-ordinates of C_2 and D are found similarly, but to find D_1 it is necessary to solve a pair of simultaneous equations relating to the lines DD_1 and C_2D_1. Figure 13.4*b* shows the general case for such a solution.

From the diagram it can be seen that the slope of the chord A is equal to the mean of the slopes of the tangents

$$y_{m,n} - y_{m,n-1} = (x_{m,n} - x_{m,n-1}) \tan \phi_A = (x_{m,n} - x_{m,n-1}) \tan \tfrac{1}{2} (\phi_{m,n} + \phi_{m,n-1}) \tag{13.40a}$$

The slope of the chord B is equal to the mean of the slopes of the tangents to the slip line of the other family, which of course is orthogonal to the first.

$$\phi_B = 180 - \tfrac{1}{2} (90 + \phi_{m,n} + 90 + \phi_{m-1,n}) = 90 - \tfrac{1}{2} (\phi_{m,n} + \phi_{m-1,n})$$

$$y_{m-1,n} - y_{m,n} = (x_{mn} - x_{m-1,n}) \tan \phi_B$$

$$y_{m,n} - y_{m-1,n} = -(x_{m,n} - x_{m-1,n}) \cot \tfrac{1}{2} (\phi_{m,n} + \phi_{m-1,n}) \tag{13.40b}$$

It may be noted that these equations correspond to those given by Hill (ref.13.1, p. 141).

In this way the co-ordinates of all the points in the quadrant can be found. The calculations are performed in sequence for small increments of angle along each slip-line, starting with C–C_3. After each calculation the question is asked whether y has reached the value $h/2$, the platen surface. When this is reached, the next line D–D_1, etc. is calculated. The field is completed when x reaches the value $b/2$, at the centre line.

Finally, as in manual drawing, the lines in the dead zone must be erased (Figure 10.3). This is most conveniently done by an instruction to the plotter to start at the centre-point F and to ignore all points above AF.

A suitable flow chart for the calculations is shown in Figure 13.5. It will be seen that the logic follows quite closely that of the manual construction, p. 260.

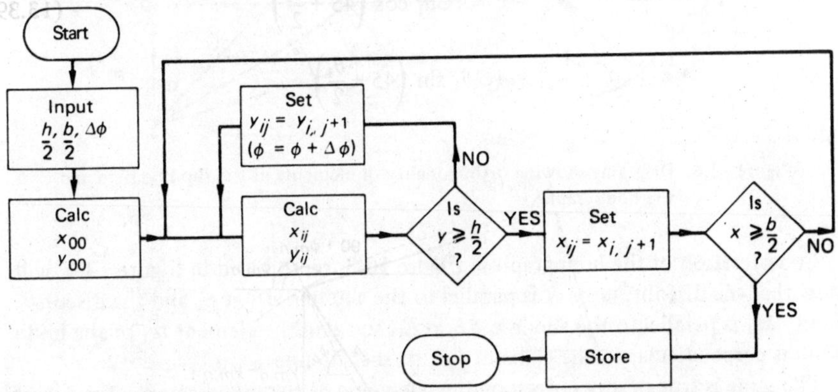

Figure 13.5. A flow chart for drawing a slip-line field for compression.

Obviously, once the program has been written (and checked) the fields can be drawn for any values of b and h. In this particular example the results are well known, but for this reason it is useful to start with this field to gain familiarity with the procedure.

13.4.2 *Drawing the hodograph*

The hodograph for this solution is also well known (see Figure 10.3). It can be drawn in exactly the same way as the slip-line field, by using the geometric relationships between the nodal points, starting at f going to c_3 and then completing the lines from e and d.

A more general method, which does not require a prior knowledge of the hodograph, utilises the fact that all elements of the hodograph are normal to the corresponding elements of the slip-line field.

Care should be taken that this does not cause confusion, since in Chapter 5 and elsewhere in this book hodographs have been drawn on the basis that there can be no change in the velocity component normal to a slip line. It follows that when there is a velocity discontinuity on a slip line, it must be parallel to that line. For example, the velocity discontinuity al in Figure 5.10 is parallel to the slip line AB at the point L. Correspondingly in Figure 13.6, the velocity discontinuity am is parallel to the slip line at M. However, the element of the hodograph line between l and m is perpendicular to both al and am, to the degree of accuracy with which the straight line ml represents the true curved line, therefore this element lm is perpendicular to the element of the slip line LM. The orthogonality is of course exact if the element is infinitesimally small.

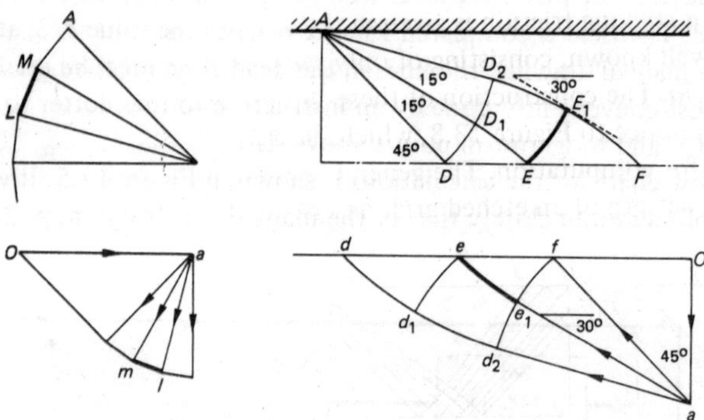

Figure 13.6. Diagrams showing orthogonality of elements of the slip line field and the hodograph.

Consideration of the hodograph in Figure 10.3, reproduced in Figure 13.6, will show that the discontinuity af is parallel to the slip line AF at F, and the discontinuity ae_1 is parallel to the slip line AF at E_1, so again the element fe_1 of the hodograph is perpendicular to the element FE_1 of the slip line.

The same is true of any corresponding elements of the two diagrams, for example ee_1 and EE_1, when there is not a finite discontinuity in velocity from one side of the slip line to the other. The lines actually representing a finite velocity discontinuity, such as am, af, ae, or ad_2 remain parallel to the slip lines.

This property can be utilised in producing the hodograph from the computer solution of the slip-line field. The hodograph is started with the co-ordinates $(0, -1)$ at a and $(-1, 0)$ at f, and the point e_1 is found from the condition that

fe_1 is perpendicular to FE_1 and the discontinuity ae_1 is parallel to AF at E, giving a pair of simultaneous equations.

Having located e_1, the construction proceeds by noting that e_1e is perpendicular to E_1E and e has the y co-ordinate zero. The point d_1 is next found using the conditions that ed_1 is perpendicular to ED_1 and d_2d_1 is perpendicular to D_2D_1. The co-ordinates are then determined from trigonometric simultaneous equations similar to equations 13.40a and b.

This technique is slightly more difficult to use manually than the one described in Chapter 5 for drawing hodographs, but it is necessary for the numerical solutions. It also has the advantage of distinguishing clearly between slip lines that carry a real finite velocity discontinuity and those where the assumed discontinuity is only a function of the mesh size. It is therefore fundamentally more satisfactory.

The hodograph is completed when all the nodal points of the slip-line field have been used. The final exit velocity can be printed out as Od.

Two further examples will be discussed to show the advantage of the computer method. One allows the friction to vary for given dimensions and the other follows the progressive deformation.

13.4.3 *Slip-line fields for all possible friction coefficients, applied to compression with width/height ratio 2:1*

The slip-line field for sticking friction is given in the previous section and the field for zero friction is well known, consisting of only two lines in the quadrant, as shown in Figure 13.7*a*. The construction of these slip-line fields by computer can be followed with reference to Figure 13.8, which shows a typical field together with the flow chart used for computation. The general form of the field was suggested by the deformation pattern of an etched grid, described in the next section.

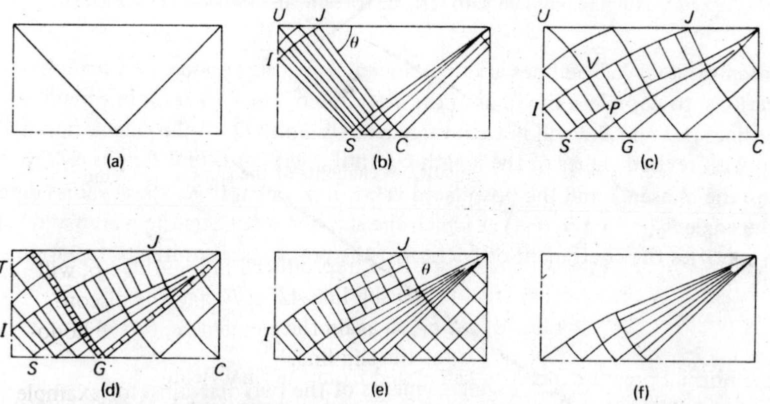

Figure 13.7 Slip-line fields for various average friction values[12.4].

A value λ for the angle of the arc IV, equal to SP, is first assumed, and an arbitrary location of the point I is selected. The line IS must be straight and at 45° to the two centre lines, so the arcs IV and SP and the line $UVPG$ can be drawn. To continue the field, it is postulated that the angle γ_1 of the arc VJ_0 (=PA) is equal to

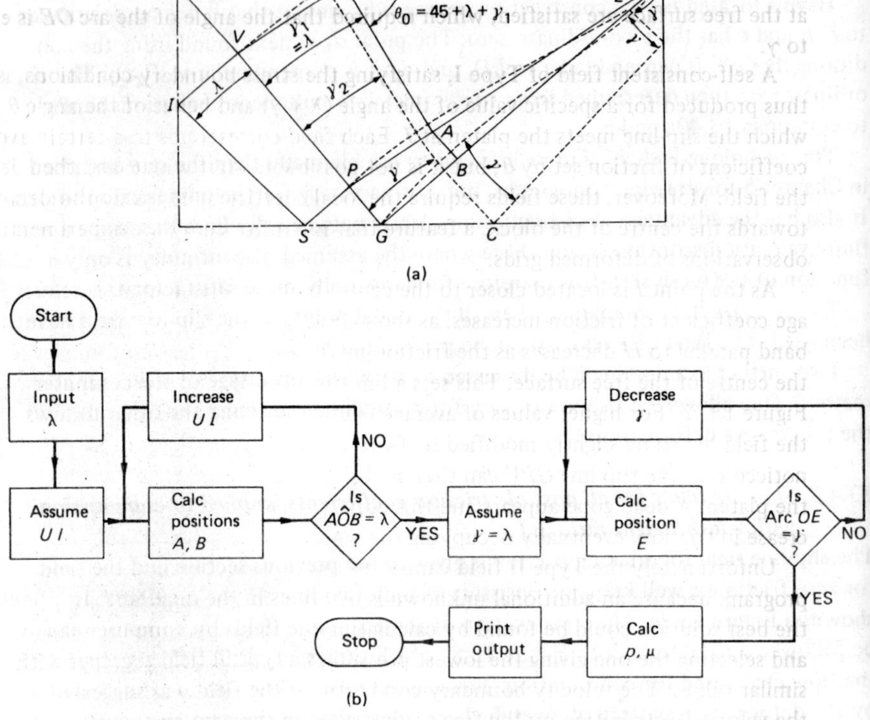

Figure 13.8. A typical slip-line field (a), for computer solution of a plane-strain forging problem with variable friction, and the flow chart used (b).

the chosen value of λ. The lines are then drawn, using the condition of orthogonal intersections, to complete the field in the direction of the corner O. In general the straight lines AO and BO will not converge exactly on to O, so the calculation is iterated with revised values of the length UI until a self-consistent field is established based on the chosen λ and the postulated value of γ_1 equal to λ. These values determine the angle $\theta_0(= 45 + \gamma_1 + \lambda)$ at which the slip line intersects the platen and this in turn specifies the coefficient of friction at the point J_0, according to equation 5.10,

$$\cos 2\theta_0 = \mu q/k$$

since the normal pressure q can be calculated from the final field.

Now it is evident that there is no justification for setting $\gamma = \lambda$, though they will be similar in magnitude. A correction is therefore needed to this preliminary field, as can also be seen from the fact that the β line BG is concave towards the horizontal centre line, which implies a pressure p_G significantly greater than k. A field such as that shown dotted and passing through J is more likely, giving $p = k$ at both G and C. This can be regarded as a modification of the preliminary field with a reduced value of γ and the necessary introduction of a reflex curvature along AO and BO.

The next step of the program therefore allows γ to decrease and an iteration is continued until the boundary conditions of 45° intersection on the centre line and at the free surface are satisfied, which required that the angle of the arc OE is equal to γ.

A self-consistent field of Type I, satisfying the stress boundary-conditions, is thus produced for a specific value of the angle $(\lambda + \gamma)$ and hence of the angle θ at which the slip-line meets the platen at J. Each field corresponds to a certain average coefficient of friction set by θ, but it is not possible to start with θ and then deduce the field. Moreover, these fields require the local coefficient of friction to decrease towards the centre of the block, a feature that is confirmed by the experimental observations of deformed grids.

As the point I is located closer to the centre-point of the block, the deduced average coefficient of friction increases, as shown in Figures 13.7*a*–*c*. The discontinuity band parallel to IJ decreases as the friction increases, and the point C moves towards the centre of the free surface. This sets a limit to the fields of this type, as shown in Figure 13.7*c*. For higher values of average friction coefficient, above about 0·18, the field has to be slightly modified to Type II, as shown in Figure 13.7*d*. It will be noticed that the slip line GPV can then no longer pass through the centre-point of the platen. A dead zone appears and this continues to grow in size with further increase in friction, eventually occupying the whole platen surface in Figure 13.7*f*.

Unfortunately the Type II field cannot be drawn uniquely by the computer program, because an additional unknown is introduced, the length TI. In principle, the best solution could be found by calculating the mean pressure for various fields and selecting the one giving the lowest pressure, but all these fields predict closely similar values. The velocity boundary-conditions could also in principle be used, but the velocity distribution on the free surface is unknown, except that the mean value is determined by constancy of volume. It is therefore necessary in this example to take some arbitrary criterion, such as the assumption that the size of the dead zone increases directly with the friction angle θ. The fields shown in Figures 13.7*d*–*f* were drawn in this way. A more accurate technique is given by Collins[13·21].

This example shows that there is a steady variation of the fields over the whole range of friction, with one transition in type. The fields could have been drawn manually, but require a trial-and-error approach to produce each self-consistent field. It would be very tedious to obtain such fields without the use of a computer.

13.4.4 *Application to non-steady conditions. Progressive deformation in forging with sticking friction*

It is possible to draw the slip-line field and hodograph for any average values of friction, as shown above, but the fields are complex. In considering progressive deformation, it is therefore simpler to restrict attention to the sticking friction condition, as shown in Figure 13.9.

The velocity of any element can be obtained from the hodograph and hence its displacement after a small increment of deformation. For the purposes of calculation the whole zone can be divided into four different areas. In the dead zone A for example, all elements move with the velocity of the platen, while in the rigid zone B all move uniformly outwards. This incidentally accords with the sur-

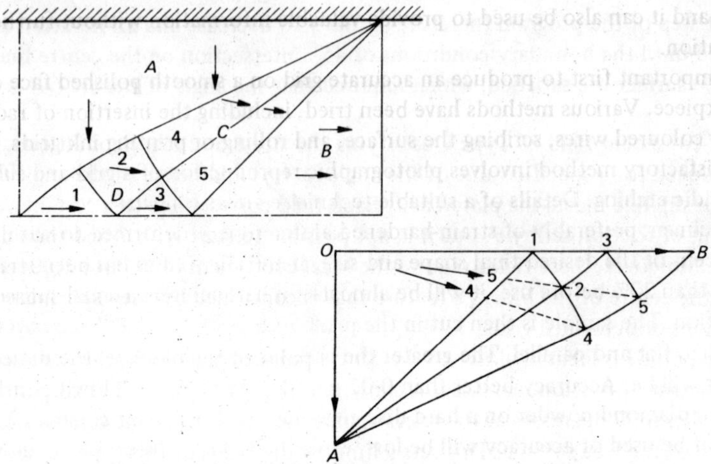

Figure 13.9. Division of the workpiece into zones where the deformation patterns differ.

prising practical result that under very high friction there is no barrelling, though moderately high friction produces marked barrelling. In the regions *C* and *D* the velocity changes rapidly.

The displacement of each point is used to form a new matrix of co-ordinate positions, from which the calculation is repeated. The next increment of deformation is then made and next positions calculated, and so on until the total desired deformation is achieved.

The results can be interpreted directly as displacements or as local strains. The velocities can also be differentiated to give strain rates as in §13.3.3. Figure 13.10 shows the calculated distortion of a grid surface after 2% deformation and after 11 successive 2% steps to the same total reduction. The final result agrees fairly well with an experimental measurement, but strain-hardening in the real material further broadens the zone of intense shear and it is not possible to establish full sticking friction over the whole surface in practice.

13.4.5 *Construction of a slip-line field from a distorted grid. Experimental technique*

The computations described above require a starting field that can then be developed in various ways. An adaptation of the visioplasticity method is very useful for this

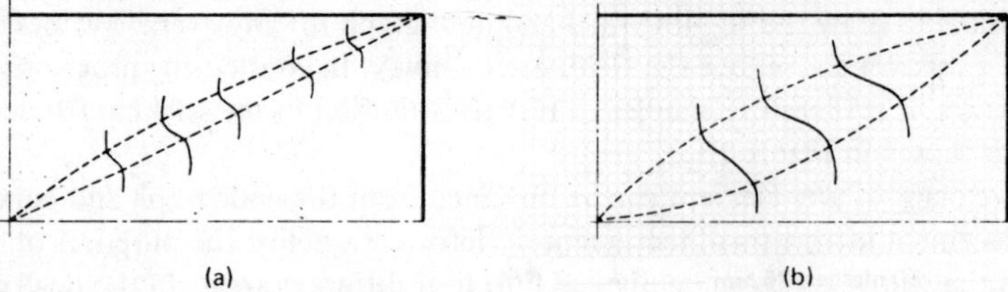

Figure 13.10. Calculated distortion of a grid after forging (a) with 2% compression (b) with 11 steps.

purpose and it can also be used to provide valuable information without further computation.

It is important first to produce an accurate grid on a smooth polished face of the workpiece. Various methods have been tried, including the insertion of radioactive or coloured wires, scribing the surface, and rolling or printing ink grids. The most satisfactory method involves photographic reproduction of a grid and subsequent acidic etching. Details of a suitable technique are as follows:

A specimen, preferably of strain-hardened aluminium, is deformed to within a few per cent of the desired final shape and size. If the aluminium has been strained by more than 20% before use, it will be almost non-hardening in a small subsequent deformation. The sample is then cut in the plane to be examined and the two faces are polished flat and parallel. The greater the care taken at this stage, the better the result will be. Accuracy better than 0·01 mm should be sought. Final polishing with 1 μm diamond powder on a hard disc gives a good finish. Soft polishing pads should not be used or accuracy will be lost, especially at the edges.

One of these faces is then prepared for the photographic process. It is degreased by immersion in 5% NaOH solution and then in 35% HNO_3, after which it is washed in distilled water and dried at room temperature. A mixture of commercial 'metal-etching reagent' or 'photo-resist' emulsion and thinner, at about 2:1 ratio, is then spread thinly and uniformly over the surface. This is best done by spinning the specimen, if it is reasonably symmetrical, at say 250 rad/m for 30 seconds, or by blowing a moderate stream of air across an irregular surface. This should be done in a dark room under yellow or red light.

The coated specimen is then dried in dust-free air at room temperature for 15 m and baked in an oven at 120°C for a further 10 m. When it has cooled to room temperature a photographic negative of the grid is placed on it and held in tight contact. The emulsion is then exposed to a 300 watt ultraviolet lamp, through the grid negative, for 3 m.

The emulsion is next developed in a commercial developer for 2 m, washed and dried at 120°C again for 10 m. It can then be handled, with considerable care, in open daylight again. When it is cool, the exposed metal in the grid pattern is gently swabbed with wet cotton wool soaked in 10% NaOH solution. The fine lines of

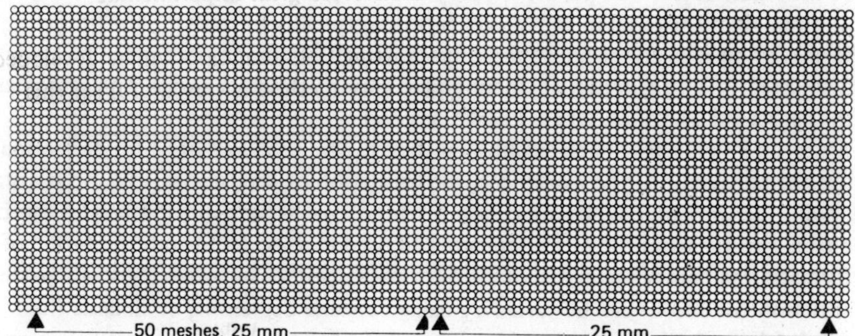

Figure 13.11. A grid suitable for use in the photo-resist etching technique. The diameter of the etched mesh should be about 0·5 mm.

hardened emulsion are still very delicate and must not be damaged at this stage. When the etching is judged to be sufficiently deep, the specimen is washed with water and dried in warm air.

Either circular- or square-mesh grids may be used. Circles deform into ellipses whose major and minor diameters give a direct measure of the local strain though the principal axes may rotate. The centres of the circles can be used as reference points for a computer input matrix. A useful grid is reproduced in Figure 13.11.

When the specimen has been prepared, it is carefully reassembled with its polished companion and held in close contact during a deformation of about 4%. It is found that this allows the deformed elliptical meshes to be measured satisfactorily without introducing large enough overall strain to distort the slip-line field pattern unduly. If the etched surface is held tightly against the counterface, using a suitable jig in plane strain or by close fitting in an axisymmetric die, the highly-strained regions distort and mark the polished counterface, giving an easily-recognised pattern of the most highly-strained regions.

The deformed grid is examined either by photographic enlargement or preferably at 100x magnification in a digitised projection microscope. The major and minor axes are measured and their directions recorded for each mesh. These directions, for moderate deformation, coincide with the directions of principal strain increment. They are plotted on a large-scale (20x) map and short lines at 45° to each axis are added. These show the local directions of maximum shear strain, assumed to coincide with the directions of maximum shear-strain rate and maximum shear stress, for the isotropic aluminium.

An experimental slip-line field can then be constructed by superimposing a square grid of comparable mesh size and transferring these directions. A line is first drawn in any square, somewhere well inside the boundaries of the specimen, right across the square in the direction of maximum shear-stress. This line is then continued right across the next square, changing direction according to the 45° lines from the nearest ellipse. The process is continued until a boundary is reached, and the line is then completed in the opposite direction, after returning to the starting point. Sufficient other lines are then drawn in, using the same method, until the slip-line pattern is evident.

Usually the slip-lines so drawn do not obey all the rules of slip-line field geometry over the whole surface. In particular, they are likely to diverge near a singularity and not to remain orthogonal in high-strain regions, probably because of the finite deformation needed to recognise the ellipses. Nevertheless they can often be smoothed and adjusted to conform to ideal incremental conditions quite easily. The agreement with theoretical solutions, where known, is good.

Once such fields have been established they can be used as a basis for deductions about the behaviour of the workpiece, the formation of dead zones, etc., or as a starting point for computer solutions as in § 13.4.3. Local strains and strain rates can be deduced directly from the measurements and used as a model for interpretation of deformation in harder materials, or even for prediction of hot-working strain patterns.

Friction conditions can be varied by using benzene degreasing or silicone fluid for high friction and PTFE tape or lead or indium foil for low friction. Lanolin and other common lubricants can of course also be used. An advantage in some circumstances of using metallic foils is that the shear strength is known; for indium or lead on work-hardened aluminium, the shear strength ratios are respectively about 0·05 and 0·27.

13.5 Upper-bound solutions

Simple upper-bound solutions have been described in Chapter 5 and their value for rapid calculations is shown in later chapters, for example on p. 188. The further development of this method to more complex problems and also to axisymmetric deformation has been facilitated by the concept of unit deforming regions, introduced by Kudo[13.5]. The workpiece is divided into suitable rectangular regions and each of these is then sub-divided into rigid triangles or sometimes into other shapes.

The usual assumptions are made, namely that the metal is non-hardening, rigid-plastic, and obeys the von Mises yield criterion. It is also assumed that flow follows the Levy-Mises stress/strain-rate relationships, equations 13.18a and b.

The limit theorem is applied, which states that the true rate at which work is done by the applied tractions is not larger than the internal rate of energy dissipation, $\dot{E}$, derived from any admissible velocity field. Now for plane strain, pp. 55, 95,

$$\dot{E} = \sqrt{2}\, k \int_V \sqrt{(\dot{\epsilon}_x^{\,2} + \dot{\epsilon}_y^{\,2} + \tfrac{1}{2}\,\dot{\gamma}_{xy}^{2})}\, \mathrm{d}V + \int_S \tau\, u \,\mathrm{d}S \qquad (13.41)$$

but if we assume rigid triangles separated by shear lines the first term, which relates to the volume of continuously deforming metal, is zero. The shear stress τ in the second term is either zero for a frictionless interface or equal to k for an internal shear line or a sticking interface. Thus for a plane-strain specimen of width w and breadth b

$$\dot{E} = \Sigma\, kus \,.\, w \qquad (13.42)$$

and this is equated to the rate of performing work by the external force at velocity V, so

$$\bar{p} \,.\, A \,.\, V = \bar{p} \,.\, V \,.\, wb = \Sigma\, kus \,.\, w \qquad (13.43)$$

as in equation 5.32.

Now consider a unit rectangular region in the deforming metal, as in Figure 13.12. The surface 14 represents the tool face, descending with velocity V. Surfaces 12 and 23 offer rigid constraint but 34 can move outwards with no resistance.

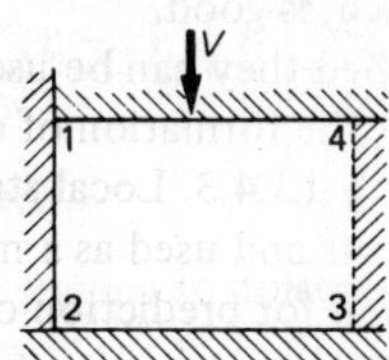

Figure 13.12. A unit deforming zone

The problem is then to find a pattern of internal velocity discontinuities that is compatible with the imposed velocity conditions and gives as low an energy dissipation as possible. Subdivision into rigid triangles is usually the best solution. Two basic patterns emerge and can be considered in detail.

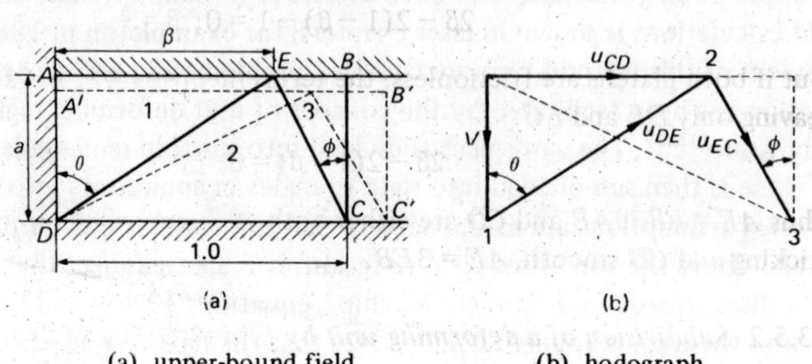

(a) upper-bound field (b) hodograph

Figure 13.13. Subdivision of a deforming unit zone into three rigid triangles[13.5]

13.5.1 ***Subdivision of a deforming unit by two shear lines (Type I)***

Suppose the rectangular zone has a breadth 1·0 and a depth a, and that it is divided by two lines meeting at a distance β from the fixed end. The block deforms, infinitesimally, to $A'B'C'D$ and the hodograph shown in Figure 13.13*b* is constructed according to the procedure described in Chapter 5.

Then, calculating the lengths of the shear lines from Figure 13.13*a* and the velocity discontinuities on them from Figure 13.13*b*, and assuming sticking friction everywhere,

$$\Sigma\, kus = k\{AEu_{AE} + DEu_{DE} + [DCu_{DC}] + ECu_{EC} + EBu_{EB}\}$$

$$= k\left\{\beta.0 + \frac{\beta}{\sin\theta}\cdot\frac{V}{\cos\theta} + [1\,.\,V\tan\theta] + \frac{1-\beta}{\sin\phi}\cdot\frac{V}{\cos\phi} + (1-\beta)\,V(\tan\theta + \tan\phi)\right\}$$

$$= kV\left\{\frac{\beta}{\sin\theta\,\cos\theta} + [\tan\theta] + \frac{1-\beta}{\sin\phi\,\cos\phi} + (1-\beta)(\tan\theta + \tan\phi)\right\} \tag{13.44}$$

$$\text{Now } \tan\theta = \frac{\beta}{a};\ \tan\phi = \frac{1-\beta}{a};\ \sin\theta = \frac{\beta}{\sqrt{a^2+\beta^2}};\ \cos\theta = \frac{a}{\sqrt{a^2+\beta^2}}$$

$$\sin\phi = \frac{1-\beta}{\sqrt{(1-\beta)^2 + a^2}};\ \cos\phi = \frac{a}{\sqrt{(1-\beta)^2 + a^2}}$$

$$\Sigma\, kus = kV\left\{\underset{DE}{\frac{a^2+\beta^2}{a}} + \underset{DC}{\left[\frac{\beta}{a}\right]} + \underset{EC}{\frac{(1-\beta)^2 + a^2}{a}} + \underset{EB}{\frac{(1-\beta)}{a}}\right\} \tag{13.45}$$

Now to find the minimum dissipation of energy this expression is differentiated w.r.t. β and equated to zero

$$\frac{\mathrm{d}}{\mathrm{d}\beta}(\Sigma\, kus) = \frac{kV}{a}\{2\beta + [1] - 2(1-\beta) - 1\} = 0 \tag{13.46}$$

giving $\beta = \frac{1}{2}$ (a)

for sticking friction on all surfaces.

If the lower platen *DC* has zero friction the term DCu_{DC} is multiplied by zero and the corresponding terms in the square brackets in each equation are zero. Then

$$2\beta - 2(1-\beta) - 1 = 0; \; \beta = \tfrac{3}{4} \qquad \text{(b)}$$

But if both platens are frictionless, the terms involving *AE*, *EB* and *DC* are all zero, leaving only *DE* and *EC*:

$$2\beta - 2(1-\beta) = 0; \; \beta = \tfrac{1}{2} \qquad \text{(c)}$$

Thus *AE* = *EB* if *AB* and *CD* are either both sticking or both smooth, but if *AB* is sticking and *CD* smooth, *AE* = 3*EB*.

13.5.2 *Subdivision of a deforming unit by four shear lines (Type II)*

The detailed solution for this field follows the same pattern and is left for an example. The final equation for β is

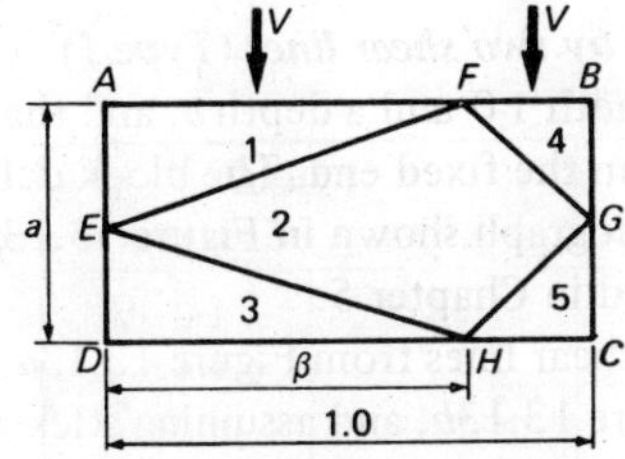

(a) upper-bound field (b) hodograph

Figure 13.14. subdivision of a deforming unit zone into five rigid zones.

$$\frac{d}{d\beta}(\Sigma\, kus) = \frac{2kV}{a}\{\underset{EF,EH}{2\beta} - \underset{FG,HG}{2(1-\beta)} + \underset{FB}{(1-2\beta)} - \underset{HC}{[2(1-\beta)]}\} = 0 \qquad (13.47)$$

when *FB* and *HC* are both sticking, $\beta = \frac{3}{4}$ *AF* = 3*FB* (a)

FB is sticking and *HC* smooth, $\beta = \frac{1}{2}$ *AF* = *FB* (b)

FB and *HC* are both smooth, $\beta = \frac{1}{2}$ *AF* = *FB* (c)

13.5.3 *Application to simple forging*

Each half of the forging in Figure 13.15 can be considered as a single zone. The centre line forms the boundary *AD* of Figure 13.13 and the block is free to move

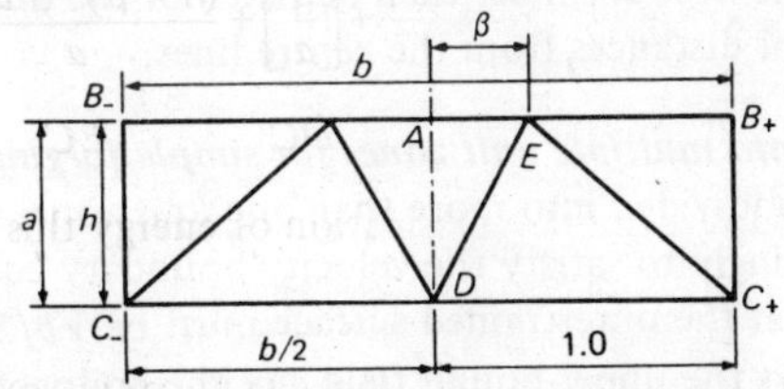

Figure 13.15. A simple forging divided into two symmetrical unit zones[13.5]

outwards at BC but must descend with the platen AB. If the unit zone is divided by two shear lines and sticking friction is assumed, $\beta = \frac{1}{2}$(equation 13.46a). Substituting this into equation 13.45 and equating the rate of dissipation of energy by internal shear to the rate of performing work by the external force on the zone of width $AB = 1$,

$$\bar{p} \, . \, V = \Sigma \, kus = \frac{kV}{a}\{a^2 + \beta^2 + \beta + (1-\beta)^2 + a^2 + (1-\beta)\}$$

$$\bar{p} = \frac{k}{a}\{2a^2 + \tfrac{3}{2}\} \qquad (13.48)$$

For zero friction on both platens, the only non-zero terms are contributed by shear on DE and EC, and β is again $\frac{1}{2}$.

$$\bar{p} = \frac{k}{a}\{2a^2 + \tfrac{1}{2}\} \qquad (13.49)$$

Similar calculations can be performed for the 5-block subdivision (Type II). The results are tabulated in Table 13.1.

Table 13.1
Average indentation pressures $p/2k$

b/h	a	Type I Sticking	Type I Frictionless	Type II Sticking	Type II Frictionless
2	1	1·75	1·25	1·38	1·0
4	$\frac{1}{2}$	2·0	1·0	2·0	1·25
8	$\frac{1}{4}$	3·25	1·25	3·6	2·12

It is of interest to compare these results with those obtained by direct application of simple upper-bound theory. For example, with division of a 2:1 block by two lines at 45°, $\bar{p}/2k = 1{\cdot}5$ for sticking friction and 1·0 for zero friction. Hill[13.1] has given a formula based on slip-line field solutions for this problem. At $b/h = 2$ with sticking friction

$$\frac{p}{2k} = \tfrac{3}{4} + \frac{b}{4h} = 1{\cdot}25 \qquad (13.50)$$

For frictionless compression, of course, $p = 2k$ for all b/h ratios greater than 1; subject to the minor increase for non-integral values (see p. 79).

Table 13.1 also shows that $\bar{p}$ increases rapidly as the width/height ratio increases. It is, in fact, preferable to divide the workpiece into multiple blocks, each with an internal shear pattern. Account must then be taken of the different velocities of the blocks at different distances from the centre lines.

13.5.4 *Subdivision into multiple unit zones for simple forging*

If the workpiece is subdivided into more than one unit zone, the hodograph must be drawn correspondingly to satisfy the velocity boundary-conditions. For constant volume, the velocity at the unrestrained surface must be $Vb/2h$.

Figure 13.16 shows the upper-bound field and the hodograph for two adjacent

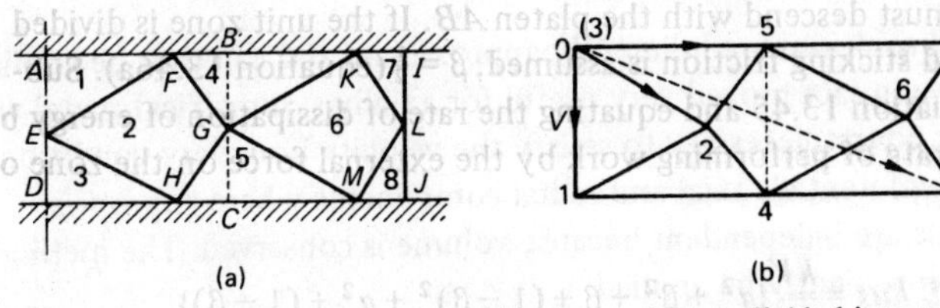

Figure 13.16. A simple compression with the workpiece divided into two unit zones on each side of the centre line.

unit zones. The velocity discontinuities are the same in each unit, but the velocity along *FB* is less than that along *KI*. The solution follows the same pattern as for a single zone, as shown in § 13.5.2, leading to the solution

$$\bar{p} = \frac{k}{2ab}\{4a^2 + \tfrac{5}{4}b^2\} = 1{\cdot}25 \tag{13.51}$$

If both platens are frictionless then again there is no energy dissipated except at the shear lines, and the single and double zones both give the same pressure.

Detailed calculations of the optimum shear-line configuration and zone size have been made, but it suffices for most purposes to divide the block into zones that are approximately square; $na = b/2$.

13.5.5 *Application to more complex conditions*

The method can also be applied to problems in which the metal can flow in more than one direction, for example in plane-strain extrusion-forging. Some knowledge of slip-line field solutions for related problems, or some experimental evidence from visioplasticity is useful in suggesting suitable divisions and shear patterns. Figure 13.17*b*, for example, resembles the slip-line field for forging with an orifice in one platen (p. 271).

Various configurations have been examined by Kudo and others.

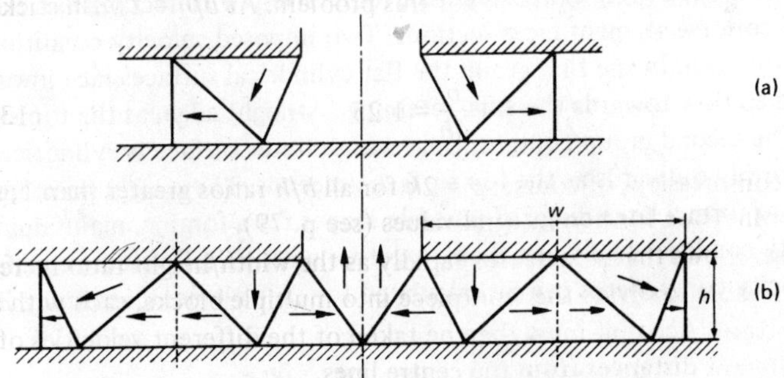

Figure 13.17. Upper-bound unit zones and shear lines for extrusion-forging[13.5].

13.5.6 *Application of upper-bound solutions to axial symmetry*[13.6]

In practical problems, it is much more common to find axially-symmetric conditions than plane–strain. This has always set a serious limitation to rigorous slip-line field theory, though approximations can be made[1.9, 13.18].

Upper-bound solutions are inherently approximate and can thus be applied more freely. They have in fact proved very useful for axially-symmetrical metal-working[(13.7)]. It is of course necessary to satisfy the velocity boundary-conditions, taking account of circumferential, axial and radial components where appropriate, though only two of these are independent because volume is conserved. The method just described can conveniently be applied.

13.5.7 *Unit zone upper-bound solutions for axial symmetry*

The solutions described in §13.5.1–13.5.5 for plane–strain conditions can be adapted to axial symmetry by supposing that the diagrams represent diametral sections of the workpiece. Similar divisions into rectangular or approximately square unit sections are madę, but each then represents a cylindrical or tubular volume. When the shearing force on a line within such a unit is calculated, the area used is the conical area generated by rotating the line about the axis. The velocity conditions also relate to shearing on this conical surface, but have to be taken as average values. As the size of the unit zone decreases, the assumed velocities become more accurate, but the complexity of the solution increases.

A more convenient analytical method has also been introduced by Kudo[(13.6)], which considers only the unit ring-shaped zones of rectangular section, without subdividing them. The shearing is assumed to occur between these rings only, which otherwise deform homogeneously, so the energy is governed by the surface areas of the rings and the velocities of sliding at the boundaries. With suitable choice of the hypothetical rings, this method can give results of useful accuracy, but it also rapidly becomes lengthy for any configurations that depart from a simple geometry.

13.5.8 *Curved-element upper-bound solutions*

The concept of rigid zones with boundary shearing has been extended by Bramley[(13.8)] to include triangular and curved sections of rings, as shown in Figure 13.18.

Four basic types of ring are considered, respectively with rectangular, triangular, convex and concave segment cross-sections. Two imposed velocity conditions are postulated for each. In the first group the flat cylindrical surfaces face inwards and are assumed to flow towards the axis, retaining a straight edge, as the top surface descends. The second group contains rings whose outer surface is cylindrical and moves uniformly outwards as the top surface descends. These eight basic units can be built up to form a large range of possible shapes for a forging, maintaining admissible velocity conditions between them, together with overall volume constancy.

The velocity components can be calculated for each element, assuming incompressibility:

$$\dot{\epsilon}_r + \dot{\epsilon}_\theta + \dot{\epsilon}_z = \frac{\partial u}{\partial r} + \frac{u}{r} + \frac{\partial v}{\partial z} = 0 \qquad (13.52)$$

The resultant velocity must be parallel to the boundary surface, and the solution for a triangular element of apex angle α is

$$u = -\frac{r+1}{2}\cot\alpha;\ v = 1 + \frac{zr}{2}\cot\alpha \qquad (13.53)$$

if u is assumed to be independent of z.

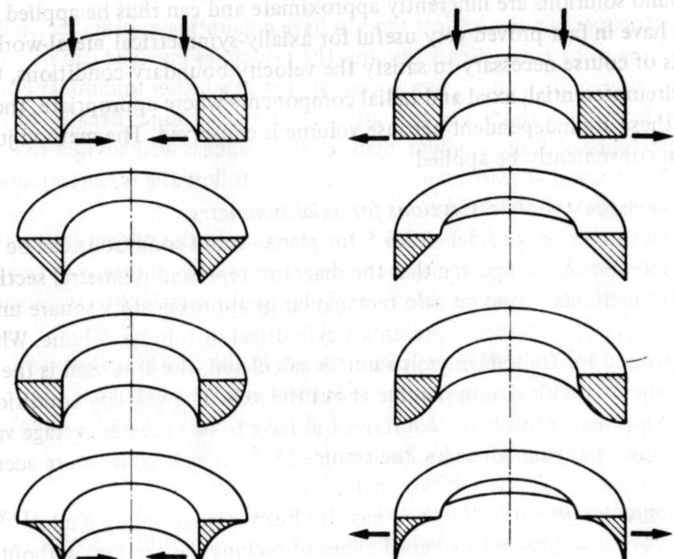

Figure 13.18. Eight basic ring-shaped units for forging analysis[13.8]

Similar expressions have been obtained for the other sections, except for the concave ones, which require some approximation.

The energy dissipated on each boundary and within each zone can then be calculated and the results are summed for the whole assemblage of units, as in equation 13.41, with a shear velocity u_d at each discontinuity.

$$\dot{E} = \frac{\sqrt{2}}{3} Y \int_V \sqrt{(\dot{\epsilon}_r^2 + \dot{\epsilon}_\theta^2 + \dot{\epsilon}_z^2 + \tfrac{1}{2}\dot{\gamma}_{rz}^2)}\, dV + \int_S \tau u_d dS \qquad (13.54)$$

The maximum force at the end of the forging can thus be found, but the calculation involves the solution of a large number of simultaneous equations, so a computer has to be used.

13.6 Finite-element analysis for elastic-plastic deformation

Although finite-element analysis has become well established in advanced design of elastically-deforming structures[13.9] over the past twenty years, it has only recently been applied to metalworking. In its fully detailed form it requires extensive computer facilities and is likely to remain for a considerable time confined to academic institutions, but it may have important and far-reaching consequences. Already views have been expressed that, in one or other of its forms, it will eventually supersede slip-line field and upper-bound techniques.

The main reason is that it is much more versatile and the degree of complexity that can be introduced is limited only by the size of the computer available. Stress, strain, strain-rate and temperature distributions can be predicted on a fine scale, and hardness, ductility and grain homogeneity can be deduced.

The research papers on the subject tend to have a daunting appearance, and this is not an area to be entered casually, but the principles involved can be enunciated without undue difficulty. They appear at first in a rather unfamiliar guise, but can be understood without a deep mathematical background. Because of the potential importance of finite-element analysis, the reader is well advised to make the effort to understand at least enough to be able to follow the research publications and to envisage its application.

This section should be considered as an introduction to the subject rather than as a detailed handbook, though it may be hoped that a reader thoroughly familiar with computer methods would be able to build up useful solutions to metalworking problems.

The background for finite-element plasticity lies in the elastic analysis which has been widely used with striking successes in mechanical and civil engineering. Detailed descriptions are available, notably by Zienkiewicz[(13.9)], and a valuable simplified account has been given by Paulsen[(13.10)]

13.6.1 *Fundamentals of finite-element elastic analysis*

For simplicity we shall deal with a two-dimensional problem, but choosing plane stress, not plane strain.

The basis of the method is very simple. The whole body is divided into small triangular elements, superimposed on a co-ordinate grid system. The objective is then to calculate the deflections of each element due to forces applied at the nodes, or corners of the triangles.

It is, of course, necessary that the forces and deflections for any one element are compatible with those of its neighbours. A set of simultaneous equations is therefore produced, the number depending on the number of elements into which the body is divided. The accuracy clearly increases with diminishing element size. The complexity also increases, consequently so does the computing time, but the fundamental equations remain simple.

The first step is to write expressions for the displacements of the nodes of any one element in terms of x and y co-ordinates. These are then converted into strains in the element and, by simple elastic modulus relationships, into stresses. The stiffness of each element is defined from these equations, relating the displacements at the three nodes to the forces acting there. The compatibility of each element with its neighbour then provides an expression for the stiffness of the whole body.

The external forces are of course defined by the boundary conditions of the problem, and the corresponding deflections and stress and strain vectors for every element follow from the detailed stiffness relationships.

Because the body has to be divided into a large number of triangular elements in order to obtain sufficient precision, a matrix notation is used, essentially as a shorthand way of writing down the component equations. No detailed manipulation of the matrices is required, since the computer readily operates sub-routines for transposition and inversion as required.

13.6.2 *Expressions for the displacements of the nodes of one triangular element*

The element is in principle able to move and to distort, but it must maintain

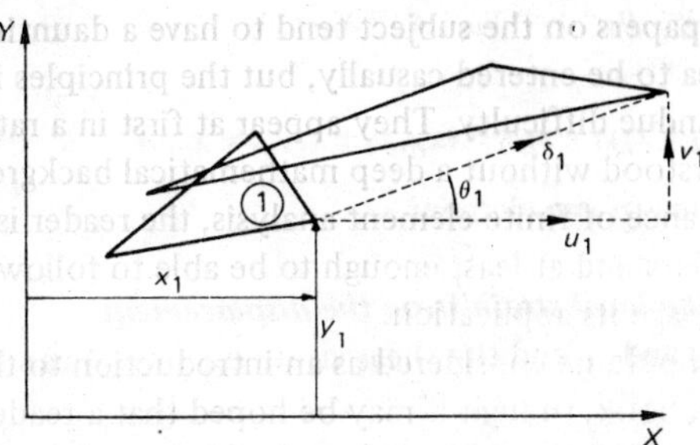

Figure 13.19. Displacement vectors u and v

strain compatibility between adjacent elements. This is possible if the displacement varies linearly between the nodes and strain is assumed to be constant throughout each element.

The displacement δ of any point within the element can be specified by its two components u and v, respectively parallel to the x and y directions. Each of these is assumed to be a linear function of the co-ordinates x and y of the point (x, y):

$$\begin{aligned} u &= n_1 + n_2 x + n_3 y \\ v &= n_4 + n_5 x + n_6 y \end{aligned} \tag{13.58}$$

It should be noted that u and v are linear distances, not velocities, here; n_1, n_2, etc. are simply constants relating to the chosen element. Thus at the corner 1 in Figure 13.19,

$$\begin{aligned} u_1 &= n_1 + n_2 x_1 + n_3 y_1 \\ v_1 &= n_4 + n_5 x_1 + n_6 y_1 \end{aligned} \tag{13.56}$$

In more general terms, if any triangular element in the body extends between the nodal points, i, j and m,

$$\begin{aligned} u_i &= n_1 + n_2 x_i + n_3 y_i \\ v_i &= n_4 + n_5 x_i + n_6 y_i \end{aligned} \tag{13.57}$$

Similarly u_j, v_j and u_m, v_m can be written in terms of the same constants and the respective co-ordinates x_j, y_j and x_m, y_m. Any other element would require six new constants.

The six equations for the element i, j, m can be solved for $n_1, n_2 \ldots n_6$ in terms of the co-ordinates. This is a lengthy procedure, but gives the following result:

$$u = \left[\frac{a_i u_i + a_j u_j + a_m u_m}{2\Delta}\right] + \left[\frac{b_i u_i + b_j u_j + b_m u_m}{2\Delta}\right] x + \left[\frac{c_i u_i + c_j u_j + c_m u_m}{2\Delta}\right] y \tag{13.58a}$$

$$v = \left[\frac{a_i v_i + a_j v_j + a_m v_m}{2\Delta}\right] + \left[\frac{b_i v_i + b_j v_j + b_m v_m}{2\Delta}\right] x + \left[\frac{c_i v_i + c_j v_j + c_m v_m}{2\Delta}\right] y$$

where $a_i = x_j y_m; -x_m y_j;\ b_i = y_j - y_m;\ c_i = x_m - x_j$
$a_j = x_m y_i - x_i y_m$, etc.

and $\Delta = x_j y_m + x_m y_i + x_i y_i + x_i y_j - (x_m y_j + x_i y_m + x_j y_i)$

which is actually the area of the element.

13.6.3 *Evaluation of the local strain from the displacement*

The direction strains ϵ_x and ϵ_y and the shear strain ϵ_{xy} $(=\gamma_{xy})$ can be written in terms of the displacements:

$$\epsilon_x = \frac{\partial u}{\partial x},\ \epsilon_y = \frac{\partial v}{\partial y},\ \epsilon_{xy} = \frac{\partial v}{\partial x} + \frac{\partial u}{\partial y} \tag{13.59}$$

(It is easy to see that an element of length δx, if displaced u at one end, will be displaced $u + \frac{\partial u}{\partial x}\delta x$ at the other and so suffer a strain $\frac{\partial u}{\partial x}$.)

Then by differentiating equation 13.58a, for example, with respect to x:

$$\epsilon_x = \frac{\partial u}{\partial x} = \frac{b_i u_i + b_j u_j + b_m u_m}{2\Delta}$$

and equation 13.58b with respect to y:

$$\epsilon_y = \frac{\partial v}{\partial y} = \frac{c_i v_i + c_j v_j + c_m v_m}{2\Delta} \tag{13.60}$$

$$\epsilon_{xy} = \frac{\partial v}{\partial x} + \frac{\partial u}{\partial y} = \frac{b_i v_i + b_j v_j + b_m v_m + c_i u_i + c_j v_j + c_m v_m}{2\Delta}$$

At this stage it is convenient to introduce the matrix notation, but only an elementary knowledge of matrix algebra is needed. For those with such a knowledge the next section can be omitted (equations 13.61–13.65).

A *matrix* is composed of elements arranged in *rows* and *columns*; thus the matrix [a] may be

$$[a] = \begin{bmatrix} a_{11} & a_{12} & a_{13} & a_{14} \\ a_{21} & a_{22} & a_{23} & a_{24} \\ a_{31} & a_{32} & a_{33} & a_{34} \end{bmatrix} \tag{13.61}$$

This has three rows and four columns, but a matrix can contain any number of rows or columns. It is always written in square brackets.

A special form of matrix has only one column. It is then known as a *vector* and it is written in curly brackets, for example

$$\{b\} = \begin{Bmatrix} b_1 \\ b_2 \\ b_3 \end{Bmatrix} \tag{13.62}$$

To multiply a matrix by a constant, each element is multiplied individually. Thus

$$n\{b\} = \begin{Bmatrix} nb_1 \\ nb_2 \\ nb_3 \end{Bmatrix} \tag{13.63}$$

Multiplication of a matrix by another follows a special rule. Thus if $[a]\,[b] = [c]$, for example,

$$\begin{bmatrix} a_{11} & a_{12} & a_{13} \\ a_{21} & a_{22} & a_{23} \\ a_{31} & a_{32} & a_{33} \end{bmatrix} \begin{bmatrix} b_{11} & b_{12} & b_{13} \\ b_{21} & b_{22} & b_{23} \\ b_{31} & b_{32} & b_{33} \end{bmatrix} = \begin{bmatrix} c_{11} & c_{12} & c_{13} \\ c_{21} & c_{22} & c_{23} \\ c_{31} & c_{32} & c_{33} \end{bmatrix} \tag{13.64}$$

The significance of say c_{23} is that it represents the sum of the products of the individual components of row 2 of the first matrix and column 3 of the second.

Thus

$$c_{23} = a_{21}\,b_{13} + a_{22}\,b_{23} + a_{23}\,b_{33} \tag{13.65}$$

Similarly

$$c_{32} = a_{31}\,b_{12} + a_{32}\,b_{22} + a_{33}\,b_{32}$$

It will be noticed that the number of rows in [b] must be the same as the number of columns in $[a]$ for multiplication.

Now reverting to equation 13.60, the convenient format is

$$\begin{aligned} 2\Delta\,.\,\epsilon_x &= b_i u_i + 0 \quad + b_j u_j + 0 \quad + b_m u_m \; + 0 \\ 2\Delta\,.\,\epsilon_y &= 0 \quad + c_i v_i + 0 \quad + c_j v_j + 0 \quad\quad + c_m v_m \\ 2\Delta\,.\,\epsilon_{xy} &= c_i u_i + b_i v_i + c_j u_j + b_j v_j + c_m u_m \; + b_m v_m \end{aligned} \tag{13.66}$$

Comparison with equations 13.63 and 13.64 shows that this can be written in matrix form as

$$\begin{bmatrix} b_i & 0 & b_j & 0 & b_m & 0 \\ 0 & c_i & 0 & c_j & 0 & c_m \\ c_i & b_i & c_j & b_j & c_m & b_m \end{bmatrix} \begin{Bmatrix} u_i \\ u_j \\ u_m \\ v_i \\ v_j \\ v_m \end{Bmatrix} = 2\Delta \begin{Bmatrix} \epsilon_x \\ \epsilon_y \\ \epsilon_{xy} \end{Bmatrix} \tag{13.67}$$

or

$$\} = [B]\,\{\delta\} \tag{13.68}$$

wher $\{\epsilon\}$ is the strain vector $\begin{Bmatrix} \epsilon_x \\ \epsilon_y \\ \epsilon_{xy} \end{Bmatrix}$ and $\{\delta\}$ is the displacement vector containing all the individual displacements of the corners of the element, in the x and y directions. The strain is thus related to the displacement through the matrix $[B]$, which consists entirely of a set of numbers, such as $b_i = y_j - y_m$ and the multiplier $1/2\Delta$ composed of positioa co-ordinates of the corners of the triangular element.

13.6.4 *Evaluation of the local stress from the strain*

The next step is to relate the strain ϵ to the stress σ. In one dimension, Hooke's Law suffices, $\epsilon_x = \sigma_x/E$, but in two dimensions account must also be taken of the lateral contraction $-\nu\epsilon_x$ due to a tensile stress σ_x, ν being Poisson's ratio, usually about 0·3, Thus

$$\epsilon_x = \frac{\sigma_x}{E} - \frac{\nu\sigma_y}{E}$$

$$\epsilon_y = \frac{\sigma_y}{E} - \frac{\nu\sigma_x}{E}$$

whence $\sigma_x = \dfrac{E}{1-\nu^2}\,\epsilon_x + \dfrac{\nu E}{1-\nu^2}\,\epsilon_y$

The shear stress is given by

$$\tau_{xy} = \frac{E}{2(1+\nu)}\,\epsilon_{xy}$$

since in pure shear $\dfrac{\epsilon_{xy}}{2} = \dfrac{\gamma_{xy}}{2} = \epsilon_x$ (from Mohr's circle) and $\sigma_x = \tau_{xy}$, $\sigma_y = -\tau_{xy}$.

These equations can be written:

$$\sigma_x = \frac{E}{1-\nu^2}\epsilon_x + \frac{E\nu}{1-\nu^2}\epsilon_y + 0 \tag{13.69}$$

$$\sigma_y = \frac{E\nu}{1-\nu^2}\epsilon_x + \frac{E}{1-\nu^2}\epsilon_y + 0$$

$$\tau_{xy} = 0 \qquad + 0 \qquad + \frac{E}{2(1+\nu)}\epsilon_{xy}$$

which is again a matrix product

$$\begin{Bmatrix} \sigma_x \\ \sigma_y \\ \tau_{xy} \end{Bmatrix} = \begin{bmatrix} \dfrac{E}{1-\nu^2} + \dfrac{\nu E}{1-\nu^2} + 0 \\ \dfrac{E\gamma}{1-\nu^2} + \dfrac{E}{1-\nu^2} + 0 \\ 0 + 0 + \dfrac{E}{2(1+\nu)} \end{bmatrix} \begin{Bmatrix} \epsilon_x \\ \epsilon_y \\ \epsilon_{xy} \end{Bmatrix}$$

$$\text{or } \{\sigma\} = [D]\,\{\epsilon\} \tag{13.70}$$

The matrix $[D]$ depends entirely on the Young's Modulus, assumed isotropic, and the Poisson's Ratio.

Since the strains are directly related to the displacements in equation 13.68, the stresses can be written

$$\{\sigma\} = [B]\ [D]\ \{\delta\} \tag{13.71}$$

The stresses are of course directly dependent upon the applied forces, so a relationship can be developed that allows the displacement vector to be determined, still for the individual element, from the applied force vector.

13.6.5. *Displacement under force. The stiffness matrix for an element*

If the force applied at the corner i of the element is f_i, it will be linearly related to the displacement u_i, since the system is elastic, as in a spring or a beam:

$$f_i = cu_i$$

It can be shown that a similar relationship is valid in general elastic deformation. In matrix form, this is

$$\{f\} = [c]\ \{u_i\} \tag{13.72}$$

if it is assumed that all forces act at the nodes of the mesh of triangular elements. Any distributed forces acting along the sides of the elements are replaced by their equivalent concentrated forces at the nodes.

Now the stresses are found by dividing the force acting at each of two nodes by the projected area of the element, for example $(x_j - x_i)\ t$, where t is the thickness of the element, constant according to the original assumption. In general terms:

$$2\{f\} = t\,[A]\ \{\sigma\} \tag{13.73}$$

where $[A]$ is a matrix dependent upon the differences of the positional co-ordinates of the nodes; in fact it is the *transpose* of the matrix $[B]$, excluding the multiplier $\frac{1}{2}\Delta$. (The transpose of a matrix contains simply the same matrix components written with rows and columns transposed.) Thus

$$\begin{bmatrix} a_{11} & a_{12} \\ a_{21} & a_{22} \end{bmatrix}^T = \begin{bmatrix} a_{11} & a_{21} \\ a_{12} & a_{22} \end{bmatrix}$$

This operation is readily performed on command by a computer sub-routine program.

Thus
$$\{f\} = [B]^T\ [D]\ [B]\ t\Delta\{\delta\} \tag{13.74}$$

or simply
$$\{f\} = [k]\ \{\delta\}$$

Since $[k]$ directly relates the deflection vector to the force vector inducing the displacements at the corners of the triangular element, it is known as the stiffness matrix for the element.

13.6.6. *The stiffness matrix for the whole system*

For the whole system, a similar relationship can be assembled

$$\{F\} = [K]\{\delta\} \tag{13.75}$$

where F represents the external forces acting on each node of the triangular-mesh network and $\{\delta\}$ represents the displacements at each node. Many of the forces may in practice be equal to zero.

13.6.7; *Method of solution of an elastic problem*

Having divided the body into triangular elements, the co-ordinates of the nodes are read into the computer and the components of the individual stiffness matrices of elements N are evaluated to compile the element stiffness matrices $[k]_N$

$$[k]_N = [B]_N^T\,[D]_N\,[B]_N\,t\Delta \tag{13.76}$$

These separate stiffnesses are then assembled to give the master matrix $[K]$.

The forces are known, so the deflections can be calculated from equation 13.75, written in the form

$$\{\delta\} = [K]^{-1}\{F\} \tag{13.77}$$

where $[K]^{-1}$ denotes the *inverse* of $[K]$, again an operation performed by a subroutine.

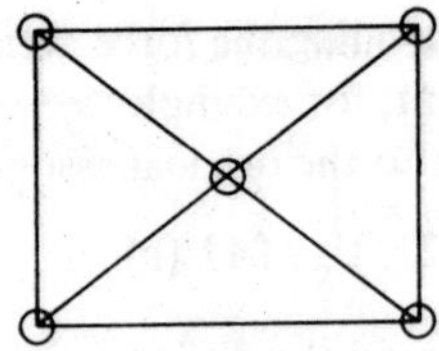

Even a very simple loading system with four elements involves five nodes and consequently ten vertical and horizontal displacement values. A 10 x 10 stiffness matrix is involved. Many real problems contain 100 to 1,000 elements. Obviously a computer is required for these solutions, and this type of analysis therefore needs some knowledge of a computer language, usually FORTRAN. Most undergraduate courses now include basic FORTRAN programming and good introductory textbooks are available, including self-study manuals.

Further development of elastic finite-element analysis is important in design work and has relevance to metalworking in assessing the stresses in elastically-deformed dies for example, but it is complex, often involving three-dimensional elements, anistropy and even composite materials. It need not concern us here.

13.6.8. *Finite-element analysis in plastic deformation*

In plastic deformation, the displacements are also related to the forces acting on small elements and can be derived from stiffness matrices[13 11]. An assumption is made that the relationship is again linear, though this is less satisfactory for plastic deformation, and methods are now available for making better approximations. There are however, important differences: the forces are considered in terms of the uniform stresses acting on the faces of the elements, not as the discrete forces at the nodes; and only incremental deformations are relevant.

For simplicity we shall deal only with two-dimensional conditions, but, as in slip-line field theory, plane strain rather than plane stress is selected.

The forces are divided into two categories, body forces F and surface tractions T. The master relationship, analogous to equation 13.75 is

$$[\Delta F] + [\Delta T] = [K]\ [\Delta U] \tag{13.78}$$

since it is the increments in the forces that cause the increment ΔU in displacement.

The stiffness matrix $[K]$ can no longer be found from the co-ordinates $[B]$ and the elastic constants $[D]$, but it can be determined by an application of the principle of virtual work. For this purpose, a hypothetical change $\delta(\Delta u_i)$ in the increment of displacement Δu_i is postulated, and the work done by the external forces F_i and T_i is equated to the work done by internal shearing to complete this change in deformation.

Now the work done by a force in moving a distance parallel to its line of action is simply the product of the force and the distance (as in equation 4.8 for example). So, for an element of original area S_0, the work done by the surface traction over an area $\mathrm{d}S$ is

$$\frac{T_i}{S_0}\,\delta(\Delta u_i)\,\mathrm{d}S$$

and the work done over the whole new area S is

$$W_T = \int_S \frac{T_i}{S_0}\,\delta(\Delta u_i)\,\mathrm{d}S \tag{13.79}$$

(It may be noted that some authors define T_i as a force per unit of original area, but it is clearer if S_0 is retained so that T_i is strictly defined as a force.) Correspondingly, the work done by the body forces throughout the volume is

$$W_F = \int_V \frac{F_i}{V_0}\,\delta(\Delta u_i)\,\mathrm{d}V \tag{13.80}$$

(again, F_i is sometimes defined as the force per unit of mass, so the expression then contains the density ρ.) Together, these two expressions are to be equated to the internal work, using the well-known result that the work done internally per unit volume is equal to the integral of stress with respect to strain, as in equation 4.6:

$$W_I = \int_V s_{ij}\,\delta(\Delta\epsilon_{ij})\,\mathrm{d}V, \tag{13.81}$$

where $\Delta\epsilon_{ij}$ is the incremental linear strain and s_{ij} is the stress initially applied. Thus the virtual work equation, as applied to the original configuration, is

$$\int_V s_{ij}\,\delta(\Delta\epsilon_{ij})\,dV = \int_S \frac{T_i}{S_0}\,\delta(\Delta u_i)\,dS + \int_V \frac{F_i}{V_0}\,\delta(\Delta u_i)\,dV \qquad (13.82)$$

A similar equation can be set up for the virtual work required to produce a further hypothetical change in the increment of deformation, starting with the material that has undergone the major increment of deformation Δu_i. The first expression, for the virtual work before the deformation, is then subtracted to give an equation for the incremental virtual work, related to the deformation increment itself. The forces required to produce a hypothetical change after the increment will be greater than before and may be written $F_i' = F_i + \Delta F_i$ and $T_i' = T_i + \Delta T_i$. Thus if the stress is now σ_{ij} and the incremental strain is ΔE_{ij},

$$\int_V \sigma_{ij}\,\delta(\Delta E_{ij})\,dV = \int_S \frac{T_i'}{S_0}\,\delta(\Delta u_i)\,dS + \int_V \frac{F_i'}{V_0}\,\delta(\Delta u_i)\,dV \qquad (13.83)$$

Care must be taken to specify the quantities with the correct reference, since in general the element, and hence the axes, will rotate during the deformation increment.

It is to be expected that the incremental strain ΔE_{ij} will contain both linear and non-linear terms, for example

$$\Delta E_{xx} = \frac{\partial \Delta u}{\partial x} + \tfrac{1}{2}\left(\frac{\partial \Delta u}{\partial x}\right)^2 + \tfrac{1}{2}\left(\frac{\partial \Delta v}{\partial x}\right)^2 \qquad (13.84a)$$

$$2\Delta E_{xy} = \frac{\partial \Delta u}{\partial y} + \frac{\partial \Delta v}{\partial x} + \frac{\partial \Delta u}{\partial x}\cdot\frac{\partial \Delta u}{\partial y} + \frac{\partial \Delta v}{\partial x}\cdot\frac{\partial \Delta v}{\partial y} \qquad (13.84b)$$

For convenience the linear terms can be written $\Delta\epsilon_{ij}$ and the non-linear term $\Delta\eta_{ij}$. Thus $\Delta E_{ij} = \Delta\epsilon_{ij} + \Delta\eta_{ij}$. We have in fact introduced *strain tensor* notation, but this, like the matrix notation, can be regarded simply as a helpful shorthand method of writing the equations. The suffixes stand for x and y, and also z when relevant, and the expressions are assembled in symmetrical matrix form:

$$[\epsilon_{ij}] = \begin{bmatrix} \epsilon_{xx} & \epsilon_{xy} \\ \epsilon_{yx} & \epsilon_{yy} \end{bmatrix}; \quad [\eta_{ij}] = \begin{bmatrix} \eta_{xx} & \eta_{xy} \\ \eta_{yx} & \eta_{yy} \end{bmatrix} \qquad (13.85)$$

Similarly for the stresses

$$[\sigma_{ij}] = \begin{bmatrix} \sigma_{xx} & \sigma_{xy} \\ \sigma_{yx} & \sigma_{yy} \end{bmatrix}$$

The *stress tensor* σ_{ij} in equation 13.83 has two constituent stresses, the stress s_{ij} originally applied and the stress increment t_{ij} measured in the deformed, and rotated, system. Both are referred to the original unit of area, by dividing the respective forces by S_0.

$$\sigma_{ij} = s_{ij} + t_{ij}$$

Thus the incremental virtual work relationship is found by subtracting the pre-increment virtual work (13.82) from the post-increment virtual work (13.83), neglecting the third order term $t_{ij}\,\delta(u_{ij})$,

$$\delta\left[\int_V (s_{ij}\,\Delta\eta_{ij} + \tfrac{1}{2}\,t_{ij}\,\Delta\epsilon_{ij})\,\mathrm{d}V\right] = \delta\left[\int_V \frac{\Delta F_i\,\Delta u_i}{V_0}\,\mathrm{d}V + \int_S \frac{\Delta T_i}{S_0}\,\Delta u_i \mathrm{d}S\right] \tag{13.86}$$

The solution of this equation becomes meaningful only if the differentation can be performed on a continuous function. It is therefore necessary to find a set of continuous weighting functions α_{ij} such that, for any one of the discrete elements, the displacement $\Delta u_j(x, y)$ forms part of a continuous displacement field. These weighting functions are used to multiply each of the discrete incremental displacements. Thus for the jth element of the continuous field

$$\Delta u_j(x, y) = \alpha_{ij}(x, y)\,\Delta u_i \tag{13.87}$$

where Δu_i is the discrete nodal incremental displacement.

Now consider the components of equation 13.86 separately. $\displaystyle\int_V \frac{\Delta F_j}{V_0}\,\Delta u_j\,\mathrm{d}V$

depends on Δu_j, so, using equation 13.87, it is written in the continuous function form of the discrete increments Δu_i. Thus

$$\int_V \frac{\Delta F_j}{V_0}\,\Delta u_j\,\mathrm{d}V = \left[\int_V \frac{\Delta F_j}{V_0}\,\alpha_{ij}\,\mathrm{d}V\right]\Delta u_i$$

$$= \Delta f_j\,.\,\Delta u_i \tag{13.88}$$

where Δf_j has been introduced to represent the integral.

Similarly

$$\int_S \frac{\Delta T_j}{S_0}\,u_j\,\mathrm{d}S = \left[\int_S \frac{T_j}{S_0}\,\alpha_{ij}\,\mathrm{d}S\right]\Delta u_i$$

$$= \Delta t_j\,.\,\Delta u_i \tag{13.89}$$

Now for the terms on the left of equation 13.86, using 13.84,

$$\Delta\eta_{ij} \text{ involves } \Delta\eta_{xx} = \tfrac{1}{2}\left(\frac{\partial \Delta u}{\partial x}\right)^2 + \tfrac{1}{2}\left(\frac{\partial \Delta v}{\partial x}\right)^2$$

$$\Delta\eta_{xy} = \tfrac{1}{2}\left(\frac{\partial \Delta u}{\partial x}\cdot\frac{\partial \Delta u}{\partial y} + \frac{\partial \Delta v}{\partial u}\cdot\frac{\partial \Delta v}{\partial y}\right)$$

which can be written $$\Delta\eta_{ij} = \left[\frac{\partial\alpha_{ik}}{\partial p} \cdot \frac{\partial\alpha_{kj}}{\partial p}\right] \Delta u_i \, . \, \Delta u_j \qquad (13.90)$$

and

$$\delta \int_V s_{ij}\,\Delta\eta_{ij}\,\mathrm{d}V = \delta \int_V \left[s_{pq}\,\frac{\partial\alpha_{ik}}{\partial p} \cdot \frac{\partial\alpha_{kj}}{\partial q}\right] \Delta\, u_i \Delta u_j\,\mathrm{d}V \qquad (13.91)$$

In a similar way

$$\Delta\epsilon_{ij} \text{ involves } \Delta\epsilon_{xx} = \frac{\partial\Delta u}{\partial x}$$

$$2\Delta\epsilon_{xy} = \frac{\partial\Delta u}{\partial y} + \frac{\partial\Delta v}{\partial x}$$

which can be written

$$2\Delta\epsilon_{ij} = \left[\frac{\partial\alpha_{ip}}{\partial q} + \frac{\partial\alpha_{iq}}{\partial p}\right] \Delta u_i \, . \, \Delta u_j \qquad (13.92)$$

This can be combined with a constitutive relationship between the stress t_{ij} and the strain increment Δ_{kl} for the material. Usually this is assumed to be linear, as in the Levy-Mises equations 13.11,

$$t_{ji} = C_{ijkl}\Delta\epsilon_{kl} \qquad (13.93)$$

Collecting together equations, 13.91, 13.92, 13.93 and 13.88, 13.89, the incremental virtual work equation 13.86 can be reformulated:

$$\delta\left[\int_V s_{pq}\,\frac{\partial\alpha_{ik}}{\partial p} \cdot \frac{\partial\alpha_{kj}}{\partial q} \cdot \Delta u_i\,\Delta u_j\,\mathrm{d}V + \tfrac{1}{4}\int_V \left(\frac{\partial\alpha_{ip}}{\partial q} + \frac{\partial\alpha_{iq}}{\partial p}\right) C_{pqst}\,\frac{\partial\alpha_{js}}{\partial t} + \frac{\partial\alpha_{jt}}{\partial s}\right.$$
$$\left. \Delta u_i \Delta u_j\,\mathrm{d}V\right] = \Delta f_j\,\Delta u_i + \Delta t_j \Delta u_i \qquad (13.94)$$

The first term on the left-hand side depends upon the initial stress field acting in the continuum and on the non-linear part of the strain tensor. It is governed by the geometric configuration and can be regarded as a stiffness k_G linking the displacement Δu_j with the force

$$\delta \int_V s_{ij}\,\Delta\eta_{ij}\,\mathrm{d}V = \delta \int_V \left(s_{pq}\,\frac{\partial\alpha_{ik}}{\partial p} \cdot \frac{\partial\alpha_{kj}}{\partial_q}\,\Delta u_i\right) \Delta u_j\,\mathrm{d}V = k_G\,\Delta u_j \qquad (13.95)$$

Similarly, the second term on the left-hand side can be regarded as a stiffness k_C depending on the constitutive equation introducing material properties (13.93).

$$\delta \int_V \tfrac{1}{2}\,t_{ij}\,\Delta\epsilon_{ij}\,\mathrm{d}V = \delta\left[\tfrac{1}{4}\int_V \left\{\left(\frac{\partial\alpha_{ip}}{\partial q} + \frac{\partial\alpha_{iq}}{\partial p}\right) C_{pqst}\left(\frac{\partial\alpha_{js}}{\partial t} + \frac{\partial\alpha_{it}}{\partial s}\right) \Delta u_i\right\} \Delta u_j\,\mathrm{d}V\right.$$
$$= k_C\,\Delta u_j$$

Thus

$$k_G\ \Delta u_j + k_C\ \Delta u_j = k_{ij}\ \Delta u_j = \Delta f_j + \Delta t_j \tag{13.96}$$

where k_{ij} is the overall stiffness, including geometric factors and the material response. It is, of course, a matrix.

This equation can be solved because all the variables are known at the end of the ith increment. ΔF_j and ΔT_j are the force increments that produced the ith increment so Δf_j and Δt_j can be calculated, and the stiffness matrix is determined by the nodal co-ordinates of the system and the stress/strain increment relationship. The discrete nodal displacement for the next increment Δu_j can thus be calculated.

The procedure is lengthy and obviously requires a computer, but the principle is not complicated. The stiffness matrix is determined for each finite element, and a series of simultaneous equations is set up for the compatibility of all the elements in the system. The master equation corresponding to 13.78 is

$$[K]\ [\Delta U] = [K_C + K_G]\ [\Delta U] = [\Delta F] + [\Delta T] \tag{13.97}$$

The solution is further complicated if both elastic and plastic deformation are included, because account must then be taken of the separate regimes and of the transition between them, governed by the von Mises criterion of yielding, which in tensor notation is, as in equation 3.22,

$$\tfrac{1}{2}s'_{ij}s'_{ij} = k^2 = \left(\frac{S}{2}\right)^2 \tag{13.98}$$

where k is a constant, the shear yield stress of the material.

This account is intended to provide the background understanding of the application of finite element theory to elastic and plastic problems, but further study of the original papers on the subject is necessary before the reader embarks on programs of his own. A large computer is needed.

The method is undoubtedly powerful and is likely to increase in popularity as sub-routine programs become available. It has been applied to problems involving the elastic–plastic boundaries in plane-strain drawing, to temperature-rise calculations in forging, and to related stress and strain distributions.

13.7 Application of variational calculus

The calculus of variations can be applied to metalworking in several ways, including the upper-bound type of analysis. Two important methods will be considered briefly; the weighted residuals approach and a matrix integral function.

13.7.1. *The general method of weighted residuals*

This mathematical technique has been available for a long time, but has found fairly extensive applications in engineering only during the past decade [13.12]. It offers an approximate method of solution for changes in distributed systems. First, a reasonable guess is made on the basis of previous experience and knowledge of related problems, and this is then progressively improved until an acceptable accuracy is obtained.

The input is a system of differential or integral/differential equations with associated constitutive relationships and boundary conditions. Given also a set of initial conditions, a trial solution is assumed. This may, for example, specify a functional dependence upon positional co-ordinates but include undetermined functions of time. The latter can then be determined by requiring the trial solution to satisfy the original differential equations to some specified approximation.

Suppose that a differential equation for displacement u as a function of co-ordinates x and time t is given as

$$N(u) - \frac{\partial u}{\partial t} = 0 \text{ for } t > 0 \text{ and } x \text{ in a 3-dimensional volume } V \text{ with boundary } S \tag{13.99}$$

$N(u)$ contains only spatial derivatives of u.

For the initial and boundary conditions:

$$u(x, 0) = u_0 x \text{ in the continuum } V$$

$$u(x, t) = f_s(x, t) \text{ on the surface } S$$

A trial solution is assumed, satisfying the boundary conditions but not necessarily the differential equations. This can take the form

$$u^*(x, t) = u_s(x, t) + \sum_{i=1}^{N} C_i(t)\, u_i(x, t) \tag{13.100}$$

where the approximate discrete functions u_i are prescribed and satisfy the boundary conditions on the surface

$$u_s = f_s, \quad u_i = 0$$

These equations can be written in the form of *residuals*, which are of course zero if the solution is exact.

$$R(u^*) = N(u^*) - \frac{\partial u^*}{\partial t} \tag{13.101}$$

$$R_0(u^*) = u_0 X - u_s(x, 0) - \sum_{i=1}^{N} C_i(0)\, u_i(x, 0)$$

The difference between these residuals is a measure of the extent to which u^* satisfies the differential equations and the initial conditions respectively.

A weighting function w_j is prescribed so that the integral of the residuals over the whole volume is zero, or as small a value as is acceptable. If u^* is exact, the solutions are valid regardless of the weighting functions, but this would be a special case. The weighting functions may be chosen in various ways, by reduction to ordinary differential equations, by the method of moments, by the principle of least squares, by collocation $w_j = \delta(x_j - x)$, or by a unified method due to Crandall[13.13].

13.7.2. ***Weighted-residuals method in plasticity***

This method has been shown to provide another powerful technique for solving metalworking problems.

Suppose that an unknown continuous function is to be found that satisfies the differential equations

$$\begin{aligned} D_1\,[f(x)] &= f_1\,(x) \text{ in the continuum } V \\ D_2\,[f(x)] &= f_2\,(x) \text{ at the boundary } S \end{aligned} \tag{13.102}$$

where D_1 and D_2 are differential operators and f_1 and f_2 are given functions of the position co-ordinates.

Now a function $f^*(x)$ is postulated in terms of discrete values of the co-ordinates. It must satisfy both continuity and stress equations for the deforming body.

$$f^*(x) = \sum_{i=1} A_i\,\phi_i(x) \tag{13.103}$$

Then the error, or residual, will be

$$E(x) = \Sigma A_i D_1\,[\phi_i(x)] - f(x) \tag{13.104}$$

which depends on the form of the function ϕ and the choice of parameters A_i.

The next step is to choose these values so that the error is reduced to a desired minimum. This can be done, in principle, by selecting weighting factors w_i such that, as nearly as possible

$$\int_V w_i(x)\,E(x)\,\mathrm{d}V = 0 \tag{13.105}$$

The most convenient way of doing this is to use the method of least squares, namely

$$\int_V E^2(x)\,\mathrm{d}V = \text{minimum}$$

Since the variable parameters are A_i, this is equivalent to

$$\frac{\partial}{\partial A_i}\int_V E^2(x)\mathrm{d}V = 0$$

or

$$\int_V \frac{\partial}{\partial A_i}\,[E(x)] \cdot E(x)\,\mathrm{d}V = 0 \tag{13.106}$$

from which it can be seen that

$$w_i(x) = \frac{\partial}{\partial A_i}\cdot[E(x)] \tag{13.107}$$

13.7.3. *Application of weighted residuals to axisymmetric extrusion*(13.14)

The basic continuity equation for an incompressible material in axial symmetry is (as in equation 13.52)

$$\dot{\epsilon}_r + \dot{\epsilon}_\theta + \dot{\epsilon}_z = \frac{\partial \dot{u}}{\partial r} + \frac{\dot{u}}{r} + \frac{\partial \dot{v}}{\partial z} = 0 \tag{13.108}$$

and the stress equilibrium equations are

$$\frac{\partial \sigma_r}{\partial r} + \frac{\partial \tau_{rz}}{\partial z} + \frac{\sigma_r - \sigma_\theta}{r} = 0$$

$$\frac{\partial \tau_{rz}}{\partial r} + \frac{\partial \sigma_z}{\partial z} + \frac{\tau_{rz}}{r} = 0 \tag{13.109}$$

There are two unknowns in the strain-rate equation, so a scalar flow function ψ can be chosen, such that

$$\dot{u} = \frac{1}{r}\frac{\partial \psi}{\partial z}; \quad \dot{v} = -\frac{1}{r}\frac{\partial \psi}{\partial r}$$

satisfying equation 13.52.

The four unknown stresses require two independent functions ϕ and χ to be specified satisfying the two equations 13.109. These may be chosen as

$$\sigma_r = \frac{1}{r}\frac{\partial^2 \phi}{\partial z^2} + \frac{\chi}{r}$$

$$\sigma_\theta = \frac{\partial \chi}{\partial r}$$

$$\sigma_z = \frac{1}{r}\frac{\partial^2 \phi}{\partial r^2} \tag{13.110}$$

$$\tau_{rz} = -\frac{1}{r}\frac{\partial^2 \phi}{\partial r \partial z}$$

The six unknowns are in this way reduced to three functions, which are expressed as polynomials whose order is a compromise between accuracy and computing time.

In extrusion, the following flow function has been suggested

$$\psi = \psi_0 \frac{r^2}{r^{*2}} + (r - r^*)^2 \, \Sigma\Sigma A_{ij} r^{2i+2} z^{j+1} \tag{13.111}$$

where $r^*(z)$ is the profile of the die.

The expression must give $v_r = 0$ at $r = 0$ and v_z constant at $z = 0$. Also $\psi = \psi_0 =$ constant when $r = r^*$.

Suitable stress functions are

$$\phi = \sum_{k=0}^{K} \sum_{l=0}^{L} B_{kl} r^{2k+3} z^l \tag{13.112}$$

$$\chi = \sum_{p=0}^{P} \sum_{q=0}^{Q} C_{pq} r^{2p+1} z^q$$

The best values of the parameters A_{ij}, B_{kl}, C_{pq} have then to be chosen, by the least squares method.

The equation for comparison is the Levy-Mises flow rule (13.18b), which for the radial direction is

$$\sigma_r - \sigma_m = \frac{1}{\lambda}\dot{\epsilon}_r$$

So the error E_r at any point N in the deformation zone is

$$(E_r)_N = (\sigma_r - \sigma_m) - \frac{1}{\lambda}\dot{\epsilon}_r \qquad (13.113)$$

and over the whole volume the least squares equation is applied:

$$E_r^2 = \sum_N (E_r)_N^2 \geqslant 0 \qquad (13.114)$$

Similar flow-rule equations govern the circumferential, axial and shear-strain rates, giving error functions E_θ, E_z, E_γ

A total error function is thus built up:

$$E_T = \sum E_i w_i \qquad (13.115)$$

where the weighting factors are found by differentiating with respect to A_{ij}, B_{kl} and C_{pq}, according to equation 13.107.

The whole system of functions is very complex because each of the four functions depends on A, B and C, while σ_r, σ_θ, σ_z and τ_{rz} are functions of r, z, B and C, and $\dot{\epsilon}_r$ is a function of r, z and A.

The procedure itself is lengthy, and if the flow rules are not linear it will require several iterations. Nevertheless, the calculation is fully numerical and complete solutions have been obtained. Detailed stress distributions through the deformation zone have been plotted for extrusion and for drawing[13.14].

13.7.4. *A matrix functional method*[13.15]

As mentioned in § 13.6, the finite element method rapidly becomes complex and requires extensive computer storage as well as calculation time. If the yield stress varies with position because of strain hardening, and with direction because of anistropy, the problem can no longer be treated as linear and the complexity multiplies, especially if large deformations are included.

A variational approach can be used to reduce the computation and storage. This has some resemblance to the unit-zone upper-bound method and to the weighted residuals method.

Making the usual assumptions that the material is rigid-plastic and that elastic deformation can be ignored, the rate of performing work can be described in terms of three components, as in equations 13.80–13.82.

The rate of working due to homogeneous deformation can be written in terms of the generalised stress $\bar{\sigma}$ and strain rate $\dot{\bar{\epsilon}}$ (see pp. 55, 351)

$$\frac{\mathrm{d}(W_H)}{\mathrm{d}t} = \int_V \bar{\sigma}\,\dot{\bar{\epsilon}}\,\mathrm{d}V \tag{13.116}$$

These are defined in equations 13.16–13.18.

The rate of working due to inhomogeneous deformation is assumed to be expressible in a similar form but with a Lagranian multiplier to take account of the redundant work factor.

$$\frac{\mathrm{d}(W_I)}{\mathrm{d}t} = \int_V \lambda(\dot{\epsilon}_1 + \dot{\epsilon}_2 + \dot{\epsilon}_3)\,\mathrm{d}V \tag{13.117}$$

The rate of working due to the surface traction T with a displacement rate $\dot{u}$ is

$$\frac{\mathrm{d}(W_S)}{\mathrm{d}t} = \int_S T\dot{u}\,\mathrm{d}S \tag{13.118}$$

Then a functional, or integral expression Φ can be defined, in a way analogous to that of §13.7.2, giving the residuals

$$\Phi = \int_V \bar{\sigma}\dot{\bar{\epsilon}}\,\mathrm{d}V + \int_V \lambda(\dot{\epsilon}_1 + \dot{\epsilon}_2 + \dot{\epsilon}_3)\,\mathrm{d}V - \int_S T\dot{u}\,\mathrm{d}S \tag{13.119}$$

The true solution will reduce this to a minimum, so

$$\frac{\mathrm{d}\Phi}{\mathrm{d}\lambda} = 0 \tag{13.120}$$

In this solution, the velocity fields must satisfy the velocity boundary conditions on the free surface.

To formulate the problem, the workpiece is divided into elements connected at their nodal points, as in other numerical methods, and the integral (functional) Φ is approximated by a function ϕ of the discrete nodal point velocities u_i and the multipliers λ_i for each element. The problem is thus reduced to finding a set of u and λ values that give a minimum value of ϕ.

The analysis is still lengthy, but the method is flexible. It is possible to include variation of flow stress with strain and even to make allowance for anisotropy, but an iterative process is then necessary.

One of the important deficiencies of variational methods in general is that rigid zones can not be calculated. It is however possible to find the boundary of any region where the strain is below a set limit and to eliminate all nodal points within this region from the calculation. They can then be reintroduced for the next increment of deformation if required.

This matrix method has been applied in detail to a cold heading problem[13.15, 13.16].

13.8. Numerical computation of stress-analysis solutions

One of the important advantages of using a computer for metalworking problems is that progressive solutions can be readily obtained once the initial equations and conditions have been set up. This has already been discussed with reference to slip-line fields (§13.4) and to finite-element analysis (§13.6). It can also be utilised in solutions obtained by the stress-analysis technique, sometimes known as the *slab method*[13.17].

Several simple solutions of this type have been described in Chapter 4 and later chapters. A slab of infinitesimal thickness is selected and the equations representing the force equilibrium for the slab are compiled, assuming that the deformation is homogeneous within the slab and that one of the major directions is a principal stress direction. The resulting differential equation is then solved with appropriate boundary conditions.

To apply this method to more complex deformation patterns, such as are found in closed-die forging, the whole workpiece is divided into unit deformation zones. These are not optimised, as they are in the upper-bound solutions, but are chosen only so that flow is reasonably uniform in each one. Figure 13.20 shows an example.

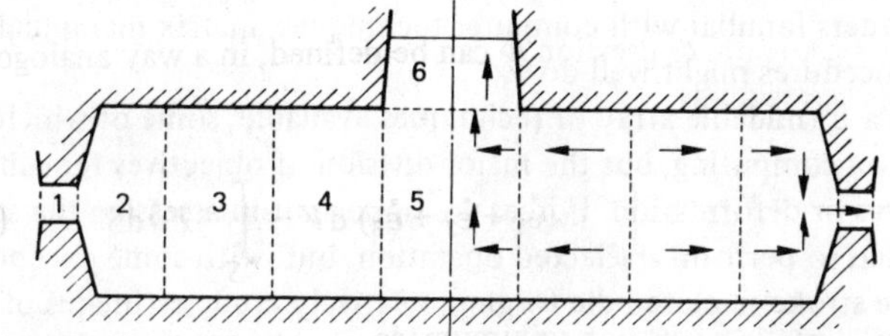

Figure 13.20. Division of a closed-die forging workpiece into unit zones.

The force balance is evaluated for each zone, starting at a free surface and maintaining continuity of the stress distribution. Different expressions are obtained for the different types of zone, for example zones 2 and 3 which in flow is respectively convergent and parallel.

Since the metal can flow outwards into the flash gutter (1) and upwards into the stem (6) until the stem is filled, two sets of calculations must be undertaken, starting at the free surfaces of 1 and 6 respectively. The flow divide will then be found by the condition that the stresses must balance on each side of the neutral surface. In fact of course, as we have seen in Chapter 10, the flow divide will not be a simple plane, but it can be regarded as such to the approximation required here.

Evidently the flow pattern will change in stages as deformation develops. At first the metal flows smoothly outwards and to a lesser extent upwards, but as the metal enters the peripheral convergent zone the flow into the stem will be enhanced. When the stem cavity is filled, flow is possible only through the flash gutter, and if deformation is continued there is a rapid rise in the force required.

The analysis must therefore be done in small steps, revising the boundary conditions after each step. It is of course important to observe the volume constancy condition throughout. At some critical stages the flow mode may change and the correct choice must be made, on the criterion of minimum energy, which is equivalent to finding the minimum axial stress.

A complete analysis of this type[13.17] has been presented for the axisymmetric closed-die forging shown in Figure 13.20. The whole procedure is again elaborate, though the principle is simple. Further detail can however be introduced, to make allowance for variation in flow stress with strain, or differential friction. It should also be possible to include local strain-rate and temperature variations and their influence on yield stress. While this method can give very good predictions of the load variation with deformation, it is inherently incapable of predicting shape changes such as barrelling in open-die forging. Perhaps the most important application is in prediction of approximate tool stresses so that the likelihood of die fracture can be reduced.

13.9 Assessment of the current state of metalworking theory

This chapter is intended to show the range of development that has recently taken place in metalworking theory and to present the underlying principles in sufficient detail to be readily understood. It does not aim, as earlier chapters have done, to equip the reader immediately to undertaken solutions to new metalworking problems, though readers familiar with computer techniques, matrix manipulation and optimisation procedures might well do so.

There is now a formidable array of techniques available, some of which make heavy demands on computing, but the major division of objectives remains to determine either forces or deformation. Forces are important in assessing the size of equipment needed to perform a selected operation, but, with some exceptions such as the end of the stroke in closed-die forging, relatively crude estimates of total load will usually suffice. The newer methods can be more valuable in providing information on a finer scale about the local stresses acting on tools, so that these can be designed with adequate strength and toughness. Immediately therefore, the analyses are linked with the material properties of dies and punches. This close connection with metallurgy is even more evident in problems involving the deformation and flow. The general assumptions of rigid-plastic, non-hardening and isotropic material provide necessary simplifications for the first approach to real problems and many useful conclusions can be reached by suitable use of slip-line field theory, for example. Nevertheless, especially in non-steady processes, the constitutive relationships between stress and strain, or stress and strain-rate have to be taken into account for individual materials as soon as more detailed flow patterns are needed. It is of course also essential to know the fracture properties of the workpiece when considering the possibility of defects, such as surface cracks or internal bursts, being produced. The computer methods thus differ in two significant ways from the earlier theoretical analyses.They are capable of providing more detailed information about local stresses, strains and strain-rates, but in order for these to be significant it is no longer sufficient to consider the idealised universal material of the more classical theory.

It is perhaps too early to make a reliable assessment of the relative merits of the various methods that have been proposed in recent years and are still being developed, but some evaluation can be attempted.

There seems little doubt that the basic type of stress analysis described in Chapter 4 remains generally useful for a wide variety of quasi-steady processes. This depends on the force equilibrium of infinitesimal slabs and is essentially concerned with forces and surface stress-distributions. In rather more complex guise, it can be used for progressive deformation (§ 13.8) and in that form can give predictions of major flow velocities, for example in extrusion forging. It does however suffer from its neglect of redundant deformation. In some examples, such as in simple extrusion (§ 8.8.2) or indentation, the redundant work factor is large and the load is consequently grossly underestimated. For the same reason the stress-analysis method is inherently incapable of giving information about internal distortion. It can, however, easily include strain hardening.

Slip-line field theory is mathematically precise and takes full account of redundant deformation, so it is in principle capable of giving very accurate and detailed information, provided that its assumptions are met by the material. For uniformly strain-hardened aluminium, as one example, very close correlation between predicted and observed dead zones, rigid zones and deformation patterns can be obtained after small increments of deformation. In special instances, particularly with heavily-rolled sheet, the assumption of isotropy is invalid, but slip-line field theory does not have great relevance to the sheet-forming processes, which are primarily governed by the stress equilibrium and geometric conditions. Neglect of elastic deformation is again of minor importance in most heavy-working processes. There remain the two important assumptions of homogeneity and absence of strain hardening. These become particularly relevant when the slip-line field theory is extended to progressive deformation (§ 13.4.4). In cold working, local strain-hardening in regions of larger strain tends to inhibit further deformation in the same areas, and to spread the deformation zone into adjoining regions of less-hardened material. The general effect is therefore that the observed deformation occupies significantly broader bands than theory predicts, though at very high strain rates it may be possible for adiabatic shear heating to reduce the effective flow stress and thus to confine the strain to very narrow bands, even leading to fracture. During hot working similar concentrations of shear may arise, but the dominant inhomogeneity is introduced by inevitable temperature variations in all real billets.

Probably the most severe limitation on slip-line field theory was originally the requirement for plane–strain deformation, but useful extensions of the subject have now been made[13.18]. Very few real processes involve true plane–strain conditions. The rolling of thin flat strip approximates closely to it, but the redundant work factor is so small for this process that stress analysis gives good results. Tube ironing or drawing of a thin-walled tube in a close pass, where the wall thickness is reduced with little change in diameter, is another example, because the circumferential strain is nearly zero. It is possible to regard the outer annuli in many axisymmetric workpieces as thin-walled tubes and thus to approximate to plane–strain conditions. This is of course not valid in regions near the axis, but these often contribute little to the total force. Useful results can thus be obtained by constructing a slip-line field for a diametral section of an axisymmetric billet as though it were deforming in plane–strain and then making allowance for the conical areas in calculating forces. More accurate analysis is possible at the expense of considerably greater complexity[13·22].

If these limitations are borne in mind, a valuable insight into the deformation modes and the stress distribution can be obtained from slip-line fields. It is even possible to indicate where defects are likely to occur, either in the form of cracks or profile irregularities, such as barrelling or the extrusion defect (p. 272). Further exploration of the inclusion of strain-hardening and of the applicability of slip-line fields to more complicated geometric forms would considerably extend their range of use, already widened by the technique of drawing fields by a computer.[13·21]

Visioplasticity is a very valuable adjunct to slip-line field theory in this respect. Relatively simple model experiments can provide detailed information about the deformation patterns and also about the ways in which they are likely to be influenced by strain-hardening and by departure from plane-strain conditions. Apart from this, it is an important technique in its own right, and useful strain and strain-rate distributions can be deduced from model experiments.

The Moiré fringe method[13.19] and the photoplastic method[13.20] should also be mentioned in this context, though they have not been discussed in detail here. These experimental techniques also provide information about the local strains, but they are less precise than the etching method described in §13.4.5.

Upper-bound solutions in the simple form shown in §5.10 are useful for rapid estimation of approximate working loads. It is always quicker to calculate new results by this method than to use the stress analysis, though the latter is preferable when a solution is known, for example in wire drawing, and only the reduction ratio or the friction is to be varied. The upper bounds do however include redundant work inherently, whereas allowance has to be made for it in the stress equations, usually by a redundant work factor, as in equation 6.51.

The unit-zone upper-bound methods are more complex, but can be applied to quite complicated geometric shapes, especially when curved elements are used. Some skill is required in choosing appropriate zones, but the methods are simpler than the finite-element analyses and can almost be reduced to routine computer procedures. They appear to be the best methods for calculating forces for plane–strain, axisymmetric and possibly unsymmetrical conditions when the tools do not have simple shapes. It seems likely that some forms of upper-bound technique will prove useful for a variety of metalworking problems, especially as verified computer sub-routines and linking programs become available. The results are, however, primarily concerned with calculation of working loads and to a lesser extent with overall flow, rather than with detailed distributions of stress and strain.

The finite-element and variational methods are in principle very powerful and can give detailed results, including strain, strain-rate and temperature influences on local material properties. They do however require extensive computer facilities, as well as considerable expertise. It seems likely that they will remain the province of academic research laboratories, though their results may well have important industrial significance.

References

13.1 Hill, R., *The mathematical theory of plasticity*, Oxford University Press, 1956, p. 140.

13.2 Thomson, E. G., Yang, C. T. and Kobayashi, S., *Mechanics of plastic deformation in metal processing*, Macmillan, 1965.

13.3 Medrano, R. E. and Gillis, P. P., 'Visioplasticity techniques in axisymmetric extrusion', *J. Strain Anal.* 7, 170–177, 1972.

13.4 Rowe, G. W., Li, T. F. and Farmer, L. E., 'A study of deformation in simple forging with variable finite friction', *Proc. III N.Amer. Metalwkg. Res. Conf.*, Pittsburg, 1975, pp. 85–99.

13.5 Kudo, H. 'An upper-bound approach to plane-strain forging and extrusion –
I *Int. J. Mech. Sci.* **1**, 57–83, 1960.
II *Int. J. Mech. Sci.* **1**, 229–252, 366–8, 1960.

13.6 Johnson, W., and Kudo, H., 'Use of upper-bound solutions for determination of temperature distributions in fast hot rolling and axi-symmetric extrusion processes,' *Int. J. Mech. Sci.*, **1**, 175–191, 1960.

13.7 Avitzur, B., *Metal forming processes and analysis*, McGraw-Hill, 1968.

13.8 McDermott, R. P. and Bramley, A. N., 'Forging analysis – a new approach', *Proc. 2nd N. Amer. Metalwkg. Res. Conf.*, Madison 1974, pp. 35–47.

13.9 Zienkiewicz, O. C., *The finite element method in structural and continuum mechanics*, McGraw-Hill, 1967.

13.10 Paulsen, W. C., 'Finite element analysis. Pts 1–3', *Machine Des.*, Sept. 30, 46–52, Oct. 14, 146–150, Oct. 28, 90–94, 1971.

13.11 Gordon, J. L. and Weinstein, A. S., 'A finite element analysis of the plane–strain drawing problem', *Proc. 2nd N. Amer. Metalwkg. Res. Conf.*, 194–208, 1974.

13.12 Finlayson, B. A. and Scriven, L. E., 'The method of weighted residuals – a review', *App. Mech. Rev.*, **19**, 735–748, 1966.

13.13 Crandall, S. H., *Engineering Analysis*, McGraw-Hill, 1956.

13.14 Steck, E., *Numerische Behandlung von Verfahren der Umformtechnik*, Bericht 22, Inst. Umf. Univ. Stuttgart, 1971.

13.15 Lee, C. H. and Kobayashi, S., 'New solutions to rigid-plastic deformation problems using a matrix method', *J. Engg. Ind. Trans A.S.M.E.*, **95**, 865–870, 1973.

13.16 Shah, S. N. and Kobayashi, S., 'Rigid-plastic analysis of cold heading by a matrix method', *Proc. 15th Mach. Tool Des. Res. Conf.* Birmingham 1974, 603–610.

13.17 Altan, T., 'Computer simulation to predict load, stress and metal flow in an axisymmetric closed-die forging' in *Metal Forming*, ed. Hoffmanner, A. L., 325–347, Plenum Press, 1971.

13.18 Johnson, W. and Mellor, P. B., *Engineering Plasticity*, van Nostrand Reinhold, 1973.

13.19 Kato, K., Murota, T. and Jimma, T., 'Improvement of Moiré and grid methods of plastic strain analysis and their application to extrusion', *Bull. Jap. Soc. Mech. Engrs.* **12**, 32–42, 1969.

13.20 Nisida, M., Hondo, M. and Hasunuma, T., 'Studies of plastic deformation by the photoplastic method', *Proc. 6th Jap. Nat. Cong. App. Mech.*, 137–140, 1956.

13.21 Dewhurst, P. and Collins, L. F., 'A matrix technique for constructing slip-line field solutions to a class of plane-strain plasticity problems', *Int. J. Num. Meth. in Eng.* 7, 357–378, 1973.

13.22 Cox, A. D., Eason, G. and Hopkins, H. G., 'Axially-symmetric plastic deformation in soils', *Phil. Trans. Roy. Soc.* **A254**, 1–45, 1961.

Author and Subject Index

(Principal references in bold type)